INTRODUCTION TO SUPERCONDUCTIVITY

Second Edition

Michael Tinkham
Rumford Professor of Physics
and
Gordon McKay Professor of Applied Physics
Harvard University

McGraw-Hill, Inc.
New York St. Louis San Francisco Auckland Bogotá Caracas
Lisbon London Madrid Mexico City Milan Montreal New Delhi
San Juan Singapore Sydney Tokyo Toronto

INTRODUCTION TO SUPERCONDUCTIVITY
International Editions 1996

1 2 3 4 5 6 7 8 9 0 CWP UPE 9 8 7 6

This book was set in Times Roman by Keyword Publishing Services.
The editors were Jack Shira and Eleanor Castellano;
the production supervisor was Elizabeth J. Strange.
The cover was designed by Amy Becker.

Library of Congress Cataloging-in-Publication Data

Tinkham, Michael.
Introduction to superconductivity / Michael Tinkham. – 2nd ed.
p. cm. – (International series in pure and applied physics)
Includes index.
ISBN 0-07-064878-6
1. Superconductivity. I. Title. II. Series.
QC611.92.T56 1996
537.6'23–dc20 95-22378

When ordering this title, use ISBN 0-07-114782-9

Printed in Singapore

ABOUT THE AUTHOR

A native of Wisconsin, **Michael Tinkham** received an A. B. degree from Ripon College and his M. S. and Ph.D. from MIT. After a postdoctoral year at the Clarendon Laboratory in Oxford, he spent 11 years teaching at the University of California, Berkeley, before moving to Harvard in 1966, where he is now the Rumford Professor of Physics and Gordon McKay Professor of Applied Physics. Over the years, he has spent sabbatical leaves at MIT and at the University of Paris, Orsay, as a Guggenheim Fellow; at the Cavendish Laboratory in Cambridge University as an NSF Senior Postdoctoral Fellow; at the Institute for Theory of Condensed Matter in Karlsruhe, Germany, as a Humboldt Prize Fellow; at the University of California in Berkeley as a Visiting Miller Professor; and as a Visiting Professor at the Technical University of Delft, the Netherlands.

He is a Member of the National Academy of Sciences and a Fellow of the American Academy of Arts and Sciences, of the Amercian Physical Society, and of the American Association for the Advancement of Science. Honors from the American Physical Society include the Richtmyer lectureship and the Buckley Solid State Physics Prize for his research on the electromagnetic properties of superconductors. In 1976 he was awarded an honorary Sc.D. from Ripon College. He has also served on the US National Committee of IUPAP and as chairman of the Fritz London Award Committee.

Author of over 200 research publications, he has written three previous books: *Group Theory and Quantum Mechanics, Superconductivity*, and the first edition of *Introduction to Superconductivity*, which has been translated into Russian, Japanese, and Chinese.

CONTENTS

PREFACE

The first edition was written after the surge of activity in the 1950s and 1960s, in which the Bardeen-Cooper-Schrieffer (BCS) and Ginzburg-Landau (GL) theories were assimilated and used to develop what was referred to in the first paragraph of that edition as a "remarkably complete and satisfactory picture" of superconducting phenomena. It was written with the aim of presenting that picture to its readers, and it has performed that function well during the subsequent 20 years. Very little in the first edition has proved to be incorrect or misleading. Why, then, make the very substantial effort to prepare a second edition?

The major impetus has, of course, come from the discovery of high-temperature superconductivity, and the immense body of activity which this has spawned. This phenomenon is not only of great interest for its own sake, but it has also provided us with a broader perspective for understanding the properties of the classic superconductors as well. Another impetus was my desire to include an introductory survey of the many developments in classic superconductivity over the past two decades. Inevitably, I have given some extra emphasis to areas in which my own group has been active, such as Josephson junctions, single-electron tunneling effects, and nonequilibrium superconductivity, but these are also topics of broad interest. Finally, I wanted to take advantage of the 20 years of teaching experience with the first edition and to try to produce a new edition which is somewhat more "user friendly" for the beginner in the field.

Rather than reviewing the whole book, let me simply highlight the changes from the first edition. First, before attacking the intimidating BCS theory, I have introduced a new Chap. 2, which, drawing on the latter part of the old Chap. 3, treats example applications of the elementary London theory of the electrodynamics of superconductors, including diamagnetic screening, the intermediate state, and high-frequency absorption. For many practical purposes, this is the most important level of understanding, and it seems counterproductive to delay discussing these examples until after BCS, as was done in the first edition.

Second, although the old BCS chapter has been renumbered as Chap. 3, it is largely unchanged except for the addition of a new Sec. 3.11 on the penetration depth. This new section collects the essential material from the first part of the old Chap. 3 on the implications of the BCS nonlocal electrodynamics for the determination of the effective *local* penetration depth used throughout the rest of the book. At the same time, the less essential mathematical details of this discussion are either relegated to the appendix or omitted entirely, because the *conceptual* importance of nonlocality is now well established; but nonlocality is not relevant to the high-temperature superconductors or the dirty materials of technical importance for magnet wire. Instead, there is some new discussion of how λ is measured experimentally and other new material introducing the perpendicular penetration depth $\lambda_\perp$ of thin films.

The first really major change has been to replace the old Chap. 6 on the Josephson effect by two entirely new chapters. In this expanded and modernized treatment, Chap. 6 covers the more basic principles and applications, making use of the popular "tilted-washboard" picture for discussing the RCSJ model, including *I-V* characteristics of both under- and overdamped junctions. It also discusses Shapiro steps, photon-assisted tunneling, quantum interference, and sine-Gordon solitons in long junctions. This is followed by a modernized discussion of applications of Josephson junctions to SQUIDs, an entirely new discussion of *arrays* of Josephson junctions, and a brief discussion of superconducting tunnel junctions as high-frequency detectors.

The new Chap. 7 deals with the many special properties of the small, low-capacitance junctions made possible by modern nanofabrication techniques. In such junctions, the charging energy of single electrons is important and may dominate the Josephson coupling energy, so that the particle number becomes a better quantum number than the phase φ. This chapter deals with such topics as the Coulomb blockade, the single-electron tunneling transistor, and macroscopic quantum tunneling, as well as the importance of the damping of high-resistance junctions by low-impedance leads. Although these topics are currently the focus of intense research activity and can considerably deepen one's understanding of the phase-number uncertainty relation in superconductors, this chapter can be skimmed over in a first course on superconductivity.

The second major change is the introduction of an entirely new Chap. 9, dealing primarily with the high-temperature superconductors. Because the fundamental mechanism responsible for the high T_c remains to be identified with certainty, we sidestep this question and emphasize instead the many properties of these materials which can be understood in the framework of the classic Lawrence-Doniach model of layered superconductors. These include the magnetic anisotropy and the implications of the flux-line-lattice melting transition for the resistive transition. Although its applicability is not restricted to the high-temperature superconductors, we next review the Larkin-Ovchinnikov model of collective pinning, including a short discussion of flux creep in this model and also in the vortex-glass and Boson glass models. The chapter concludes with a discussion of anomalous properties of the high-temperature superconductors which cannot be understood in terms of standard *s*-wave BCS superconductivity but, instead, suggest *d*-wave pairing.

The third major change is the addition of the new Chap. 11, dealing with nonequilibrium superconductivity, using a simplified version of the Schmid-Schön formalism to discuss the many implications of quasi-particle disequilibrium in driven superconductors. This chapter includes discussions of the *enhancement* of superconductivity by microwave radiation, by quasi-particle tunneling, and by dynamic nonequilibrium effects associated with a time-dependent energy gap. Also discussed here are phase-slip centers and the interconversion of normal and supercurrent by Andreev reflection at *NS* interfaces. The latter is a relatively old subject which, applied to more general configurations, has recently enjoyed a resurgence of interest.

The other five chapters have been left largely unchanged, not because they could not be improved but, rather, because doing so would have unduly, perhaps indefinitely, delayed completion of this new edition. Nontheless, some changes were made. A number of new references were added to reflect the progress made in the intervening years. Also, brief discussions of a number of new topics were inserted, such as the Kosterlitz-Thouless resistive transition in two-dimensional superconductors.

While the expansion of the text which was required to include new developments will make it more useful as a reference for researchers, it also makes the book too long to be covered completely in a one-semester course. On the other hand, the instructors in such a course can take advantage of this plethora of material to pick topics to their own taste, leaving other topics to be pursued in individual study, perhaps leading to the preparation of a term paper.

McGraw-Hill and I would like to thank the following reviewers for their many helpful comments and suggestions in the early stages of the development of the second edition: Alex de Lozanne, Philip Duxbury, Richard S. Newrock, John Ruvalds, Mark Rzchowski, and Dale Van Harlingen.

Finally, I am pleased to acknowledge the assistance of many other colleagues in encouraging my efforts, in helping to guide the focus of the revisions, and in providing generous assistance in improving the quality of the presentation. I am uniquely indebted to Rick Newrock for his careful and speedy reading of chapter after chapter and for his extensive detailed criticism of the manuscript as it neared final form, which spurred me on to make many improvements and clarifications. It is, of course, impossible to acknowledge all those who have helped in many ways over the many years (and I apologize in advance to all those inadvertently omitted), but I should at least mention (in alphabetical order) Ryogo Aoki, Mac Beasley, Chuck Black, Greg Blonder, John Clarke, Dick Ferrell, Michael Flatté, Pierre-Gilles de Gennes, Rolf Glover, Ashraf Hanna, Jack Hergenrother, Marco Iansiti, Mark Itzler, Charlie Johnson, Teun Klapwijk, Kostya Likharev, Chris Lobb, Hans Mooij, David Nelson, Miguel Octavio, Dan Prober, Dan Ralph, Mark Rzchowski, Albert Schmid, Gerd Schön, Bill Skocpol, Mark Tuominen, and Valerii Vinokur. Of course, I cheerfully accept full responsiblity for any errors or misunderstandings which may appear despite their assistance!

Michael Tinkham

SUGGESTIONS FOR USING THIS BOOK

The first edition of this book was sufficiently slender so that most of its contents could be covered in a briskly paced one-semester course. This is no longer the case after the expansion required to bring the second edition up to date. The best choice of topics to cover will depend on the length of the course, the level of students, and the interest of the instructor. The following suggestions are offered as guidance in structuring a one-semester introductory course at the beginning graduate or advanced undergraduate level based on the material in this book.

The historical overview in Chap. 1 should be read primarily to provide a bird's-eye view of the subject for orientation. The discussion of electrodynamics in Sec. 1.3 can be used in an introductory lecture, with the rest of the material essentially deferred until treated carefully in later chapters. Chapter 2 presents a systematic treatment of the electrodynamics of classic superconductors, the property that gives the subject much of its interest and importance, at the simple, but very useful, level of the London equations. It should be covered carefully, except that Sec. 2.4 can be omitted to save time.

Although the discussion of the BCS theory in Chap. 3 is kept as simple as possible without loss of rigor, it is still the most technically intimidating chapter. Because of its importance, it should all be examined at whatever depth seems appropriate to the class. In an elementary course, one might focus on the key *results*: the ground state, the energy gap, the density of states, electron tunneling, and the penetration depth, skimming lightly over the rest.

Chapters 4 and 5 on the Ginzburg-Landau theory and type II superconductors are central and should be covered carefully, except for Secs. 5.7 and 5.8, which can be skimmed. Likewise, the Josephson effect, treated in Chap. 6, is fundamentally very important, but the more specialized Secs. 6.6 and 6.7 can be skimmed or omitted in a first course. The special features of very small Josephson junctions treated in Chap. 7 are of considerable research interest at the present time, but this chapter can be omitted or skimmed in a first course, unless it is of special interest to the instructor. Similarly, much of the discussion of fluctuation effects in Chap. 8 is of somewhat specialized interest, but Secs. 8.1 and 8.2 on fluctuation-induced electrical resistance should be covered, and Secs. 8.3 through 8.5 should at least be skimmed. With the intense current interest in high-temperature superconductivity, much of Chap. 9 should be covered, at least at the qualitative level. The material from Secs. 9.6.3 through 9.8 is of more specialized interest and might be skimmed or omitted.

Finally, the material in Chaps. 10 and 11 considerably enriches our understanding of superconductivity by considering such topics as dirty superconductors, gapless superconductors, and time-dependent and nonequilibrium regimes of superconductivity. Unfortunately, time limitations in a short course will probably only allow these topics to be skimmed over lightly for the general ideas, with special attention to Sec. 10.1.

In summary, the more advanced and specialized material, which can be skimmed or omitted in a one-semester course, is that in Chaps. 7, 10 and 11, and in some of the latter parts in Chaps. 3, 5, 6, 8, and 9.

PREFACE TO THE FIRST EDITION

This book has evolved from a set of lecture notes originally written for a graduate course at Harvard University during the fall term of 1969. They were subsequently rewritten during a sabbatical leave at the Cavendish laboratory in 1971–1972 and during a repeat of the course in 1973.

The objective of the lectures, and of this book, is to provide an up-to-date introduction to the intriguing subject of superconductivity and some of its potential applications. The emphasis is on the rich array of phenomena and how they may be understood in the simplest possible way. Consequently, the use of thermal Green functions has been completely avoided, despite their fashionability and undeniable power in the hands of skilled theorists. Rather, the power of phenomenological theory in giving insight is emphasized, and microscopic theory is often narrowly directed to the task of computing the coefficients in phenomenological equations. It is hoped that this emphasis will make the treatment more palatable to the experimentalist, and also complement the more generous coverage of the formal theoretical aspects of the subject in most books presently available. Finally, the author was motivated by the hope that if the theoretical techniques were kept as elementary as possible, the work might have more value to undergraduates and technologists with incomplete backgrounds in theoretical physics.

In a sense this book forms an updated and greatly expanded version of the Les Houches lectures of the author, written in 1961. However, so much development of the subject has occurred in the intervening years that these notes were really rewritten (twice) from start to finish. In the process, the author has drawn frequently on the excellent book of de Gennes, *Superconductivity in Metals and Alloys*, and on the two-volume treatise *Superconductivity* edited by Parks. There is little in the book which has not been published previously in some form, but some topics—particularly fluctuation effects—have developed too recently to have appeared in previous books.

No attempt has been made to give an exhaustive or definitive treatment. Such a treatment required the two-volume Parks treatise mentioned above. Rather, the author has chosen to introduce the reader to a selection of topics which reflect his own focus on the electrodynamic properties of superconductors, which, after all, give the subject its unique interest. The time limitation of a semester lecture course provided unrelenting discipline in limiting the number of topics and the depth of treatment.

The book starts with an introductory survey which lays out the ground to be covered in the book, and gives some of the milestones in the historical development of the subject. The reader is advised to treat this as an overview only, intended to introduce concepts and language, with the detailed explanations to be developed in subsequent chapters. He definitely should not puzzle over issues which are only sketchily introduced at this point.

The second chapter is devoted to "basic BCS," the microscopic theory developed by Bardeen, Cooper, and Schrieffer to explain the superconducting state. This theory is placed at the beginning because no serious discussion of superconductivity is possible without concepts derived from the theory. Unfortunately, this chapter has by far the most forbidding formal nature of any part of the book, but this should not be allowed to discourage the reader. Little use of the mathematical details will be made in the following chapters, and so this chapter can be skimmed for the general ideas (which are summarized in the concluding section), and referred to later if more detailed understanding of some particular point is required.

With Chap. 3, we move into the phenomenological level of treatment, which characterizes the rest of the book. First, the implications of the nonlocal electrodynamics in determining the effective penetration depth of a magnetic field into bulk and thin film superconductors are explored, the thorough discussion of the latter topic reflecting a historical interest of the author. A simplified discussion is then given of the intermediate state, in which superconducting and normal material coexist in the presence of a magnetic field.

Chapter 4 develops the Ginzburg-Landau theory from the same phenomenological point of view used by the original authors. The theory is then applied to an extensive catalog of classic problems: domain-wall energy, critical-current density, fluxoid quantization, critical fields of films and foils, the upper critical field H_{c2}, the Abrikosov vortex state, and the surface nucleation field H_{c3}. The concepts treated here underlie the subjects treated in the following chapters, in addition to illustrating the power of the Ginzburg-Landau approach.

In Chap. 5, the magnetic properties of type II superconductors are developed in some detail. After the equilibrium flux density has been worked out, attention is focused on the creep and flow of the flux under the influence of transport currents. In this way, insight is obtained into the considerations which limit potential applications of type II superconductors in high-field magnets. The chapter concludes with a discussion of the factors governing the design of superconducting magnets to cope with time-varying fields, including the use of twisted multicore composite conductors to minimize ac losses while maintaining thermal stability.

Chapter 6 is devoted to the Josephson effect and macroscopic quantum phenomena. These subjects represent some of the purest and most fundamental aspects of superconductivity, yet also provide the basis for sensitive instruments which have revolutionized electromagnetic measurements. Both aspects are reflected in the treatment given; in particular, the detailed discussion of practical SQUID magnetometers is the first to appear in a textbook.

Although for years it was thought that the effects of thermodynamic fluctuations were unobservably small in superconductors, the advent of the superconducting dectectors just mentioned has made it possible to observe such effects both above and below T_c. Chapter 7 surveys these phenomena in both electrical conductivity and diamagnetism. For example, it is shown how fluctuation effects put a limit (though an astronomical one) on the lifetime of "persistent" currents below T_c, and how they also give rise to "precursors" of superconductivity above T_c. Because this subject has flowered since the date of the Parks treatise, this book is the first containing a thorough discussion of this interesting and informative new aspect of superconductivity.

The final chapter is devoted to introductory discussions of three topics: the Bogoliubov method, gapless superconductivity, and time-dependent Ginzburg-Landau theory. These topics go beyond the elementary Ginzburg-Landau phenomenology and bring in more microscopic considerations. Yet the basic concepts and conclusions have been drawn inevitably into the discussions of the topics treated earlier; morover, taken together, they lay the groundwork for work going on at the present frontiers of research. Hence, it seems fitting to close the book with a peek at these topics, where the last word is by no means in.

Finally, the author is pleased to thank the reviewers of the manuscript for constructive suggestions; the detailed reading of the final manuscript by Dr. Richard Harris is especially appreciated. The comments of students who have used the notes also were particularly helpful. The speedy and accurate typing of Miss Patricia McCarthy in preparing the final manuscript was an invaluable incentive to continued progress. More generally, the author wants to thank his numerous students, colleagues, and collaborators, especially in Berkeley, Orsay, Harvard, and the other Cambridge, for making his exploration of superconductivity the pleasure it has been. Although it would be impossible to list them all here, I cannot close this Preface without explicitly acknowledging numerous seminal discussions over the years with M. R. Beasley, J. Clarke, P. G. de Gennes, R. A. Ferrell, and R. E. Glover III. If this book serves to initiate others into the fascination I have found in this subject, it will have well served its intended purpose.

Michael Tinkham

INTRODUCTION TO SUPERCONDUCTIVITY

CHAPTER 1

HISTORICAL OVERVIEW

Superconductivity was discovered in 1911 by H. Kamerlingh Onnes[1] in Leiden, just 3 years after he had first liquefied helium, which gave him the refrigeration technique required to reach temperatures of a few degrees Kelvin. For decades, a fundamental understanding of this phenomenon eluded the many scientists who were working in the field. Then, in the 1950s and 1960s, a remarkably complete and satisfactory theoretical picture of the classic superconductors emerged. This situation was overturned and the subject was revitalized in 1986, when a new class of high-temperature superconductors was discovered by Bednorz and Müller.[2] These new superconductors seem to obey the same general phenomenology as the classic superconductors, but the basic microscopic mechanism remains an open and contentious question at the time of this writing.

The purpose of this book is to introduce the reader to the field of superconductivity, which remains fascinating after more than 80 years of investigation. To retard early obsolescence, we shall emphasize the aspects which seem to be reasonably securely understood at the present time.

The goal of this introductory chapter is primarily to give some historical perspective to the evolution of the subject. All detailed discussion is deferred to later chapters, where the topics are examined again in much greater depth. We start by reviewing the basic observed electrodynamic phenomena and their early

[1]H. Kamerlingh Onnes, *Leiden Comm.* **120b**, **122b**, **124c** (1911).

[2]G. Bednorz and K. A. Müller, *Z. Phys.* **B64**, 189 (1986).

phenomenological description by the Londons. We then briefly sketch the subsequent evolution of the concepts which are central to our present understanding. This quasi-historical review of the development of the subject is probably too terse to be fully understood on the first reading. Rather, it is intended to provide a quick overview to help orient the reader while reading subsequent chapters, in which the ideas are developed in sufficient detail to be self-contained. In fact, some readers have found this survey more useful to highlight the major points *after* working through the details in subsequent chapters.

1.1 THE BASIC PHENOMENA

What Kamerlingh Onnes observed was that the electrical resistance of various metals such as mercury, lead, and tin disappeared completely in a small temperature range at a critical temperature T_c, which is characteristic of the material. The complete disappearance of resistance is most sensitively demonstrated by experiments with persistent currents in superconducting rings, as shown schematically in Fig. 1.1. Once set up, such currents have been observed to flow without measurable decrease for a year, and a lower bound of some 10^5 years for their characteristic decay time has been established by using nuclear resonance to detect any slight decrease in the field produced by the circulating current. In fact, we shall see that under many circumstances we expect absolutely no change in field or current to occur in times less than $10^{10^{10}}$ years! Thus, *perfect conductivity* is the first traditional hallmark of superconductivity. It is also the prerequisite for most potential applications, such as high-current transmission lines or high-field magnets.

The next hallmark to be discovered was *perfect diamagnetism*, found in 1933 by Meissner and Ochsenfeld.[3,4] They found that not only a magnetic field is *excluded* from entering a superconductor (see Fig. 1.2), as might appear to be

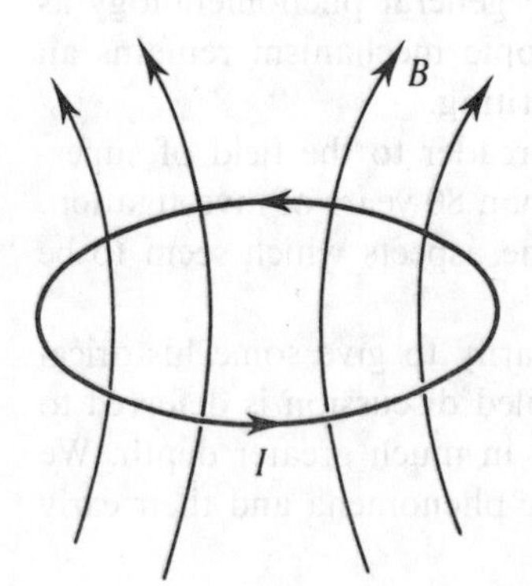

FIGURE 1.1
Schematic diagram of persistent current experiment.

[3]W. Meissner and R. Ochsenfeld, *Naturwissenschaften* **21**, 787 (1933).

[4]Actually, the diamagnetism is perfect only for *bulk* samples, since the field does penetrate a finite distance λ, typically approximately 500 Å.

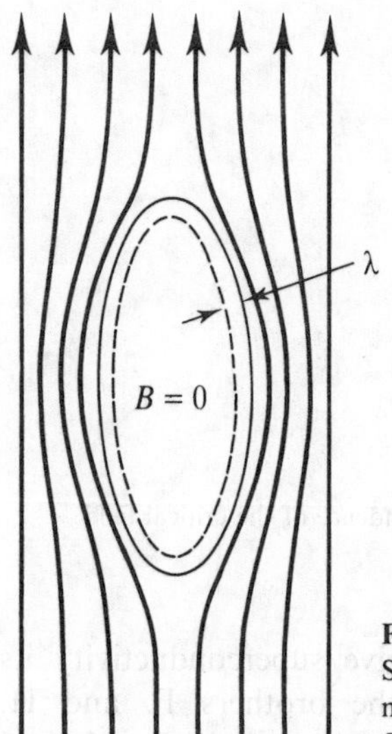

FIGURE 1.2
Schematic diagram of exclusion of magnetic flux from interior of massive superconductor. λ is the penetration depth, typically only 500 Å.

explained by perfect conductivity, but also that a field in an originally normal sample is *expelled* as it is cooled through T_c. This certainly could *not* be explained by perfect conductivity, which would tend to trap flux *in*. The existence of such a reversible *Meissner effect* implies that superconductivity will be destroyed by a critical magnetic field H_c, which is related thermodynamically to the free-energy difference between the normal and superconducting states in zero field, the so-called condensation energy of the superconducting state. More precisely, this *thermodynamic critical field* H_c is determined by equating the energy $H^2/8\pi$ per unit volume, associated with holding the field out against the magnetic pressure, with the condensation energy. That is,

$$\frac{H_c^2(T)}{8\pi} = f_n(T) - f_s(T) \tag{1.1}$$

where f_n and f_s are the Helmholtz free energies per unit volume in the respective phases in zero field. It was found empirically that $H_c(T)$ is quite well approximated by a parabolic law

$$H_c(T) \approx H_c(0)[1 - (T/T_c)^2] \tag{1.2}$$

illustrated in Fig. 1.3. While the transition in zero field at T_c is of second order, the transition in the presence of a field is of first order since there is a discontinuous change in the thermodynamic state of the system and an associated latent heat.

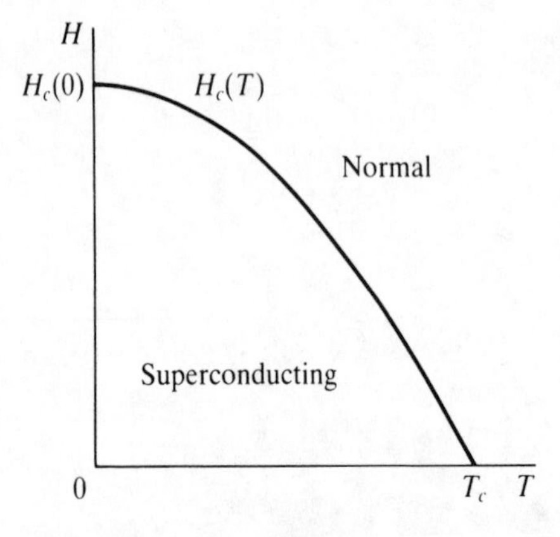

FIGURE 1.3
Temperature dependence of the critical field.

1.2 THE LONDON EQUATIONS

These two basic electrodynamic properties, which give superconductivity its unique interest, were well described in 1935 by the brothers F. and H. London,[5] who proposed two equations to govern the microscopic electric and magnetic fields

$$\mathbf{E} = \frac{\partial}{\partial t}(\Lambda \mathbf{J}_s) \tag{1.3}$$

$$\mathbf{h} = -c \text{ curl } (\Lambda \mathbf{J}_s) \tag{1.4}$$

where

$$\Lambda = \frac{4\pi\lambda^2}{c^2} = \frac{m}{n_s e^2} \tag{1.5}$$

is a phenomenological parameter. It was expected that n_s, the *number density of superconducting electrons*, would vary continuously from zero at T_c to a limiting value of the order of n, the density of conduction electrons, at $T \ll T_c$. In (1.4), we introduce our notational convention of using **h** to denote the value of the flux density on a microscopic scale, reserving **B** to denote a macroscopic average value. Although notational symmetry would suggest using **e** for the microscopic local value of **E** in the same way, to avoid constant confusion with the charge e of the electron, we shall do so only in the few cases[6] where it is really useful. These notational conventions are discussed further in the appendix.

[5]F. and H. London, *Proc. Roy. Soc.* (London) **A149**, 71 (1935).

[6]The fundamental basis for our notational asymmetry in treating **E** and **B** is in the Maxwell equations curl $\mathbf{h} = 4\pi\mathbf{J}/c$ and curl $\mathbf{e} = -(1/c)\partial\mathbf{h}/\partial t$. Superconductors in equilibrium can have nonzero $\mathbf{J}_s$, as described by the London equations, causing **h** to vary on the scale of λ. But in equilibrium, or even steady state, $\partial\mathbf{h}/\partial t = 0$, so that **e** is zero, or at least constant in space, so the use of both **e** and **E** offers no advantage. The distinction is useful only in discussing time-dependent phenomena such as motion of flux-bearing vortices in type II superconductors.

The first of these equations (1.3) describes perfect conductivity since any electric field *accelerates* the superconducting electrons rather than simply sustaining their velocity against resistance as described in Ohm's law in a normal conductor. The second London equation (1.4), when combined with the Maxwell equation curl $\mathbf{h} = 4\pi\mathbf{J}/c$, leads to

$$\nabla^2\mathbf{h} = \frac{\mathbf{h}}{\lambda^2} \tag{1.6}$$

This implies that a magnetic field is exponentially screened from the interior of a sample with penetration depth λ, i.e., the Meissner effect. Thus, the parameter λ is operationally defined as a penetration depth; empirically, the temperature dependence of λ is found to be approximately described by

$$\lambda(T) \approx \lambda(0)[1 - (T/T_c)^4]^{-1/2} \tag{1.7}$$

The implications of the London equations are illustrated much more thoroughly in Chap. 2.

A simple, but unsound, "derivation" of (1.3) can be given by computing the response to a uniform electric field of a perfect normal conductor, i.e., a free-electron gas with mean free path $\ell = \infty$. In that case, $d(m\mathbf{v})/dt = e\mathbf{E}$, and since $\mathbf{J} = ne\mathbf{v}$, (1.3) follows. But this computation is not rigorous for the spatially nonuniform fields in the penetration depth, for which (1.3) and (1.4) are most useful. The fault is that the response of an electron gas to electric fields is nonlocal; i.e., the current at a point is determined by the electric field averaged over a region of radius $\sim\ell$ about that point. Consequently, only fields that are uniform over a region of this size give a full response; in particular, the conductivity becomes *infinite* as $\ell \to \infty$ *only* for fields filling all space. Since we are dealing here with an interface between a region with field and one with no field, it is clear that even for $\ell = \infty$, the effective conductivity would remain finite. For the case of a high-frequency current, this corresponds to the extreme anomalous limit of the normal skin effect, in which the surface resistance remains finite even as $\ell \to \infty$.

A more profound motivation for the London equations is the quantum one, emphasizing use of the vector potential $\mathbf{A}$, given by F. London[7] himself. Noting that the canonical momentum $\mathbf{p}$ is $(m\mathbf{v} + e\mathbf{A}/c)$, and arguing that in the absence of an applied field we would expect the ground state to have zero net momentum (as shown in a theorem[8] of Bloch), we are led to the relation for the local average velocity in the presence of the field

$$\langle \mathbf{v}_s \rangle = \frac{-e\mathbf{A}}{mc}$$

---

[7] F. London, *Superfluids*, vol. I, Wiley, New York, 1950.

[8] This theorem is apparently unpublished, though famous. See p. 143 of the preceding reference.

This will hold if we postulate that for some reason the wavefunction of the superconducting electrons is "rigid" and retains its ground-state property that $\langle \mathbf{p} \rangle = 0$. Denoting the number density of electrons participating in this rigid ground state by n_s, we then have

$$\mathbf{J}_s = n_s e \langle \mathbf{v}_s \rangle = \frac{-n_s e^2 \mathbf{A}}{mc} = \frac{-\mathbf{A}}{\Lambda c} \tag{1.8}$$

Taking the time derivative of both sides yields (1.3) and taking the curl leads to (1.4). Thus, (1.8) contains both London equations in a compact and suggestive form.[9]

This argument of London leaves open the actual value of n_s, but a natural upper limit is provided by the total density of conduction electrons n. If this is inserted in (1.5), we obtain

$$\lambda_L(0) = \left(\frac{mc^2}{4\pi n e^2} \right)^{1/2} \tag{1.9}$$

The notation here is chosen to indicate that this is an ideal theoretical limit as $T \to 0$. Note that n_s is expected to decrease continuously to zero as $T \to T_c$, causing $\lambda(T)$ to diverge at T_c as described by (1.7). Careful comparisons of the rf penetration depths of samples in the normal and superconducting states have shown that the superconducting penetration depths λ are always larger than $\lambda_L(0)$, even after an extrapolation of the data to $T = 0$. The quantitative explanation of this excess penetration depth required introduction of an additional concept by Pippard: the coherence length ξ_0.

1.3 THE PIPPARD NONLOCAL ELECTRODYNAMICS

Pippard[10] introduced the coherence length while proposing a nonlocal generalization of the London equation (1.8). This was done in analogy to Chambers's nonlocal generalization[11] of Ohm's law from $\mathbf{J}(\mathbf{r}) = \sigma \mathbf{E}(\mathbf{r})$ to

$$\mathbf{J}(\mathbf{r}) = \frac{3\sigma}{4\pi\ell} \int \frac{\mathbf{R}[\mathbf{R} \cdot \mathbf{E}(\mathbf{r}')] e^{-R/\ell}}{R^4} \, d\mathbf{r}'$$

---

[9] Since (1.8) is evidently not gauge-invariant, it will only be correct for a particular gauge choice. This choice, known as the London gauge, is specified by requiring that div $\mathbf{A} = 0$ (so that div $\mathbf{J} = 0$), that the normal component of $\mathbf{A}$ over the surface be related to any supercurrent through the surface by (1.8), and that $\mathbf{A} \to 0$ in the interior of bulk samples.

[10] A. B. Pippard, *Proc. Roy. Soc.* (London) **A216**, 547 (1953).

[11] This approach of Chambers is discussed, e.g., in J. M. Ziman, *Principles of the Theory of Solids*, Cambridge University Press, New York (1964), p. 242.

where $\mathbf{R} = \mathbf{r} - \mathbf{r}'$; this formula takes into account the fact that the current at a point $\mathbf{r}$ depends on $\mathbf{E}(\mathbf{r}')$ throughout a volume of radius $\sim\ell$ about $\mathbf{r}$. Pippard argued that the superconducting wavefunction should have a similar characteristic dimension ξ_0 which could be estimated by an uncertainty-principle argument, as follows: Only electrons within $\sim kT_c$ of the Fermi energy can play a major role in a phenomenon which sets in at T_c, and these electrons have a momentum range $\Delta p \approx kT_c/v_F$, where v_F, is the Fermi velocity. Thus,

$$\Delta x \gtrsim \hbar/\Delta p \approx \hbar v_F/kT_c$$

leading to the definition of a characteristic length

$$\xi_0 = a\frac{\hbar v_F}{kT_c} \tag{1.10}$$

where a is a numerical constant of order unity, to be determined. For typical elemental superconductors such as tin and aluminum, $\xi_0 \gg \lambda_L(0)$. If ξ_0 represents the smallest size of a wave packet that the superconducting charge carriers can form, then one would expect a weakened supercurrent response to a vector potential $\mathbf{A}(\mathbf{r})$ which did not maintain its full value over a volume of radius $\sim\xi_0$ about the point of interest. Thus, ξ_0 plays a role analogous to the mean free path ℓ in the nonlocal electrodynamics of normal metals. Of course, if the ordinary mean free path is less than ξ_0, one might expect a further reduction in the response to an applied field.

Collecting these ideas into a concrete form, Pippard proposed replacement of (1.8) by

$$\mathbf{J}_s(\mathbf{r}) = -\frac{3}{4\pi\xi_0\Lambda c}\int\frac{\mathbf{R}[\mathbf{R}\cdot\mathbf{A}(\mathbf{r}')]}{R^4}e^{-R/\xi}d\mathbf{r}' \tag{1.11}$$

where again $\mathbf{R} = \mathbf{r} - \mathbf{r}'$ and the coherence length ξ in the presence of scattering was assumed to be related to that of pure material ξ_0 by

$$\frac{1}{\xi} = \frac{1}{\xi_0} + \frac{1}{\ell} \tag{1.12}$$

Using (1.11), Pippard found[12] that he could fit the experimental data on both tin and aluminum by the choice of a single parameter $a = 0.15$ in (1.10). [We shall see in Chap. 3 that the microscopic theory of Bardeen, Cooper, and Schrieffer[13] (BCS) confirms this form, with the numerical constant $a = 0.18$.] For both metals, λ is considerably larger than $\lambda_L(0)$ because $\mathbf{A}(\mathbf{r})$ decreases sharply over a distance $\lambda \ll \xi_0$, giving a weakened supercurrent response, and hence an increased field penetration. Moreover, the increase of λ with the decreasing mean free path predicted by (1.11) and (1.12) was consistent with data on a series

---

[12] T. E. Faber and A. B. Pippard, *Proc. Roy. Soc.* (London) **A231**, 336 (1955).

[13] J. Bardeen, L. N. Cooper, and J. R. Schrieffer, *Phys. Rev.* **108**, 1175 (1957).

of tin-indium alloys with a varying mean free path. Thus, Pippard's nonlocal electrodynamic equation (1.11) not only fitted the experimental data, but it also anticipated the form of electrodynamics found several years later from the microscopic theory.

1.4 THE ENERGY GAP AND THE BCS THEORY

The next step in the evolution of our understanding of superconductors was the establishment of the existence of an energy gap Δ, of order kT_c, between the ground state and the quasi-particle excitations of the system. This concept had been suggested earlier by Daunt and Mendelssohn[14] to explain the observed absence of thermoelectric effects, and it had been postulated theoretically by various workers.[15,16] However, the first quantitative experimental evidence arose from precise measurements of the specific heat of superconductors by Corak et al.[17] These measurements showed that the electronic specific heat well below T_c was dominated by an exponential dependence so that

$$C_{es} \approx \gamma T_c a e^{-bT_c/T} \tag{1.13}$$

where the normal-state electronic specific heat is $C_{en} = \gamma T$, and a and b are numerical constants. Such an exponential dependence, with b found to be $\sim$1.5, implies a minimum excitation energy per particle of $\sim 1.5kT_c$.

At about the same time, measurements of electromagnetic absorption in the region of $\hbar\omega \sim kT_c$ were first carried out. Using millimeter-microwave techniques, Biondi et al.[18] reached this region in aluminum, which has a low $T_c \approx 1.2$ K and hence a small gap, but they were not able to carry the measurements to temperatures much below T_c. Working from the far-infrared side as well as from the microwave side, Glover and Tinkham[19] were able to make a more complete study of thin lead films at temperatures far below $T_c \approx 7.2$ K. These measurements and similar ones on tin films could be interpreted quite convincingly in terms of an energy gap of 3 to 4 times kT_c. This result was consistent with the calorimetric one if excitations always were produced in pairs, as would be expected if they obeyed Fermi statistics. The spectroscopic measurement gives

[14] J. G. Daunt and K. Mendelssohn, *Proc. Roy. Soc.* (London) **A185**, 225 (1946).

[15] See, e.g., V. L. Ginzburg, *Fortschr. Phys.* **1**, 101 (1953) and references cited therein.

[16] J. Bardeen, "Theory of Superconductivity," in S. Flügge (ed.), *Handbuch der Physik*, vol. XV, Springer Verlag, Berlin (1956), pp. 303–310. (This article showed explicitly that an energy gap would account for the Pippard nonlocal electrodynamics.)

[17] W. S. Corak, B. B. Goodman, C. B. Satterthwaite, and A. Wexler, *Phys. Rev.* **96**, 1442 (1954); **102**, 656 (1956).

[18] M. A. Biondi, M. P. Garfunkel, and A. O. McCoubrey, *Phys. Rev.* **102**, 1427 (1956).

[19] R. E. Glover and M. Tinkham, *Phys, Rev.* **104**, 844 (1956); **108**, 243 (1957).

the minimum total energy E_g required to create the pair of excitations; the thermal one measures the energy $E_g/2$ per statistically independent particle.

At this point, Bardeen, Cooper, and Schrieffer[20] (BCS) produced their epoch-making pairing theory of superconductivity, which forms the subject of Chap. 3. In the BCS theory, it was shown that even a weak attractive interaction between electrons, such as that caused in second order by the electron-phonon interaction, causes an instability of the ordinary Fermi-sea ground state of the electron gas with respect to the formation of bound pairs of electrons occupying states with equal and opposite momentum and spin. These so-called *Cooper pairs* have a spatial extension of order ξ_0 and, crudely speaking, comprise the superconducting charge carriers anticipated in the phenomenological theories.

One of the key predictions of this theory was that a minimum energy $E_g = 2\Delta(T)$ should be required to break a pair, creating two quasi-particle excitations. This $\Delta(T)$ was predicted to increase from zero at T_c to a limiting value

$$E_g(0) = 2\Delta(0) = 3.528kT_c \tag{1.14}$$

for $T \ll T_c$. Not only did this result agree with the measured gap widths, but the BCS prediction for the shape of the absorption edge above $\hbar\omega_g = E_g$ was also in quantitative agreement with the data of Glover and Tinkham. This agreement provided one of the most decisive early verifications of the microscopic theory.

1.5 THE GINZBURG-LANDAU THEORY

Although a considerable body of work followed the appearance of the BCS theory, serving to substantiate its predictions for various processes such as nuclear relaxation and ultrasonic attenuation in which the energy gap and excitation spectrum play a key role, the most exciting developments of the ensuing decade came in another direction. This direction is epitomized by the Ginzburg-Landau (GL) theory of superconductivity, which concentrates entirely on the superconducting electrons rather than on excitations, and was actually proposed in 1950, 7 years before BCS. Ginzburg and Landau[21] introduced a complex pseudowavefunction ψ as an order parameter within Landau's general theory of second-order phase transitions. This ψ describes the superconducting electrons, and the local density of superconducting electrons (as defined in the London equations) was given by

$$n_s = |\psi(x)|^2 \tag{1.15}$$

[20]J. Bardeen, L. N. Cooper, and J. R. Schrieffer, *Phys. Rev.* **108**, 1175 (1957).

[21]V. L. Ginzburg and L. D. Landau, *Zh. Eksperim. i Teor. Fiz.* **20**, 1064 (1950).

Then, using a variational principle and working from an assumed series expansion of the free energy in powers of ψ and $\nabla\psi$ with expansion coefficients α and β, they derived the following differential equation for ψ:

$$\frac{1}{2m^*}\left(\frac{\hbar}{i}\nabla - \frac{e^*}{c}A\right)^2 \psi + \beta|\psi|^2\psi = -\alpha(T)\psi \tag{1.16}$$

Note that this is analogous to the Schrödinger equation for a free particle, but with a nonlinear term. The corresponding equation for the supercurrent

$$\mathbf{J}_s = \frac{e^*\hbar}{i2m^*}(\psi^*\nabla\psi - \psi\nabla\psi^*) - \frac{e^{*2}}{m^*c}|\psi|^2\,\mathbf{A} \tag{1.17}$$

was also the same as the usual quantum-mechanical current expression for particles of charge e^* and mass m^*. With this formalism they were able to treat two features which were beyond the scope of the London theory, namely: (1) nonlinear effects of fields strong enough to change n_s (or $|\psi|^2$) and (2) the spatial variation of n_s. A major early triumph of the theory was in handling the so-called intermediate state of superconductors (discussed in Chap. 2), in which superconducting and normal domains coexist in the presence of $H \approx H_c$. The interface between two such domains is shown schematically in Fig. 1.4.

When first proposed, the theory appeared rather phenomenological, and its importance was not generally appreciated, especially in the western literature. However, in 1959, Gor'kov[22] was able to show that the GL theory was, in fact, a limiting form of the microscopic theory of BCS (suitably generalized to deal with spatially varying situations), valid near T_c, in which ψ is directly proportional to the gap parameter Δ. More physically, ψ can be thought of as the wavefunction of the center-of-mass motion of the Cooper pairs. The GL theory is now universally accepted as a masterstroke of physical intuition which embodies in a simple way

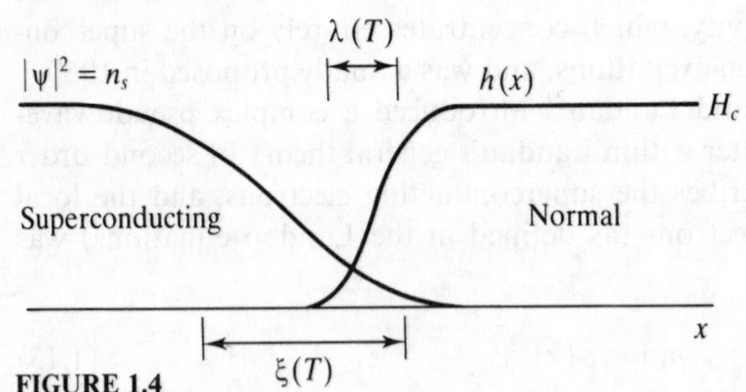

FIGURE 1.4
Interface between superconducting and normal domains in the intermediate state.

[22] L. P. Gor'kov, *Zh. Eksperim. i Teor. Fiz.* **36**, 1918 (1959) [*Sov. Phys.—JETP* **9**, 1364 (1959)].

the macroscopic quantum-mechanical nature of the superconducting state that is crucial for understanding its unique electrodynamic properties.

The GL theory introduces a characteristic length, now usually called the *GL coherence length*,

$$\xi(T) = \frac{\hbar}{|2m^*\alpha(T)|^{1/2}} \tag{1.18}$$

which characterizes the distance over which $\psi(\mathbf{r})$ can vary without undue energy increase. In a pure superconductor far below T_c, $\xi(T) \approx \xi_0$, the (temperature-independent) Pippard coherence length; near T_c, however, $\xi(T)$ diverges as $(T_c - T)^{-1/2}$ since α vanishes as $(T - T_c)$. Thus, these two "coherence lengths" are related but distinct quantities.

The ratio of the two characteristic lengths defines the GL parameter

$$\kappa = \frac{\lambda}{\xi} \tag{1.19}$$

Since λ also diverges as $(T_c - T)^{-1/2}$ near T_c, this dimensionless ratio is approximately independent of temperature. For typical classic pure superconductors, $\lambda \approx 500$ Å and $\xi \approx 3{,}000$ Å, so $\kappa \ll 1$. In this case, one can show[23] (see Chap. 4) that there is a positive surface energy associated with a domain wall between normal and superconducting material. This positive surface energy stabilizes a domain pattern in the intermediate state, with a scale of subdivision intermediate between the microscopic length ξ and the macroscopic sample size.

1.6 TYPE II SUPERCONDUCTORS

In 1957 (the same year as BCS), Abrikosov[24] published a remarkably significant paper, almost overlooked at the time, in which he investigated what would happen in GL theory if κ were large instead of small, i.e., if $\xi < \lambda$, rather than the reverse. Reversing the argument cited above, this should lead to a *negative* surface energy, so that the process of subdivision into domains would proceed until it is limited by the *microscopic* length ξ, below which the gradient energy term would become excessive. Because this behavior is so radically different from the classic intermediate-state behavior described earlier, Abrikosov called these *type II superconductors* to distinguish them from the earlier *type I* variety. He showed that the exact breakpoint between the two regimes was at $\kappa = 1/\sqrt{2}$. For materials with $\kappa > 1/\sqrt{2}$, he found that instead of discontinuous breakdown of superconductivity in a first-order transition at H_c, there was a continuous increase in flux penetration starting at a lower critical field H_{c1} and reaching $B = H$ at an upper

[23] The physical reason is that there is an interfacial layer of thickness $\sim(\xi - \lambda)$ which pays the energetic cost of excluding the magnetic field without enjoying the full condensation energy of the superconducting state.

[24] A. A. Abrikosov, *Zh. Eksperim. i Teor. Fiz.* **32**, 1442 (1957) [*Sov. Phys.—JETP* **5**, 1174 (1957)].

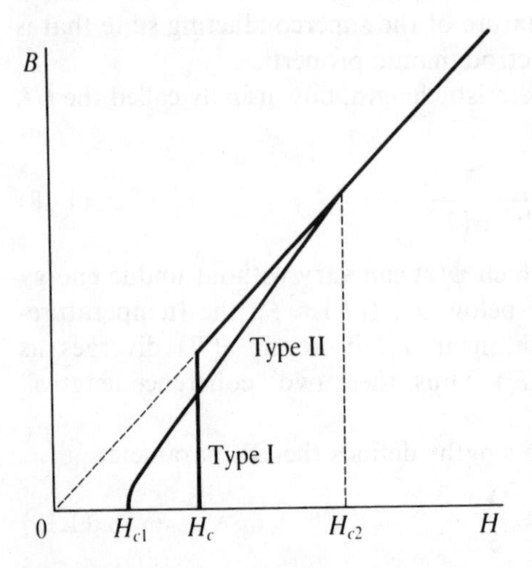

FIGURE 1.5
Comparison of flux penetration behavior of type I and type II superconductors with the same thermodynamic critical field H_c. $H_{c2} = \sqrt{2}\kappa H_c$. The ratio of B/H_{c2} from this plot also gives the approximate variation of R/R_n, where R is the electrical resistance for the case of negligible pinning, and R_n is the normal-state resistance.

critical field H_{c2}, as shown schematically in Fig. 1.5. Because of the partial flux penetration, the diamagnetic energy cost of holding the field out is less, so H_{c2} (which turns out to be given by $\sqrt{2}\kappa H_c$) can be much greater than the thermodynamic critical field H_c (at which nothing special happens). This property has made possible high-field superconducting magnets.

Another result of Abrikosov's analysis was that, in the so-called *mixed state*, or *Schubnikov phase*, between H_{c1} and H_{c2}, the flux should not penetrate in laminar domains but, rather, in a regular array of flux tubes, each carrying a quantum of flux

$$\Phi_0 = \frac{hc}{2e} = 2.07 \times 10^{-7}\text{G} - \text{cm}^2 \qquad (1.20)$$

Within each unit cell of the array, there is a vortex of supercurrent concentrating the flux toward the vortex center. Although Abrikosov predicted a square array, it was later shown, upon correcting a numerical error, that a triangular array should have a slightly lower free energy. This vortex array was first demonstrated experimentally by a magnetic decoration technique coupled with electron microscopy.[25] More recently, scanning tunneling microscope measurements[26] have not only

[25] U. Essmann and H. Träuble, *Phys. Lett.* **24A**, 526 (1967).

[26] H. F. Hess et al. *Phys. Rev. Lett.* **62**, 214 (1989); *Phys. Rev. Lett.* **64**, 2711 (1990).

confirmed the existence of the vortex array, but they have also made possible detailed measurements of the density of electronic states in the quasi-normal core at the center of each vortex. Of course, random inhomogeneities in the underlying material lead to "pinning" of vortices at favorable locations, so that in some cases one finds a glasslike pattern of flux tubes.

We have already noted that type II superconductors are not perfectly diamagnetic, and since $|\psi|^2$ turns out to go to zero in the centers of the vortices, we are not surprised to find that there is no energy gap in the cores. Thus, we are led to ask whether the first hallmark—perfect conductivity—is also lost. The answer is a bit equivocal, and the details are the subject of ongoing research. In the presence of a transport current, the flux tubes experience a so-called *Lorentz force* $\mathbf{J} \times \Phi_0/c$ per unit length (analogous to the macroscopic force density $\mathbf{J} \times \mathbf{B}/c$) tending to make them move sideways, in which case a longitudinal "resistive" voltage is induced. In an ideal homogeneous material, Bardeen and Stephen[27] showed that this flux motion is resisted only by a viscous drag, and that type II superconductors should show a resistance comparable to that in the normal state, only reduced by a factor $\sim B/H_{c2}$. In real materials, however, there is always some inhomogeneity to pin the flux, so that there is essentially no resistance until a finite current is reached, such that the Lorentz force exceeds the pinning force. In superconducting magnet wire, the pinning is deliberately made strong enough to give large critical currents.

1.7 PHASE, JOSEPHSON TUNNELING, AND FLUXOID QUANTIZATION

Faced with these fallen hallmarks, one might well ask what really is the essential universal characteristic of the superconducting state. The answer is the existence of the many-particle condensate wavefunction $\psi(\mathbf{r})$, which has amplitude and phase and which maintains phase coherence over macroscopic distances. This condensate is analogous to, but not identical to, the familiar Bose-Einstein condensate, with Cooper pairs of electrons replacing the single bosons which condense in superfluid helium, for example.

Since the phase and particle number are conjugate variables, reflecting complementary aspects of the wave-particle dualism, there is an uncertainty relation

$$\Delta N \, \Delta\varphi \gtrsim 1 \tag{1.21}$$

which limits the precision with which N and φ can be simultaneously known. However, since $N \sim 10^{22}$ in a macroscopic sample, both N and φ can be known to within small fractional uncertainties, and the phase may be treated as a semiclassical variable. As we shall see in Chap. 7, however, this is not the case in very small mesoscopic structures.

[27] J. Bardeen and M. J. Stephen, *Phys. Rev.* **140**, A1197 (1965).

The physical significance of the phase degree of freedom was first emphasized in the work of Josephson,[28] who predicted that pairs should be able to tunnel between two superconductors even at *zero* voltage difference, giving a supercurrent density

$$J = J_c \sin(\varphi_1 - \varphi_2) \tag{1.22}$$

where J_c is a constant and φ_i is the phase of ψ in the ith superconductor at the tunnel junction. He also predicted that a voltage difference V_{12} between the electrodes would cause the phase difference to increase with time as $2eV_{12}t/\hbar$, so that the current would oscillate with frequency $\omega = 2eV_{12}/\hbar$. Although originally received with some skepticism, these predictions have been extremely thoroughly verified. Subsequently, Josephson junctions have been utilized in ultrasensitive voltmeters and magnetometers, and in making the most accurate available measurements of the ratio of fundamental constants h/e. In fact, the standard volt is now *defined* in terms of the frequency of the ac Josephson current.

The most basic implication of the existence of a phase factor in $\psi(\mathbf{r}) \equiv |\psi(\mathbf{r})| e^{i\varphi(\mathbf{r})}$, however, is operative in the simple case of a superconducting ring. In that case, the single-valuedness of ψ requires that $\varphi(\mathbf{r})$ return to itself (modulo 2π) on going once around the ring on any path. Just as the corresponding condition in an atom leads to the quantization of orbital angular momentum in integral multiples of $\hbar$, here this condition requires that the *fluxoid* Φ' take on only integral multiples of $\Phi_0 = hc/2e$. The fluxoid is a quantity introduced by F. London, which can be written

$$\Phi' = \Phi + \frac{m^* c}{e^{*2}} \oint \frac{\mathbf{J}_s \cdot d\mathbf{s}}{|\psi|^2} \tag{1.23}$$

where $\Phi = \oint \mathbf{A} \cdot d\mathbf{s}$ is the ordinary magnetic *flux* through the integration loop. Since the current sustaining the flux in the ring flows only in a layer of thickness $\sim\lambda$ on the inner surface of the ring, if the ring is thick compared to λ, the path of integration can be taken deeper inside the wall of the ring, where $J_s = 0$. Then (1.23) implies that $\Phi = \Phi'$, so that the flux itself has the quantized value $n\Phi_0$. This property was demonstrated experimentally[29] in 1961. When J_s is not small, as in the vortices in a type II superconductor, both terms in (1.23) may be equally important, and the value of the flux Φ inside a given contour is itself unrestricted; only the flux*oid* Φ', directly related to the line integral of $d\varphi$, always has precise quantum values.

---

[28] B. D. Josephson, *Phys. Lett.* **1**, 251 (1962).

[29] B. S. Deaver and W. M. Fairbank, *Phys. Rev. Lett.* 7, 43 (1961); R. Doll and M. Näbauer, *Phys. Rev. Lett.* 7, 51 (1961).

1.8 FLUCTUATIONS AND NONEQUILIBRIUM EFFECTS

The preceding discussion has been simplified by an implicit assumption that superconductors will always be found in the lowest-energy eigenfunction of the GL equation. While this is indeed the *most probable* single possibility, the presence of the thermal energy $\sim kT$ implies that the system will fluctuate into other low-lying states with a finite probability. At temperatures *below* T_c, the fluctuations which are most prominent are those which allow finite resistance to appear even at currents below the nominal critical current, which is defined as the maximum current before fluctuations of *zero* energy allow a resistive voltage to appear. The other side of the coin is that *above* T_c fluctuations cause some vestiges of superconductivity to remain. These were first observed by Glover,[30] who found that the conductivity of amorphous films of superconductors diverges as $(T - T_c)^{-1}$ as one approaches T_c from above. This "Curie-Weiss" form of temperature dependence with an appropriate coefficient was also predicted theoretically at about the same time. Somewhat later, the corresponding effect was also observed[31] in the diamagnetic susceptibility of pure bulk samples. In this case, the basic divergence is as $(T - T_c)^{-1/2}$.

These measurements and the associated theory show that in principle the effects of the superconducting interaction persist to arbitrarily high temperatures but that in practice a fairly strong cutoff sets in at about $2T_c$. Thus, there is not only some resistance below T_c but also some superconductivity above T_c, although the apparently abrupt switchover observed by Kamerlingh Onnes is still a good working approximation for most purposes. The superconducting transition in the classic superconductors is much sharper than other second-order phase transitions, such as those in magnetic materials, because the coherence length ξ_0 is much larger than the interatomic distance, so that each electron interacts with many others. However, this is not the case in the high-temperature superconductors, where the coherence length is comparable to atomic dimensions, leading to much more prominent fluctuation effects.

In addition to phenomena in which condensate states other than the ground state are explored by thermal fluctuations in the context of thermal equilibrium, *nonequilibrium* regimes have also been studied, as described in Chap. 11. In the simplest examples, energy is fed in from an external source to drive the quasi-particle population out of equilibrium. In this case, one can distinguish two classes of nonequilibrium, involving energy and charge, respectively, as codified by Schmid and Schön.[32] The former category includes regimes in which T_c can actually be raised by as much as a factor of 2, but the energy gap Δ is not raised above its equilibrium value at $T = 0$.

---

[30]R. E. Glover, III, *Phys. Lett.* **25A**, 542 (1967).

[31]J. P. Gollub, M. R. Beasley, R. S. Newbower, and M. Tinkham, *Phys. Rev. Lett.* **22**, 1288 (1969).

[32]A. Schmid and G. Schön, *J. Low Temp. Phys.* **20**, 207 (1975).

1.9 HIGH-TEMPERATURE SUPERCONDUCTIVITY

Finally, we mention the discovery by Bednorz and Müller[33] in 1986 of high-temperature superconductivity in layered materials dominated by copper oxide planes. Materials of this sort have subsequently been discovered with T_c well over 100 K. This has opened the way to a broader range of practical applications than for the classic superconductors because cooling by liquid helium is not required. Realization of these applications has not been immediate, however, because these oxide materials are hard to fabricate in useful forms and have low electron densities compared to conventional metals. Resistance-causing fluctuations are much more prominent than in the classic superconductors for several reasons: (1) Operation at higher temperatures inevitably makes thermal fluctuations more important; (2) using (1.10), the high T_c and the low Fermi velocity stemming from the low electron density implies a short coherence length, which allows easier fluctuations; and (3) the anisotropy induced by the weak interplanar coupling reduces the integrity of vortex lines, so that pinning is less effective. Despite these practical obstacles, the intellectual challenge of understanding these unusual materials has motivated an enormous volume of research, and remarkable progress has been made.

At the time of this writing, the basic physical mechanism responsible for the high T_c is not yet clear. It *is* clear that a two-electron pairing is involved, but the nature of the pairing (*s* wave vs. *d* wave) remains controversial, although very recent experiments seem to favor *d* wave pairing. Nonetheless, the magnetic properties, including the melting of the flux-line lattice giving easily measurable resistance over a substantial range of fields below $H_{c2}(T)$, can be addressed quite satisfactorily within the framework of a model of a layered superconductor with Josephson coupling between the layers. Such a model was introduced by Lawrence and Doniach[34] more than 20 years ago to describe layered low-temperature superconductors. Accordingly, our treatment of high-temperature superconductivity in Chap. 9 will focus primarily on the many implications of the Lawrence-Doniach model for these materials. We shall also include a brief introduction to the Larkin-Ovchinnikov[35] theory of collective pinning and flux creep, and to other topics that are not specifically tied to high-temperature superconductivity but which have been illuminated by the intensive theoretical activity on all aspects of superconductivity brought on by its discovery. Finally, we shall briefly review the experimental evidence in favor of unconventional pairing in these materials.

With this quick overview behind us, we now proceed to a more detailed discussion of the topics mentioned in this outline.

[33] G. Bednorz and K. A. Müller, *Z. Phys.* **B64**, 189 (1986).

[34] W. E. Lawrence and S. Doniach, in E. Kanda (ed.), *Proc. 12th Int. Conf. Low Temp. Phys.*, Kyoto, Japan (1970) [Keigaku, Tokyo, (1971)], p. 361.

[35] A. I. Larkin and Yu. V. Ovchinnikov, *J. Low Temp. Phys.* **34**, 409 (1979).

CHAPTER 2

INTRODUCTION TO ELECTRODYNAMICS OF SUPERCONDUCTORS

In this chapter, we shall work through a number of illustrative examples to get a feeling for the electrodynamic behavior of the classic type I superconductors at the level of the London equations, simply taking the penetration depth $\lambda(T)$ and the critical field $H_c(T)$ as given parameters. This treatment, which avoids the details of the Ginzburg-Landau (GL) theory and the underlying BCS theory, is a good approximation for the cases treated because the GL theory reduces to the London theory when n_s can be taken to have its equilibrium value everywhere, and because the BCS theory introduces qualitative new features *only* at frequencies above the energy gap. In fact, this simple treatment is all that is needed for many applications.

We start by reviewing the physics described by the London equations (1.3) and (1.4) in terms of a simple model. We then illustrate their implications by treating the simple example of a superconducting slab in a parallel dc magnetic field and the effect of the self-field of a current in a wire. Next, we treat the coexistence of superconductivity and normal metal in the so-called *intermediate state* of a type I superconductor in a magnetic field that is comparable to H_c. Finally, we introduce the complex ac conductivity of superconductors and use it to discuss the absorption of electromagnetic radiation at frequencies below the energy gap frequency.

2.1 THE LONDON EQUATIONS

When we introduced the London equations in Sec. 1.2, we pointed out that a classical "derivation" in terms of electrons with infinite mean free path could not be justified in any rigorous way. Nonetheless, such an approach gives a useful physical feeling for the phenomena they describe, and we shall use it here to motivate the form of the equations and also to estimate the value of the penetration depth λ.

In the standard Drude model for electrical conductivity, one applies classical mechanics to the electron motion and writes

$$m\, d\mathbf{v}/dt = e\mathbf{E} - m\mathbf{v}/\tau$$

Here $\mathbf{v}$ is the average or "drift" velocity of the electrons, and τ is a phenomenological relaxation time describing the time it would take the scattering from defects to bring the drift velocity of the electrons to zero. In a normal metal, the competition described by this equation between the scattering and the acceleration by $\mathbf{E}$ leads to a steady-state drift velocity $\mathbf{v} = e\mathbf{E}\tau/m$. If there are n conduction electrons per unit volume, this produces an electric current density $\mathbf{J} = ne\mathbf{v} = (ne^2\tau/m)\mathbf{E} = \sigma\mathbf{E}$, i.e., Ohm's law.

Insofar as it is possible to describe the perfect conductivity of a superconductor by postulating that a certain density n_s of its electrons act as if there were no scattering term (by letting their τ_s go to infinity), Ohm's law is replaced by an *accelerative supercurrent*. That is, we have $d\mathbf{v}_s/dt = e\mathbf{E}/m$, so that the total supercurrent $\mathbf{J}_s$ is governed by

$$d\mathbf{J}_s/dt = (n_s e^2/m)\mathbf{E} = \mathbf{E}/\Lambda = (c^2/4\pi\lambda^2)\mathbf{E} \qquad (1.3')$$

which is equivalent to the first London equation (1.3) together with the definition (1.5) of the parameters Λ and λ. (Our notational convention is to use λ to denote the phenomenological penetration depth involving the phenomenological parameter n_s, reserving λ_L to denote the specific theoretical limiting value for a pure superconductor with local electrodynamics, as defined in the BCS theory.)

Taking the time derivative of the Maxwell equation $\nabla \times \mathbf{h} = 4\pi\mathbf{J}/c$, inserting (1.3′), and then eliminating $\partial\mathbf{h}/\partial t$ using the other Maxwell equation $\nabla \times \mathbf{E} = -(1/c)\,\partial\mathbf{h}/\partial t$, we obtain

$$-\nabla \times \nabla \times \mathbf{E} = \nabla^2\mathbf{E} = \mathbf{E}/\lambda^2 \qquad (2.1a)$$

after using a vector identity. These results apply only to time-varying electric fields since (1.3′) would imply a current accelerating to infinity in response to a strictly dc electric field. Equation (2.1*a*) shows that such a time-dependent electric field is screened out exponentially in a distance λ. By further use of the Maxwell equations, we can show that this result implies that *time-varying magnetic* fields are also screened in this same distance λ.

The essential content of the second London equation (1.4) (which *cannot* be "derived" by a classical argument of this sort) is that this screening of magnetic fields also applies to *time-independent* magnetic fields, as is required to describe the Meissner effect. This can be seen in detail by taking the curl of both sides of the Maxwell equation $\nabla \times \mathbf{h} = (4\pi/c)\mathbf{J}$ and by substituting $-\mathbf{h}/c\Lambda$ from (1.4) for $\nabla \times \mathbf{J}$. (Recall that our notational convention is that $\mathbf{h}$ represents the microscopic value of the magnetic flux density, whereas $\mathbf{B}$ denotes its macroscopic average.) After noting that $\nabla \cdot \mathbf{h} = 0$ by Maxwell's equations, we see that

$$\nabla^2 \mathbf{h} = (1/\lambda^2)\mathbf{h} \tag{2.1b}$$

with

$$\lambda^2 = mc^2/4\pi n_s e^2 \tag{2.2}$$

as stated in (1.5). Equations (2.1*b*) and (2.2) can be considered the operational definition of the phenomenological superconducting electron density n_s in terms of the measurable quantity λ.

2.2 SCREENING OF A STATIC MAGNETIC FIELD

The most important consequence of the London equations (1.3) and (1.4), or the derived equations (2.1*a*,*b*), is that electromagnetic fields are screened from the interior of a bulk superconductor in a characteristic penetration depth λ, given by (2.2). For example, a particular solution of (2.1*b*) describes a magnetic field h, parallel to the surface, which decreases exponentially into the interior of a bulk superconductor as

$$h(x) = h(0)e^{-x/\lambda},$$

where x is measured in from the surface. Equation (2.1*a*) shows that a time-varying *electric* field is screened in the same way.

If one inserts the effective density of conduction electrons n (as determined, e.g., from surface impedance measurements of the electromagnetic skin depth in the normal state) for n_s in (2.2), the predicted value of λ is found to be ~200 Å in typical classic metallic conductors. The experimental values of λ for pure samples are typically 500 Å for $T \ll T_c$. As noted in Sec. 1.3, this quantitative discrepancy results from the nonlocal electrodynamics in pure superconductors, which will be discussed further in connection with the BCS theory in Chap. 3. In superconductors with short electronic mean free path ("dirty superconductors") or short coherence length (such as the high-temperature superconductors), the electrodynamics becomes local, as in the London theory. These materials have larger values of λ (typically up to 1500 Å) corresponding to a much smaller value of the phenomenological parameter n_s.

In all these cases, as one approaches the second-order phase transition at T_c, $n_s \to 0$ continuously; as a result, $\lambda(T)$ diverges as $T \to T_c$. The BCS theory will account for this temperature dependence, which depends in detail on the ratios of the characteristic lengths λ, ξ, and ℓ but is qualitatively the same in all cases. A frequently used empirical temperature dependence is

$$\lambda(T) \approx \lambda(0)/(1 - t^4)^{1/2} \tag{2.3}$$

In terms of (2.2), this corresponds to n_s going to zero at T_c as $(1 - t^4)$, where $t = T/T_c$ is referred to as the *reduced temperature*. This dependence is often called the *two-fluid temperature dependence*, referring to an early model of Gorter and Casimir,[1] which interpreted the thermodynamics of superconductors in terms of coexisting fluids of normal and "condensed" or superconducting electrons. This model related the measured T dependences of the specific heat and the critical field, and gave this dependence for the density of condensed electrons.

2.2.1 Flat Slab in Parallel Magnetic Field

A classic and important example of the application of the London equations is the case of a flat superconducting slab of finite thickness d in an applied parallel magnetic field H_a. Solving (2.1b) with the boundary conditions that $h = H_a$ at the two surfaces at $x = \pm d/2$, one obtains a superposition of exponentials penetrating from both sides, so that

$$h = H_a \frac{\cosh(x/\lambda)}{\cosh(d/2\lambda)} \tag{2.4}$$

This shows that h is reduced to a minimum value $H_a/\cosh(d/2\lambda)$ at the midplane of the slab. Averaged over the sample thickness d, one finds

$$B \equiv \bar{h} \equiv H_a + 4\pi M = H_a \frac{2\lambda}{d} \tanh \frac{d}{2\lambda} \tag{2.5}$$

It is clear from (2.5) that when $d \gg \lambda$, $B \to 0$ and $M \to -H_a/4\pi$. This is the *Meissner effect* limit of perfect diamagnetism of bulk superconductors. On the other hand, when $d \ll \lambda$, series expansion of $\tanh x \approx x - x^3/3 + \cdots$ shows that $B \to H_a(1 - d^2/12\lambda^2)$, so that

$$M \to -(H_a/4\pi)(d^2/12\lambda^2) \tag{2.5a}$$

Some of the earliest experimental determinations of λ were made by comparing magnetization measurements on thin films with the results of calculations of this sort.

We focus particularly on the magnetization M because of its key role in determining the observed critical field. The reason is that the superconducting state becomes energetically unfavorable above the magnetic field H_m at which the added magnetic energy associated with the diamagnetic response in the superconducting state becomes greater than its initial advantage in free energy in zero field. That is, H_m is determined by the relation

$$(F_n - F_s)\,|_{H=0} = -\int_0^{H_m} M(H)\, dH \tag{2.6}$$

---

[1] C. J. Gorter and H. B. G. Casimir, *Phys. Z.* **35**, 963 (1934); *Physica* **1**, 306 (1934).

For the Meissner case, where $M = -H/4\pi$, this maximum field H_m is *called* the *thermodynamic critical field* for this reason, and is given the special symbol H_c. From (2.6) it is clear that

$$H_c^2/8\pi \equiv (F_n - F_s)|_{H=0} \tag{2.6a}$$

as anticipated in (1.1).

For the case of a thin film that is parallel to the applied field, we insert (2.5*a*) into (2.6), and see that the critical field is increased from H_c to

$$H_{c\|} = \sqrt{12} H_c \lambda / d \tag{2.6b}$$

[Although this result is qualitatively correct, the full GL theory (see Chap. 4) replaces the factor $\sqrt{12}$ by $\sqrt{24}$.] To give a numerical example, a 100 Å thick film of tin, which has $\lambda \approx$ 500 Å and $H_c \approx 300$ Œ, will have $H_{c\|} \approx 7,500$ Œ $\gg H_c$. This great increase in the critical magnetic field of a thin film in a parallel magnetic field is an important thermodynamic consequence of the fact that its magnetization M is greatly reduced below the Meissner value.

2.2.2 Critical Current of Wire

Consider a long superconducting wire of circular cross section with radius $a \gg \lambda$, carrying a current I. This current produces a circumferential self-field at the surface of the wire of magnitude $H = 2I/ca$. When this field reaches the critical field H_c, it will destroy the superconductivity. (This is the so-called *Silsbee criterion.*) Thus, the critical current will be $I_c = caH_c/2$, which scales with the *perimeter*, not the cross-sectional area, of the wire. This suggests that the current flows only in a surface layer of constant thickness. It can be confirmed analytically by application of the London and Maxwell equations in this geometry that this is so, and that the thickness of the surface layer is λ. Since the cross-sectional area of this surface layer will be $2\pi a\lambda$, the critical current *density* J_c will be $I_c/2\pi a\lambda$, namely,

$$J_c = \frac{c}{4\pi}\frac{H_c}{\lambda} \tag{2.7}$$

Although this argument in terms of the critical field cannot be used to show it, more general energetic arguments indicate that this value of J_c also holds for wires that are much thinner than λ, where the current density is nearly uniform and I_c *is* proportional to the cross-sectional area. [Again, the full GL theory (see Chap. 4) gives a result differing by a numerical factor of $(2/3)^{3/2} \sim 0.5$.] Putting typical numerical values $H_c = 500$ Œ and $\lambda = 500$ Å in (2.7), one finds that J_c is typically of order 10^8 A/cm^2, a very large value indeed.

2.3 TYPE I SUPERCONDUCTORS IN STRONG MAGNETIC FIELDS: THE INTERMEDIATE STATE

We now consider the effect of fields strong enough to destroy superconductivity, rather than simply induce screening currents to keep the field out of the interior of the sample. The effect of such fields depends on the *shape* of the sample. The simplest case is that of a long, thin cylinder or sheet, parallel to the field, such as the slab treated in the previous section, because in this case the field everywhere along the surface is just equal to the applied field H_a. For other geometries, in which the demagnetizing factor of the sample is *not* zero, the field over part of the surface will exceed the applied field, as is illustrated in Fig. 2.1, causing some normal regions to appear while H_a is still less than H_c.

Let us now consider the relevant free energies in some detail, restricting our attention at first to the simple case of zero demagnetizing factor. When the sample (of volume V) is normal, the total Helmholtz free energy is given by

$$F_n = Vf_{n0} + V\frac{H_a^2}{8\pi} + V_{\text{ext}}\frac{H_a^2}{8\pi} \tag{2.8}$$

where f_{n0} is the free-energy density in the normal state in the absence of the field, and the terms in H_a^2 denote the energy of the field inside and outside the sample, respectively. When the sample is superconducting, the Meissner effect excludes the field from the interior, so that

$$F_s = Vf_{s0} + V_{\text{ext}}\frac{H_a^2}{8\pi} \tag{2.9}$$

where f_{s0} is the free-energy density in the superconducting state. (We assume macroscopic sample dimensions, so that it is permissible to ignore the effects of field penetration and currents in a layer λ deep on the surface.) Taking the difference, we have

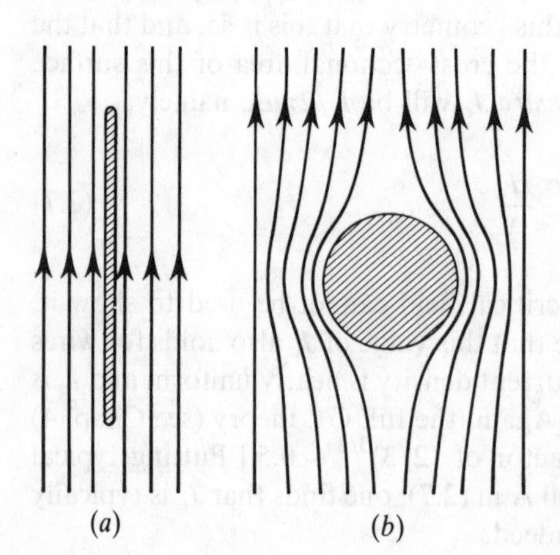

FIGURE 2.1
Contrast of exterior-field pattern (*a*) when demagnetizing coefficient is nearly zero and (*b*) when it is $\frac{1}{3}$ for a sphere. In (*b*) the equatorial field is three halves the applied field for the case of full Meissner effect, which is shown.

$$F_n - F_s = V(f_{n0} - f_{s0}) + \frac{VH_a^2}{8\pi}$$
$$= V\left(\frac{H_c^2}{8\pi}\right) + V\left(\frac{H_a^2}{8\pi}\right) \tag{2.10}$$

In the second form, we have used the defining relation for the thermodynamic critical field H_c

$$f_{n0} - f_{s0} = \frac{H_c^2}{8\pi} \tag{2.11}$$

In particular, when $H_a = H_c$, (2.10) becomes

$$F_n - F_s|_{H_c} = V\left(\frac{H_c^2}{4\pi}\right) \tag{2.12}$$

Thus, at the transition from the superconducting to the normal state, the free energy F increases by $H_c^2/4\pi$ per unit volume. Where does the energy come from? It comes from the energy source maintaining the constant field doing work against the back emf (electromotive force) induced as the flux enters the sample, as can be shown by an elementary computation.

The reason for the awkward necessity of considering the energy of the source of the field is that we carried out the preceding discussion in terms of the Helmholtz free energy. This free energy is appropriate for situations in which **B** is held constant rather than **H** because if **B** is constant, there is no induced emf and no energy input from the current generator. The appropriate thermodynamic potential for the case of constant **H** is the Gibbs free energy G. This differs from F by the term $-V(BH/4\pi)$, which essentially accounts automatically for the work[2] done by the generator. Thus, we consider the Gibbs free-energy density

$$g = f - \frac{hH}{4\pi} \tag{2.13}$$

This leads to

$$G_n = Vf_{n0} - \frac{VH_a^2}{8\pi} - \frac{V_{\text{ext}}H_a^2}{8\pi} \tag{2.14}$$

since **h** = **B** = **H** in the normal state and outside the sample, whereas

$$G_s = Vf_{s0} - \frac{V_{\text{ext}}H_a^2}{8\pi} \tag{2.15}$$

[2]For a discussion of magnetic work, see, e.g., F. Reif, *Fundamentals of Statistical and Thermal Physics*, McGraw-Hill, New York (1965), pp. 439–444.

since $\mathbf{h} = \mathbf{B} = 0$ in the superconducting state. Taking the difference yields

$$G_n - G_s = V(f_{n0} - f_{s0}) - \frac{VH_a^2}{8\pi} \tag{2.16}$$

Because the requirement for phase equilibrium is the equality of the normal and superconducting values of the appropriate thermodynamic potential, which is G for the case of fixed $\mathbf{H}$, (2.16) together with the definition of H_c given in (2.11) imply that $H_a = H_c$ is the condition for coexistence of superconducting and normal phases in equilibrium.

2.3.1 Nonzero Demagnetizing Factor

The preceding discussion was, of course, an idealization for the sake of simplicity. Any real sample will have a nonzero demagnetizing factor, which will cause the field at the surface to be different from H_a, the uniform applied field at large distances from the sample. As a concrete example, let us treat the case of a spherical superconducting sample of radius R. On a macroscopic scale, we still have $\mathbf{B} = 0$ inside the superconductor, at least for $H_a \ll H_c$. Outside, the field satisfies

$$\nabla \cdot \mathbf{B} = \nabla \times \mathbf{B} = \nabla^2 \mathbf{B} = 0 \tag{2.17a}$$

with boundary conditions

$$\mathbf{B} \to \mathbf{H}_a \quad \text{as } r \to \infty \tag{2.17b}$$

$$B_n = 0 \quad \text{at } r = R \tag{2.17c}$$

where B_n is the normal component of $\mathbf{B}$. This is a standard boundary-value problem, with exterior solution

$$\mathbf{B} = \mathbf{H}_a + \frac{H_a R^3}{2} \nabla \left(\frac{\cos\theta}{r^2} \right) \tag{2.18}$$

where θ is the polar angle measured from the direction of $\mathbf{H}_a$. It can readily be verified that (2.18) satisfies all the conditions of (2.17), including $B_n = 0$. Similarly, a direct calculation shows that the surface tangential component of $\mathbf{B}$ is

$$(B_\theta)_R = \tfrac{3}{2} H_a \sin\theta \tag{2.19}$$

Note that this exceeds H_a over an equatorial band of angles from $\theta = 42°$ to $138°$. At the equator, $B_\theta = 3H_a/2$, so that the equatorial field reaches H_c as soon as H_a reaches $2H_c/3$. Therefore, for even slightly higher H_a, certain regions of the sphere must go normal. Still, the whole sphere cannot go normal since if it did, the diamagnetism would disappear completely, leaving $H = H_a \approx 2H_c/3$ everywhere, a value insufficient to keep the superconductivity from reappearing. Thus, for fields in the range

$$\frac{2H_c}{3} < H_a < H_c$$

there must be a coexistence of superconducting and normal regions, which, following historical usage, is called *the intermediate state*. The size of the superconducting and normal regions depends on the value of the (positive) interfacial energy. If this energy were zero (or negative), the subdivision could be finer than λ and the Meissner effect would not be observed.[3]

Generalizing to other ellipsoidial shapes (for which alone a demagnetizing factor is well defined), we expect an intermediate state whenever the applied field lies in range

$$1 - \eta < \frac{H_a}{H_c} < 1 \tag{2.20}$$

The demagnetizing factor η as defined here ranges from zero for the limit of a long, thin cylinder or thin plate in a parallel field to $\frac{1}{3}$ for a sphere to $\frac{1}{2}$ for a cylinder in a transverse field and finally to unity for an infinite flat slab in a perpendicular field. Because the slab in a perpendicular field *always* shows the intermediate state, we treat that limiting case in the greatest detail. It is also the configuration in which most experimental studies of the intermediate state have been carried out.

2.3.2 Intermediate State in a Flat Slab

Let us consider an infinite flat slab of thickness $d \gg \lambda$ in a perpendicular field, a problem first treated in a classic paper of Landau.[4] In this case, it is appropriate to consider the average *flux* per unit area in the slab to be fixed by the source at a value B_a, which will equal the external field far enough from the slab that any inhomogeneities induced by the slab will average out. (See Fig. 2.2) Since $h = 0$ in the superconducting regions, the fraction ρ_n of normal material must then be related to the flux density h_n in it by

$$\rho_n = \frac{B_a}{h_n} \tag{2.21}$$

Neglecting surface-energy effects, $h_n = H_c$, as expected from (2.16), but there will be corrections to this value.

To find the scale and shape of the superconducting and normal regions, we need to consider surface energies in the energy balance. We group these into two terms: F_1, arising from the interfaces within the slab between superconducting (S) and normal (N) domains, and F_2, dominantly dependent on the behavior near the interface between the sample and the space outside.

Since our discussion so far is limited to situations in which the superconductivity is constant in space, we must anticipate our later development of the Ginzburg-Landau (GL) theory (which can deal with spatial gradients) by

---

[3]F. London, *Superfluids*, vol. I, Wiley, New York (1950), p. 128.

[4]L. D. Landau, *Phys. Z. Sowjet*, **11**, 129 (1937).

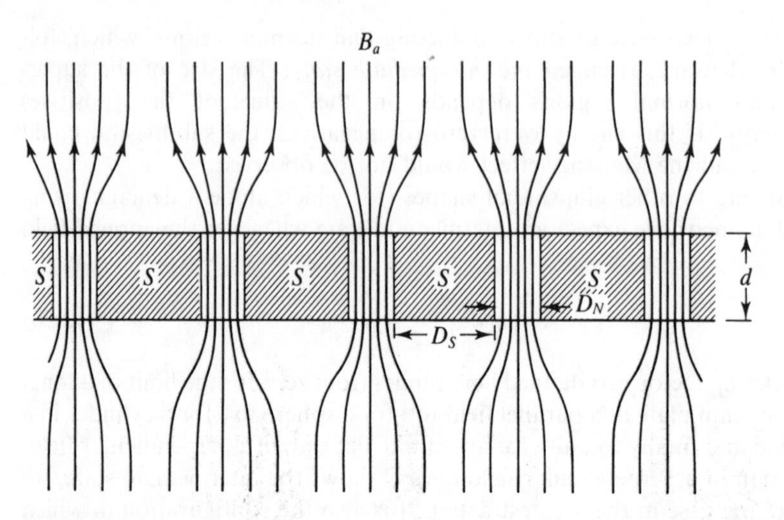

FIGURE 2.2
Schematic diagram showing magnetic flux channeling through the normal laminae in the intermediate state of a type I superconductor. Flux density is B_a at large distances and zero or $h_n(\approx H_c)$ in the cross section of the slab. The normal regions are macroscopic, in contrast to the vortices in a type II superconductor, which contain only a single quantum of flux.

introducing a phenomenological surface-energy term associated with the NS interface. This is usually expressed in terms of a length, which we shall denote[5] $\delta(T)$, such that the additional energy per unit area of interface is

$$\gamma = \frac{H_c^2}{8\pi}\delta \tag{2.22}$$

Since it will turn out that $\delta \approx \xi - \lambda$, δ is positive and of order 10^{-5} to 10^{-4} cm for typical type I superconductors. If the surface energy were negative, we could not have a stable equilibrium of macroscopic volumes of the two phases. Instead, the interfaces would proliferate to gain negative surface energy. This is essentially what we shall find happens in type II superconductors. For the present, then, we take γ to be positive.

Given a positive interface energy γ, the domain walls will assume a configuration of minimum area, all other considerations being equal. In particular, given a pattern of normal regions at the surface (which determines the energy F_2 mentioned earlier), the domain walls will to a good approximation run straight through the slab perpendicular to the surface, since any other choice would increase F_1. It is less clear what the two-dimensional domain pattern should be since it involves optimizing the trade-off between interface energy within the sample (F_1) and field energy just outside the sample (F_2). No rigorous solutions have been found in any

[5]Another common notation is $\Delta(T)$; we avoid this to prevent confusion with the energy gap.

generality; rather, various models are compared. From such studies[6] it has become clear that the free-energy differences are small between even radically different geometries (such as laminae and tubes of normal material), so long as the scale of the structure is optimized in each case. This suggests that many different configurations will be found, depending on the exact experimental conditions and the sample quality, and this expectation is borne out in practice.[7]

Because of its analytical simplicity, and because it is representative of actual observed structures, we shall concentrate on an analysis of the laminar model of the intermediate state. In this model, there is a one-dimensional array of alternating N and S domains, of thickness D_n and D_s, with period $D = D_n + D_s$, as illustrated in Fig. 2.2. The interface energy F_1 per unit area of the slab is then readily seen to be

$$F_1 = \frac{2d\gamma}{D} = \frac{2\ d\delta}{D}\frac{H_c^2}{8\pi} \tag{2.23}$$

By itself, this term favors a very coarse structure, but the exterior term F_2 works in the opposite direction, as we now show.

Although numerical calculations of F_2 were carried out by Landau and his coworkers, we shall content ourselves with a simple physical argument which gives quite similar results. The dominant contribution to F_2 is the energy of the nonuniform external magnetic field outside the domain structure of the intermediate state relative to that of the uniform field which is there if the sample is in the normal state or has an infinitely finely divided domain pattern. At the surface, the average energy density of the field is

$$\frac{\rho_n h_n^2}{8\pi}$$

since a fraction ρ_n of the volume has a field h_n, whereas, using (2.21), the energy density of the uniform field is

$$\frac{B_a^2}{8\pi} = \frac{\rho_n^2 h_n^2}{8\pi}$$

Thus, the average *excess* energy density at the surface due to the domain structure is

$$\frac{(\rho_n - \rho_n^2)h_n^2}{8\pi} = \frac{\rho_n\,\rho_s h_n^2}{8\pi} \tag{2.24}$$

---

[6]See, e.g., E. R. Andrew, *Proc. Roy. Soc.* (London) **A194**, 98 (1948); R. N. Goren and M. Tinkham, *J. Low Temp. Phys.*, **5**, 465 (1971).

[7]For photographs, see, e.g., R. P. Huebener, *Magnetic Flux Structures in Superconductors*, Springer (1979), Chap. 2; J. D. Livingston and W. DeSorbo, "The Intermediate State in Type I Superconductors," in R. D. Parks (ed.), *Superconductivity*, Dekker, New York (1969), p. 1235; A. C. Rose Innes and E. H. Rhoderick, *Introduction to Superconductivity*, 2nd ed., Pergamon, Oxford (1978); T. E. Faber, *Proc. Roy. Soc.* (London) **A248**, 460 (1958).

where $\rho_s = 1 - \rho_n$ is the superconducting fraction. Above the surface, the field inhomogeneity leading to this excess energy will be substantially reduced in a "healing length" L, which will be of the order of the lesser of the lengths D_n and D_s. A convenient mathematical form embodying this observation is

$$L = (D_n^{-1} + D_s^{-1})^{-1} = \frac{D}{\rho_n^{-1} + \rho_s^{-1}} = D\rho_s \rho_n$$

Approximating F_2 by the excess energy density (2.24) out to a distance L on either side of the slab, we have

$$F_2 = \frac{2\rho_n^2 \rho_s^2 D h_n^2}{8\pi} \tag{2.25}$$

If we now minimize the sum of F_1 and F_2 with respect to D, we find

$$D = \frac{(d\delta)^{1/2} H_c}{\rho_n \rho_s h_n} \approx \frac{(d\delta)^{1/2}}{\rho_n \rho_s} \tag{2.26}$$

as the period of the domain structure. Note that its order of magnitude is set by the geometric mean of a macroscopic dimension, the sample thickness d, and a microscopic dimension, the domain-wall thickness δ. For typical values, $D \approx 10^{-2}$ cm. Another characteristic feature is that the number of domains becomes small (i.e., D becomes large) when either ρ_n or ρ_s is small, i.e., near $B_a = 0$ or H_c.

Such domain patterns have been observed experimentally[8] by such varied techniques as: (1) moving a tiny magnetoresistive or Hall-effect probe over the surface, (2) making powder patterns with either ferromagnetic or superconducting (diamagnetic) powders which outline the flux-bearing regions, and (3) using the Faraday magneto-optic effect in magnetic glasses in contact with the surface. Orderly laminar patterns are favored if the magnetic field is applied at an angle to the normal, causing laminae aligned with the field direction to have less domain-wall area and hence lower energy. From measurements on such structures, values of the surface-energy parameter δ have been obtained which are in satisfactory agreement with theoretical expectations based on the GL theory. In fact, these measurements played an important role in establishing that theory in the first place.

In addition to determining the scale of the domain structure, the surface energy also depresses the critical field in the intermediate state to a value H_{cI}, which is somewhat below H_c, the critical field for the case of zero demagnetizing factor. (Do not confuse this H_{cI} for a type I superconductor in the intermediate state with the H_{c1} of a type II superconductor.) We may estimate the size of this effect by computing the surface energy $F_1 + F_2$ with the optimized domain size D

[8] Access to this extensive literature is provided by the review of J. D. Livingston and W. DeSorbo, in R. D. Parks (ed.), *Superconductivity*, Dekker, New York (1969), Chap. 21.

given by (2.26) and by adding it to the volume energy terms, appropriately weighted with ρ_n or ρ_s. The resulting average free energy per unit volume of sample is

$$f_I = \rho_s f_{s0} + \rho_n\left(f_{s0} + \frac{H_c^2}{8\pi} + \frac{h_n^2}{8\pi}\right) + \frac{F_1 + F_2}{d}$$

$$= f_{s0} + \rho_n \frac{H_c^2}{8\pi} + \frac{B_a^2}{\rho_n 8\pi} + 4(1-\rho_n)\left(\frac{\delta}{d}\right)^{1/2} \frac{H_c B_a}{8\pi} \quad (2.27)$$

We note first that if we neglect the surface terms, f_I has its minimum when $\rho_n = B_a/H_c$, or $h_n = H_c$, as expected in that case. When the surface energy is included, f_I has its minimum when

$$\rho_n = \left(\frac{B_a}{H_c}\right)\left[1 - 4\left(\frac{\delta}{d}\right)^{1/2}\left(\frac{B_a}{H_c}\right)\right]^{-1/2} \quad (2.28)$$

At low fields this starts out as (B_a/H_c), but the correction becomes more important at higher fields. Since we define H_{cI} as the value of B_a for which $\rho_s \to 0$ or $\rho_n \to 1$, we have, upon solving by the quadratic formula,

$$H_{cI} = H_c\left[\left(1 + \frac{4\delta}{d}\right)^{1/2} - 2\left(\frac{\delta}{d}\right)^{1/2}\right]$$

$$\approx H_c\left[1 - 2\left(\frac{\delta}{d}\right)^{1/2}\right] \qquad d \gg \delta \quad (2.29)$$

The numerical coefficient of the correction term depends on this particular detailed model, with all its approximations, but the general form of the result seems to hold for quite a variety of models.

In concluding our discussion of this very simplified model, we note that the flux density h_n in the normal regions is given by B_a/ρ_n. Using (2.28), we see that h_n decreases from H_c to H_{cI} as the applied field is increased from zero to its critical value. Thus, it is generally true that the field in the normal regions is somewhat less than H_c. Although this result may appear paradoxical, it simply reflects the role of the surface energies neglected in zeroth-order energy arguments, which consider only terms that are proportional to the volume.

REFINEMENTS. The preceding discussion of a simplified model outlines the major features of the intermediate state in a way which is certainly semiquantitatively correct in its predictions of domain size and of H_{cI}. However, it does not deal with one important qualitative feature, namely, the spreading out of the flux before it leaves the sample. This can occur most simply by having the normal domains fan out near the surface, or by more complex branching or corrugation of the domains. All these refinements increase the interface energy somewhat in order to decrease the field energy by a greater amount.

In his original treatment of the problem, Landau took into account the fanning out of the normal domains within his laminar model. The result of his numerical calculations was to replace the factor $\rho_n \rho_s h_n/H_c = \rho_s B_a/H_c$ in the denominator of (2.26) by a computed function $\phi(B_a/H_c)$ which has a qualitatively similar dependence on the applied field. Thus, this refinement has little effect on the domain size. Because of the fanning out at the surface, the flux density in the normal regions at the surface is less than the interior value, as is partially anticipated by the result of our simple model that $h_n < H_c$. It is also possible to estimate the depression of H_{cI} below H_c by considering the stability of a single isolated superconducting domain (the last one, say), taking into account the fact that the surface tension of the curved interface with the surrounding normal material effectively helps the magnetic field to destroy the last bit of superconductivity.

Curiously, Landau seems to have been somewhat unclear on the stability of this interface and he proposed a second model,[9] in which the normal domains were assumed to branch into two, repeating as necessary, in order to spread out the flux at the surface without having the flux density in the normal regions fall below H_c. Subsequent work has shown that for samples of reasonable thickness, even a single branching would raise the free energy above that of the unbranched model because the reduction in field energy is less than the increase in interface energy. However, features resembling a Landau branching structure have been reported by Solomon and Harris[10] in lead, which has a particularly low surface-energy parameter.

Extensive experimental observations by Faber[11] showed that a complex maze structure of corrugated normal domains was often seen. Presumably, thin normal laminae, flat in the interior, develop a corrugation of increasing amplitude as they approach the surface. This accomplishes the effective dispersal of the emerging flux over a band whose width is equal to the amplitude of the corrugation in a way which appears to be more economical of interface energy than is the branching model of Landau. Obviously, such corrugations affect the interpretation of observed domain sizes in terms of a surface-energy parameter.

It should also be mentioned that flux spots or tubes, rather than laminae, may be observed under suitable circumstances. For example, Landau pointed out that normal tubes should be more favorable at low flux density, whereas superconducting tubes should be more favorable for the last superconducting material near H_c. Experiments[12] of Träuble and Essman have revealed a regular array of flux spots in lead foils in a perpendicular field, whereas Kirchner[13] has observed both flux spots and laminalike "meanders" in rather similar samples. The

[9] L. D. Landau, *Nature* **141**, 688 (1938); *J. Phys. U.S.S.R.* **7**, 99 (1943).

[10] P. R. Solomon and R. E. Harris, *Phys. Rev.* **B3**, 2969 (1971).

[11] T. E. Faber, *Proc. Roy. Soc.* (London) **A248**, 460 (1958).

[12] H. Träuble and U. Essmann, *Phys. Stat. Sol.* **25**, 395 (1968).

[13] H. Kirchner, *Phys. Lett.* **26A**, 651 (1968).

evolution of the flux pattern with increasing field from flux tubes to corrugations, then branches, and finally into superconducting tubes at high fields is particularly clearly demonstrated by motion pictures taken by various groups[14] using the magneto-optic technique. Evidently, the richness of the phenomena observed in the intermediate state poses a severe challenge to any complete theoretical understanding. Yet another dimension of complexity is added in the time-dependent phenomena of the dynamic intermediate state, but we shall not go into that aspect here.

2.3.3 Intermediate State of a Sphere

To illustrate the application of our results in a more general geometry, we now return to the case of the sphere. As found previously, the intermediate state will exist when $\frac{2}{3} < H_a/H_c < 1$. In this range, the volume of the sphere is subdivided into S and N laminae, which fan out near the surface and may branch or become corrugated, but we shall ignore these refinements in the present discussion. Moreover, we shall assume that the radius of the sphere is large enough compared to the domain-wall thickness δ so that we can ignore the difference between H_{cI} and H_c. Then the flux density in the N laminae is always exactly H_c, and the normal fraction ρ_n is B/H_c, where $\mathbf{B}$ is the average of $\mathbf{h}(\mathbf{r})$ over the laminar structure. In the macroscopic Maxwell equations, this average serves for $\mathbf{B}$ everywhere inside the sphere. On the other hand, the magnitude of the Maxwell $\mathbf{H}$ in the sphere throughout the intermediate state is just H_c. This follows since $H = h = H_c$ in the normal laminae, and the tangential component of $\mathbf{H}$ is continuous across the interface between laminae since the only currents there are internal ones associated with the medium in thermodynamic equilibrium. Thus, as is the case in more familiar examples, the macroscopic fields inside the sphere are uniform, whereas those outside are the sum of the applied field plus a dipole field, namely,

$$\mathbf{B} = \mathbf{H} = \mathbf{H}_a + \frac{H_1 R^3}{2} \nabla \left(\frac{\cos\theta}{r^2} \right) \tag{2.30}$$

This has the same form as the expression (2.18) which we found to hold in the linear regime before $2H_c/3$; in that case, the parameter H_1 was chosen to equal H_a, so as to match $B_n = 0$ at $r = R$. In the intermediate state, $B_n \neq 0$. Rather, we determine H_1 by equating the internal and external values of B_n and of H_{tang}:

$$B_n = B \cos\theta = H_a \cos\theta - H_1 \cos\theta \tag{2.31}$$

[14] See, e.g., P. R. Solomon and R. E. Harris, *Proc. 12th Intl. Conf. on Low Temp. Phys.*, Kyoto, Japan (1970), p. 475.

$$H_{\text{tang}} = H_c \sin\theta = H_a \sin\theta + \frac{1}{2} H_1 \sin\theta \tag{2.32}$$

Solving, we find $H_1 = \frac{2(H_c - B)}{3}$, so that

$$B = 3H_a - 2H_c \qquad \frac{2}{3} \le \frac{H_a}{H_c} \le 1 \tag{2.33}$$

Thus, the magnetic induction of the sphere increases linearly from zero to H_c as the applied field H_a increases from $2H_c/3$ to H_c, as depicted in Fig. 2.3.

Because B_n is continuous, B can be measured external to the sphere by measuring B at the pole, $\theta = 0$. Similarly, the continuity of H_{tang} implies that the internal value of H can be measured externally by measuring the equatorial surface field, $B_{\text{equat}} = H_{\text{equat}}$. The predicted dependence of this quantity is also shown in Fig. 2.3. Experimental data on clean samples actually follow these predictions quite well.

2.4 INTERMEDIATE STATE ABOVE CRITICAL CURRENT OF A SUPERCONDUCTING WIRE

As our final example of dc electrodynamics of type I superconductors, we now discuss the appearance of resistance in a superconducting wire above its critical current. Consider a wire of radius a carrying a current I. By Maxwell's equation, the magnetic field at the surface of the wire is $2I/ca$. When this equals H_c, the wire can no longer be entirely superconducting. As noted in Sec. 2.2.2, this defines a critical current

$$I_c = \frac{H_c ca}{2} \tag{2.34}$$

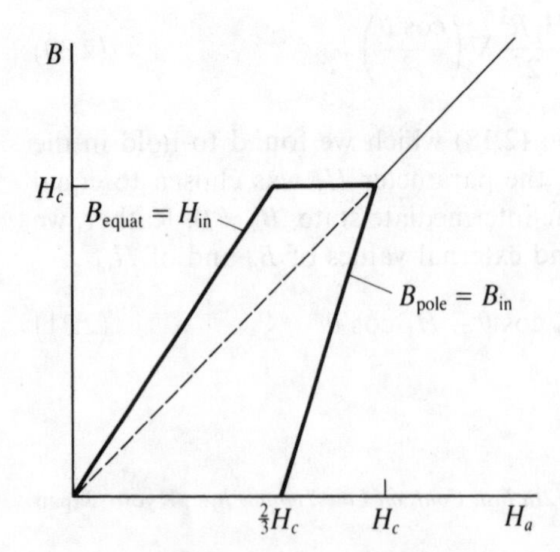

FIGURE 2.3
Internal values of B and H in a superconducting sphere in an applied field H_a. As indicated, these can be measured externally by measuring the surface field B at the pole and the equator, respectively. The sphere is in the intermediate state for $2H_c/3 < H_a < H_c$.

based on Silsbee's rule that the critical current cannot exceed that which produces a critical magnetic field at the superconductor. (The critical current may be much *less* than is given by this criterion, especially if the thickness of the superconductor is much less than λ.) If $I > I_c$, then the surface field exceeds H_c, and the surface (at least) must become normal.

But if a surface layer were to go normal and to leave a fully superconducting core, the current would all go through the core, leading to a still greater field at *its* surface, which would a fortiori be greater than H_c. Thus, no stable configuration exists with a solid superconducting core surrounded by normal material. What if the sample went entirely normal? In this case, the current density J would be uniform across the cross section, leading to

$$H(r) = \frac{2Ir}{ca^2}$$

Since this drops below H_c as $r \to 0$, the core could not be wholly normal either.

These observations suggest a core region (of radius $r_1 < a$) in an intermediate state, surrounded by a normal layer which also carries current. The latter requires a longitudinal electric field, which is compatible with an intermediate-state structure, so long as its layers are oriented transverse to the axis.

The nature of the intermediate-state structure is dictated by the requirement that, neglecting surface energies, $H(r) = H_c$ for $r \leq r_1$. Since $H(r) = 2I(r)/cr$, where $I(r)$ is the total current inside radius r, we need $I(r) = crH_c/2$. This requires a current density

$$J(r) = \frac{1}{2\pi r}\frac{dI}{dr} = \frac{cH_c}{4\pi r} \tag{2.35}$$

Yet, the longitudinal electric field E is independent of r, as can be seen since curl $\mathbf{E} = -(1/c)(\partial \mathbf{B}/\partial t) = 0$ if we assume the structure is stable in time. These requirements are approximately reconciled by the configuration shown in Fig. 2.4, first proposed by F. London,[15] in which the fractional path length (parallel to the axis of the wire) of resistive material is r/r_1. If the normal resistivity is ρ, this leads to

$$J(r) = \frac{Er_1}{\rho r}$$

for $r < r_1$. Combining this with (2.35), we see that

$$r_1 = \frac{\rho c H_c}{4\pi E} \tag{2.36}$$

[15] F. London, "Une Conception Nouvelle de la Supraconductibilite," *Act. Sci. et Ind.*, no. 458, Hermann & Cie., Paris (1937). A more accessible discussion may be found on p. 120 in London's book *Superfluids*, vol. I, Wiley, New York (1950).

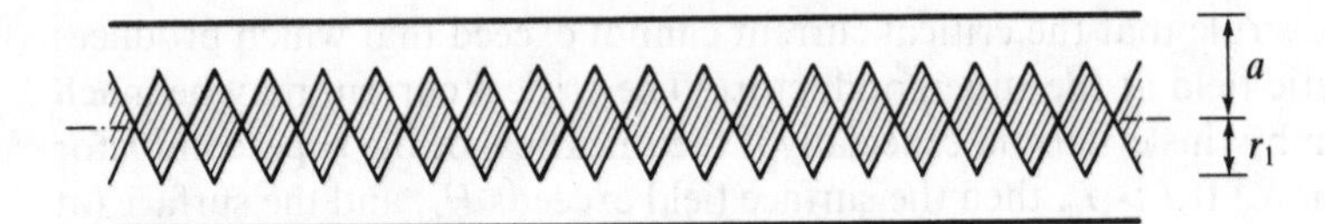

FIGURE 2.4
London's model of the intermediate-state structure in a wire carrying a current in excess of I_c. The shaded region is superconducting. The core radius r_1 is a at I_c and ideally approaches zero only asymptotically as $I \to \infty$.

Since the current inside the core I_1 must generate a field H_c at the surface of the core, we have

$$I_1 = \frac{cr_1 H_c}{2} = \frac{c^2 H_c^2 \rho}{8\pi E}$$

The current in the outer normal layer is

$$I_2 = \frac{E}{\rho}\pi(a^2 - r_1^2) = \frac{\pi a^2 E}{\rho} - \frac{c^2 H_c^2 \rho}{16\pi E}$$

Adding this to I_1, we obtain the total current in the wire

$$I = \frac{\pi a^2 E}{\rho} + \frac{c^2 H_c^2 \rho}{16\pi E} \tag{2.37}$$

Solving the quadratic equation for $E(I)$, and using (2.34), we find

$$E = \frac{\rho I}{2\pi a^2}\left\{1 \pm \left[1 - \left(\frac{I_c}{I}\right)^2\right]^{1/2}\right\} \tag{2.38}$$

The plus sign must be chosen if E is to increase with an increase of I, as required for stability. Note also that in the normal state $E = \rho I/\pi a^2$, by the usual Ohm's law relation. Thus, we can write our results in terms of an apparent fractional resistance

$$\frac{R}{R_n} = \begin{cases} 0 & I < I_c \\ \dfrac{1}{2}\left\{1 + \left[1 - \left(\dfrac{I_c}{I}\right)^2\right]^{1/2}\right\} & I > I_c \end{cases} \tag{2.39}$$

From this we see that half the resistance appears discontinuously at I_c, at which point the intermediate-state pattern suddenly fills the entire wire. With further increase of current, the resistance increases continuously as the intermediate-state region shrinks to a smaller and smaller central core, the asymptotic behavior being $r_1/a = I_c/2I$. In principle, some superconducting material will continue to exist in the core for all finite currents. In practice, however, the Joule heating above I_c makes it hard to carry out an isothermal experiment to confirm this property in detail.

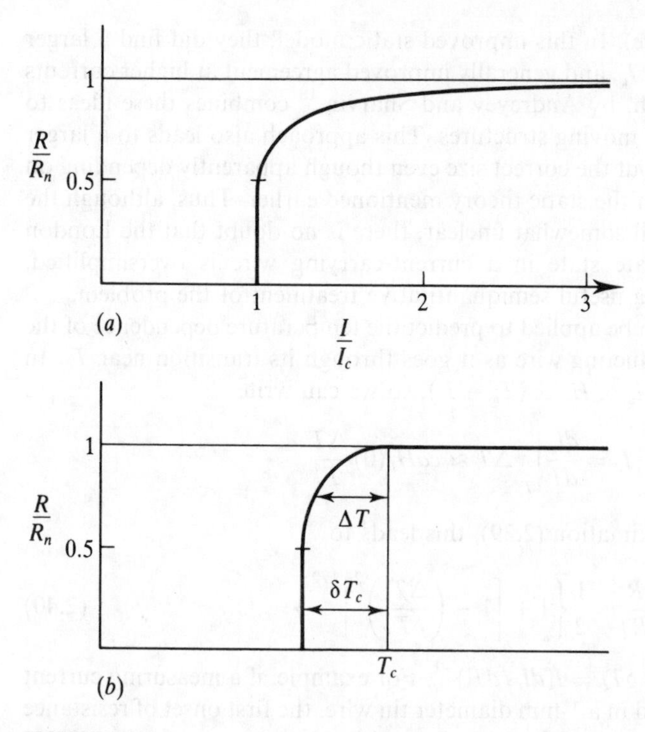

FIGURE 2.5
Resistance of a wire in the intermediate state. (*a*) Current dependence at constant temperature. (*b*) Temperature dependence at constant current, showing the broadening and depression of the apparent transition temperature. The parameter $\delta T_c = I(dI_c/dT)^{-1}$.

Experimental data are in good qualitative agreement with the theoretical result (2.39), which is plotted in Fig. 2.5, but there are quantitative discrepancies. In particular, the discontinuous jump in resistance typically goes from zero to $0.7\text{–}0.8R_n$, rather than to $\frac{1}{2}R_n$, as predicted by the simple theory.[16] This has led to a number of reexaminations of the London model. For example, Gorter[17] considered a dynamic model, with continually moving phase boundaries. On the other hand, Baird and Mukherjee[16] have made more detailed numerical studies of static models similar to London's. They were able to find the optimum ratio of domain period to wire diameter ($\sim$0.7), and the curved domain-wall profiles on which $H = H_c$ and which come closer than the London model to truly satisfying the condition that $H = H_c$ throughout (rather than doing so only on the average

[16]D. C. Baird and B. K. Mukherjee, *Phys. Lett.* **25A**, 137 (1967), and references cited therein. See also B. K. Mukherjee, J. F. Allen, and D. C. Baird, *Proc. 11th Intl. Conf. on Low Temp. Phys.*, St. Andrews (1968), p. 827.

[17]C. J. Gorter, *Physica* **23**, 45 (1957).

over the domain structure). In this improved static model, they did find a larger jump in R/R_n (to 0.69) at I_c, and generally improved agreement at higher currents as well. A third approach, by Andreyev and Sharvin,[18] combines these ideas to consider a broad class of moving structures. This approach also leads to a larger jump in R/R_n at I_c of about the correct size even though apparently depending on different parameters from the static theory mentioned earlier. Thus, although the theoretical position is still somewhat unclear, there is no doubt that the London theory of the intermediate state in a current-carrying wire is oversimplified. Nonetheless, it provides a useful semiquantitative treatment of the problem.

This theory can also be applied to predict the temperature dependence of the resistance of a superconducting wire as it goes through its transition near T_c. In this temperature region, $I_c \propto H_c \propto (T_c - T)$, so we can write

$$I_c = \left.\frac{dI_c}{dT}\right|_{T_c} \Delta T \approx caH_c(0)\frac{\Delta T}{T_c}$$

With the London approximation (2.39), this leads to

$$\frac{R}{R_n} = \frac{1}{2}\left\{1 + \left[1 - \left(\frac{\Delta T}{\delta T_c}\right)^2\right]^{1/2}\right\} \tag{2.40}$$

where $\Delta T = T_c - T$, and $\delta T_c = I(dI_c/dT)^{-1}$. For example, if a measuring current of 1 A (ampere) were used in a 1-mm diameter tin wire, the first onset of resistance would occur about 0.03K below T_c. Thus, in a critical-temperature measurement, I must be kept small enough so that the δT_c from this source is negligible compared to the intrinsic breadth of the transition as limited by sample inhomogeneity. The shape of the resistive transition due to finite current is illustrated in Fig. 2.5*b*.

There is also a current-induced intermediate state in thin-film superconductors. Although the geometry is much less simple to handle theoretically because of edge effects, this configuration has the advantage that the intermediate-state structure can be viewed by a magneto-optic technique. Experiments by Huebener and collaborators[19] have shown that the resistance increases in discrete increments, each associated with the appearance of an additional channel for the motion of magnetic-flux tubes across the strip. The fact that the flux pattern is moving can be demonstrated by the observation of an induced voltage in another adjacent superconducting film in a thin-film sandwich. In these experiments, a time of flight of the order of 10^{-3} sec could be inferred from noise-spectrum measurements. In other experiments, such as those of L. Rinderer, and those of

[18]A. F. Andreyev and Yu. V. Sharvin, *Zh. Eksperim. i Teor. Fiz.* **53**, 1499 (1967); see also A. F. Andreyev, *Proc. 11th Intl. Conf. on Low Temp. Phys.*, St. Andrews (1968), p. 831.

[19]R. P. Huebener and R. T. Kampwirth, *Solid State Comm.* **10**, 1289 (1972); R. P. Huebener and D. E. Gallus, *Phys. Rev.* **B7**, 4089 (1973).

Solomon and Harris cited earlier, motion pictures have been taken of the moving domain patterns under situations in which the motion is much slower. These experimental results confirm that some sort of time-dependent structure is characteristic of resistive regimes in superconductors. We shall return to this point in connection with dissipative effects in type II superconductors in Chap. 5.

2.5 HIGH-FREQUENCY ELECTRODYNAMICS

In the *static* examples treated in earlier sections of this chapter, the superconductor has been described entirely in terms of a lossless diamagnetic response, except for the completely normal domains created in response to strong fields and currents. Most practical applications of superconductivity, however, involve ac currents, whether at low frequencies in power lines or at high frequencies in microwave and computer applications, and superconductors always show finite dissipation when carrying alternating currents. The reason for this is simple. According to the first London equation, a *time-varying* supercurrent requires an electric field $\mathbf{E}$ to accelerate and decelerate the superconducting electrons. This electric field also acts on the so-called "normal" electrons (really thermal excitations from the superconducting ground state, as we shall see in Chap. 3), which scatter from impurities, and can be described by Ohm's law. In this section, we introduce the so-called *two-fluid model*, which describes the electrodynamics that results from the superposition of the response of the "superconducting" and "normal" electron fluids to alternating electromagnetic fields. Although this model is, of course, an oversimplification, it is the standard working approximation for understanding electrical losses in superconductors, so that dissipation can be anticipated and minimized in applications such as microwave resonators. The validity of the model is restricted, however, to frequencies below the energy-gap frequency, since above that frequency additional loss mechanisms set in and the dissipation approaches that in the normal state.

2.5.1 Complex Conductivity in Two-Fluid Approximation

In the Drude model, introduced in Sec. 2.1, the drift velocity of the electron gas, as governed by Newton's second law, obeys the equation

$$m\, d\mathbf{v}/dt = e\mathbf{E} - m\mathbf{v}/\tau \tag{2.41}$$

In a general two-fluid model, one assumes that the total electron density n can be divided into two parts: the density of superconducting electrons is n_s and that of normal electrons is n_n, and they have different relaxation times τ_s and τ_n in (2.41). If one crudely models the behavior of the superconducting electrons simply by assuming $\tau_s = \infty$, as we did in motivating the first London equation, the last term drops out, and we obtain the first London equation (1.3) in the form $d\mathbf{J}_s/dt = (n_s e^2/m)\mathbf{E} = (c^2/4\pi\lambda^2)\mathbf{E}$. The normal electrons give a parallel ohmic conduction channel, with $\mathbf{J}_n = (n_n e^2 \tau_n / m)\mathbf{E}$, provided that $\omega \ll 1/\tau_n$ (as is typically the case even at microwave frequencies).

So long as we are interested only in the *linear* response of the superconductor, it is very convenient to Fourier analyze the applied electric field and to treat the response to each frequency separately. According to (2.41), the ac response of either parallel channel $(i = n, s)$ to a field $Ee^{i\omega t}$ is described by a complex conductivity

$$\sigma_i(\omega) \equiv \sigma_{1i}(\omega) - i\sigma_{2i}(\omega) = (n_i e^2 \tau_i / m)/(1 + i\omega\tau_i) \tag{2.42}$$

whose real and imaginary parts are

$$\sigma_{1i}(\omega) = \sigma_{0i}/(1 + \omega^2 \tau_i^2) \tag{2.43a}$$

$$\sigma_{2i}(\omega) = \sigma_{0i}\omega\tau_i/(1 + \omega^2 \tau_i^2) \tag{2.43b}$$

where $\sigma_{0i} \equiv n_i e^2 \tau_i / m$. If we now describe the superconducting electrons by letting τ_s increase continuously to ∞, we see that the $\sigma_{1s}(\omega)$ curve becomes higher and higher at $\omega = 0$ but cuts off at a lower and lower frequency, retaining a constant area under the curve. Eventually, it shrinks to a δ function $\sigma_{1s}(\omega) = (\pi/2)(n_s e^2/m)\delta(\omega)$, the strength of which can be confirmed by an elementary integration of $\int \sigma_{1s}(\omega)\, d\omega$ over positive frequencies. In this same limiting process, $\sigma_{2s}(\omega)$ approaches the limit $n_s e^2/m\omega$, which is equivalent to the first London equation. For both the normal and superconducting components, the area under the $\sigma_{1i}(\omega)$ curve has the value $(\pi/2)n_i e^2/m$, proportional to n_i and independent of τ_i, as required by the general quantum mechanical *oscillator strength sum rule*, although we have used a classical model. In the historical Gorter-Casimir two-fluid model, it was thought that $n_n \sim t^4$ and $n_s \sim (1 - t^4)$, where $t = T/T_c$. Such a dependence is consistent with the often used empirical approximation that $\lambda \sim (1 - t^4)^{-1/2}$. The modern picture, described in a later chapter, is more complicated, but the temperature dependences are qualitatively similar.

Since this treatment applies only to frequencies below the energy gap, it is usually possible to assume that frequencies are also low enough so that $\omega\tau_n \ll 1$. In that case, the combined response of the two fluids to an electric field reduces simply to

$$\sigma_1(\omega) = (\pi n_s e^2/2m)\delta(\omega) + n_n e^2 \tau_n / m \tag{2.44a}$$

$$\sigma_2(\omega) = n_s e^2/m\omega \tag{2.44b}$$

This approximation is qualitatively very useful, despite its obvious limitations. The essential importance of the normal fluid contribution is that it provides *nonzero* dissipation in superconductors *at all nonzero frequencies*. For example, this accounts for the finite Q in superconducting microwave cavities, as we shall see in detail subsequently.

2.5.2 High-Frequency Dissipation in Superconductors

The ideal dc properties of superconductors are governed by the flow of lossless supercurrent within an equilibrium state of the system. We now examine the response of a superconductor to a high-frequency current, in which the normal electrons give a finite amount of dissipation because the supercurrent response is no longer a zero-impedance shunt. To proceed, we use the approximate two-fluid expressions (2.44) for the complex conductivity. From these expressions we see that at any nonzero frequency, the conductivity has a real part $\sigma_1 = n_n e^2 \tau_n / m$ and an imaginary part $\sigma_2 = n_s e^2 / m\omega$, where n_n and n_s are the densities of normal and superconducting electrons, respectively, in this model description.

It may be helpful to use a circuit analogy, in which σ_1 is the conductance $1/R$ of a resistive channel in parallel with an inductive channel of admittance $1/i\omega L$. This circuit has a characteristic frequency, $\omega_0 = R/L$, below which the dominant current flow is in the lossless inductive channel and above which the resistive channel dominates. Returning to the superconducting case, the ratio of currents in the two channels is

$$\frac{J_s}{J_n} = \frac{n_s e^2 / m\omega}{n_n e^2 \tau_n / m} = \frac{n_s}{n_n \omega \tau_n} \tag{2.45}$$

Thus, the crossover frequency will be $\omega \approx (n_s/n_n)(1/\tau_n)$.[20] Since τ_n is typically $\sim 10^{-12}$ sec, the crossover frequency will be typically $\sim 10^{11}$ Hz (Hertz) if n_s/n_n is of order unity. In the historic two-fluid approximation, the temperature dependence of this ratio is given by $n_s/n_n \approx (1 - t^4)/t^4$. In a more modern version based on the energy gap Δ in the BCS theory, $n_n \sim e^{-\Delta/kT}$, which has the same qualitative behavior as t^4 but goes to zero *exponentially* at low temperatures instead of as a power law. We conclude that at frequencies below the high-microwave range, most of the current will be carried as a supercurrent, but there will be *nonzero* dissipation from the normal component for any nonzero frequency, even one that is orders of magnitude below the crossover frequency.

To make this argument more quantitative, we start by noting that realistic experimental arrangements usually involve a *current bias* as opposed to a *voltage bias* because superconductors have much lower impedance than typical sources of electrical energy, causing the external source to be the current-limiting element in the circuit. Given an imposed ac current density J, the power dissipated per unit volume is $\rho J^2 = \text{Re}\ (1/\sigma) J^2 = [\sigma_1/(\sigma_1^2 + \sigma_2^2)] J^2 \approx (\sigma_1/\sigma_2^2) J^2$, where in the last step we have used the fact that typically $\sigma_1 \ll \sigma_2$. Despite its simplicity, this argument is generally sound and leads to two important and correct conclusions. First, the

---

[20] Actually, in a BCS superconductor with $\omega_g \tau \ll 1$, a better approximation to the crossover frequency is $n\omega_g/n_n$. Here $\omega_g = \Delta/\hbar$ is the energy-gap frequency. This does not change the order of magnitude estimate of the crossover frequency.

frequency dependence of the dissipation is $\sim\omega^2$ because of the factor $1/\sigma_2^2$. Physically, this reflects the fact that the dissipation is $\sigma_1 E^2$, and that the electric field E required to accelerate a given amplitude supercurrent must rise as ω because the acceleration of the electrons must be accomplished in a shorter time period $\sim 1/\omega$. The second important conclusion is that the dissipation is proportional to σ_1, i.e., the density of normal electrons, since they provide the mechanism by which the electric field dissipates energy. [It should be remembered that these conclusions only hold for frequencies with photon energies that are lower than the BCS energy gap, which is $\sim kT_c$. Above that frequency ($\sim 10^{11} - 10^{12}$ Hz), there is little difference between the superconducting and normal-state dissipation properties.]

As a concrete example, we consider the surface resistance and absorptivity of a superconducting surface such as those forming the walls of a microwave cavity. For simplicity, we consider the response to an incoming plane wave at normal incidence. Because the metal impedance is so low compared to free space, the incident wave is almost perfectly reflected, forming a standing wave with a maximum of H at the surface. The amplitude of the oscillating H at the surface is twice the incident amplitude H_{inc} because of the constructive superposition of incident and reflected waves at the surface. Applying the Maxwell equation curl $\mathbf{H} = 4\pi\mathbf{J}/c$, the discontinuity between $h = 2H_{\text{inc}}$ outside the surface and $h = 0$ inside implies a surface *sheet* current density $\mathcal{J} = cH_{\text{inc}}/2\pi$ flowing in the skin layer of depth δ. (This is an example of the *current bias* mentioned in the preceding paragraph.) The dissipated power per unit area is then $\mathcal{J}^2 R_s$, where R_s is the *surface resistance*, i.e, the resistance per square of the surface layer of thickness δ. By solving the skin-depth problem for a general complex conductivity, we find $\delta = c[2\pi\omega(|\sigma| + \sigma_2)]^{-1/2}$, and then

$$R_s = \delta^{-1}\,\text{Re}\,(1/\sigma) = \delta^{-1}\sigma_1/|\sigma|^2 \approx \delta^{-1}\sigma_1/\sigma_2^2 \tag{2.46}$$

The physical significance of this surface resistance is made more evident by using it to compute the absorptivity $\mathcal{A}$ of the surface, i.e., the fraction of the incident electromagnetic energy which is absorbed. Using the magnitude of the Poynting vector to obtain the incoming power, we obtain the following:

$$\mathcal{A} = \frac{P_{\text{abs}}}{P_{\text{inc}}} = \frac{\mathcal{J}^2 R_s}{cH_{\text{inc}}^2/4\pi} = \frac{c}{\pi} R_s \tag{2.47}$$

Thus, the surface resistance is a direct measure of the absorptivity of the surface.

Applying (2.47) to the normal state of a metal, where $\sigma_1 = \sigma_n$ and $\sigma_2 = 0$, we find the familiar result

$$\mathcal{A}_n = (2\omega/\pi\sigma_n)^{1/2} \tag{2.47a}$$

In the superconducting state, the corresponding result is

$$\mathcal{A}_s \approx (2\sigma_1\omega^{1/2})/(\sigma_2^{3/2}\pi^{1/2}) \propto \omega^2\sigma_1 \tag{2.47b}$$

in agreement with our general result for the dependence on frequency and σ_1. The different frequency dependences of (2.47*a,b*) are of practical importance. Since the normal absorptivity rises only as $\omega^{1/2}$ whereas the superconducting absorptivity rises as ω^2, the difference between them narrows as the frequency increases, even well below the energy-gap frequency, above which $\mathscr{A}_s$ rapidly rises to equal $\mathscr{A}_n$.

Since these absorptivities are typically less than 10^{-3} and can be as small as 10^{-10} in superconductors, they are difficult to measure in a single reflection. However, in a resonant microwave cavity with superconducting walls, the radiation is reflected many times as it bounces around in the cavity, thus magnifying the absorption. Experimentally, the level of dissipation in a resonator is described by its Q or *quality factor*, which is defined as the energy stored divided by the energy dissipated per radian (i.e., a time interval of $1/\omega$). This Q also describes the width Δf of the resonant response of the cavity since it can be shown that $\Delta f/f = 1/Q$.

We can approximately relate Q to $\mathscr{A}$ by the steps indicated in the following equation:

$$Q = \frac{\text{stored energy}}{\text{loss per radian}} = \frac{(H^2/8\pi)V}{(c/4\pi\omega)H^2\mathscr{A}S} = \frac{\omega}{2c}\frac{V}{S}\frac{1}{\mathscr{A}} \approx \frac{L}{\lambda}\frac{1}{\mathscr{A}} \approx \frac{1}{\mathscr{A}} \tag{2.48}$$

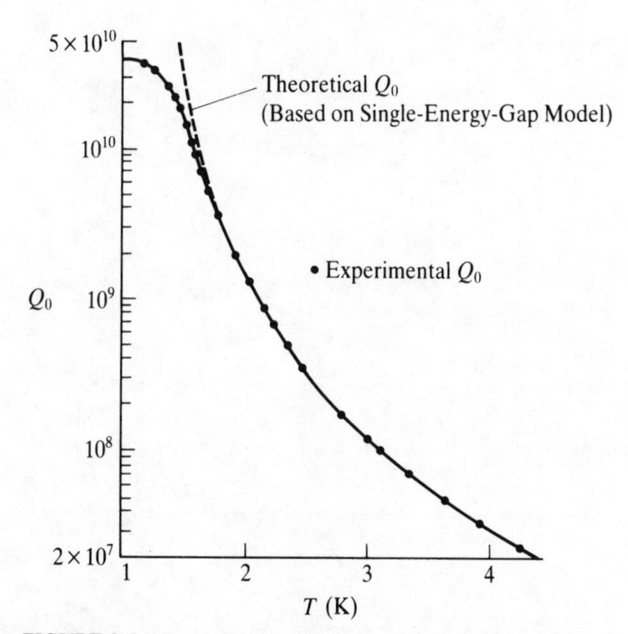

FIGURE 2.6
Temperature dependence of Q_0 for a 11.2-GHz niobium cavity. [*After Turneaure and Weissman*, J. Appl. Phys. **39**, *4417 (1968)*.]

In (2.48), V is the volume of the cavity, S is its surface area, $L = V/S$ is a typical linear dimension of the cavity, and λ is the wavelength of the radiation. The last step is appropriate only if the cavity is operating in its lowest mode, so that its linear dimension is of the order of the wavelength; at best, (2.48) is only roughly valid. Despite the crude approximations, (2.48) provides a useful guide to understanding the relation between absorptivity and cavity Q. With superconducting cavities, Q values as high as 10^{10} have been achieved, indicating extremely low absorption compared to normal metals, for which typically $Q \leq 10^4$. The temperature dependence of the Q of a niobium cavity is shown in Fig 2.6. The exponentially rapid variation reflects the freezing out of normal electrons, causing $n_n(T)$ to fall approximately as $e^{-\Delta/kT}$, where Δ is the BCS energy gap. Such high-Q cavities are of technical importance in applications as very narrow-band microwave filters. For example, a Q of 10^{10} implies a frequency resolution of 1 Hz in a cavity resonant at 10^{10} Hz.

CHAPTER 3

THE BCS THEORY

Although this book emphasizes the phenomenological rather than the microscopic theory of superconductivity, it seems appropriate to put "basic BCS" early in the treatment because this theory[1] revolutionized the quality of our understanding of superconductivity. It deepens our understanding of the phenomenological theory by providing the framework for *calculating* the parameters of that theory, such as H_c and λ, as well as showing how to treat phenomena which require a more microscopic approach, such as the energy gap and coherence factors.

Unfortunately, putting this survey of the BCS theory so early in the book has a pedagogical disadvantage: It is one of the more forbidding chapters, largely because the technique of second quantization is used. To minimize this problem, we have kept the treatment as simple as possible and give some explanation of the method to assist a novice to follow the presentation. It is also worth emphasizing that the phenomenological theory, to which the rest of the book is largely devoted, was generally developed to explain experimental observations independently of, and prior to, its "deduction" from microscopic theory. The phenomenological theory can be studied, appreciated, and used in practice with only a limited "cultural" understanding of the microscopic theory. Thus, readers are urged not to allow themselves to get bogged down in this chapter. If this threatens, skim quickly ahead to Sec. 3.11 on the theory of the penetration depth and

[1] J. Bardeen, L. N. Cooper, and J. R. Schrieffer, *Phys. Rev.* **108**, 1175 (1957).

the concluding summary. Then in reading later chapters, simply refer back to this chapter when necessary to clarify specific points.

3.1 COOPER PAIRS

The basic idea that even a weak attraction can bind pairs of electrons into a bound state was presented by Cooper[2] in 1956. He showed that the Fermi sea of electrons is unstable against the formation of at least one bound pair, regardless of how weak the interaction is, so long as it is attractive. This result is a consequence of the Fermi statistics and of the existence of the Fermi-sea background, since it is well known that binding does not ordinarily occur in the two-body problem in three dimensions until the strength of the potential exceeds a finite threshold value.

To see how this binding comes about, we consider a simple model of two electrons added to a Fermi sea at $T = 0$, with the stipulation that the extra electrons interact with each other but not with those in the sea, except via the exclusion principle. Thus, we seek a two-particle wavefunction. By the general arguments of Bloch, we expect the lowest-energy state to have zero total momentum, so that the two electrons must have equal and opposite momenta. This suggests building up an orbital wavefunction of the sort

$$\psi_0(\mathbf{r}_1, \mathbf{r}_2) = \sum_{\mathbf{k}} g_{\mathbf{k}} e^{i\mathbf{k}\cdot\mathbf{r}_1} e^{-i\mathbf{k}\cdot\mathbf{r}_2}$$

Taking into account the antisymmetry of the total wavefunction with respect to exchange of the two electrons, ψ_0 is converted either to a sum of products of cos $\mathbf{k} \cdot (\mathbf{r}_1 - \mathbf{r}_2)$ with the antisymmetric singlet spin function $(\alpha_1\beta_2 - \beta_1\alpha_2)$ or to a sum of products of sin $\mathbf{k} \cdot (\mathbf{r}_1 - \mathbf{r}_2)$ with one of the symmetric triplet spin functions $(\alpha_1\alpha_2, \alpha_1\beta_2 + \beta_1\alpha_2, \beta_1\beta_2)$. (In these expressions, α_1 refers to the "up" spin state of particle 1, whereas β_1 refers to its "down" state.) Anticipating an attractive interaction, we expect the singlet coupling to have lower energy because the cosinusoidal dependence of its orbital wavefunction on $(\mathbf{r}_1 - \mathbf{r}_2)$ gives a larger probability amplitude for the electrons to be near each other. Thus, we consider a two-electron singlet wavefunction of the form

$$\psi_0(\mathbf{r}_1 - \mathbf{r}_2) = \left[\sum_{k>k_F} g_{\mathbf{k}} \cos \mathbf{k} \cdot (\mathbf{r}_1 - \mathbf{r}_2)\right] (\alpha_1\beta_2 - \beta_1\alpha_2) \tag{3.1}$$

By inserting (3.1) into the Schrödinger equation of the problem, one can show that the weighting coefficients $g_{\mathbf{k}}$ and the energy eigenvalue E are to be determined by solving

$$(E - 2\epsilon_{\mathbf{k}})g_{\mathbf{k}} = \sum_{k'>k_F} V_{\mathbf{k}\mathbf{k}'} g_{\mathbf{k}'} \tag{3.2}$$

[2] L. N. Cooper, *Phys. Rev.* **104**, 1189 (1956).

In this expression, the $\epsilon_{\mathbf{k}}$ are unperturbed plane-wave energies, and the $V_{\mathbf{kk}'}$ are the matrix elements of the interaction potential

$$V_{\mathbf{kk}'} = \Omega^{-1} \int V(\mathbf{r}) e^{i(\mathbf{k}'-\mathbf{k})\cdot\mathbf{r}} d\mathbf{r} \tag{3.3}$$

where $\mathbf{r}$ is the distance between the two electrons and Ω is the normalization volume. This $V_{\mathbf{kk}'}$ characterizes the strength of the potential for scattering a pair of electrons with momenta $(\mathbf{k}', -\mathbf{k}')$ to momenta $(\mathbf{k}, -\mathbf{k})$. If a set of $g_{\mathbf{k}}$ satisfying (3.2) with $E < 2E_F$ can be found, then a bound-pair state exists.

Since it is hard to analyze this situation for general $V_{\mathbf{kk}'}$, Cooper introduced the very serviceable approximation that all $V_{\mathbf{kk}'} = -V$ for $\mathbf{k}$ states out to a cutoff energy $\hbar\omega_c$ away from E_F, and that $V_{\mathbf{kk}'} = 0$ beyond $\hbar\omega_c$. Then the right-hand side of (3.2) is a constant, independent of $\mathbf{k}$, and we have

$$g_{\mathbf{k}} = V \frac{\sum g_{\mathbf{k}'}}{2\epsilon_{\mathbf{k}} - E} \tag{3.4}$$

Summing both sides and canceling $\sum g_{\mathbf{k}}$, we obtain

$$\frac{1}{V} = \sum_{k>k_F} (2\epsilon_{\mathbf{k}} - E)^{-1} \tag{3.5}$$

When we replace the summation by an integration, with $N(0)$ denoting the density of states at the Fermi level for electrons of one spin orientation, this becomes

$$\frac{1}{V} = N(0) \int_{E_F}^{E_F+\hbar\omega_c} \frac{d\epsilon}{2\epsilon - E} = \frac{1}{2} N(0) \ln \frac{2E_F - E + 2\hbar\omega_c}{2E_F - E}$$

In most classic superconductors, it is found that $N(0)V < 0.3$. This allows use of the so-called *weak-coupling approximation*, valid for $N(0)V \ll 1$, in which the solution to the preceding equation can be written as

$$E \approx 2E_F - 2\hbar\omega_c e^{-2/N(0)V} \tag{3.6}$$

Thus, indeed, there *is* a bound state with negative energy with respect to the Fermi surface made up entirely of electrons with $k > k_F$, i.e., with kinetic energy in excess of E_F. The contribution to the energy of the attractive potential outweighs this excess kinetic energy, leading to binding regardless of how small V is. Note that the form of the binding energy is not analytic at $V = 0$; i.e., it cannot be expanded in powers of V. As a result, it cannot be obtained by perturbation theory, a fact that greatly delayed the genesis of the theory.

Returning to the wavefunction, we see that the dependence on the relative coordinate $\mathbf{r} = \mathbf{r}_1 - \mathbf{r}_2$ is proportional to

$$\sum_{k>k_F} \frac{\cos \mathbf{k}\cdot\mathbf{r}}{2\xi_{\mathbf{k}} + E'}$$

where we have gone over to energies measured from the Fermi energy, so that

$$\xi_{\mathbf{k}} = \epsilon_{\mathbf{k}} - E_F \quad \text{and} \quad E' = 2E_F - E > 0 \tag{3.7}$$

(Because of the sign change, E' is now the *binding* energy relative to $2E_F$.) Since $g_\mathbf{k}$ depends only on $\xi_\mathbf{k}$, this solution has spherical symmetry; hence, it is an S state as well as a singlet spin state. Note that the weighting factor $(2\xi_\mathbf{k} + E')^{-1}$ has its maximum value $1/E'$ when $\xi_\mathbf{k} = 0$, i.e., for electrons *at* the Fermi level, and that it falls off with higher values of $\xi_\mathbf{k}$. Thus, the electron states within a range of energy $\sim E'$ above E_F are those most strongly involved in forming the bound state. Since $E' \ll \hbar\omega_c$ for $N(0)V < 1$, this shows that the detailed behavior of $V_{\mathbf{kk}'}$ out around $\hbar\omega_c$ will not have any great effect on the result. This fact gives us some justification for making such a crude approximation to $V_{\mathbf{kk}'}$. A second consequence of this small range of energy states is that, by the uncertainty-principle argument of Pippard cited earlier, it implies that the size of the bound pair is not less than $\sim \hbar v_F/E'$. Since kT_c turns out to be of the order of E', this implies that the size of the Cooper pair state is $\sim\xi_0 = a\hbar v_F/kT_c$, much larger than the interparticle distance. Thus, the pairs are highly overlapping.

3.2 ORIGIN OF THE ATTRACTIVE INTERACTION

We now must examine the origin of the negative $V_{\mathbf{kk}'}$ needed for superconductivity. If we take the bare Coulomb interaction $V(\mathbf{r}) = e^2/r$ and carry out the computation of $V(\mathbf{q})$

$$V(\mathbf{q}) = V(\mathbf{k} - \mathbf{k}') = V_{\mathbf{kk}'} = \Omega^{-1} \int V(\mathbf{r}) e^{i\mathbf{q}\cdot\mathbf{r}} d\mathbf{r}$$

we find

$$V(\mathbf{q}) = \frac{4\pi e^2}{\Omega q^2} = \frac{4\pi e^2}{q^2} \tag{3.8}$$

where the last equality holds for unit normalization volume Ω. Evidently, this $V(\mathbf{q})$ is always positive.

Now, if we take into account the dielectric function $\epsilon(\mathbf{q}, \omega)$ of the medium, $V(\mathbf{q})$ is reduced by a factor $\epsilon^{-1}(\mathbf{q}, \omega)$. The most obvious ingredient in $\epsilon(\mathbf{q}, \omega)$ is the screening effect[3] of the conduction electrons. This introduces a screening length $1/k_s \approx 1$ Å. In the Fermi-Thomas approximation ϵ is given by $\epsilon = 1 + k_s^2/q^2$, so that

$$V(\mathbf{q}) = \frac{4\pi e^2}{q^2 + k_s^2} \tag{3.9}$$

Thus, the electronic screening has eliminated the divergence at $\mathbf{q} = 0$, but it still leaves a positive $V_{\mathbf{kk}'}$. Hence, no superconductivity would result.

[3]See, e.g., J. M. Ziman, *Principles of the Theory of Solids*, Cambridge University Press, New York (1964), Chap. 5.

Negative terms come in only when one takes the motion of the ion cores into account. The physical idea is that the first electron polarizes the medium by attracting positive ions; these excess positive ions in turn attract the second electron, giving an effective attractive interaction between the electrons. If this attraction is strong enough to override the repulsive screened Coulomb interaction, it gives rise to a net attractive interaction, and superconductivity results. Historically, the importance of the electron-lattice interaction in explaining superconductivity was first suggested by Fröhlich[4] in 1950. This suggestion was confirmed experimentally by the discovery[5] of the *isotope effect*, i.e., the proportionality of T_c and H_c to $M^{-1/2}$ for isotopes of the same element.

Since these lattice deformations are resisted by the same stiffness that makes a solid elastic, it is clear that the characteristic vibrational, or phonon, frequencies will play a role. (For the electronically screened Coulomb interaction, the characteristic frequency is the plasma frequency, which is so high that we can assume instantaneous response.) From momentum conservation, we can see that if an electron is scattered from $\mathbf{k}$ to $\mathbf{k}'$, the relevant phonon must carry the momentum $\mathbf{q} = \mathbf{k} - \mathbf{k}'$, and the characteristic frequency must then be the phonon frequency $\omega_{\mathbf{q}}$. As a result, it is plausible that the phonon contribution to the screening function be proportional to $(\omega^2 - \omega_{\mathbf{q}}^2)^{-1}$. Evidently, this resonance denominator gives a *negative* sign if $\omega < \omega_{\mathbf{q}}$, corresponding to the physical argument above; for higher frequencies, i.e., electron energy differences larger than $\hbar\omega_{\mathbf{q}}$, the interaction becomes repulsive. Thus, the cutoff energy $\hbar\omega_c$ of Cooper's attractive matrix element $-V$ is expected to be of the order of the Debye energy $\hbar\omega_D = k\Theta_D$, which characterizes the cutoff of the phonon spectrum.

Early analyses of the best way to treat the coupled electron-phonon system were given by Fröhlich[6] and by Bardeen and Pines.[7] The first attempt to test the theoretical criterion for superconductivity systematically throughout the periodic table was a calculation by Pines.[8] He used the "jellium" model, in which the solid is approximated by a fluid of electrons and point ions, with complete neglect of crystal structure and Brillouin zone effects as well as of the finite ion-core size. As shown in the book of de Gennes,[9] e.g., the jellium model in a certain approximation leads to

$$V(\mathbf{q}, \omega) = \frac{4\pi e^2}{q^2 + k_s^2} + \frac{4\pi e^2}{q^2 + k_s^2} \frac{\omega_{\mathbf{q}}^2}{\omega^2 - \omega_{\mathbf{q}}^2} \tag{3.10}$$

[4]H. Fröhlich, *Phys. Rev.* **79**, 845 (1950).

[5]E. Maxwell, *Phys. Rev.* **78**, 477 (1950); C. A. Reynolds, B. Serin, W. H. Wright, and L. B. Nesbitt, *Phys. Rev.* **78**, 487 (1950).

[6]H. Fröhlich, *Proc. Roy. Soc.* (London) **A215**, 291 (1952).

[7]J. Bardeen and D. Pines, *Phys. Rev.* **99**, 1140 (1955).

[8]D. Pines, *Phys. Rev.* **109**, 280 (1958).

[9]P. G. de Gennes, *Superconductivity in Metals and Alloys*, W. A. Benjamin, New York (1966), reprinted by Addison-Wesley, Reading, MA, 1989, p. 102.

The first term is the screened Coulomb repulsion, whereas the second term is the phonon-mediated interaction, which is attractive for $\omega < \omega_{\mathbf{q}}$. Unfortunately, (3.10) is too simplified to be of much use as a criterion for superconductivity since it reduces to zero for $\omega = 0$, and it is always negative for $\omega < \omega_{\mathbf{q}}$ regardless of material parameters. It does, however, illustrate that the phonon-mediated interaction is of the same order of magnitude as the direct one, so that the concept of achieving a net, negative interaction matrix element in this way is not unreasonable. In fact, a rather quantitative account of superconducting properties of many specific superconductors has been given by careful microscopic calculations of the band structure and electron-phonon coupling. This work has been discussed in an extensive review by Carbotte,[10] and experimental tests of the role of phonons are cited below in Sec. 3.8.5.

Although the phonon-mediated attraction is the basis for superconductivity in the classic superconductors, it is important to recognize that the BCS pairing model requires only an attractive interaction giving a matrix element that can be approximated as $-V$ over a range of energies near E_F. Different pairing interactions, involving the exchange of bosons other than phonons, may well be responsible for superconductivity in some of the more exotic organic, heavy fermion, and high-temperature superconductors. In this case, the electron pairing may have *p*-wave or *d*-wave character, rather than the *s*-wave form assumed here; in fact, as will be discussed in Sec. 9.9, there is considerable evidence for *d*-wave pairing in the high-temperature superconductor YBCO. Nonetheless, the macroscopic phenomenology of the resulting superconducting state is changed only in detail by these differences. Hence, the basic BCS model, treated here, underlies the understanding of all superconductors, even the exotic materials for which significant generalizations are required.

3.3 THE BCS GROUND STATE

Having seen that the Fermi sea is unstable against the formation of a bound Cooper pair when the net interaction is attractive, clearly we must then expect pairs to condense until an equilibrium point is reached. This will occur when the state of the system is so greatly changed from the Fermi sea (because of the large number of bound pairs) that the binding energy for an additional pair has gone to zero. Evidently, it would not be easy to handle such a complicated state unless an ingenious mathematical form could be found. Such a form was provided by the BCS wavefunction.

When we write down wavefunctions for more than two electrons, the manner of handling the antisymmetry used above for a single Cooper pair becomes quite awkward, and it is convenient to replace it by a scheme of $N \times N$ *Slater*

---

[10] J. P. Carbotte, *Revs. Mod. Phys.* **62**, 1027 (1990).

determinants to specify N-electron antisymmetrized product functions. The Slater determinants in turn are more compactly expressed using the language of second quantization, in which the occupied states (including spin index) are specified by the use of "creation operators" such as $c^*_{\mathbf{k}\uparrow}$, which creates an electron of momentum $\mathbf{k}$ and spin up.[11] It is also necessary to introduce annihilation operators $c_{\mathbf{k}\uparrow}$ which empty the corresponding state. In this notation, the singlet wavefunction discussed earlier is written

$$|\psi_0\rangle = \sum_{k>k_F} g_{\mathbf{k}} c^*_{\mathbf{k}\uparrow} c^*_{-\mathbf{k}\downarrow} |F\rangle \tag{3.11}$$

where $|F\rangle$ represents the Fermi sea with all states filled up to k_F. This form makes it obvious that pairs of time-reversed states are always occupied together, a feature that Anderson[12] showed is maintained in the case of dirty superconductors, where a generalized pairing scheme is needed since $\mathbf{k}$ is no longer a good quantum number. One may verify that (3.11) is equivalent to the form (3.1) given above for singlet pairing by summing the two 2×2 Slater determinants with the (equal) coefficients $g_{\mathbf{k}}$ and $g_{-\mathbf{k}}$.

Since electrons obey Fermi statistics, the creation and annihilation operators introduced earlier obey the characteristic anticommutation relations of fermion operators

$$[c_{\mathbf{k}\sigma}, c^*_{\mathbf{k}'\sigma'}]_+ \equiv c_{\mathbf{k}\sigma} c^*_{\mathbf{k}'\sigma'} + c^*_{\mathbf{k}'\sigma'} c_{\mathbf{k}\sigma} = \delta_{\mathbf{k}\mathbf{k}'}\delta_{\sigma\sigma'}$$

$$[c_{\mathbf{k}\sigma}, c_{\mathbf{k}'\sigma'}]_+ = [c^*_{\mathbf{k}\sigma}, c^*_{\mathbf{k}'\sigma'}]_+ = 0 \tag{3.12}$$

where σ refers to the spin index. The particle number operator $n_{\mathbf{k}\sigma}$ is defined by

$$n_{\mathbf{k}\sigma} = c^*_{\mathbf{k}\sigma} c_{\mathbf{k}\sigma} \tag{3.13}$$

which has an eigenvalue of unity when operating on an occupied state and gives zero when operating on an empty state. For our purposes, only elementary manipulations using these rules will be required in carrying out applications of this formalism. We use it simply as a compact notation for dealing with many-electron wavefunctions and operators which act on them.

We approach the BCS wavefunction by observing that the most general N-electron wavefunction expressed in terms of momentum eigenfunctions and with the Cooper pairing built in is

$$|\psi_N\rangle = \sum g(\mathbf{k}_i, \ldots, \mathbf{k}_l) c^*_{\mathbf{k}_i\uparrow} c^*_{-\mathbf{k}_i\downarrow} \cdots c^*_{\mathbf{k}_l\uparrow} c^*_{-\mathbf{k}_l\downarrow} |\phi_0\rangle$$

where $|\phi_0\rangle$ is the vacuum state with no particles present, $\mathbf{k}_i$ and $\mathbf{k}_l$ designate the first and last of the M $\mathbf{k}$ values in the band which are occupied in a given term in the sum, and g specifies the weight with which the product of this set of $N/2$ pairs

[11] We retain the original notation of BCS, but note that $c^*_{\mathbf{k}\uparrow}$ is written as $c^+_{\mathbf{k}\uparrow}$ in more modern notation.

[12] P. W. Anderson, *J. Phys. Chem. Solids* **11**, 26 (1959).

of creation operators appears. The sum runs over all **k** values in the band. Since there are

$$\frac{M!}{[M-(N/2)]!(N/2)!} \approx 10^{(10^{20})}$$

ways of choosing the $N/2$ states for pair occupancy, there will be that many terms in the sum and that many of the $g(\mathbf{k}, \ldots)$ to determine. This is obviously hopeless. BCS argued that with so many particles involved it would be a good approximation to use a Hartree self-consistent field or *mean-field* approach, in which the occupancy of each state **k** is taken to depend solely on the *average* occupancy of other states. In its simplest form, this relaxes the constraint on the total number of particles being N since occupancies are treated only statistically. However, because the number of particles is huge, no serious error is made by working with a system in which only $\bar{N}$ is fixed. Essentially we work in a grand canonical ensemble.

BCS took as their form for the ground state

$$|\psi_G\rangle = \prod_{\mathbf{k}=\mathbf{k}_1,\ldots,\mathbf{k}_M} (u_\mathbf{k} + v_\mathbf{k} c^*_{\mathbf{k}\uparrow} c^*_{-\mathbf{k}\downarrow})|\phi_0\rangle \tag{3.14}$$

where $|u_\mathbf{k}|^2 + |v_\mathbf{k}|^2 = 1$. This form implies that the probability of the pair $(\mathbf{k}\uparrow, -\mathbf{k}\downarrow)$ being occupied is $|v_\mathbf{k}|^2$, whereas the probability that it is unoccupied is $|u_\mathbf{k}|^2 = 1 - |v_\mathbf{k}|^2$. For simplicity, we can consider $u_\mathbf{k}$ and $v_\mathbf{k}$ all real, but it will soon prove important to let them differ by a phase factor $e^{i\varphi}$, where φ is independent of k, and will turn out to be the phase of the macroscopic condensate wavefunction. Evidently, this $|\psi_G\rangle$ can be expressed as a sum

$$|\psi_G\rangle = \sum_N \lambda_N |\psi_N\rangle \tag{3.15}$$

where each term represents the part of the expansion of the product form (3.14) containing $N/2$ pairs. [These $|\psi_N\rangle$ are special cases of the general form mentioned earlier, in which $g(\mathbf{k}, \ldots)$ is given by $\Pi_\mathbf{k} u_\mathbf{k} \Pi_{\mathbf{k}'} v_{\mathbf{k}'}$, where **k** runs over the $(M - N/2)$ unoccupied pair states and **k**′ runs over the $N/2$ occupied pair states.] If all the $u_\mathbf{k}$ and $v_\mathbf{k}$ are finite, there is a finite probability of any N from 0 to $2M$, and $|\psi_G\rangle$ is not an eigenstate of the number operator. However, the values of $|\lambda_N|^2$ are very sharply peaked about the average value

$$\bar{N} = \sum_\mathbf{k} 2|v_\mathbf{k}|^2 \tag{3.16}$$

As an illustration of the formal manipulation of these second quantized forms, let us run through the mechanics of how $\bar{N}$ is calculated, although (3.16) is obviously correct in view of the physical significance of $v_\mathbf{k}$, mentioned earlier. To start,

$$\bar{N} = \langle N_{op} \rangle = \left\langle \sum_{\mathbf{k},\sigma} n_{\mathbf{k}\sigma} \right\rangle = \langle \psi_G | \sum_{\mathbf{k}} (c^*_{\mathbf{k}\uparrow} c_{\mathbf{k}\uparrow} + c^*_{\mathbf{k}\downarrow} c_{\mathbf{k}\downarrow}) | \psi_G \rangle$$

$$= 2 \sum_{\mathbf{k}} \langle \psi_G | c^*_{\mathbf{k}\uparrow} c_{\mathbf{k}\uparrow} | \psi_G \rangle$$

since the electrons all occur in pairs with antiparallel spin. Putting in $|\psi_G\rangle$ explicitly yields

$$\bar{N} = 2 \sum_{\mathbf{k}} \langle \phi_0 | (u^*_{\mathbf{k}} + v^*_{\mathbf{k}} c_{-\mathbf{k}\downarrow} c_{\mathbf{k}\uparrow}) c^*_{\mathbf{k}\uparrow} c_{\mathbf{k}\uparrow} (u_{\mathbf{k}} + v_{\mathbf{k}} c^*_{\mathbf{k}\uparrow} c^*_{-\mathbf{k}\downarrow})$$

$$\times \prod_{\mathbf{l} \neq \mathbf{k}} (u^*_{\mathbf{l}} + v^*_{\mathbf{l}} c_{-\mathbf{l}\downarrow} c_{\mathbf{l}\uparrow})(u_{\mathbf{l}} + v_{\mathbf{l}} c^*_{\mathbf{l}\uparrow} c^*_{-\mathbf{l}\downarrow}) | \phi_0 \rangle$$

In writing this, we have used the property that $\langle A\phi|\psi\rangle = \langle \phi|A^\dagger|\psi\rangle$, and that the adjoint of a product of operators is the product of the adjoints in reverse order. Also, we have been able to rearrange the order of factors in the products to group together all those concerning a given pair state **k** or **l** because, by (3.12), commutation of even numbers of dissimilar Fermi operators introduces no sign change. As we proceed to evaluate this expression, we may think of $|\phi_0\rangle$ as the product of vacuum states for each **k** value. This enables the factor relating to each pair to be evaluated separately. Multiplying out the factor for $\mathbf{l} \neq \mathbf{k}$, we have

$$|u_{\mathbf{l}}|^2 + u^*_{\mathbf{l}} v_{\mathbf{l}} c^*_{\mathbf{l}\uparrow} c^*_{-\mathbf{l}\downarrow} + v^*_{\mathbf{l}} u_{\mathbf{l}} c_{-\mathbf{l}\downarrow} c_{\mathbf{l}\uparrow} + |v_{\mathbf{l}}|^2 c_{-\mathbf{l}\downarrow} c_{\mathbf{l}\uparrow} c^*_{\mathbf{l}\uparrow} c^*_{-\mathbf{l}\downarrow}$$

When we take the $\langle \phi_0| \quad |\phi_0\rangle$ matrix element of this expression, the middle two terms give zero since they change the occupancy of the **l**th pair. The last term creates and then annihilates the pair, leading to a factor of unity. [More carefully, using the commutation relations (3.12), the operators in the last term can be transformed by successive binary interchanges to $-c_{\mathbf{l}\uparrow} c_{-\mathbf{l}\downarrow} c^*_{\mathbf{l}\uparrow} c^*_{-\mathbf{l}\downarrow}$, $+(c_{\mathbf{l}\uparrow} c^*_{\mathbf{l}\uparrow})(c_{-\mathbf{l}\downarrow} c^*_{-\mathbf{l}\downarrow})$, and $+(1 - c^*_{\mathbf{l}\uparrow} c_{\mathbf{l}\uparrow})(1 - c^*_{-\mathbf{l}\downarrow} c_{-\mathbf{l}\downarrow})$, both factors of which give unity when operating on $|\phi_0\rangle$.] Thus, each factor for $\mathbf{l} \neq \mathbf{k}$ simply reduces to $|u_{\mathbf{l}}|^2 + |v_{\mathbf{l}}|^2 = 1$. When the same procedure is followed in the product with $\mathbf{l} = \mathbf{k}$, the cross terms in $u_{\mathbf{k}} v_{\mathbf{k}}$ still drop out since the extra factor $c^*_{\mathbf{k}\uparrow} c_{\mathbf{k}\uparrow}$ leaves particle conservation unaffected. Moreover, because $c_{\mathbf{k}\uparrow}|\phi_0\rangle$ gives zero, the $|u_{\mathbf{k}}|^2$ term also drops out, leaving simply $|v_{\mathbf{k}}|^2$. Thus, we recover (3.16) as anticipated.

To estimate the sharpness of the peak at $\bar{N}$, one needs to evaluate

$$\langle (N - \bar{N})^2 \rangle = \langle N^2 - 2N\bar{N} + \bar{N}^2 \rangle = \langle N^2 \rangle - \bar{N}^2$$

Carrying through a calculation similar to that above, one finds

$$\langle (N - \bar{N})^2 \rangle = 4 \sum_{\mathbf{k}} u^2_{\mathbf{k}} v^2_{\mathbf{k}}$$

This is nonzero unless the occupancy cuts off discontinuously. In fact, it will turn out that $v_{\mathbf{k}}$ goes from 1 to 0, and $u_{\mathbf{k}}$ goes from 0 to 1 in an energy range of $\sim kT_c$, so that the sum is $\sim (T_c/T_F)\bar{N}$. Also, note that both $\bar{N}$ and $\langle (N - \bar{N})^2 \rangle$ scale in proportion to the volume if one compares systems of various sizes but the same particle density. (This follows because the number of **k** values in a given energy range is proportional to the volume.) Accordingly,

$$\delta N_{\text{rms}} = \langle (N - \bar{N})^2 \rangle^{1/2} \approx (T_c/T_F)^{1/2} \bar{N}^{1/2} \approx 10^9 \tag{3.17a}$$

while the fractional uncertainty is

$$\frac{\delta N_{\text{rms}}}{\bar{N}} \approx \frac{(T_c/T_F)^{1/2}}{\bar{N}^{1/2}} \approx 10^{-13} \tag{3.17b}$$

Thus, as is typical of many-particle statistical situations, as $N \to \infty$, the absolute fluctuations become large, but the fractional fluctuations approach zero.

Although for macroscopic samples we can usually ignore exact particle-number conservation, it is useful to note that we can project out the N-particle part of the BCS ground state, if necessary, by a method used by P. W. Anderson.[13] We write $|\psi_G\rangle$, associating an arbitrary phase factor $e^{i\varphi}$ with the creation of each pair, as

$$|\psi_\varphi\rangle = \prod_{\mathbf{k}} (|u_k| + |v_k| e^{i\varphi} c^*_{\mathbf{k}\uparrow} c^*_{-\mathbf{k}\downarrow}) |\phi_0\rangle \tag{3.18a}$$

When we carry out the multiplication over **k**, it produces many terms, which can be grouped into a sum of the sort described by (3.15). The members of the $|\psi_N\rangle$ term are identified by a common phase factor $e^{iN\varphi/2}$, where $N/2$ is the number of pairs in an N-particle state. (N must be an even number since only pairs are included in the wavefunction.) We can then project out this $|\psi_N\rangle$ by simply multiplying by $e^{-iN\varphi/2}$ and integrating on φ over 2π, since this gives zero except for those terms in the expansion of the product in (3.18*a*) in which there are precisely $N/2$ factors of $e^{i\varphi}$, each of which is associated with the creation of a pair. That is,

$$|\psi_N\rangle = \int_0^{2\pi} d\varphi e^{-iN\varphi/2} \prod_{\mathbf{k}} (|u_{\mathbf{k}}| + |v_{\mathbf{k}}| e^{i\varphi} c^*_{\mathbf{k}\uparrow} c^*_{-\mathbf{k}\downarrow}) |\phi_0\rangle = \int_0^{2\pi} d\varphi e^{-iN\varphi/2} |\psi_\varphi\rangle \tag{3.18b}$$

By integrating over all values of φ, i.e., by making φ completely uncertain, we have enforced a precise specification of the number N. On the other hand, with φ fixed as in (3.18*a*), we have seen that $\delta N \approx 10^9$. These results illustrate the uncertainty relation

$$\Delta N \, \Delta\varphi \gtrsim 1 \tag{3.19}$$

The relation between phase and number will be discussed in greater depth in Sec. 7.3, which deals with mesoscopic systems. These effects become crucial when dealing with small particles, where the large Coulomb energy resulting from any excess electrons forces ΔN to be small.

---

[13] P. W. Anderson, "The Josephson Effect and Quantum Coherence Measurements in Superconductors and Superfluids," in C. J. Gorter (ed.), *Progress in Low Temp. Phys.* vol. 5, Wiley, New York (1967), p. 5.

There is an instructive analogy to (3.19) in the case of electromagnetic radiation. In order to have a semiclassical electric field **E** with well-defined phase and amplitude, one must have enough photons (as in a laser), so that one can tolerate a superposition of states with various numbers of photons present.

3.4 VARIATIONAL METHOD

We have studied the structure of ψ_G with some care to bring out some of its interesting features. Now we must actually make it explicit by finding appropriate values for the $u_{\mathbf{k}}$ and $v_{\mathbf{k}}$. Our first approach is a variational calculation, as was used in the original BCS paper. Later we shall discuss another technique which leads to the same conclusions but in somewhat more modern form.

3.4.1 Determination of the Coefficients

We make the calculation by using the so-called *pairing hamiltonian* or *reduced hamiltonian*

$$\mathscr{H} = \sum_{\mathbf{k}\sigma} \epsilon_{\mathbf{k}} n_{\mathbf{k}\sigma} + \sum_{\mathbf{kl}} V_{\mathbf{kl}} c^*_{\mathbf{k}\uparrow} c^*_{-\mathbf{k}\downarrow} c_{-\mathbf{l}\downarrow} c_{\mathbf{l}\uparrow} \tag{3.20}$$

presuming that it includes the terms that are decisive for superconductivity, although it omits many other terms which involve electrons not paired as $(\mathbf{k}\uparrow, -\mathbf{k}\downarrow)$. Such terms have zero expectation value in the BCS ground-state wavefunction but may be important in other applications. To regulate the mean number of particles $\bar{N}$, we include a term $-\mu N_{\text{op}}$, where μ is the chemical potential (or Fermi energy) and N_{op} is the particle-number operator. We then minimize the expectation value of the sum by setting

$$\delta\langle\psi_G|\mathscr{H} - \mu N_{\text{op}}|\psi_G\rangle = 0$$

The inclusion of $-\mu N_{\text{op}}$ is mathematically equivalent to taking the zero of kinetic energy to be μ (or E_F). So. more explicitly, we set

$$\delta\left\langle \psi_G \middle| \sum_{\mathbf{k}\sigma} \xi_{\mathbf{k}} n_{\mathbf{k}\sigma} + \sum_{\mathbf{kl}} V_{\mathbf{kl}} c^*_{\mathbf{k}\uparrow} c^*_{-\mathbf{k}\downarrow} c_{-\mathbf{l}\downarrow} c_{\mathbf{l}\uparrow} \middle| \psi_G \right\rangle = 0$$

where, as before, $\xi_{\mathbf{k}} = \epsilon_{\mathbf{k}} - \mu$ is the single-particle energy relative to the Fermi energy. By the method of calculation used previously to find $\bar{N}$, we see at once that the first term yields

$$\langle KE - \mu N\rangle = 2\sum_{\mathbf{k}} \xi_{\mathbf{k}} |v_{\mathbf{k}}|^2 \tag{3.21}$$

Similarly, the interaction term gives

$$\langle V\rangle = \sum_{\mathbf{kl}} V_{\mathbf{kl}} u_{\mathbf{k}} v^*_{\mathbf{k}} u^*_{\mathbf{l}} v_{\mathbf{l}} \tag{3.22}$$

as can be seen by direct calculation. Alternatively, it can be seen by inspection by noting that the term $V_{\mathbf{kl}}$ scatters from a state with $(\mathbf{l}\uparrow, -\mathbf{l}\downarrow)$ to one with $(\mathbf{k}\uparrow, -\mathbf{k}\downarrow)$. This requires the initial state to have the **l** pair occupied and the **k** pair empty and vice versa for the final state. The probability *amplitude* for such an initial state is $u_{\mathbf{k}}v_{\mathbf{l}}$ and for the final state it is $v_{\mathbf{k}}^*u_{\mathbf{l}}^*$, thus leading to the preceding result. We should note perhaps that $V_{\mathbf{kl}}$ contributes nothing to the energy in the normal state. This is obvious at $T = 0$ since states are either 100 percent occupied or empty, so that the product of the probabilities of being full and empty in zero. At $T > 0$, the Fermi distribution does not cut off sharply, and so one might think there would be a nonzero contribution. However, in the normal state, the various Slater determinants representing specific electron occupation numbers are superimposed with random relative phase, so that the appropriate products of probability amplitudes (corresponding to $u_{\mathbf{k}}v_{\mathbf{k}}^*u_{\mathbf{l}}^*v_{\mathbf{l}}$ in the ordered BCS state) average to zero. Hence, these scattering terms make no contribution to the average energy in the normal state.

Combining (3.21) and (3.22), and for simplicity taking $u_{\mathbf{k}}$ and $v_{\mathbf{k}}$ to be real, we have

$$\langle\psi_G|\mathcal{H} - \mu N_{\mathrm{op}}|\psi_G\rangle = 2\sum_{\mathbf{k}} \xi_{\mathbf{k}} v_{\mathbf{k}}^2 + \sum_{\mathbf{kl}} V_{\mathbf{kl}} u_{\mathbf{k}} v_{\mathbf{k}} u_{\mathbf{l}} v_{\mathbf{l}} \tag{3.23}$$

which is to be minimized subject to the constraint that $u_{\mathbf{k}}^2 + v_{\mathbf{k}}^2 = 1$. This constraint is conveniently imposed by letting

$$u_{\mathbf{k}} = \sin\theta_{\mathbf{k}} \quad \text{and} \quad v_{\mathbf{k}} = \cos\theta_{\mathbf{k}} \tag{3.24}$$

Then, after using elementary trigonometric identities, the right member of (3.23) can be written

$$\sum_{\mathbf{k}} \xi_{\mathbf{k}}(1 + \cos 2\theta_{\mathbf{k}}) + \tfrac{1}{4}\sum_{\mathbf{kl}} V_{\mathbf{kl}} \sin 2\theta_{\mathbf{k}} \sin 2\theta_{\mathbf{l}}$$

whence

$$\frac{\partial}{\partial\theta_{\mathbf{k}}}\langle\psi_G|\mathcal{H} - \mu N_{\mathrm{op}}|\psi_G\rangle = 0 = -2\xi_{\mathbf{k}} \sin 2\theta_{\mathbf{k}} + \sum_{\mathbf{l}} V_{\mathbf{kl}} \cos 2\theta_{\mathbf{k}} \sin 2\theta_{\mathbf{l}} \tag{3.25}$$

(The extra factor of 2 enters in the second sum because both **k** and **l** indices run over any given value $\mathbf{k}'$.) Thus,

$$\tan 2\theta_{\mathbf{k}} = \frac{\sum_{\mathbf{l}} V_{\mathbf{kl}} \sin 2\theta_{\mathbf{l}}}{2\xi_{\mathbf{k}}} \tag{3.26}$$

Now we *define* the quantities

$$\Delta_{\mathbf{k}} = -\sum_{\mathbf{l}} V_{\mathbf{kl}} u_{\mathbf{l}} v_{\mathbf{l}} = -\tfrac{1}{2}\sum_{\mathbf{l}} V_{\mathbf{kl}} \sin 2\theta_{\mathbf{l}} \tag{3.27}$$

and

$$E_{\mathbf{k}} = (\Delta_{\mathbf{k}}^2 + \xi_{\mathbf{k}}^2)^{1/2} \tag{3.28}$$

This $E_{\mathbf{k}}$ will soon be shown to be the excitation energy of a quasi-particle of momentum $\hbar\mathbf{k}$, while $\Delta_{\mathbf{k}}$ will be essentially independent of $\mathbf{k}$, and hence is the mimimum excitation energy, or energy gap. It will also become the *order parameter* in the phenomenological theory, having a phase factor $e^{i\varphi}$, where φ is the relative phase of $u_{\mathbf{k}}$ and $v_{\mathbf{k}}$ as in (3.18*a*).

With these two definitions, (3.26) becomes

$$\tan 2\theta_{\mathbf{k}} = -\frac{\Delta_{\mathbf{k}}}{\xi_{\mathbf{k}}} \tag{3.29a}$$

so that

$$2u_{\mathbf{k}}v_{\mathbf{k}} = \sin 2\theta_{\mathbf{k}} = \frac{\Delta_{\mathbf{k}}}{E_{\mathbf{k}}} \tag{3.29b}$$

and

$$v_{\mathbf{k}}^2 - u_{\mathbf{k}}^2 = \cos 2\theta_{\mathbf{k}} = -\frac{\xi_{\mathbf{k}}}{E_{\mathbf{k}}} \tag{3.29c}$$

The choice of signs for the sine and cosine [only their relative sign is fixed by (3.29*a*)] gives the occupation number $v_{\mathbf{k}}^2 \to 0$ as $\xi_{\mathbf{k}} \to \infty$, as is required for a reasonable solution.

We can now substitute (3.29*b*) back into (3.27) to evaluate $\Delta_{\mathbf{k}}$, leading to the condition for self-consistency

$$\Delta_{\mathbf{k}} = -\frac{1}{2}\sum_{\mathbf{l}} \frac{\Delta_{\mathbf{l}}}{E_{\mathbf{l}}} V_{\mathbf{kl}} = -\frac{1}{2}\sum_{\mathbf{l}} \frac{\Delta_{\mathbf{l}}}{(\Delta_{\mathbf{l}}^2 + \xi_{\mathbf{l}}^2)^{1/2}} V_{\mathbf{kl}} \tag{3.30}$$

We note first the trivial solution in which $\Delta_{\mathbf{k}} = 0$, so that $v_{\mathbf{k}} = 1$ for $\xi_{\mathbf{k}} < 0$, and $v_{\mathbf{k}} = 0$ for $\xi_{\mathbf{k}} > 0$. The associated $|\psi\rangle$ is just the single Slater determinant with all states up to $\mathbf{k}_F$ occupied, the normal Fermi sea at $T = 0$. But we expect a nontrivial solution with lower energy if $V_{\mathbf{kl}}$ is negative. We retain the model of $V_{\mathbf{kl}}$ used by Cooper and by BCS, namely,

$$V_{\mathbf{kl}} = \begin{cases} -V & \text{if } |\xi_{\mathbf{k}}| \text{ and } |\xi_{\mathbf{l}}| \leq \hbar\omega_c \\ 0 & \text{otherwise} \end{cases} \tag{3.31}$$

with V being a positive constant. (Our theoretical discussion of $V_{\mathbf{kl}}$ actually suggests that the relevant energy is $|\xi_{\mathbf{k}} - \xi_{\mathbf{l}}|$, the energy change of the electron in the scattering process, but to get a simple solution, we need to make the stronger restriction that $|\xi_{\mathbf{k}}|$ and $|\xi_{\mathbf{l}}|$ separately are smaller than $\hbar\omega_c$.) Inserting this $V_{\mathbf{kl}}$ in (3.30), we find that it is satisfied by

$$\Delta_{\mathbf{k}} = \begin{cases} \Delta & \text{for } |\xi_{\mathbf{k}}| < \hbar\omega_c \\ 0 & \text{for } |\xi_{\mathbf{k}}| > \hbar\omega_c \end{cases} \tag{3.32}$$

Since in this model $\Delta_{\mathbf{k}} = \Delta$ is actually independent of $\mathbf{k}$, we may cancel it from both sides of (3.30), and our condition for self-consistency then reads

$$1 = \frac{V}{2}\sum_{\mathbf{k}} \frac{1}{E_{\mathbf{k}}} \tag{3.33}$$

Upon replacing the summation by an integration from $-\hbar\omega_c$ to $\hbar\omega_c$, and using the symmetry of $\pm\xi$ values, this becomes

$$\frac{1}{N(0)V} = \int_0^{\hbar\omega_c} \frac{d\xi}{(\Delta^2+\xi^2)^{1/2}} = \sinh^{-1}\frac{\hbar\omega_c}{\Delta} \tag{3.33a}$$

Thus
$$\Delta = \frac{\hbar\omega_c}{\sinh\,[1/N(0)V]} \approx 2\hbar\omega_c e^{-1/N(0)V} \tag{3.34}$$

where the last step is justified in the weak-coupling limit $N(0)V \ll 1$. Since it will turn out that $N(0)V$ is typically $\lesssim 0.3$ the approximate equality in (3.34) is typically good to 1 percent.

Having found Δ, we may simply compute the coefficients $u_{\mathbf{k}}$ and $v_{\mathbf{k}}$ which specify the optimum BCS wavefunction. A convenient approach is to start with (3.29*c*) and the normalization condition $u_{\mathbf{k}}^2 + v_{\mathbf{k}}^2 = 1$. In this way, we find that the fractional occupation number $v_{\mathbf{k}}^2$ is given by

$$v_{\mathbf{k}}^2 = \frac{1}{2}\left(1-\frac{\xi_{\mathbf{k}}}{E_{\mathbf{k}}}\right) = \frac{1}{2}\left[1-\frac{\xi_{\mathbf{k}}}{(\Delta^2+\xi_{\mathbf{k}}^2)^{1/2}}\right]$$

whereas
$$u_{\mathbf{k}}^2 = \frac{1}{2}\left(1+\frac{\xi_{\mathbf{k}}}{E_{\mathbf{k}}}\right) = 1 - v_{\mathbf{k}}^2 \tag{3.35}$$

A plot of $v_{\mathbf{k}}^2$ is shown in Fig. 3.1. Note that $v_{\mathbf{k}}^2$ approaches unity well below the Fermi energy and zero well above, rather like the Fermi function that is appropriate to normal metals at finite temperatures. In fact, there is a startling resemblance between $v_{\mathbf{k}}^2$ for the BCS ground state at $T=0$ and the normal-metal Fermi function at $T=T_c$, also plotted in Fig. 3.1 for comparison purposes. From this comparison we see that, contrary to the early ideas of Fröhlich, Bardeen, and others, the change in the metal on cooling from T_c to $T=0$ cannot be usefully described in terms of changes in the occupation numbers of one-electron

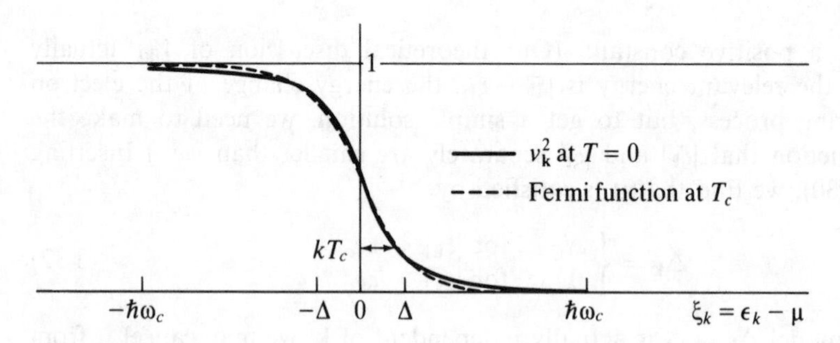

FIGURE 3.1
Plot of BCS occupation fraction $v_{\mathbf{k}}^2$ $(=1-u_{\mathbf{k}}^2)$ vs. electron energy measured from the chemical potential (Fermi energy). To make the cutoffs at $\pm\hbar\omega_c$ visible, the plot has been made for a strong-coupling superconductor with $N(0)V = 0.43$. For comparison, the Fermi function for the normal state at T_c is also shown on the same scale using the BCS relation $\Delta(0) = 1.76kT_c$.

momentum eigenstates. In particular, no gap opens up in **k** space. Rather, the disorder associated with partial occupation of these states with *random* phases is being replaced by a *single* quantum state of the system, in which more or less the same set of many-body states with various one-electron occupancies are now superposed with a *fixed* phase relation.

A further remark about our result for $v_{\mathbf{k}}^2$ is that it falls off as $1/\xi_{\mathbf{k}}^2$ for $\xi_{\mathbf{k}} \gg \Delta$, the same dependence as we found earlier for $g_{\mathbf{k}}^2$ in our simple treatment of a single Cooper pair. In fact, apart from the asymmetry introduced by the artificial restriction of keeping the Fermi sea undisturbed, that simple calculation gives quite an accurate idea of how the energy-lowering correlated-pair state is formed. Finally, we note from (3.35) that Δ is the characteristic energy determining the range of **k** values involved in forming the Cooper pairs. Since Δ will turn out to be $1.76kT_c$ at $T=0$, this reinforces our expectation of a characteristic size $\sim\xi_0 \sim \hbar v_F/\Delta \sim \hbar v_F/kT_c$.

3.4.2 Evaluation of Ground-State Energy

With $|\psi_G\rangle$ determined, we now calculate its energy, to show that it is indeed lower than the Fermi-sea state. From (3.23), using (3.27) and (3.35), we have

$$\langle\psi_G|\mathcal{H} - \mu N_{\text{op}}|\psi_G\rangle = \sum_{\mathbf{k}}\left(\xi_{\mathbf{k}} - \frac{\xi_{\mathbf{k}}^2}{E_{\mathbf{k}}}\right) - \frac{\Delta^2}{V}$$

As noted earlier, the normal state at $T=0$ corresponds to the BCS state with $\Delta=0$, in which case $E_{\mathbf{k}} = |\xi_{\mathbf{k}}|$. Thus,

$$\langle\psi_n|\mathcal{H} - \mu N_{\text{op}}|\psi_n\rangle = \sum_{|\mathbf{k}|<k_F} 2\xi_{\mathbf{k}}$$

the terms for $|\mathbf{k}| > k_F$ giving zero since $E_{\mathbf{k}} = \xi_{\mathbf{k}}$. Thus, the difference in these energies is

$$\begin{aligned}\langle E\rangle_s - \langle E\rangle_n &= \sum_{|\mathbf{k}|>k_F}\left(\xi_{\mathbf{k}} - \frac{\xi_{\mathbf{k}}^2}{E_{\mathbf{k}}}\right) + \sum_{|\mathbf{k}|<k_F}\left(-\xi_{\mathbf{k}} - \frac{\xi_{\mathbf{k}}^2}{E_{\mathbf{k}}}\right) - \frac{\Delta^2}{V}\\ &= 2\sum_{|\mathbf{k}|>k_F}\left(\xi_{\mathbf{k}} - \frac{\xi_{\mathbf{k}}^2}{E_{\mathbf{k}}}\right) - \frac{\Delta^2}{V}\end{aligned}$$

by symmetry about the Fermi energy. In this formula, the sum expresses the change in kinetic energy, whereas the term $-\Delta^2/V$ is the change in potential energy. Going over to the continuum approximation, carrying out the integration on ξ from 0 to $\hbar\omega_c$, and using the weak-coupling-limit relation (3.34), we find

$$\langle E\rangle_s - \langle E\rangle_n = \left[\frac{\Delta^2}{V} - \frac{1}{2}N(0)\Delta^2\right] - \frac{\Delta^2}{V}$$

We have kept the kinetic-energy difference together inside the brackets to make explicit the cancellation of its leading term against the attractive potential-energy term. Thus, the *net* energy lowering is down by a factor of $N(0)V/2 \approx 0.1$ from the increase in kinetic energy or the decrease in potential energy separately. Introducing the thermodynamic symbol $U(T)$ for the internal energy of the system, and anticipating that $\Delta(T)$ is temperature dependent, we have as our final result

$$U_s(0) - U_n(0) = -\frac{1}{2}N(0)\Delta^2(0) \tag{3.36}$$

This is the *condensation energy* at $T = 0$, which must by definition equal $H_c^2(0)/8\pi$, where $H_c(T)$ is the thermodynamic critical field.

3.4.3 Isotope Effect

Since (3.34) shows that Δ is proportional to $\hbar\omega_c$, and since ω_c for different isotopes of the same element should be proportional to $M^{-1/2}$, it follows from (3.36) that the thermodynamic critical field H_c should scale as $M^{-\alpha}$, with $\alpha \approx \frac{1}{2}$. This dependence of H_c (or T_c) on the isotopic mass is called the *isotope effect*. Note that this argument requires the assumption that $N(0)$ and V are unchanged from one isotope to the other. Since $N(0)$ is a purely electronic quantity, it might be expected to be independent of M, but it is less clear for the parameter V, which is determined jointly by the electrons and phonons. In fact, although the isotope-shift exponent α is quite accurately $\frac{1}{2}$ for a few classic superconductors, such as lead, experiments on other materials have shown that it can range all the way to zero or even change sign. Such variations can be accounted for theoretically[14] if one uses a more detailed theory of the interaction than we have been able to give here, but as usual a priori calculations are difficult because they depend on details of the material. It is perhaps more useful to invert the procedure and to try to use the measured isotope effect to gain information about the interaction, as was done for the classic superconductors by McMillan.[15]

More recently, the discovery of the high-temperature superconductors has renewed interest in using the isotope effect to try to elucidate the puzzle of the mechanism responsible for the high transition temperature. Since these materials contain several different types of atoms, many different types of isotopic substitution can be studied. One can even differentially substitute a given type of atom on different types of sites in the crystal. These investigations[16] have shown the importance of the phonon spectrum and revealed certain regularities, but no definitive conclusions about mechanisms had been reached at the time of this writing.

---

[14] P. Morel and P. W. Anderson, *Phys. Rev.* **125**, 1263 (1962); J. W. Garland, Jr., *Phys. Rev. Lett.* **11**, 114 (1963)

[15] W. L. McMillan, *Phys. Rev.* **167**, 331 (1968).

[16] See, e.g., the extensive review by J. P. Franck in D. M. Ginsberg (ed.), *Physical Properties of High Temperature Superconductors IV*, World Scientific, Singapore (1994), pp. 189–293.

3.5 SOLUTION BY CANONICAL TRANSFORMATION

The variational method used in the original BCS treatment, which we have just sketched, is a direct approach for calculating the condensation energy of the superconducting ground state relative to the normal state. It is somewhat clumsy, however, though workable, in dealing with excited states. In this section, we outline another approach, closer to the more sophisticated modern methods, which is well suited to handle excitations. This alternate method is also a self-consistent field method, but no appeal to a variational calculation is required.

We start with the observation that the characteristic BCS pair-interaction hamiltonian will lead to a ground state which is some phase-coherent superposition of many-body states with pairs of Bloch states $(\mathbf{k}\uparrow, -\mathbf{k}\downarrow)$ occupied or unoccupied as units. Because of the coherence, operators such as $c_{-\mathbf{k}\downarrow}c_{\mathbf{k}\uparrow}$ can have nonzero expectation values $b_\mathbf{k}$ in such a state, rather than averaging to zero as in a normal metal, where the phases are random. Moreover, because of the large numbers of particles involved, the fluctuations about these expectation values should be small. This suggests that it will be useful to express such a product of operators formally as

$$c_{-\mathbf{k}\downarrow}c_{\mathbf{k}\uparrow} = b_\mathbf{k} + (c_{-\mathbf{k}\downarrow}c_{\mathbf{k}\uparrow} - b_\mathbf{k}) \tag{3.37}$$

and subsequently neglect quantities which are bilinear in the presumably small fluctuation term in parentheses. If we follow this procedure with our pairing hamiltonian (3.20), we obtain the so-called *model-hamiltonian*

$$\mathscr{H}_M = \sum_{\mathbf{k}\sigma} \xi_\mathbf{k} c^*_{\mathbf{k}\sigma}c_{\mathbf{k}\sigma} + \sum_{\mathbf{kl}} V_{\mathbf{kl}}(c^*_{\mathbf{k}\uparrow}c^*_{-\mathbf{k}\downarrow}b_\mathbf{l} + b^*_\mathbf{k}c_{-\mathbf{l}\downarrow}c_{\mathbf{l}\uparrow} - b^*_\mathbf{k}b_\mathbf{l}) \tag{3.38}$$

where the $b_\mathbf{k}$ are to be determined self-consistently, so that

$$b_\mathbf{k} = \langle c_{-\mathbf{k}\downarrow}c_{\mathbf{k}\uparrow}\rangle_{\text{av}} \tag{3.39}$$

Note that in gaining the simplicity of eliminating quartic terms in the $c_\mathbf{k}$'s from the hamiltonian, we have thrown it into an approximate form which does not conserve particle number. Rather, there are now terms which create or destroy pairs of particles. This is analogous to the situation noted earlier in which the simple BCS product wavefunction with fixed phase contained many different numbers of particles. Only by integrating over the phase φ were we able to set an exact particle number. The corresponding situation here is that we have assigned a definite phase to $b_\mathbf{k}$. In any case, as before, we can handle this situation by introducing the chemical potential μ so as to fix $\bar{N}$ at any desired value.

Now to proceed with the solution, we define

$$\Delta_\mathbf{k} = -\sum_\mathbf{l} V_{\mathbf{kl}}b_\mathbf{l} = -\sum_\mathbf{l} V_{\mathbf{kl}}\langle c_{-\mathbf{l}\downarrow}c_{\mathbf{l}\uparrow}\rangle \tag{3.40}$$

This definition is evidently very analogous to the one given in (3.27), and it will turn out to give the gap in the energy spectrum. In terms of $\Delta_{\mathbf{k}}$, the model hamiltonian becomes (after relabeling some subscripts)

$$\mathscr{H}_M = \sum_{\mathbf{k}\sigma} \xi_{\mathbf{k}} c^*_{\mathbf{k}\sigma} c_{\mathbf{k}\sigma} - \sum_{\mathbf{k}} (\Delta_{\mathbf{k}} c^*_{\mathbf{k}\uparrow} c^*_{-\mathbf{k}\downarrow} + \Delta^*_{\mathbf{k}} c_{-\mathbf{k}\downarrow} c_{\mathbf{k}\uparrow} - \Delta_{\mathbf{k}} b^*_{\mathbf{k}}) \tag{3.41}$$

which is a sum of terms, each bilinear in the pair of operators corresponding to the partners in a Cooper pair. Such a hamiltonian can be diagonalized by a suitable linear transformation to define new Fermi operators $\gamma_{\mathbf{k}}$. As shown independently by Bogoliubov[17] and by Valatin,[18] the appropriate transformation is specified by

$$\begin{aligned} c_{\mathbf{k}\uparrow} &= u^*_{\mathbf{k}} \gamma_{\mathbf{k}0} + v_{\mathbf{k}} \gamma^*_{\mathbf{k}1} \\ c^*_{-\mathbf{k}\downarrow} &= -v^*_{\mathbf{k}} \gamma_{\mathbf{k}0} + u_{\mathbf{k}} \gamma^*_{\mathbf{k}1} \end{aligned} \tag{3.42}$$

where the numerical coefficients $u_{\mathbf{k}}$ and $v_{\mathbf{k}}$ satisfy $|u_{\mathbf{k}}|^2 + |v_{\mathbf{k}}|^2 = 1$. Note that $\gamma_{\mathbf{k}0}$ participates in destroying an electron with $\mathbf{k}\uparrow$ or creating one with $-\mathbf{k}\downarrow$; in both cases, the net effect is to decrease the system momentum by $\mathbf{k}$ and to reduce S_z by $\hbar/2$. The operator $\gamma^*_{\mathbf{k}1}$ has similar properties, so $\gamma_{\mathbf{k}1}$ itself decreases the system momentum by $-\mathbf{k}$ (i.e., increases it by $\mathbf{k}$) and has the net effect of increasing S_z.

Substituting these new operators (3.42) into the model hamiltonian (3.41), and carrying out the indicated products taking into account the noncommutivity of the operators, we obtain

$$\begin{aligned} \mathscr{H}_M = \sum_{\mathbf{k}} \xi_{\mathbf{k}} [(|u^2_{\mathbf{k}}| - |v_{\mathbf{k}}|^2)(\gamma^*_{\mathbf{k}0}\gamma_{\mathbf{k}0} + \gamma^*_{\mathbf{k}1}\gamma_{\mathbf{k}1}) + 2|v_{\mathbf{k}}|^2 + 2u^*_{\mathbf{k}} v^*_{\mathbf{k}} \gamma_{\mathbf{k}1}\gamma_{\mathbf{k}0} \\ + 2u_{\mathbf{k}} v_{\mathbf{k}} \gamma^*_{\mathbf{k}0}\gamma^*_{\mathbf{k}1}] + \sum_{\mathbf{k}} [(\Delta_{\mathbf{k}} u_{\mathbf{k}} v^*_{\mathbf{k}} + \Delta^*_{\mathbf{k}} u^*_{\mathbf{k}} v_{\mathbf{k}})(\gamma^*_{\mathbf{k}0}\gamma_{\mathbf{k}0} + \gamma^*_{\mathbf{k}1}\gamma_{\mathbf{k}1} - 1) \\ + (\Delta_{\mathbf{k}} v^{*2}_{\mathbf{k}} - \Delta^*_{\mathbf{k}} u^{*2}_{\mathbf{k}})\gamma_{\mathbf{k}1}\gamma_{\mathbf{k}0} + (\Delta^*_{\mathbf{k}} v^2_{\mathbf{k}} - \Delta_{\mathbf{k}} u^2_{\mathbf{k}})\gamma^*_{\mathbf{k}0}\gamma^*_{\mathbf{k}1} + \Delta_{\mathbf{k}} b^*_{\mathbf{k}}] \end{aligned} \tag{3.43}$$

Now, if we choose $u_{\mathbf{k}}$ and $v_{\mathbf{k}}$ so that the coefficients of $\gamma_{\mathbf{k}1}\gamma_{\mathbf{k}0}$ and $\gamma^*_{\mathbf{k}0}\gamma^*_{\mathbf{k}1}$ vanish, the hamiltonian is diagonalized; i.e., it is carried into a form containing only constants plus terms proportional to the occupation numbers $\gamma^*_{\mathbf{k}}\gamma_{\mathbf{k}}$. The coefficients of both undesired terms are zero if

$$2\xi_{\mathbf{k}} u_{\mathbf{k}} v_{\mathbf{k}} + \Delta^*_{\mathbf{k}} v^2_{\mathbf{k}} - \Delta_{\mathbf{k}} u^2_{\mathbf{k}} = 0$$

---

[17] N. N. Bogoliubov, *Nuovo Cimento* **7**, 794 (1958); *Zh. Eksperim. i Teor. Fiz.* **34**, 58 (1958) [*Soviet Phys.—JETP* **7**, 41 (1958)].

[18] J. G. Valatin, *Nuovo Cimento* **7**, 843 (1958).

When we multiply through by $\Delta_{\mathbf{k}}^*/u_{\mathbf{k}}^2$ and solve by the quadratic formula, this condition becomes

$$\frac{\Delta_{\mathbf{k}}^* v_{\mathbf{k}}}{u_{\mathbf{k}}} = (\xi_{\mathbf{k}}^2 + |\Delta_{\mathbf{k}}|^2)^{1/2} - \xi_{\mathbf{k}} \equiv E_{\mathbf{k}} - \xi_{\mathbf{k}} \tag{3.44}$$

using the definition of $E_{\mathbf{k}}$ introduced earlier. (We have chosen the positive sign of the square root so as to correspond to the stable solution of minimum rather than maximum energy.) Given the normalization requirement that $|u_{\mathbf{k}}|^2 + |v_{\mathbf{k}}|^2 = 1$, and knowing that $|v_{\mathbf{k}}/u_{\mathbf{k}}| = (E_{\mathbf{k}} - \xi_{\mathbf{k}})/|\Delta_{\mathbf{k}}|$ from (3.44), we can solve for the coefficients and find

$$|v_{\mathbf{k}}|^2 = 1 - |u_{\mathbf{k}}|^2 = \frac{1}{2}\left(1 - \frac{\xi_{\mathbf{k}}}{E_{\mathbf{k}}}\right) \tag{3.35$'$}$$

in exact agreement with (3.35), our variationally obtained result.

Although the phases of $u_{\mathbf{k}}$, $v_{\mathbf{k}}$, and $\Delta_{\mathbf{k}}$ are individually arbitrary, they are related by (3.44) since $\Delta_{\mathbf{k}}^* v_{\mathbf{k}}/u_{\mathbf{k}}$ is real. That is, the phase of $v_{\mathbf{k}}$ relative to $u_{\mathbf{k}}$ must be the phase of $\Delta_{\mathbf{k}}$. There is no loss in generality in choosing all the $u_{\mathbf{k}}$ to be real and positive. If we do so, $v_{\mathbf{k}}$ and $\Delta_{\mathbf{k}}$ must have the same phase.

3.5.1 Excitation Energies and the Energy Gap

With $u_{\mathbf{k}}$ and $v_{\mathbf{k}}$ chosen so as to diagonalize the model hamiltonian (3.43), the remaining terms reduce to

$$\mathscr{H}_M = \sum_{\mathbf{k}} (\xi_{\mathbf{k}} - E_{\mathbf{k}} + \Delta_{\mathbf{k}} b_{\mathbf{k}}^*) + \sum_{\mathbf{k}} E_{\mathbf{k}} (\gamma_{\mathbf{k}0}^* \gamma_{\mathbf{k}0} + \gamma_{\mathbf{k}1}^* \gamma_{\mathbf{k}1}) \tag{3.45}$$

The first sum is a constant, which differs from the corresponding sum for the normal state at $T = 0$ ($E_{\mathbf{k}} = |\xi_{\mathbf{k}}|$, $\Delta_{\mathbf{k}} = 0$) by exactly the condensation energy (3.36) found earlier. This is natural since the BCS ground state is the vacuum state for the quasi-particle operators $\gamma_{\mathbf{k}}$. The second sum gives the increase in energy above the ground state in terms of the number operators $\gamma_{\mathbf{k}}^* \gamma_{\mathbf{k}}$ for the $\gamma_{\mathbf{k}}$ fermions. Thus, these $\gamma_{\mathbf{k}}$ describe the elementary quasi-particle excitations of the system, which are often called *Bogoliubons*. Evidently, the energies of these excitations are just

$$E_{\mathbf{k}} = (\xi_{\mathbf{k}}^2 + |\Delta_{\mathbf{k}}|^2)^{1/2} \tag{3.46}$$

Thus, as we had anticipated, $\Delta_{\mathbf{k}}$ plays the role of an *energy gap* or minimum excitation energy since even at the Fermi surface, where $\xi_{\mathbf{k}} = 0, E_{\mathbf{k}} = |\Delta_{\mathbf{k}}| > 0$. Moreover, the notation $E_{\mathbf{k}}$ has now received its justification as the energy of an elementary excitation of momentum $\hbar\mathbf{k}$.

As in our earlier variational calculation, we require self-consistency when $\langle c_{-l\downarrow} c_{l\uparrow} \rangle$ computed from our solution is inserted back into (3.40). Rewriting the $c_{\mathbf{k}}$ operators in terms of the $\gamma_{\mathbf{k}}$, and dropping off-diagonal terms in quasi-particle operators $\gamma^*_{\mathbf{k}0}\gamma^*_{\mathbf{k}1}$ and $\gamma_{\mathbf{k}1}\gamma_{\mathbf{k}0}$ (since they do not contribute to averages), we have

$$\Delta_{\mathbf{k}} = -\sum_{\mathbf{l}} V_{\mathbf{kl}}\langle c_{-\mathbf{l}\downarrow}c_{\mathbf{l}\uparrow}\rangle = -\sum_{\mathbf{l}} V_{\mathbf{kl}}u^*_{\mathbf{l}}v_{\mathbf{l}}\langle 1 - \gamma^*_{\mathbf{l}0}\gamma_{\mathbf{l}0} - \gamma^*_{\mathbf{l}1}\gamma_{\mathbf{l}1}\rangle \tag{3.47}$$

At $T = 0$, when no quasi-particles are excited, this reduces to (3.27), so that exactly the same result (3.34) for $\Delta(0)$ in terms of $\hbar\omega_c$ and $N(0)V$ follows from the canonical transformation method as from the variational one. However, the present method is much more convenient for handling the extension of the calculation to $T > 0$.

3.6 FINITE TEMPERATURES

Since we have identified $E_{\mathbf{k}}$ as the *excitation* energy of a fermion quasi-particle, it must be a positive quantity $\geq \Delta$. The probability that it is excited in thermal equilibrium is the usual Fermi function

$$f(E_{\mathbf{k}}) = (e^{\beta E_{\mathbf{k}}} + 1)^{-1} \tag{3.48}$$

where $\beta = 1/kT$. Since $E_{\mathbf{k}} \geq \Delta$, $f(E_{\mathbf{k}})$ goes to zero at $T = 0$ for *all* $\mathbf{k}$, including $|\mathbf{k}| < k_F$. [In a parallel usage of $f(E_{\mathbf{k}})$ in the normal state, it would describe excitations from the $T = 0$ Fermi sea (i.e., holes inside and electrons outside the Fermi surface) rather than its common usage simply to describe the occupation of independent electron states.] In any case,

$$\langle 1 - \gamma^*_{\mathbf{k}0}\gamma_{\mathbf{k}0} - \gamma^*_{\mathbf{k}1}\gamma_{\mathbf{k}1}\rangle = 1 - 2f(E_{\mathbf{k}})$$

so that in general (3.47) becomes

$$\begin{aligned}\Delta_{\mathbf{k}} &= -\sum_{\mathbf{l}} V_{\mathbf{kl}}u^*_{\mathbf{l}}v_{\mathbf{l}}[1 - 2f(E_{\mathbf{l}})] \\ &= -\sum_{\mathbf{l}} V_{\mathbf{kl}}\frac{\Delta_{\mathbf{l}}}{2E_{\mathbf{l}}}\tanh\frac{\beta E_{\mathbf{l}}}{2}\end{aligned} \tag{3.49}$$

Making the BCS approximation that $V_{\mathbf{kl}} = -V$, we have $\Delta_{\mathbf{k}} = \Delta_{\mathbf{l}} = \Delta$, and the self-consistency condition becomes

$$\frac{1}{V} = \frac{1}{2}\sum_{\mathbf{k}} \frac{\tanh(\beta E_{\mathbf{k}}/2)}{E_{\mathbf{k}}} \tag{3.50}$$

where, as usual, $E_{\mathbf{k}} = (\xi^2_{\mathbf{k}} + \Delta^2)^{1/2}$. Equation (3.50) determines the temperature dependence of the energy gap $\Delta(T)$.

3.6.1 Determination of T_c

The critical temperature T_c is the temperature at which $\Delta(T) \to 0$. In this case, $E_{\mathbf{k}} \to |\xi_{\mathbf{k}}|$, and the excitation spectrum becomes the same as in the normal state. Thus, T_c is found by replacing $E_{\mathbf{k}}$ with $|\xi_{\mathbf{k}}|$ in (3.50) and solving. After changing the sum to an integral, taking advantage of the symmetry of $|\xi_{\mathbf{k}}|$ about the Fermi level, and changing to a dimensionless variable of integration, we find that this condition becomes

$$\frac{1}{N(0)V} = \int_0^{\beta_c \hbar\omega_c/2} \frac{\tanh x}{x}\, dx$$

This integral can be evaluated and yields $\ln(A\beta_c\hbar\omega_c)$, where $A = 2e^{\gamma}/\pi \approx 1.13$ and γ here is Euler's constant $\gamma = 0.577\ldots$. Consequently,

$$kT_c = \beta_c^{-1} = 1.13\hbar\omega_c e^{-1/N(0)V} \tag{3.51}$$

Comparing this with (3.34), we see that

$$\frac{\Delta(0)}{kT_c} = \frac{2}{1.13} = 1.764 \tag{3.52}$$

so that the gap at $T = 0$ is indeed comparable in energy to kT_c. The numerical factor 1.76 has been tested in many experiments and found to be reasonable. That is, experimental values of 2Δ for different materials and different directions in k space generally fall in the range from $3.0kT_c$ to $4.5kT_c$, with most clustered near the BCS value of $3.5kT_c$.

3.6.2 Temperature Dependence of the Gap

Given (3.50), or its integral equivalent

$$\frac{1}{N(0)V} = \int_0^{\hbar\omega_c} \frac{\tanh \frac{1}{2}\beta(\xi^2 + \Delta^2)^{1/2}}{(\xi^2 + \Delta^2)^{1/2}}\, d\xi \tag{3.53}$$

$\Delta(T)$ can be computed numerically. For weak-coupling superconductors, in which $\hbar\omega_c/kT_c \gg 1$, $\Delta(T)/\Delta(0)$ is a universal function of T/T_c which decreases monotonically from 1 at $T = 0$ to zero at T_c, as shown in Fig. 3.2. Near $T = 0$, the temperature variation is exponentially slow since $e^{-\Delta/kT} \approx 0$, so that the hyperbolic tangent is very nearly unity and insensitive to T. Physically speaking, Δ is nearly constant until a significant number of quasi-particles are thermally excited. On the other hand, near T_c, $\Delta(T)$ drops to zero with a vertical tangent, approximately as

$$\frac{\Delta(T)}{\Delta(0)} \approx 1.74\left(1 - \frac{T}{T_c}\right)^{1/2} \qquad T \approx T_c \tag{3.54}$$

The variation of the order parameter Δ with the square root of $(T_c - T)$ is characteristic of all mean-field theories. For example, $M(T)$ has the same dependence in the molecular-field theory of ferromagnetism.

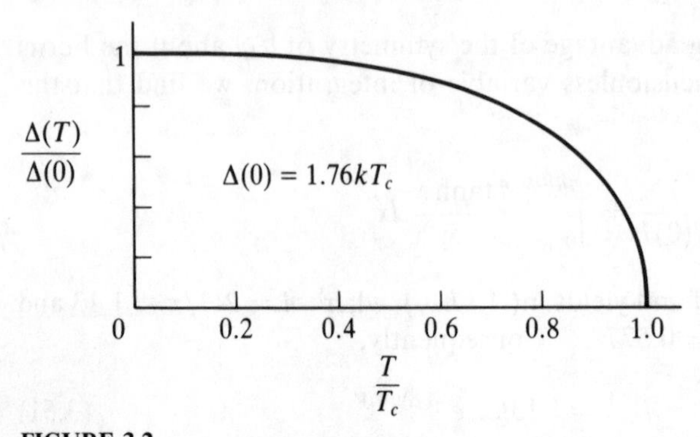

FIGURE 3.2
Temperature dependence of the energy gap in the BCS theory. Strictly speaking, this universal curve holds only in a weak-coupling limit, but it is a good approximation in most cases.

3.6.3 Thermodynamic Quantities

With $\Delta(T)$ determined, the temperature-dependent set of fermion excitation energies $E_{\mathbf{k}} = [\xi_{\mathbf{k}}^2 + \Delta(T)^2]^{1/2}$ is fixed. These energies determine the quasi-particle occupation numbers $f_{\mathbf{k}} = (1 + e^{\beta E_{\mathbf{k}}})^{-1}$, which in turn determine the electronic entropy in the usual way for a fermion gas, namely,

$$S_{es} = -2k \sum_{\mathbf{k}} [(1 - f_{\mathbf{k}}) \ln (1 - f_{\mathbf{k}}) + f_{\mathbf{k}} \ln f_{\mathbf{k}}] \tag{3.55}$$

Given $S_{es}(T)$, the specific heat can be written as

$$C_{es} = T\frac{dS_{es}}{dT} = -\beta\frac{dS_{es}}{d\beta}$$

Using (3.55), we have

$$\begin{aligned} C_{es} &= 2\beta k \sum_{\mathbf{k}} \frac{\partial f_{\mathbf{k}}}{\partial \beta} \ln \frac{f_{\mathbf{k}}}{1 - f_{\mathbf{k}}} = -2\beta^2 k \sum_{\mathbf{k}} E_{\mathbf{k}} \frac{\partial f_{\mathbf{k}}}{\partial \beta} \\ &= -2\beta^2 k \sum_{\mathbf{k}} E_{\mathbf{k}} \frac{df_{\mathbf{k}}}{d(\beta E_{\mathbf{k}})} \left(E_{\mathbf{k}} + \beta \frac{dE_{\mathbf{k}}}{d\beta} \right) \\ &= 2\beta k \sum_{\mathbf{k}} -\frac{\partial f_{\mathbf{k}}}{\partial E_{\mathbf{k}}} \left(E_{\mathbf{k}}^2 + \tfrac{1}{2}\beta \frac{d\Delta^2}{d\beta} \right) \end{aligned} \tag{3.56}$$

The first term is the usual one coming from the redistribution of quasi-particles among the various energy states as the temperature changes. The second term is more unusual and describes the effect of the temperature-dependent gap in changing the energy levels themselves.

Evidently, both terms in C_{es} will be exponentially small at $T \ll T_c$, where the minimum excitation energy Δ is much greater than kT. This accounts for the exponential form (1.13) noted earlier. Another interesting limit is very near T_c.

Then, as $\Delta(T) \to 0$, one can replace $E_{\mathbf{k}}$ by $|\xi_{\mathbf{k}}|$ in (3.56). The first term then reduces to the usual normal-state electronic specific heat

$$C_{en} = \gamma T = \frac{2\pi^2}{3} N(0)k^2 T \tag{3.57}$$

which is continuous at T_c. The second term is finite below T_c, where $d\Delta^2/dT$ is large, but it is zero above T_c, giving rise to a discontinuity ΔC is the electronic specific heat at T_c. The size of the discontinuity is readily evaluated by changing the sum to an integral, as follows:

$$\begin{aligned}\Delta C = (C_{es} - C_{en})\Big|_{T_c} &= N(0)k\beta^2 \left(\frac{d\Delta^2}{d\beta}\right) \int_{-\infty}^{\infty} \left(\frac{-\partial f}{\partial |\xi|}\right) d\xi \\ &= N(0)\left(\frac{-d\Delta^2}{dT}\right)\Big|_{T_c}\end{aligned} \tag{3.58}$$

where we have used the fact that $\partial f/\partial|\xi| = \partial f/\partial \xi$ since $\partial f/\partial \xi$ is an even function of ξ. Using the approximate form (3.54) for $\Delta(T)$, with $\Delta(0) = 1.76kT_c$, we obtain $\Delta C = 9.4N(0)k^2 T_c$. Comparing with (3.57), we find that the normalized magnitude of the discontinuity is

$$\frac{\Delta C}{C_{en}} = \frac{9.4}{2\pi^2/3} = 1.43 \tag{3.59}$$

The overall behavior of the electronic specific heat is sketched in Fig. 3.3*b*.

With $C_{es}(T)$ determined numerically from (3.56), we can integrate it to find the change in internal energy $U(T)$ as we decrease the temperature from T_c. At T_c, it must be the same as the normal value $U_{en}(0) - \frac{1}{2}\gamma T_c^2$ since the specific heat remains finite there. Thus,

$$U_{es}(T) = U_{en}(0) + \tfrac{1}{2}\gamma T_c^2 - \int_T^{T_c} C_{es}\, dT \tag{3.60}$$

From this and the entropy (3.55) we may then compute the free energy

$$F_{es}(T) = U_{es}(T) - TS_{es}(T) \tag{3.61}$$

Assuming that the effect of the superconducting transition on the lattice free energy can be neglected, the thermodynamic critical field is then determined through the relation

$$\frac{H_c^2(T)}{8\pi} = F_{en}(T) - F_{es}(T) \tag{3.62}$$

where $F_{en}(T) = U_{en}(0) - \frac{1}{2}\gamma T^2$. These various thermodynamic quantities are plotted in Fig. 3.3. A useful numerical tabulation has been given by Mühlschlegel.[19]

[19] B. Mühlschlegel, *Z. Phys.* **155**, 313 (1959).

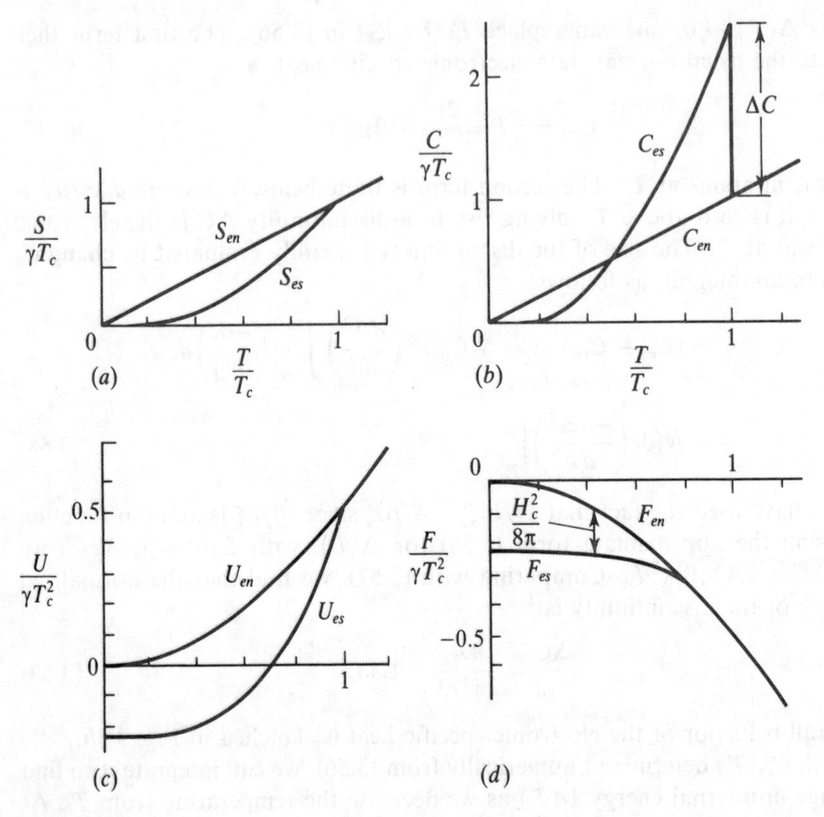

FIGURE 3.3
Comparison of thermodynamic quantities in superconducting and normal states. $U_{en}(0)$ is chosen as the zero of ordinates in (c) and (d). Because the transition is of second order, the quantities S, U, and F are continuous at T_c. Moreover, the slope of F_{es} joins continuously to that of F_{en} at T_c, since $\partial F/\partial T = -S$.

Since the critical-field curve $H_c(T)$ can be measured with greater accuracy than can typical thermodynamic quantities, it is of some importance to note that one can be derived from the other by rigorous thermodynamic computations, starting from (3.62). For example, the approximate parabolic temperature dependence of H_c quoted in (1.2) is inconsistent with an exponential variation of C_{es}, such as that quoted in (1.13), but it is consistent with a T^3 variation. Precise measurements of the deviation of $H_c(T)$ from the parabolic approximation have been used by Mapother[20] to test the BCS predictions of thermodynamic properties.

[20] D. E. Mapother, *Phys. Rev.* **126**, 2021 (1962).

3.7 STATE FUNCTIONS AND THE DENSITY OF STATES

By inverting (3.42), we find the expressions

$$\begin{aligned}\gamma_{\mathbf{k}0}^* &= u_{\mathbf{k}}^* c_{\mathbf{k}\uparrow}^* - v_{\mathbf{k}}^* c_{-\mathbf{k}\downarrow} \\ \gamma_{\mathbf{k}1}^* &= u_{\mathbf{k}}^* c_{-\mathbf{k}\downarrow}^* + v_{\mathbf{k}}^* c_{\mathbf{k}\uparrow}\end{aligned} \tag{3.63}$$

for the $\gamma_{\mathbf{k}}^*$ operators, which create quasi-particle excitations of the two spin directions from the superconducting ground state, in terms of the electron creation operators $c_{\mathbf{k}}^*$. As remarked in connection with (3.45), the superconducting ground state $|\psi_G\rangle$ is defined as the vacuum state of the γ particles, i.e., by the relations

$$\gamma_{\mathbf{k}0}|\psi_G\rangle = \gamma_{\mathbf{k}1}|\psi_G\rangle = 0 \tag{3.64}$$

The structure of $|\psi_G\rangle$ in terms of the γ particles is of no interest, but it is of interest to demonstrate that in terms of the $c_{\mathbf{k}}$, it agrees with the BCS product form (3.14). This may be verified by considering, e.g.,

$$\gamma_{\mathbf{k}0}|\psi_G\rangle = (u_{\mathbf{k}} c_{\mathbf{k}\uparrow} - v_{\mathbf{k}} c_{-\mathbf{k}\downarrow}^*) \prod_{\mathbf{l}} (u_{\mathbf{l}} + v_{\mathbf{l}} c_{\mathbf{l}\uparrow}^* c_{-\mathbf{l}\downarrow}^*)|\phi_0\rangle$$

Multiplying out the factor involving the **k**th pair, we find

$$u_{\mathbf{k}}^2 c_{\mathbf{k}\uparrow} + u_{\mathbf{k}} v_{\mathbf{k}} c_{\mathbf{k}\uparrow} c_{\mathbf{k}\uparrow}^* c_{-\mathbf{k}\downarrow}^* - v_{\mathbf{k}} u_{\mathbf{k}} c_{-\mathbf{k}\downarrow}^* - v_{\mathbf{k}}^2 c_{-\mathbf{k}\downarrow}^* c_{\mathbf{k}\uparrow}^* c_{-\mathbf{k}\downarrow}^*$$

All these terms give zero when operating on $|\phi_0\rangle$ for the following reasons: the first is zero because the annihilation operator $c_{\mathbf{k}\uparrow}$ operates on the vacuum; the next two terms cancel because $c_{\mathbf{k}\uparrow} c_{\mathbf{k}\uparrow}^*$ operating on the vacuum gives a factor of unity; the last term vanishes because it contains the same creation operator twice with no intervening annihilation operator. The case $\gamma_{\mathbf{k}1}|\psi_G\rangle$ works out similarly. Thus, (3.64) is verified, and the BCS form for the ground state is in exact agreement with that of the canonical-transformation method.

Now let us look at the excited states. For example, consider

$$\gamma_{\mathbf{k}0}^*|\psi_G\rangle = (|u_{\mathbf{k}}|^2 c_{\mathbf{k}\uparrow}^* + u_{\mathbf{k}}^* v_{\mathbf{k}} c_{\mathbf{k}\uparrow}^* c_{\mathbf{k}\uparrow}^* c_{-\mathbf{k}\downarrow}^* - v_{\mathbf{k}}^* u_{\mathbf{k}} c_{-\mathbf{k}\downarrow} - |v_{\mathbf{k}}|^2 c_{-\mathbf{k}\downarrow} c_{\mathbf{k}\uparrow}^* c_{-\mathbf{k}\downarrow}^*)$$

$$\times \prod_{\mathbf{l}\neq\mathbf{k}} (u_{\mathbf{l}} + v_{\mathbf{l}} c_{\mathbf{l}\uparrow}^* c_{-\mathbf{l}\downarrow}^*)|\phi_0\rangle$$

The two middle terms give zero for reasons mentioned earlier. The last term may be changed to $+|v_{\mathbf{k}}|^2 c_{\mathbf{k}\uparrow}^* c_{-\mathbf{k}\downarrow} c_{-\mathbf{k}\downarrow}^*$ by using the anticommutation of the fermion operators, and then the factor $c_{-\mathbf{k}\downarrow} c_{-\mathbf{k}\downarrow}^*$ may be dropped since, operating on $|\phi_0\rangle$, it gives a factor of unity. Combining the remainder with the first term, we have

$$\gamma_{\mathbf{k}0}^*|\psi_G\rangle = c_{\mathbf{k}\uparrow}^* \prod_{\mathbf{l}\neq\mathbf{k}} (u_{\mathbf{l}} + v_{\mathbf{l}} c_{\mathbf{l}\uparrow}^* c_{-\mathbf{l}\downarrow}^*)|\phi_0\rangle \tag{3.65a}$$

Similarly,

$$\gamma_{\mathbf{k}1}^*|\psi_G\rangle = c_{-\mathbf{k}\downarrow}^* \prod_{\mathbf{l}\neq\mathbf{k}} (u_\mathbf{l} + v_\mathbf{l} c_{\mathbf{l}\uparrow}^* c_{-\mathbf{l}\downarrow}^*)|\phi_0\rangle \tag{3.65b}$$

These are the excited states called *singles* in the original BCS treatment, where they were written down by inspection. They correspond to putting with certainty a single electron into one of the states of the pair $(\mathbf{k}\uparrow, -\mathbf{k}\downarrow)$, while leaving with certainty the other state of the pair empty. This effectively blocks that pair state from participation in the many-body wavefunction and increases the system energy accordingly.

The operators $\gamma_\mathbf{k}^*$ change the expectation value of the electron number in the pair of states $(\mathbf{k}\uparrow, -\mathbf{k}\downarrow)$ from $2v_\mathbf{k}^2$ in the ground state to 1 in the excited state. The change is $(1 - 2v_\mathbf{k}^2) = u_\mathbf{k}^2 - v_\mathbf{k}^2$, which ranges from -1 to $+1$ as $\xi_\mathbf{k}$ ranges from well below zero (inside Fermi surface) to well above zero (outside Fermi surface), as can be inferred[21] from the plot of $v_\mathbf{k}^2$ in Fig. 3.1. This net change results from a probability $u_\mathbf{k}^2$ of a change by $+1$ and a probability $v_\mathbf{k}^2$ of a change by -1. Such behavior would be inconsistent with exact number conservation in an isolated system. This apparent paradox is resolved by recalling that the $\gamma_\mathbf{k}^*$ operators are defined only with respect to a ground state having a definite phase of $\Delta_\mathbf{k}$, and such a state has a large uncertainty in N. If one wishes to consider excited states of an isolated system, one uses a prescription like (3.18) to project out the N-particle part *after* operating with the $\gamma_\mathbf{k}^*$ on $|\psi_G\rangle$. Note that $\gamma_\mathbf{k}^*|\psi_G\rangle$ has *no* component with an *even* number of particles. More generally, electron conservation requires that excitations always must be created or destroyed in pairs, as is expected for fermions. (It is for this reason that the *spectroscopic* gap is 2Δ, not Δ.) Considering, e.g., $\gamma_{\mathbf{k}0}^*\gamma_{\mathbf{k}'0}^*|\psi_G\rangle$, and using (3.65*a*), we see that it has a well-defined N-particle projection

$$c_{\mathbf{k}\uparrow}^* c_{\mathbf{k}'\uparrow}^* \int_0^{2\pi} d\varphi e^{-i(N-2)\varphi/2} \prod_{\mathbf{l}\neq\mathbf{k},\,\mathbf{k}'} (|u_\mathbf{l}| + |v_\mathbf{l}|e^{i\varphi} c_{\mathbf{l}\uparrow}^* c_{-\mathbf{l}\downarrow}^*)|\phi_0\rangle \tag{3.66}$$

Anticipating the discussion of tunneling experiements, in which an excited state is created by actual addition or subtraction of an electron, we see that the state resulting from $\gamma_{\mathbf{k}0}^*$ operating on an N-electron system is

$$c_{\mathbf{k}\uparrow}^* \int d\varphi e^{-iN'\varphi/2} \prod_{\mathbf{l}\neq\mathbf{k}} (|u_\mathbf{l}| + |v_\mathbf{l}|e^{i\varphi} c_{\mathbf{l}\uparrow}^* c_{-\mathbf{l}\downarrow}^*)|\varphi_0\rangle \tag{3.67}$$

where $N' = N$ if an electron is added, whereas $N' = N - 2$ if an electron is removed. Thus, we can always write out explicit expressions for these various excited states with a definite number of particles, although there is seldom any need to, except in the mesoscopic systems treated in Chap. 7. The important

[21] For more detail, see the discussion of *charge imbalance* in Sec. 11.2.

qualitative point is simply that pairs of electrons can be added to or subtracted from the condensate at will to achieve number conservation; for energy bookkeeping purposes, such electrons are at the chemical potential μ.

To avoid the need to deal with explicit wavefunctions such as (3.66) and (3.67), let us follow Josephson[22] in introducing an operator S which annihilates a Cooper pair, while S^* creates one. [For later reference in treating the Josephson effect, we note that S has the eigenvalue $e^{i\varphi}$ in a BCS state in which the phase of Δ (or of u^*v) is φ.] An equivalent operator p was also introduced by Bardeen.[23] We can then define a set of modified quasi-particle operators which definitely create either an electron or a hole, i.e., which either increase or decrease the number of electrons by one, whereas the $\gamma_{\mathbf{k}}^*$ operators of (3.63) create a linear combination of these possibilities. Thus, the two types of operators (3.63) are replaced by four:

$$\begin{aligned}
\gamma_{e\mathbf{k}0}^* &= u_{\mathbf{k}}^* c_{\mathbf{k}\uparrow}^* - v_{\mathbf{k}}^* S^* c_{-\mathbf{k}\downarrow} \\
\gamma_{h\mathbf{k}0}^* &= u_{\mathbf{k}}^* S c_{\mathbf{k}\uparrow}^* - v_{\mathbf{k}}^* c_{-\mathbf{k}\downarrow} \\
\gamma_{e\mathbf{k}1}^* &= u_{\mathbf{k}}^* c_{-\mathbf{k}\downarrow}^* + v_{\mathbf{k}}^* S^* c_{\mathbf{k}\uparrow} \\
\gamma_{h\mathbf{k}1}^* &= u_{\mathbf{k}}^* S c_{-\mathbf{k}\downarrow}^* + v_{\mathbf{k}}^* c_{\mathbf{k}\uparrow}
\end{aligned} \tag{3.68}$$

Note that the hole and electron operators are related by

$$\gamma_{h\mathbf{k}}^* = S\gamma_{e\mathbf{k}}^* \tag{3.69}$$

which corresponds to the fact that creating a holelike excitation is equivalent to annihilating a pair and creating an electronlike excitation. Thus, these are not independent excitations but, rather, the same excitation with different numbers of condensed pairs. The distinctions embodied in (3.68) are essential in calculations[24] which distinguish transfer of charge from transfer of quasi-particles.

In dealing with tunneling processes which transfer electrons from one system to another, we need to reintroduce the chemical potential explicitly since it will differ between conductors maintained at different voltages. Because all our calculations of system energy have been referred to the chemical potential μ by subtracting μN_{op}, we simply add this back in and write

$$\mathscr{H} = \mu N_{\text{op}} + E_G + \sum_{\mathbf{k}} E_{\mathbf{k}} \gamma_{\mathbf{k}}^* \gamma_{\mathbf{k}} \tag{3.70}$$

where E_G is the ground-state energy and the sum runs over all the excitations. In view of (3.70), we see that the energy to create an electronlike excitation is $E_{e\mathbf{k}} = (E_{\mathbf{k}} + \mu)$, whereas that to create a hole is $E_{h\mathbf{k}} = (E_{\mathbf{k}} - \mu)$. In an isolated

---

[22] B. D. Josephson, *Phys. Lett.* **1**, 251 (1962).

[23] J. Bardeen, *Phys. Rev. Lett.* **9**, 147 (1962)

[24] M. Tinkham, *Phys. Rev.* **B6**, 1747 (1972).

superconductor, the simplest number-conserving excitation consists of a hole and an electron, with total excitation energy

$$(E_{\mathbf{k}} + \mu) + (E_{\mathbf{k}'} - \mu) = E_{\mathbf{k}} + E_{\mathbf{k}'} \geq 2\Delta \tag{3.71}$$

On the other hand, in a tunneling process in which an electron is transferred from metal 1 to metal 2, conservation of energy requires that

$$(E_{\mathbf{k}1} - \mu_1) + (E_{\mathbf{k}'2} + \mu_2) = 0$$

so that

$$E_{\mathbf{k}1} + E_{\mathbf{k}'2} = (\mu_1 - \mu_2) = eV_{12} \tag{3.72}$$

It is of some historical interest to note that in the original form of the BCS theory it was necessary to give special treatment to *excited pairs* since the excited state formed by (3.66) with $\mathbf{k}'\uparrow$ replaced by $-\mathbf{k}\downarrow$ is not orthogonal to the ground state. As the appropriate orthogonal state is generated automatically by $\gamma^*_{\mathbf{k}1}\gamma^*_{\mathbf{k}0}|\psi_G\rangle$, however, no such special mathematical attention is required so long as excited states are expressed in terms of the $\gamma^*_{\mathbf{k}}$ operators.

3.7.1 Density of States

Now that we have seen that the quasi-particle excitations can be simply described as fermions created by the $\gamma^*_{\mathbf{k}}$, which are in one-to-one correspondence with the $c^*_{\mathbf{k}}$ of the normal metal, we can obtain the superconducting density of states $N_s(E)$ by equating

$$N_s(E)\, dE = N_n(\xi)\, d\xi$$

(Since we are thinking here of a single superconductor, we can safely revert to taking $\mu = 0$.) Because we are largely interested in energies ξ only a few millielectronvolts from the Fermi energy, we can take $N_n(\xi) = N(0)$, a constant. This leads directly to the simple result

$$\frac{N_s(E)}{N(0)} = \frac{d\xi}{dE} = \begin{cases} \dfrac{E}{(E^2 - \Delta^2)^{1/2}} & (E > \Delta) \\ 0 & (E < \Delta) \end{cases} \tag{3.73}$$

since $E^2_{\mathbf{k}} = \Delta^2 + \xi^2_{\mathbf{k}}$. Excitations with all momenta $\mathbf{k}$, even those whose $\xi_{\mathbf{k}}$ fall in the gap, have their energies raised above Δ. Moreover, we expect a divergent state density just above $E = \Delta$, as indicated in Fig. 3.4. Of course, the total number of states is conserved because of the one-to-one correspondence between the $\gamma_{\mathbf{k}}$ and the $c_{\mathbf{k}}$. The nature of this correspondence is made more explicit by Fig. 3.5, which shows the relationship between the excitation energies in the normal and superconducting states.

It is worth noting that if the BCS model is followed literally, a narrow peak in the density of states occurs at the cutoff energy $\hbar\omega_c$ because above this energy, $\Delta = 0$ and $E_{\mathbf{k}} = \xi_{\mathbf{k}}$. As a result, $N(0)\Delta^2/2\hbar\omega_c$ extra states fall in an energy range of width $\Delta^2/2\hbar\omega_c$, causing a doubling of $N(E)$ in this range. Of course, this consequence of the model is not to be taken seriously since it depends critically

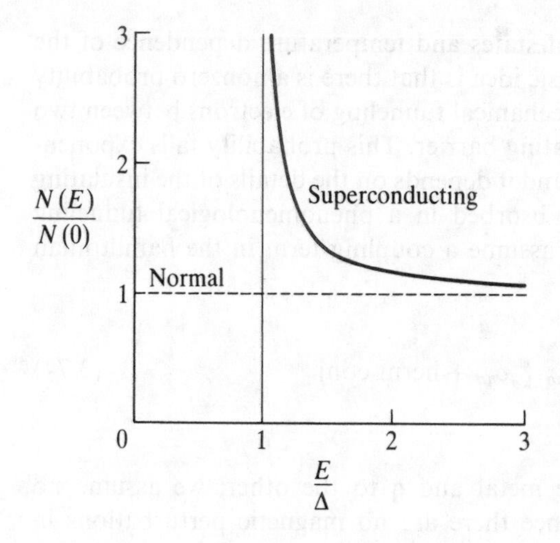

FIGURE 3.4
Density of states in superconducting compared to normal state. All **k** states whose energies fall in the gap in the normal metal are raised in energy above the gap in the superconducting state.

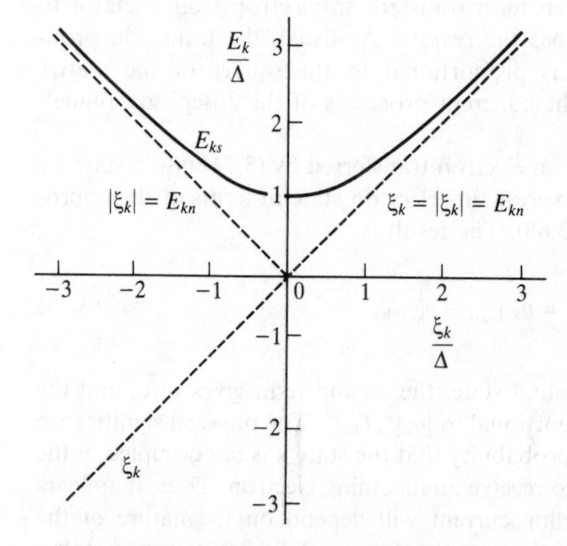

FIGURE 3.5
Energies of elementary excitations in the normal and superconducting states as functions of ξ_k, the independent-particle kinetic energy relative to the Fermi energy.

on the admittedly crude cutoff procedure. As described in the next section, a more rigorous treatment of the phonon-mediated interaction spreads this cutoff effect out over the entire energy range $\hbar\omega_c$, so that the actual departures from (3.73) are of order $(\Delta/\hbar\omega_c)^2$ or $(T_c/\Theta_D)^2$ to give about the same integrated effect.

3.8 ELECTRON TUNNELING

By far the most detailed experimental examination of the density of states is provided by electron tunneling. This technique was pioneered by Giaever,[25]

[25] I. Giaever, *Phys. Rev. Lett.* **5**, 147, 464 (1960).

who used it to confirm the density of states and temperature dependence of the energy gap predicted by BCS. The basic idea is that there is a nonzero probability of charge transfer by the quantum-mechanical tunneling of electrons between two conductors separated by a thin insulating barrier. This probability falls exponentially with the distance of separation and it depends on the details of the insulating material, but these aspects can be absorbed in a phenomenological tunneling matrix element $T_{\mathbf{kq}}$. That is, we can assume a coupling term in the hamiltonian of the form

$$\mathscr{H}_T = \sum_{\sigma\mathbf{kq}} T_{\mathbf{kq}} c^*_{\mathbf{k}\sigma} c_{\mathbf{q}\sigma} + \text{herm conj} \tag{3.74}$$

where the subscript **k** refers to one metal and **q** to the other; we assume no spin flip in the tunneling process since there are no magnetic perturbations in the problem. The explicitly written term transfers an electron from metal **q** to **k**, whereas the conjugate term does the reverse. As usual, the transition probability (and hence the current) is proportional to the square of the matrix element, so long as we exclude the coherent processes of the Josephson tunneling.

If we consider, for example, an electron transferred by (3.74) into a state $\mathbf{k}\uparrow$ in a superconductor, we must reexpress this electron state in terms of the appropriate excitations, the $\gamma_{\mathbf{k}}$, using (3.68). The result is

$$c^*_{\mathbf{k}\uparrow} = u_{\mathbf{k}} \gamma^*_{e\mathbf{k}0} + v^*_{\mathbf{k}} \gamma_{h\mathbf{k}1} \tag{3.75}$$

If the superconductor is in its ground state, the second term gives zero, and the process contributes a current proportional to $|u_{\mathbf{k}}|^2 |T_{\mathbf{kq}}|^2$. The physical significance of the factor $|u_{\mathbf{k}}|^2$ is that it is the probability that the state **k** is *not* occupied in the BCS function, and hence is able to receive an incoming electron. Thus, it appears on the face of it that the tunneling current will depend on the nature of the superconducting ground state as well as on the density of available excited states; but this turns out not to be true. As is evident from Fig. 3.5, there is another state $\mathbf{k}'$ having exactly the same energy $E_{\mathbf{k}'} = E_{\mathbf{k}}$, but with $\xi_{\mathbf{k}'} = -\xi_{\mathbf{k}}$. Using (3.75), with **k** replaced by $\mathbf{k}'$ we see that tunneling into $\mathbf{k}'$ contributes a current proportional to $|u_{\mathbf{k}'}|^2 |T_{\mathbf{k}'\mathbf{q}}|^2 = |v_{\mathbf{k}}|^2 |T_{\mathbf{k}'\mathbf{q}}|^2$ since $|u(-\xi)| = |v(\xi)|$. Making the reasonable assumption that the two matrix elements are nearly equal since **k** and $\mathbf{k}'$ are both near the same point on the Fermi surface, the total current from these two channels is proportional to $(|u_{\mathbf{k}}|^2 + |v_{\mathbf{k}}|^2)|T_{\mathbf{kq}}|^2 = |T_{\mathbf{kq}}|^2$, and the characteristic coherence factors of the superconducting wavefunction, $u_{\mathbf{k}}$ and $v_{\mathbf{k}}$, have dropped out. If we now generalize to finite temperatures, so that the quasi-particle occupation numbers $f_{\mathbf{k}}$ are nonzero, both terms of (3.75) contribute, the first as $(1 - f_{\mathbf{k}})$, the second as $f_{\mathbf{k}}$. Again, when the degenerate channels are combined, the current is simply proportional to $|T_{\mathbf{kq}}|^2$.

3.8.1 The Semiconductor Model

This disappearance of the coherence factors $u_{\mathbf{k}}$ and $v_{\mathbf{k}}$ makes it possible and convenient to reexpress the computation of the tunneling current in what is often called the *semiconductor model*. In this method, illustrated in Fig. 3.6, the normal metal is represented in the familiar elementary way as a continuous distribution of independent-particle energy states with density $N(0)$, including energies below as well as above the Fermi level. The superconductor is represented by an ordinary semiconductor with a density of independent-particle states obtained from Fig. 3.4 by adding its reflection on the negative-energy side of the chemical potential, so that it will reduce properly to the normal-metal density of states as $\Delta \to 0$. At $T = 0$, all states up to μ are filled; for $T > 0$, the occupation numbers are given by the Fermi function. It is worth noting that $f_{\mathbf{k}}$ now runs from 0 to 1, whereas in our previous convention, $f_{\mathbf{k}}$ ranged only from 0 to $\frac{1}{2}$ since $E_{\mathbf{k}} \gtrsim 0$. This difference reflects the fact that in the present model $f_{\mathbf{k}}$ measures a departure from the vacuum, whereas in the previous *excitation representation* it measured a departure from the ground state of the system.

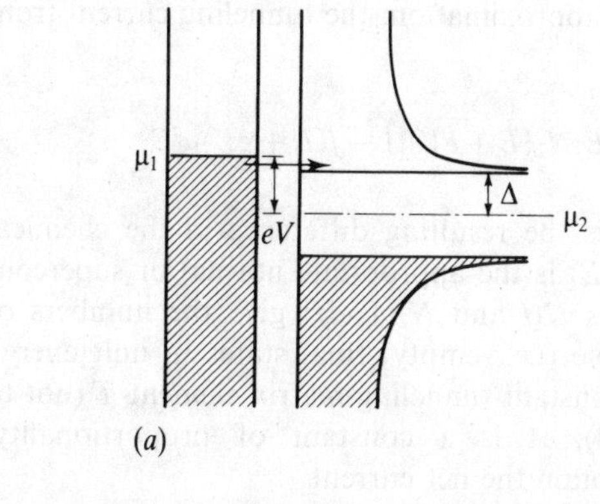

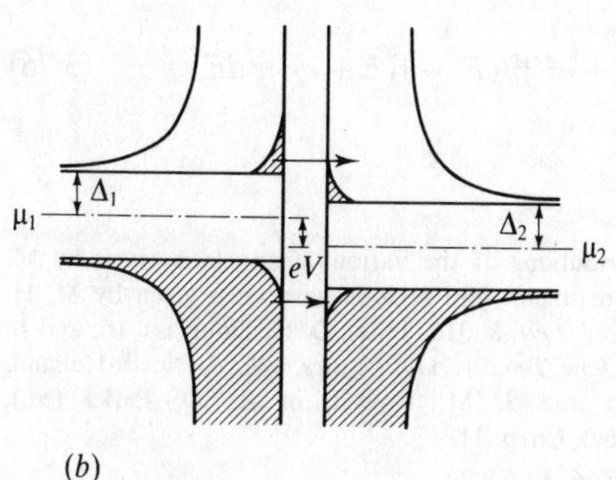

FIGURE 3.6
Example of semiconductor model description of electron tunneling. Density of states is plotted horizontally vs. energy vertically. Shading denotes states occupied by electrons. (*a*) *N-S* tunneling at $T = 0$, with bias voltage just above the conduction threshold, i.e., eV slightly exceeds the energy gap Δ. Horizontal arrow depicts electrons from the left tunneling into empty states on the right. (*b*) *S-S* tunneling at $T > 0$, with bias voltage below the threshold for conduction at $T = 0$, i.e., with $eV < \Delta_1 + \Delta_2$. Horizontal arrows depict tunneling involving thermally excited electrons or holes, respectively.

With this model, tunneling transitions are all *horizontal*, i.e., they occur at constant energy after adjusting the relative levels of μ in the two metals to account for the applied potential difference eV. This property facilitates summing up all contributions to the current in an elementary way since the various parallel channels noted earlier do not have to be considered anew in each case. Because this scheme so greatly simplifies the computations, we shall use it here to work out the tunneling characteristics of various types of junctions, and simply refer the reader to more detailed treatments which are available in the literature.[26] It should be borne in mind, however, that this technique to some extent *over*simplifies. It is sound to treat the normal metal in this way, but it is less safe to conceal the mixing of hole and electron states which is present in the superconducting state even at $T=0$. Although our previous argument for simply adding the currents from the two degenerate channels is valid for the usual case, there can be an interference effect between them which causes an oscillatory variation of the tunnel current with voltage or sample thickness known as the Tomasch effect.[27] Also, the method is inadequate for treating *charge-imbalance* regimes, where the states inside and outside the Fermi surface are not in equilibrium. (See Chap. 11.) These comments illustrate the need for caution in using the semiconductor model. Of course, this model also is inadequate for dealing with processes in which the condensed pairs play a role, since the ground state does not appear in the energy-level diagram.

Within the independent-particle approximation, the tunneling current from metal 1 to metal 2 can be written as

$$I_{1\to 2} = A\int_{-\infty}^{\infty} |T|^2 N_1(E)f(E)N_2(E+eV)[1-f(E+eV)]dE$$

where V is the applied voltage, eV is the resulting difference in the chemical potential across the junction, and $N(E)$ is the appropriate normal or superconducting density of states. The factors $N_1 f$ and $N_2(1-f)$ give the numbers of occupied initial states and of available (i.e., empty) final states in unit energy interval. This expression assumes a constant tunneling-matrix element T (not to be confused with the temperature!); A is a constant of proportionality. Subtracting the reverse current, we obtain the net current

$$I = A|T|^2\int_{-\infty}^{\infty} N_1(E)N_2(E+eV)[f(E)-f(E+eV)]\, dE \tag{3.76}$$

---

[26]A particularly explicit discussion of the contributions of the various channels is given by M. Tinkham, *Phys. Rev.* **B6**, 1747 (1972). Earlier treatments and reviews have been given by M. H. Cohen, L. M. Falicov, and J. C. Phillips, *Phys. Rev. Lett.* **8**, 316 (1962); D. H. Douglass, Jr., and L. M. Falicov, in C. J. Gorter (ed.), *Progress in Low Temperature Physics*, vol. 4, North-Holland, Amsterdam (1964), p. 97; W. L. McMillan and J. M. Rowell, in R. D. Parks (ed.), *Superconductivity*, vol. 1, Dekker, New York (1969), Chap. 11.

[27]W. J. Tomasch, *Phys. Rev. Lett.* **15**, 672 (1965); **16**, 16 (1966).

We shall now use this expression to treat a number of important cases.

3.8.2 Normal-Normal Tunneling

If both metals are normal, (3.76) becomes

$$I_{nn} = A|T|^2 N_1(0) N_2(0) \int_{-\infty}^{\infty} [f(E) - f(E + eV)] dE$$

$$= A|T|^2 N_1(0) N_2(0) eV \equiv G_{nn} V \tag{3.77}$$

so that the junction is *ohmic*; i.e., it has a well-defined conductance G_{nn}, independent of V. Note that it is also independent of the temperature.

To help reduce any lingering confusion about the relation of this semiconductor, or independent-particle, scheme to the elementary excitation scheme, let us indicate how this simple case would have been treated in the other framework. First, at $T = 0$, all $f_{\mathbf{k}} = 0$, and there are no excitations present, both metals being in their Fermi-sea ground states. Thus, any tunneling process must involve creating two excitations, a hole in one metal and an electron in the other, the sum of the two excitation energies being eV, as given by (3.72). The resulting current is

$$I = A|T|^2 \int_0^{eV} N_1(E) N_2(eV - E) dE$$

$$= A|T|^2 N_1(0) N_2(0) eV$$

exactly as found in (3.77). For $T > 0$, the current from this process is reduced by the excitations already present, which block final states, but this effect is canceled by the extra current from the tunneling of the excitations, leading to a temperature-independent result.

3.8.3 Normal-Superconductor Tunneling

A more interesting case arises if one metal is superconducting. Then Fig. 3.6*a* is relevant, and (3.76) becomes

$$I_{ns} = A|T|^2 N_1(0) \int_{-\infty}^{\infty} N_{2s}(E)[f(E) - f(E + eV)] \, dE$$

$$= \frac{G_{nn}}{e} \int_{-\infty}^{\infty} \frac{N_{2s}(E)}{N_2(0)} [f(E) - f(E + eV)] \, dE \tag{3.78}$$

In general, numerical means are required to evaluate this expression for the BCS density of states and thus to allow quantitative comparison with experiment, although the qualitative behavior is easily sketched. As indicated in Fig. 3.7*a*, at $T = 0$, there is no tunneling current until $e|V| \gtrsim \Delta$, since the chemical-potential difference must provide enough energy to create an excitation in the superconductor. The magnitude of the current is independent of the sign of V because

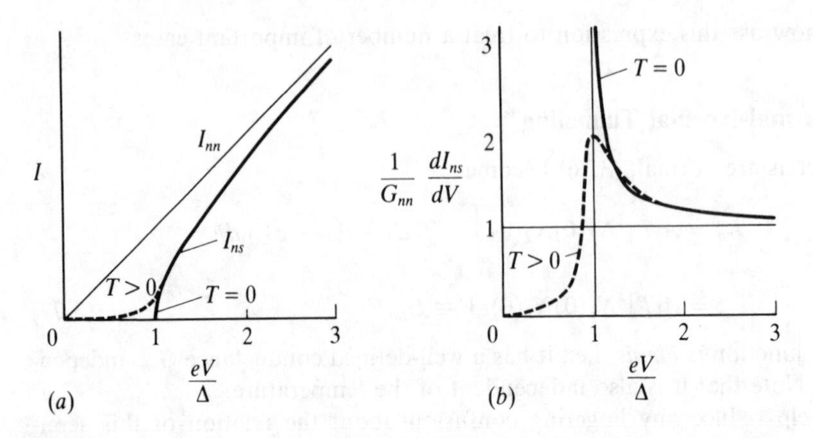

FIGURE 3.7
Characteristics of normal-superconductor tunnel junctions. (*a*) *I-V* characteristic. (*b*) Differential conductance. Solid curves refer to $T = 0$; dashed curves refer to a finite temperature.

hole and electron excitations have equal energies. For $T > 0$, the energy of excitations already present allows them to tunnel at lower voltages, giving an exponential tail of the current in the region below $eV = \Delta$.

A more direct comparison of theory and experiment can be made if one considers the differential conductance dI/dV as a function of V. From (3.78)

$$G_{ns} = \frac{dI_{ns}}{dV} = G_{nn} \int_{-\infty}^{\infty} \frac{N_{2s}(E)}{N_2(0)} \left[-\frac{\partial f(E + eV)}{\partial (eV)} \right] dE \tag{3.79}$$

Since $-[\partial f(E + eV)/\partial (eV)]$ is a bell-shaped weighting function peaked at $E = -eV$, with width $\sim 4kT$ and unit area under the curve, it is clear that as $kT \to 0$, this approaches

$$G_{ns}\Big|_{T=0} = \frac{dI_{ns}}{dV}\Big|_{T=0} = G_{nn} \frac{N_{2s}(e|V|)}{N_2(0)} \tag{3.80}$$

Thus, in the low-temperature limit, the differential conductance measures directly the density of states. At finite temperatures, as shown in Fig. 3.7*b*, the conductance measures a density of states smeared by $\sim \pm 2kT$ in energy, due to the width of the weighting function. Because this function has exponential "skirts," it turns out that the differential conductance at $V = 0$ is related exponentially to the width of the gap. In the limit $kT \ll \Delta$, this relation reduces to

$$\frac{G_{ns}}{G_{nn}}\Big|_{V=0} = \left(\frac{2\pi\Delta}{kT} \right)^{1/2} e^{-\Delta/kT} \tag{3.81}$$

3.8.4 Superconductor-Superconductor Tunneling

If both metals are superconducting, the energy-level structure is as shown in Fig. 3.6*b* and (3.76) becomes

$$I_{ss} = \frac{G_{nn}}{e}\int_{-\infty}^{\infty} \frac{N_{1s}(E)}{N_1(0)}\frac{N_{2s}(E+eV)}{N_2(0)}[f(E)-f(E+eV)]\,dE$$

$$= \frac{G_{nn}}{e}\int_{-\infty}^{\infty} \frac{|E|}{[E^2-\Delta_1^2]^{1/2}}\frac{|E+eV|}{[(E+eV)^2-\Delta_2^2]^{1/2}}[f(E)-f(E+eV)]\,dE \qquad (3.82)$$

In the second form, it is understood that the range of integration excludes values of E such that $|E| < |\Delta_1|$ and $|E+eV| < |\Delta_2|$. Again numerical integration is required to compute complete *I-V* curves. However, the qualitative features are indicated in Fig. 3.8. At $T = 0$, no current can flow until $eV = \Delta_1 + \Delta_2$; at this point, the potential difference supplies enough energy to create a hole on one side and a particle on the other. Since the density of states is infinite at the gap edges, it turns out that there is a discontinuous jump in I_{ss} at $eV = \Delta_1 + \Delta_2$, even at finite temperatures. At $T > 0$, current also flows at lower voltages because of the availability of thermally excited quasi-particles. This current rises sharply to a peak when $eV = |\Delta_1 - \Delta_2|$ because this voltage provides just the energy to allow thermally excited quasi-particles in the peak of the density of states at Δ_1, say, to tunnel into the peaked density of available states at Δ_2. The existence of this peak leads to a *negative-resistance region* $[(dI/dV) < 0]$ for $|\Delta_1 - \Delta_2| \leq eV \leq \Delta_1 + \Delta_2$. This region cannot be observed with the usual current-source arrangement since there are three possible values of V for given I and the one with $dI/dV < 0$ is unstable. A voltage source must be used to maintain stable operation. The existence of sharp features at both $|\Delta_1 - \Delta_2|$ and $\Delta_1 + \Delta_2$ allows very convenient determinations of $\Delta_1(T)$ and $\Delta_2(T)$ from the tunneling curves. The *S-S* tunneling method is superior to the *N-S* tunneling method in this

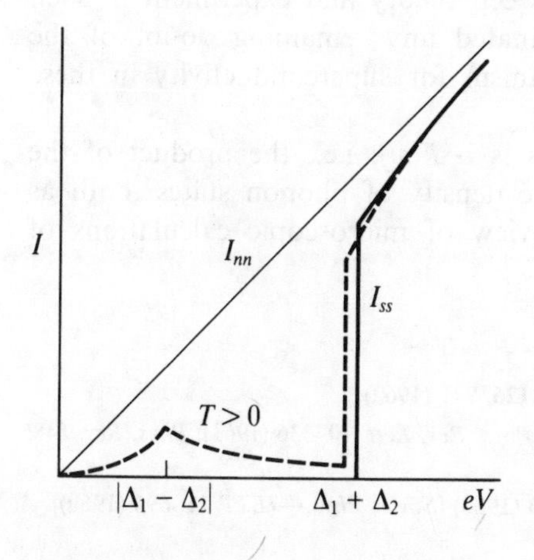

FIGURE 3.8
Superconductor-superconductor tunneling characteristic. Note that for $T > 0$ there are sharp features corresponding to both the sum and the difference of the two gap values. The peak at $|\Delta_1 - \Delta_2|$ would actually be a logarithmic singularity in the absence of gap anisotropy and level broadening due to lifetime effects.

regard because the existence of very sharply peaked densities of states at the gap edges of both materials helps to counteract the effects of thermal smearing.

3.8.5 Phonon Structure

When tunneling curves are measured in materials with strong electron-phonon coupling, structure beyond that outlined above is quite readily observed. Giaever et al.[28] noticed this first in lead and remarked that the structure occurred near energies that were characteristic of the phonon structure. This observation has since been greatly refined, from both experimental and theoretical viewpoints. The key point, made by Schrieffer, Scalapino, and Wilkins,[29] is that the observed density of states should be

$$N_s(E) = N(0)\,\mathrm{Re}\frac{E}{[E^2 - \Delta^2(E)]^{1/2}} \tag{3.83}$$

where Re indicates the real part of the expression that follows. This reduces to our earlier result (3.73) if Δ can be taken to be real and constant, as in the simple BCS model, but in the strong-coupling theory Δ becomes complex and energy dependent. The resulting energy-dependent phase of $\Delta(E)$ has physical content and is entirely distinct from the single arbitrary phase choice of Δ within the BCS approximation. The imaginary part of Δ corresponds to a damping of the quasi-particle excitations by decay with creation of real phonons; hence, it is large when $E \approx \hbar\omega_{ph}$. There is a corresponding resonant variation of Re Δ. Such behavior is, of course, hardly surprising given our model in which the interaction causing superconductivity is phonon-mediated. The simple BCS approximation suppresses these details caused by the retarded nature of the interaction [apart from the crude manifestation of a narrow peak in $N(E)$ at $\hbar\omega_c$, mentioned at the end of Sec. 3.7], but the more exact Eliashberg[30] procedure appears to give a quantitative account of the observed phenomena. This agreement between theory and experiment in such materials as lead and mercury has eliminated any remaining doubt of the correctness of the electron-phonon mechanism for superconductivity in these materials.

The central quantity in this analysis is $\alpha^2F(\omega)$, i.e., the product of the electron-phonon coupling strength and the density of phonon states, both as functions of energy. A comprehensive review of microscopic calculations of

[28] I. Giaever, H. R. Hart, and K. Megerle, *Phys. Rev.* **126**, 941 (1962).

[29] J. R. Schrieffer, D. J. Scalapino, and J. W. Wilkins, *Phys. Rev. Lett.* **10**, 336 (1963); *Phys. Rev.* **148**, 263 (1966).

[30] G. M. Eliashberg, *Zh. Eksperim. i Teor. Fiz.* **38**, 966 (1960) [*Soviet Phys.—JETP* **11**, 696 (1960)].

$\alpha^2F(\omega)$ for many materials has been given by Carbotte.[31] Working from the experimental side, McMillan and Rowell[32] were very successful in inverting the tunneling data to find the spectrum of $\alpha^2F(\omega)$. Their results generally agree well with expectations based on neutron scattering from phonons where such data are available. In fact, the tunneling data give results of superior accuracy for features such as the location of van Hove singularities in the phonon density of states, which appear as peaks in the experimentally obtained trace of the second derivative of the tunneling current.

3.9 TRANSITION PROBABILITIES AND COHERENCE EFFECTS

The effect of an external perturbation on the electrons in a metal can be expressed in terms of an interaction hamiltonian

$$\mathscr{H}_1 = \sum_{\mathbf{k}\sigma,\mathbf{k}'\sigma'} B_{\mathbf{k}'\sigma',\,\mathbf{k}\sigma}\, c^*_{\mathbf{k}'\sigma'} c_{\mathbf{k}\sigma} \tag{3.84}$$

where the $B_{\mathbf{k}'\sigma',\,\mathbf{k}\sigma}$ are matrix elements of the perturbing operator between the ordinary one-electron states of the normal metal. In the normal state, each term in this sum is independent and the square of each $B_{\mathbf{k}'\sigma',\,\mathbf{k}\sigma}$ is proportional to a corresponding transition probability. This is not the case in the superconducting state, however, because it consists of a phase-coherent superposition of occupied one-electron states. As a result, there are interference terms which are not present in the normal state.

This can be seen in detail by expanding the terms in (3.84) using the γ operators. It is then seen that the terms $c^*_{\mathbf{k}'\sigma'}c_{\mathbf{k}\sigma}$ and $c^*_{-\mathbf{k}-\sigma}c_{-\mathbf{k}'-\sigma'}$ connect the same quasi-particle states. For example,

$$c^*_{\mathbf{k}'\uparrow}c_{\mathbf{k}\uparrow} = u_{\mathbf{k}'}u^*_{\mathbf{k}}\gamma^*_{\mathbf{k}'0}\gamma_{\mathbf{k}0} - v^*_{\mathbf{k}'}v_{\mathbf{k}}\gamma^*_{\mathbf{k}1}\gamma_{\mathbf{k}'1} + u_{\mathbf{k}'}v_{\mathbf{k}}\gamma^*_{\mathbf{k}'0}\gamma^*_{\mathbf{k}1} + v^*_{\mathbf{k}'}u^*_{\mathbf{k}}\gamma_{\mathbf{k}'1}\gamma_{\mathbf{k}0} \tag{3.85a}$$

and

$$c^*_{-\mathbf{k}\downarrow}c_{-\mathbf{k}'\downarrow} = -v^*_{\mathbf{k}}v_{\mathbf{k}'}\gamma^*_{\mathbf{k}'0}\gamma_{\mathbf{k}0} + u_{\mathbf{k}}u^*_{\mathbf{k}'}\gamma^*_{\mathbf{k}1}\gamma_{\mathbf{k}'1} + u_{\mathbf{k}}v_{\mathbf{k}'}\gamma^*_{\mathbf{k}'0}\gamma^*_{\mathbf{k}1} + v^*_{\mathbf{k}}u^*_{\mathbf{k}'}\gamma_{\mathbf{k}'1}\gamma_{\mathbf{k}0} \tag{3.85b}$$

Thus, it is clear that matrix elements of these two terms in (3.84) must be added *before* squaring since they add coherently in determining the transition rate.

Fortunately, this addition can be done quite readily since one expects the coefficients $B_{\mathbf{k}'\sigma',\,\mathbf{k}\sigma}$ and $B_{-\mathbf{k}-\sigma,\,-\mathbf{k}'-\sigma'}$ to differ at most in sign because both

---

[31] J. Carbotte, *Revs. Mod. Phys.* **62**, 1027 (1990).

[32] W. L. McMillan and J. M Rowell, *Phys. Rev. Lett.* **14**, 108 (1965); see also J. M. Rowell and L. Kopf, *Phys. Rev.* **137**, A907 (1965). An excellent review is given by McMillan and Rowell in R. D. Parks (ed.), *Superconductivity*, Dekker, New York (1969), Chap. 11.

represent processes in which the momentum change of the electron is $\mathbf{k}' - \mathbf{k}$ and its spin change is $\sigma' - \sigma$. Thus, these terms can be combined as

$$B_{\mathbf{k}'\sigma',\,\mathbf{k}\sigma}(c^*_{\mathbf{k}'\sigma'}c_{\mathbf{k}\sigma} \pm c^*_{-\mathbf{k}-\sigma}c_{-\mathbf{k}'-\sigma'}) \tag{3.86}$$

where the sign choice depends on the nature of $\mathscr{H}_1$.

As shown by BCS, there are two cases. Case I is typified by the electron-phonon interaction responsible for ultrasonic attenuation. Being the interaction of the electron with a simple scalar deformation potential, it depends only on the momentum change; it is independent of the sense of $\mathbf{k}$ or σ, and the two matrix elements have the same sign and add coherently. Case II is typified by the interaction of the electron with the electromagnetic field via a term $\mathbf{p} \cdot \mathbf{A}$; since this changes sign on replacing $\mathbf{k}$ by $-\mathbf{k}$, the negative sign in (3.86) is appropriate.

Neither of these two interactions has involved the spin, so that we have had $\sigma = \sigma'$. If there *is* a spin change, as with terms of the sort I_+S_- in the hyperfine interaction of the electron with a nucleus, the sign associations are formally reversed. Following BCS, we indicate this with a factor $\Theta_{\sigma\sigma'}$, which is ± 1 for $\sigma' = \pm\sigma$. Thus, upon collecting terms and taking $u_{\mathbf{k}}$, $v_{\mathbf{k}}$, and Δ real, we see that (3.86) becomes

$$\begin{aligned} B_{\mathbf{k}'\sigma',\,\mathbf{k}\sigma}[&(u_{\mathbf{k}'}u_{\mathbf{k}} \mp v_{\mathbf{k}'}v_{\mathbf{k}})(\gamma^*_{\mathbf{k}'\sigma'}\gamma_{\mathbf{k}\sigma} \pm \Theta_{\sigma'\sigma}\gamma^*_{-\mathbf{k}-\sigma}\gamma_{-\mathbf{k}'-\sigma'}) \\ &+(v_{\mathbf{k}}u_{\mathbf{k}'} \pm u_{\mathbf{k}}v_{\mathbf{k}'})(\gamma^*_{\mathbf{k}'\sigma'}\gamma^*_{-\mathbf{k}-\sigma} \pm \Theta_{\sigma'\sigma}\gamma_{-\mathbf{k}'-\sigma'}\gamma_{\mathbf{k}\sigma})] \end{aligned} \tag{3.87}$$

(To facilitate writing this out for general spin directions, we have used a slightly modified notation, in which $\gamma_{\mathbf{k}\sigma} = \gamma_{\mathbf{k}0}$ for $\sigma = \uparrow$, and $\gamma_{\mathbf{k}\sigma} = \gamma_{-\mathbf{k}1}$ for $\sigma = \downarrow$.) From this expansion we see that the factor $\Theta_{\sigma\sigma'}$ really has no effect on the magnitude of transition probabilities; it affects only the relative phase of off-diagonal matrix elements connecting disparate states. The decisive point is whether the interaction is of case I or II, corresponding to the upper and lower signs in the coherence factors, i.e., the combinations of $u_{\mathbf{k}}$ and $v_{\mathbf{k}}$ in (3.87). For example, although the part I_zS_z of the hyperfine coupling does not flip the spin, and hence has $\Theta_{\sigma\sigma'} = +1$ rather than -1 as above for I_+S_-, both terms are governed by case II coherence factors because they are odd with respect to reversal of the spin. Generalizing from these examples, we see that cases I and II pertain to perturbations which are even and odd, respectively, under time reversal of the electronic states, which interchanges the partners in the Cooper-pairing scheme.

From (3.87) we see that in the computation of transition probabilities, the squared matrix elements $|B_{\mathbf{k}'\sigma',\,\mathbf{k}\sigma}|^2$ will be multiplied by so-called *coherence factors*, namely, $(uu' \mp vv')^2$ for the scattering of quasi-particles, and $(vu' \pm uv')^2$ for the creation or annihilation of two quasi-particles. (We have made an obvious condensation of the notation here.) With u and v given by (3.35), these coherence factors can be evaluated as explicit functions of energy. For example,

$$(uu' \mp vv')^2 = \frac{1}{4}\left\{\left[\left(1+\frac{\xi}{E}\right)\left(1+\frac{\xi'}{E'}\right)\right]^{1/2} \mp \left[\left(1-\frac{\xi}{E}\right)\left(1-\frac{\xi'}{E'}\right)\right]^{1/2}\right\}^2$$

$$= \frac{1}{4}\left\{\left(1+\frac{\xi}{E}+\frac{\xi'}{E'}+\frac{\xi\xi'}{EE'}\right)+\left(1-\frac{\xi}{E}-\frac{\xi'}{E'}+\frac{\xi\xi'}{EE'}\right)\right.$$

$$\left.\mp 2\left[\left(1-\frac{\xi^2}{E^2}\right)\left(1-\frac{\xi'^2}{E'^2}\right)\right]^{1/2}\right\} = \frac{1}{2}\left(1+\frac{\xi\xi'}{EE'} \mp \frac{\Delta^2}{EE'}\right)$$

Since E is an even function of ξ, when we sum over $\xi_{\mathbf{k}}$, terms appear in pairs such that the terms odd in ξ or ξ' cancel. Thus, effectively the coherence factor for *scattering* is

$$(uu' \mp vv')^2 = \frac{1}{2}\left(1 \mp \frac{\Delta^2}{EE'}\right) \tag{3.88a}$$

Similarly, the coherence factor for *creation* or *annihilation* of a pair of quasi-particles is

$$(vu' \pm uv')^2 = \frac{1}{2}\left(1 \pm \frac{\Delta^2}{EE'}\right) \tag{3.88b}$$

It is convenient to note that in the *semiconductor-model* sign convention, in which one of each pair of quasi-particles created or destroyed is assigned a negative energy, the coherence factors for both scattering and pair creation have the same form

$$F(\Delta, E, E') = \frac{1}{2}\left(1 \mp \frac{\Delta^2}{EE'}\right) \tag{3.89}$$

where the upper sign corresponds to case I and the lower to case II.

Evidently, the greatest effect of these coherence factors is for energies E and E' near the gap edge Δ, in which case (3.89) is either ~ 0 or ~ 1, depending on the sign. If one considers low-energy scattering processes for $\hbar\omega \ll \Delta$, no quasi-particles are created, so E and E' have the same sign. Then for case I processes like ultrasonic attentuation, $F \ll 1$, whereas for case II processes like nuclear relaxation, $F \sim 1$. The situation is reversed for high-energy processes with $\hbar\omega \gtrsim 2\,\Delta$, which create pairs of quasi-particles. Then $F \sim 1$ for case I processes and $F \ll 1$ for case II processes. Of course, if E and $E' \gg \Delta$, there is little difference between case I and II, and the superconducting coherence is unimportant.

These general considerations are made more clear by considering examples of the calculation of transition rates. Following the same line of argument used in reaching (3.76) for the case of tunneling, we expect a net transition rate between energy levels E and $E' = E + \hbar\omega$ to be proportional to

$$\alpha_s = \int |M|^2 F(\Delta, E, E+\hbar\omega) N_s(E) N_s(E+\hbar\omega) \times [f(E) - f(E+\hbar\omega)]\, dE \tag{3.90}$$

where M is the magnitude of a suitable one-electron matrix element. Since we shall always be interested in ratios to the normal-state values, we do not need to know more about the actual value of M. Upon inserting the explicit expressions for N_s and F, and simplifying, we find that (3.90) becomes

$$\alpha_s = |M|^2 N^2(0) \int_{-\infty}^{\infty} \frac{|E(E+\hbar\omega) \mp \Delta^2| [f(E) - f(E+\hbar\omega)]}{(E^2 - \Delta^2)^{1/2} [(E+\hbar\omega)^2 - \Delta^2]^{1/2}}\, dE$$

where it is understood that the regions with $|E|$ or $|E+\hbar\omega| < \Delta$ are excluded from the integration. In the normal state, $\Delta = 0$, and the corresponding expression reduces to $\alpha_n = |M|^2 N^2(0)\hbar\omega$. Thus, the desired ratio is

$$\frac{\alpha_s}{\alpha_n} = \frac{1}{\hbar\omega} \int_{-\infty}^{\infty} \frac{|E(E+\hbar\omega) \mp \Delta^2| [f(E) - f(E+\hbar\omega)]}{(E^2 - \Delta^2)^{1/2} [(E+\hbar\omega)^2 - \Delta^2]^{1/2}}\, dE \tag{3.91}$$

with the upper sign referring to case I processes and the lower to case II. We now use this general expression to treat some important specific cases.

3.9.1 Ultrasonic Attenuation

As noted earlier, the relevant matrix elements for treating the attenuation of longitudinal sound waves have case I coherence factors, i.e., the upper sign in (3.91). (We restrict our attention to longitudinal waves to avoid the complications which arise in the transverse case because currents are generated which are screened electromagnetically, giving a mixture of effects.) We note further that in typical ultrasonic experiments the sound frequency is less than 10^9 Hz, so that $\hbar\omega \lesssim 10^{-2}\Delta(0)$; also, $\hbar\omega \ll kT$. These inequalities enable us to consider only a simple low-frequency limiting case. Inspecting (3.91) in the limit as $\hbar\omega \to 0$, we see that most factors cancel, leaving

$$\frac{\alpha_s}{\alpha_n} = \lim_{\hbar\omega \to 0} \frac{1}{\hbar\omega} \int [f(E) - f(E+\hbar\omega)]\, dE$$

$$= -\int \frac{\partial f}{\partial E}\, dE$$

with the integration extending from $-\infty$ to $-\Delta$ and from Δ to ∞. Thus,

$$\begin{aligned}\frac{\alpha_s}{\alpha_n} &= f(-\infty) - f(-\Delta) + f(\Delta) - f(\infty) \\ &= 2f(\Delta) = \frac{2}{1 + e^{\Delta/kT}}\end{aligned} \tag{3.92}$$

This very simple result, combined with our previous calculation of $\Delta(T)$, predicts the behavior shown in Fig. 3.9. In particular, the infinite slope of $\Delta(T)$ at T_c causes α_s/α_n to drop with infinite slope as T is lowered below T_c. On the other hand, for $T \ll T_c$, the excess of α_s/α_n above a residual value due to nonelectronic mechanisms becomes exponentially small as the number of thermally excited quasi-particles available to absorb energy goes to zero.

When it is possible to carry experiments to low enough values of T/T_c to establish the residual attenuation level quite exactly, one can infer a value of $\Delta(T)$ from the attenuation data using (3.92). In fact, by propagating sound in different directions in single crystals, Morse and coworkers[33] were among the first to be able to get some measure of the anisotropy of the gap $\Delta_{\mathbf{k}}$ with respect to crystalline axes. This technique suffers from the fact that a given direction of sound propagation $\hat{\mathbf{k}}_s$ measures an average of $\Delta_{\mathbf{k}}$ over a disk perpendicular to $\hat{\mathbf{k}}_s$. The reason for this is that for efficient energy transfer between the sound wave and the electrons, the component of quasi-particle velocity parallel to $\hat{\mathbf{k}}_s$ must equal the sound velocity. Since $v_{\text{sound}} \ll v_{\text{electron}}$, this means that only electrons moving almost perpendicular to $\hat{\mathbf{k}}_s$ are effective in the attenuation. Nonetheless, values

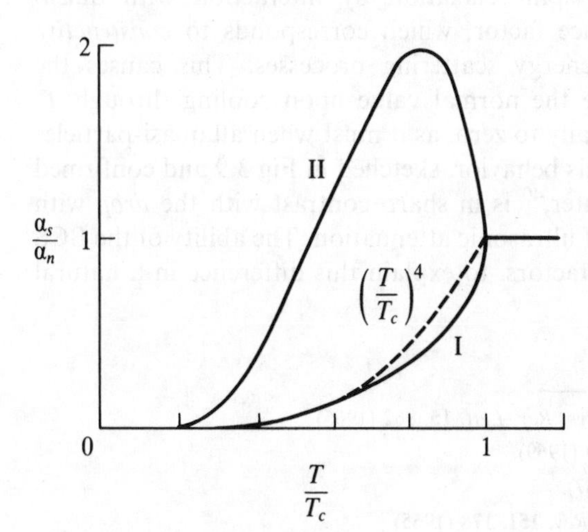

FIGURE 3.9
Temperature dependence of low-frequency absorption processes obeying case I and II coherence factors, compared with the $(T/T_c)^4$ dependence that might be expected for *all* processes from a simple two-fluid model. The curve for case I applies to ultrasonic attenuation, and it is a well-defined low-frequency limit. The curve for case II, which applies to nuclear relaxation or electromagnetic absorption, has no well-defined low-frequency limit unless gap anisotropy or level broadening is taken into account. The curve drawn here corresponds to a broadening of about $0.02\Delta(0)$.

[33] R. W. Morse, "Ultrasonic Attenuation in Metals at Low Temperatures," in K. Mendelssohn (ed.) *Progress in Cryogenics*, vol. I, Heywood, London (1959), p. 219.

of $2\Delta(0)$ in tin ranging from $3.3kT_c$ to $3.9kT_c$ have been inferred from such measurements, e.g., by Claiborne and Einspruch.[34] Phillips[35] has carefully re-examined the detailed nature of elastic scattering in anisotropic superconductors and concluded that some earlier discussions were oversimplified.

The complication of electromagnetic screening of transverse sound waves mentioned earlier can be put to good use as a means of measuring the penetration depth very near T_c, where it approaches infinity. The point is that the screening becomes essentially complete as soon as $\lambda(T)$ is less than the ultrasonic wavelength, so that the electrodynamic attenuation becomes negligible for lower temperatures, leading to a very sharp drop in attenuation within less than 1 percent of T_c. Using this effect, Fossheim[36] obtained what are probably the most reliable available values for λ of indium from measurements of α_s/α_n within a few millidegrees of T_c.

Another complication which can enter ultrasonic-attenuation measurements arises from the electronic damping of the motion of dislocations driven by the sound waves.[37] Because of dislocation pinning effects, this leads to a nonlinear response which makes definition of α_s/α_n difficult except at very low signal levels. Mason[38] showed that when these effects were taken into account, the apparently anomalous measurements on lead[39] could be reconciled with expectations based on a gap $2\Delta(0) = 4.1kT_c$, a value in reasonable agreement with results from other techniques.

3.9.2 Nuclear Relaxation

The matrix elements for nuclear-spin relaxation by interaction with quasi-particles have the case II coherence factor, which corresponds to *constructive* interference in the relevant low-energy scattering processes. This causes the relaxation rate $1/T_1$ to *rise* above the normal value upon cooling through T_c before it eventually goes exponentially to zero, as it must when all quasi-particles above the gap are "frozen out." This behavior, sketched in Fig 3.9 and confirmed experimentally by Hebel and Slichter,[40] is in sharp contrast with the *drop* with vertical tangent found in the case of ultrasonic attenuation. The ability of the BCS pairing theory, with its coherence factors, to explain this difference in a natural

[34] L. T. Claiborne and N. G. Einspruch, *Phys. Rev. Lett.* **15**, 862 (1965).

[35] W. A. Phillips, *Proc. Roy. Soc.* **A309**, 259 (1969).

[36] K. Fossheim, *Phys. Rev. Lett.* **19**, 81 (1967).

[37] B. R. Tittman and H. E. Bömmel, *Phys. Rev.* **151**, 178 (1965).

[38] W. P. Mason, *Phys. Rev.* **143**, 229 (1966).

[39] R. E. Love, R. W. Shaw, and W. A. Fate, *Phys. Rev.* **138**, A1453 (1965).

[40] L. C. Hebel and C. P. Slichter, *Phys. Rev.* **107**, 901 (1957); **113**, 1504 (1959); L. C. Hebel, *Phys. Rev.* **116**, 79 (1959).

way was one of the key triumphs which validated the theory. By way of contrast, any simple "two-fluid" model which attributed the ultrasonic attenuation and nuclear relaxation to a certain fraction of "normal electrons" would have to give the same temperature dependence for all such properties. For example, one might expect a normal-electron density $n_n/n = (T/T_c)^4$ in a model which correlated the empirical temperature variation of the penetration depth $\lambda^{-2} \sim [1 - (T/T_c)^4]$ with a similar variation of n_s/n in the London theory. As indicated by the dashed curve in Fig. 3.8, this gives a qualitative fit to the ultrasonic-attenuation result; but it cannot possibly explain the rise in $1/T_{1s}$ above $1/T_{1n}$ since that would require the number of normal electrons to exceed the total number of electrons, despite the existence of "superconducting electrons" as well.

Let us now examine this in more detail. We are again interested in the limit in which $\hbar\omega = \hbar\gamma H$ is much less than Δ and kT. However, we cannot proceed as simply as before because for the case II coherence factors $\alpha_s/\alpha_n \to \infty$ as $\omega \to 0$. However, we can still simplify the Fermi function by converting to a derivative form. Thus,

$$\frac{\alpha_s}{\alpha_n} = 2\int_\Delta^\infty \frac{E(E+\hbar\omega)+\Delta^2}{(E^2-\Delta^2)^{1/2}[(E+\hbar\omega)^2-\Delta^2]^{1/2}}\left(-\frac{\partial f}{\partial E}\right)dE \tag{3.93}$$

where the factor of 2 results from the fact that the integration over negative energies $(E+\hbar\omega) \leq -\Delta$ gives exactly the same contribution as the integration over positive energies $E \geq \Delta$. If we now try to set $\omega = 0$, we find

$$\frac{\alpha_s}{\alpha_n} = 2\int_\Delta^\infty \frac{E^2+\Delta^2}{E^2-\Delta^2}\left(-\frac{\partial f}{\partial E}\right)dE$$

which is easily seen to diverge logarithmically from the integration at Δ. If one kept ω finite, the divergence would be replaced by a factor of order $\ln(\Delta/\hbar\omega) \sim 10$ for typical values. This is still much greater than the experimentally observed rise in $1/T_1$, typically only a factor of 2, before the exponential fall at low temperatures. Thus, some other concept must be brought in to explain this quantitative discrepancy. The origin of the incipient divergence in α_s can be traced to the product of the two highly peaked, superconducting density-of-states factors in (3.93). Dropping nonsingular factors, we obtain

$$\alpha \propto \int N(E)N(E+\hbar\omega)dE \approx \int N^2(E)\,dE$$

for $\hbar\omega \ll \Delta$. Since $\int N(E)\,dE$ is conserved on going into the superconducting state, the sharp rise of $N_s(E)$ must cause $\int N_s^2\,dE$ to exceed $\int N_n^2\,dE$. An excessively high value of α results if we have overestimated the sharpness of the peak in state density by using the simple BCS form.

Two reasons have been advanced for the apparent extra breadth in the peak in the density of states. The first is that the anisotropy of the energy gap in real crystals leads to a range in $\Delta_{\mathbf{k}}$ over the Fermi surface. Thus, the peaks in $N_s(E)$ will be smeared out over a fractional energy range, typically $\frac{1}{10}$. Experiments of

Masuda[41] on aluminum with varying amounts of impurity support this explanation. He observed a bigger rise of $1/T_1$ in the dirtier samples. This is consistent with the Anderson theory of dirty superconductors, which holds that the gap should become more nearly the same for all electrons when rapid electron scattering causes electrons to sample many **k** values during the relevant time interval $\hbar/\Delta$.

The other explanation, due to Fibich,[42] is that the finite lifetime of the quasi-particle states against decay into phonons limits the sharpness of the peak, as indicated by an uncertainty-principle argument. Such a mechanism would be expected to be roughly independent of impurity but more important in strong-coupling superconductors. Some data on indium seem to support this mechanism. Thus, both mechanisms seem important in appropriate situations.

We conclude this section by noting that the high-temperature superconductors seem to show *no* Hebel-Slichter peak at all. As discussed in Sec. 9.9, this may reflect a gapless density of states associated with nonconventional pairing.

3.9.3 Electromagnetic Absorption

Since the interaction hamiltonian $\mathbf{p}\cdot\mathbf{A}$ also obeys the case II coherence factors, we can carry over the results of the previous section on nuclear relaxation without modification to describe the absorption of low-frequency electromagnetic radiation. The quantity α_s/α_n is now called σ_{1s}/σ_n since for a given E field the electromagnetic energy absorption per unit volume is $\sigma_1 E^2$, where σ_1 is the real part of the complex conductivity $\sigma_1(\omega) - i\sigma_2(\omega)$. Thus, for $\hbar\omega \ll \Delta$, we expect σ_{1s}/σ_n to rise above unity just below T_c and then to fall exponentially to zero at low temperatures. As noted earlier, such behavior is qualitatively incompatible with a simple two-fluid picture in which $n_n \leq n$.

Unlike the case of nuclear relaxation, however, it is now possible to utilize frequencies large enough to create pairs of quasi-particles. Such processes occur, in addition to the scattering processes treated already, as soon as $\hbar\omega \gtrsim 2\Delta$. In fact, at $T = 0$, there are no thermally excited quasi-particles present, and the *only* process allowing absorption of energy is the creation of pairs. With our semiconductor sign conventions, the initial state energy E must be $\leq -\Delta$, and the final state energy $E + \hbar\omega \geq \Delta$. Thus, $\sigma_1(\omega) = 0$ for $\hbar\omega < 2\Delta$, at which point there is an *absorption edge*, as shown in Fig. 3.10. The absorption can be computed by using (3.91), with the Fermi functions being either 0 or 1 at $T = 0$. Thus,

$$\left.\frac{\sigma_{1s}}{\sigma_n}\right|_{T=0} = \frac{1}{\hbar\omega}\int_{\Delta-\hbar\omega}^{-\Delta} \frac{|E(E+\hbar\omega)+\Delta^2|}{(E^2-\Delta^2)^{1/2}[(E+\hbar\omega)^2-\Delta^2]^{1/2}}\, dE \tag{3.94}$$

---

[41] Y. Masuda, *Phys. Rev.* **126**, 1271 (1962).

[42] M. Fibich, *Phys. Rev. Lett.* **14**, 561 (1965).

As shown by Mattis and Bardeen,[43] this integral can be expressed in terms of the tabulated complete elliptic integrals E and K, namely,

$$\left.\frac{\sigma_{1s}}{\sigma_n}\right|_{T=0} = \left(1 + \frac{2\Delta}{\hbar\omega}\right) E(k) - \frac{4\Delta}{\hbar\omega} K(k) \quad \hbar\omega \geq 2\Delta \tag{3.95}$$

where

$$k = \frac{\hbar\omega - 2\Delta}{\hbar\omega + 2\Delta} \tag{3.95a}$$

As shown in Fig. 3.10, σ_{1s}/σ_n rises from zero with finite slope at $\hbar\omega = 2\Delta$ and slowly approaches unity for $\hbar\omega \gg 2\Delta$. At finite temperatures, $\Delta(T) < \Delta(0)$, and also the thermally excited quasi-particles contribute absorption for $\hbar\omega < 2\Delta(T)$. Numerical calculations are needed to describe the exact behavior, but the qualitative behavior for $T > 0$ is indicated by the dashed curve. The rise as $\hbar\omega \to 0$ is the logarithmic dependence discussed in connection with $1/T_1$.

Historically speaking, the first spectroscopic measurements clearly showing the existence and width of the energy gap in superconductors well below T_c were made by Glover and Tinkham[44] with far-infrared radiation in the region of the absorption edge. These first measurements slightly preceded the appearance of the BCS theory but were soon found to be in excellent accord with (3.95). With improvements in techniques over the years,[45] the quality of data has reached the state that small deviations from the simple BCS curves observed in measure-

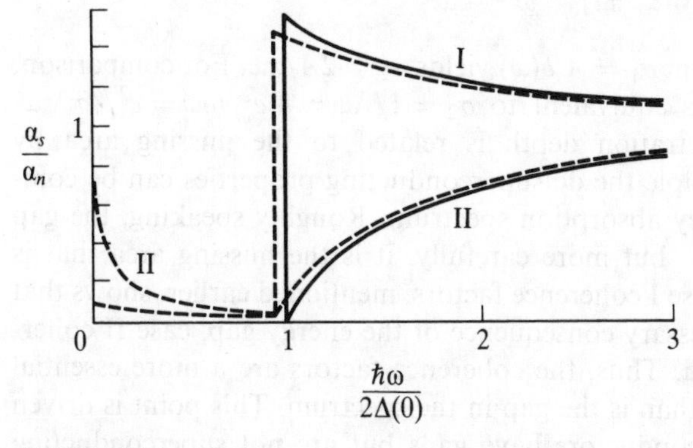

FIGURE 3.10
Frequency dependence of absorption processes obeying case I and II coherence factors at $T = 0$ (solid curves) and $T \approx \frac{1}{2} T_c$ (dashed curves).

[43] D. C. Mattis and J. Bardeen, *Phys. Rev.* **111**, 412 (1958).
[44] R. E. Glover and M. Tinkham, *Phys. Rev.* **104**, 844 (1956).
[45] D. M. Ginsberg and M. Tinkham, *Phys. Rev.* **118**, 990 (1960); L. H. Palmer and M. Tinkham, *Phys. Rev.* **165**, 588 (1968).

ments on thin lead films can be interpreted in terms of strong-coupling effects analogous to the phonon structure observed in tunneling experiments.

It is worth remarking that with case I coherence factors, α_s/α_n rises discontinuously at $\hbar\omega = 2\Delta$ to a value greater than 1, and then decreases, as shown in Fig. 3.10. One can show that the total area under the curve is conserved in this case, whereas for case II the area seems to disappear when the gap opens up. However, the oscillator-strength sum rule in the form[46]

$$\int_0^\infty \sigma_1(\omega)d\omega = \frac{\pi ne^2}{2m}$$

requires that the area under the curve of $\sigma_1(\omega)$ have the same value in the superconducting as in the normal state. Tinkham and Ferrell[47] were able to argue that the "missing area" A at finite frequencies appears as a δ function at $\omega = 0$, which physically represents the absorption of energy from a dc electric field to supply the kinetic energy of the accelerated supercurrent. The argument is based on the Kramers-Kronig relations,[48] which connect the real and imaginary parts of any causal linear-response function. Written in terms of the complex conductivity $\sigma_1 - i\sigma_2$, and with a time dependence of $e^{+i\omega t}$, they have the form

$$\sigma_1(\omega) = \frac{2}{\pi}\int_0^\infty \frac{\omega'\sigma_2(\omega')\,d\omega'}{\omega'^2 - \omega^2} + \text{const} \tag{3.96a}$$

$$\sigma_2(\omega) = -\frac{2\omega}{\pi}\int_0^\infty \frac{\sigma_1(\omega')\,d\omega'}{\omega'^2 - \omega^2} \tag{3.96b}$$

From (3.96*b*) we see that a term $\sigma_1 = A\;\delta(\omega)$ yields $\sigma_2 = 2A/\pi\omega$. For comparison, the London equation (1.3) is equivalent to $\sigma_2 = 1/\Lambda\omega = n_s e^2/m\omega = c^2/4\pi\lambda^2\omega$. Thus, we see that the penetration depth is related to the missing area by $\lambda^{-2} = 8A/c^2$, so that in principle the dc superconducting properties can be computed from the high-frequency absorption spectrum. Roughly speaking, the gap implies the superconductivity, but more carefully, it is the missing area that is important. The example of case I coherence factors, mentioned earlier, shows that the missing area is not a necessary consequence of the energy gap; case II coherence factors are also required. Thus, the coherence factors are a more essential feature of superconductivity than is the gap in the spectrum. This point is driven home by recalling that *semi*conductors have gaps but are not superconducting (because there is no missing area), whereas it has been shown that superconduc-

---

[46]See, e.g., R. Kubo, *J. Phys. Soc. Jpn.* **12**, 570 (1957).

[47]M. Tinkham and R. A. Ferrell, *Phys. Rev. Lett.* **2**, 331 (1959).

[48]For a more detailed discussion of the application of sum rule and Kramers-Kronig methods to superconducting electrodynamics, see the chapter on superconductivity by M. Tinkham in C. de Witt, B. Dreyfus, and P. G. de Gennes (eds.), *Low Temperature Physics*, Gordon and Breach, New York (1962). Also, see Sec. 2.5.1 of this book.

tors made gapless by magnetic impurities retain superconducting properties as long as there is still a missing area under the $\sigma_1(\omega)$ curve.

3.10 ELECTRODYNAMICS

The simple treatment just given of the absorption of electromagnetic fields gives no direct account of the dramatic supercurrent properties because we restricted our discussion to dissipative processes analogous to those of normal electrons. Rather than undertake the complications of a complete treatment of the response of a superconductor to an arbitrary electromagnetic field, we shall content ourselves with a treatment of the response to a static magnetic field. This response must be nondissipative, and hence is complementary to that just treated. Since the Meissner effect and related superfluid responses are nearly independent of frequency until one reaches frequencies on the order of the gap, this treatment actually is useful in a wide range of applications.

In the cases of interest, the field can be described by a classical transverse vector potential $\mathbf{A}(\mathbf{r})$ such that $\mathbf{B} = \text{curl}\,\mathbf{A}$. These are the total fields, including the effects of screening by supercurrents, which will have to be introduced in a self-consistent way. In the presence of a vector potential, it is well known that the canonical momentum of both classical and quantum mechanics contains a potential-momentum term as well as the usual kinetic momentum, so that $\mathbf{p} = m\mathbf{v} + e\mathbf{A}/c$, where e is the charge of the particle. The kinetic energy $\frac{1}{2}mv^2$ then becomes $(\mathbf{p} - e\mathbf{A}/c)^2/2m$, whereas the potential-energy expression is unchanged. Since we are interested only in calculating the linear response to weak fields, we can expand this kinetic-energy expression, keeping only terms linear in $\mathbf{A}$. With the operator replacement $\mathbf{p}_i \to -i\hbar\nabla_i$, the resulting perturbation term is

$$\mathscr{H}_1 = \frac{ie\hbar}{2mc}\sum_i(\nabla_i \cdot \mathbf{A} + \mathbf{A} \cdot \nabla_i)$$

where the sum runs over all the particles. If the vector potential is expanded in spatial Fourier components

$$\mathbf{A}(\mathbf{r}) = \sum_{\mathbf{k}} \mathbf{a}(\mathbf{q})e^{i\mathbf{q}\cdot\mathbf{r}}$$

$\mathscr{H}_1$ may be written

$$\mathscr{H}_1 = -\frac{e\hbar}{mc}\sum_{\mathbf{k},\,\mathbf{q}} \mathbf{k} \cdot \mathbf{a}(\mathbf{q})c^*_{\mathbf{k}+\mathbf{q},\sigma}c_{\mathbf{k},\,\sigma} \tag{3.97}$$

since $\mathbf{q} \cdot \mathbf{a}(\mathbf{q}) = 0$ for a transverse field, by definition. Because the dominant interaction is with the orbital motion, the spin index is unchanged. The explicit form of the $B_{\mathbf{k}'\sigma',\,\mathbf{k}\sigma}$ of (3.84) may be read out from (3.97). From the proportionality of these matrix elements to **k** we see that they are of case II, as noted earlier, and following (3.87), the interaction can be written out in full as

$$\mathscr{H}_1 = -\frac{e\hbar}{mc}\sum_{\mathbf{k},\mathbf{q}} \mathbf{k}\cdot\mathbf{a}(\mathbf{q})[u_{\mathbf{k}}u_{\mathbf{k+q}} + v_{\mathbf{k}}v_{\mathbf{k+q}})(\gamma^*_{\mathbf{k+q},0}\gamma_{\mathbf{k},0} - \gamma^*_{\mathbf{k}1}\gamma_{\mathbf{k+q},1})$$

$$+(v_{\mathbf{k}}u_{\mathbf{k+q}} - u_{\mathbf{k}}v_{\mathbf{k+q}})(\gamma^*_{\mathbf{k+q},0}\gamma^*_{\mathbf{k}1} - \gamma_{\mathbf{k+q},1}\gamma_{\mathbf{k}0})] \quad (3.98)$$

In the preceding section, we dealt with the energy absorption due to this operator. Now we consider the *current* induced by this perturbation. In the presence of a vector potential, the current consists of two terms, $\mathbf{J}_1$ and $\mathbf{J}_2$, corresponding to the two terms in $\mathbf{v} = (\mathbf{p}/m) - (e\mathbf{A}/mc)$. The second term gives the simple result

$$\mathbf{J}_2 = -\frac{ne^2}{mc}\mathbf{A} \quad (3.99)$$

which would correspond exactly to the London equation (1.8) if n could be interpreted as n_s, the number of "superconducting electrons." In fact, though, n is always the *total* electron density, and this term has the same value also in the normal state. Thus, the first term $\mathbf{J}_1$, often called the *paramagnetic* current term because it tends to cancel the diamagnetic current $\mathbf{J}_2$, must play a vital role. It is not hard to see that the **q**th Fourier component of $\mathbf{J}_1$ will be found by evaluating the operator

$$\mathbf{J}_1(\mathbf{q}) = \frac{e\hbar}{m}\sum_{\mathbf{k}} \mathbf{k}c^*_{\mathbf{k-q}}c_{\mathbf{k}} \quad (3.100)$$

which may then be expanded in terms of the γ operators to give a form analogous to (3.98).

Before proceeding with detailed calculations, it is convenient to introduce a standard notation for describing the current response to the various Fourier components of a vector potential, namely,

$$\mathbf{J}(\mathbf{q}) = -\frac{c}{4\pi}K(\mathbf{q})\mathbf{a}(\mathbf{q}) \quad (3.101)$$

For example, in the London theory, where

$$\mathbf{J}(\mathbf{r}) = -\frac{1}{c\Lambda}\mathbf{A}(\mathbf{r}) = -\frac{c}{4\pi\lambda^2}\mathbf{A}(\mathbf{r}) \quad (3.102)$$

the response is independent of **q**, and

$$K(\mathbf{q}) = K(0) = \frac{1}{\lambda^2} \quad (3.103)$$

Using the definition $\lambda_L^2(0) = mc^2/4\pi ne^2$ for the ideal limit at $T = 0$ and (3.99), we may write the response function, including both $\mathbf{J}_1$ and $\mathbf{J}_2$, as

$$K(\mathbf{q}, T) = \lambda_L^{-2}(0)[1 + \lambda_L^2(0)K_1(\mathbf{q}, T)] \quad (3.104)$$

where the preceding discussion leads us to expect that K_1 will be negative. As we are restricting our attention to isotropic systems and transverse fields, $K(\mathbf{q})$ is a function only of $|\mathbf{q}|$, and we shall normally denote it simply $K(q)$.

It is also useful to consider, in general, the relation between a q-dependent $K(q)$ and the corresponding nonlocal response in coordinate space described by a range function or kernel $F(R)$ in the expression

$$\mathbf{J}(\mathbf{r}) = C \int \frac{\mathbf{R}[\mathbf{R} \cdot \mathbf{A}(\mathbf{r}')]}{R^4} F(R)\, d\mathbf{r}' \tag{3.105}$$

where $\mathbf{R} = \mathbf{r} - \mathbf{r}'$. By inserting $\mathbf{A}(\mathbf{r}') \sim e^{i\mathbf{q}\cdot\mathbf{r}'}$, one can show that the relation between between $K(q)$ and $F(R)$ is

$$K(q) = \frac{16\pi^2 C}{3c} \int_0^\infty \left[\frac{3}{qR} j_1(qR)\right] F(R)\, dR \tag{3.106}$$

where $j_1(x) = x^{-2} \sin x - x^{-1} \cos x$ is a spherical Bessel function, such that $[3x^{-1} j_1(x)]$ is a damped oscillatory function with the value unity for $x = 0$. Thus,

$$K(0) = \frac{16\pi^2 C}{3c} \int_0^\infty F(R)\, dR \tag{3.106a}$$

depends on the integral of $F(R)$, whereas as $q \to \infty$

$$K(q) \underset{q\to\infty}{\rightarrow} \frac{16\pi^2 C}{3c} \frac{F(0)}{q} \int_0^\infty \frac{3}{x} j_1(x)\, dx = \frac{4\pi^3 C F(0)}{cq} \tag{3.106b}$$

which depends only on the value of $F(R)$ at the origin. Taking the ratio, we have

$$\frac{K(q)}{K(0)} \underset{q\to\infty}{\rightarrow} \frac{3\pi}{4qL} \tag{3.106c}$$

where $L = F^{-1}(0) \int F(R) dR$ is a measure of the range of the real-space kernel $F(R)$. These relations are very general; they apply to the relation between $\mathbf{J}$ and $\mathbf{E}$ in the normal state with $L = \ell$ and to the relation between $\mathbf{J}$ and $\mathbf{A}$ in the superconducting state with, as we shall soon see, $L \approx \xi_0$.

3.10.1 Calculation of $K(0, T)$ or $\lambda_L(T)$

Now let us proceed to calculate the temperature dependence of $K(q, T)$ for the simple limiting case $q = 0$, which corresponds to infinite wavelength. This will determine the temperature dependence of $\lambda_L(T)$, which is defined (for $\ell = \infty$) by

$$K(0, T) \equiv \frac{1}{\lambda_L^2(T)} \tag{3.107}$$

since the nonlocal theory reduces to the local London theory for fields which vary slowly in space. In other words, the temperature dependence of $K(0, T)$ is identified with the temperature dependence of n_s in the London theory. For $\mathbf{q} = 0$, obviously the coherence factor in the second term in (3.98) is zero, whereas that in the first term is unity. Moreover, for $\mathbf{q} = 0$, $\gamma^*_{\mathbf{k}+\mathbf{q},0}\gamma_{\mathbf{k}0}$ becomes the number operator $\gamma^*_{\mathbf{k}0}\gamma_{\mathbf{k}0}$. Thus, the perturbing hamiltonian (3.98) simply shifts the energies of the quasi-particle excitations, as follows:

$$E_{\mathbf{k}0} \to E_{\mathbf{k}0} - \frac{e\hbar}{mc}\mathbf{k}\cdot\mathbf{a}(0)$$
$$E_{\mathbf{k}1} \to E_{\mathbf{k}1} + \frac{e\hbar}{mc}\mathbf{k}\cdot\mathbf{a}(0) \tag{3.108}$$

Similarly, the expansion of $\mathbf{J}_1(0)$ in quasi-particle operators reduces simply to

$$\mathbf{J}_1(0) = \frac{e\hbar}{m}\sum_{\mathbf{k}}\mathbf{k}(\gamma^*_{\mathbf{k}0}\gamma_{\mathbf{k}0} - \gamma^*_{\mathbf{k}1}\gamma_{\mathbf{k}1})$$
$$= \frac{e\hbar}{m}\sum_{\mathbf{k}}\mathbf{k}(f_{\mathbf{k}0} - f_{\mathbf{k}1}) \tag{3.109}$$

where in the second line the operators have been replaced by their expectation values $f_{\mathbf{k}0}$ and $f_{\mathbf{k}1}$, the Fermi functions corresponding to the shifted energies (3.108). In the limit of small $\mathbf{a}(0)$, $f_{\mathbf{k}0} \approx f_{\mathbf{k}1}$, and the difference may be found by taking the first term in a Taylor's series expansion; i.e.,

$$f_{\mathbf{k}0} - f_{\mathbf{k}1} \approx \left(-\frac{\partial f}{\partial E_{\mathbf{k}}}\right)\frac{2e\hbar}{mc}\mathbf{k}\cdot\mathbf{a}(0)$$

Inserting this in (3.109), we have

$$\mathbf{J}_1(0) = \frac{2e^2\hbar^2}{m^2c}\sum_{\mathbf{k}}[\mathbf{a}(0)\cdot\mathbf{k}]\mathbf{k}\left(-\frac{\partial f}{\partial E_{\mathbf{k}}}\right) \tag{3.110}$$

By symmetry, $\mathbf{J}_1(0)$ is parallel to $\mathbf{a}(0)$, and the average of the square of the component of $\mathbf{k}$ along $\mathbf{J}_1$ (or $\mathbf{a}$) will be $k_F^2/3$ since $\cos^2\theta$ averages to one third over a sphere. Thus, the K_1 corresponding to (3.110) can be written as

$$K_1(0,T) = \frac{-4\pi\mathbf{J}_1(0)}{c\mathbf{a}(0)} = -\left(\frac{4\pi ne^2}{mc^2}\right)\left(\frac{4E_F}{3n}\right)\sum_{\mathbf{k}}\left(-\frac{\partial f}{\partial E_{\mathbf{k}}}\right)$$

Since $N(0) = 3n/4E_F$, this can be rewritten as

$$K_1(0,T) = -\lambda_L^{-2}(0)\int_{-\infty}^{\infty}\left(-\frac{\partial f}{\partial E}\right)d\xi$$

Inserting this in (3.104) yields the total response, including $\mathbf{J}_1$ and $\mathbf{J}_2$,

$$K(0,T) = \lambda_L^{-2}(T) = \lambda_L^{-2}(0)\left[1 - 2\int_{\Delta}^{\infty}\left(-\frac{\partial f}{\partial E}\right)\frac{E}{(E^2-\Delta^2)^{1/2}}dE\right] \tag{3.111}$$

We note first that if $\Delta = 0$ (normal state, $T \geq T_c$), then the integral reduces to $f(0) = \frac{1}{2}$, and $K(0, T \geq T_c) = 0$. This corresponds to the absence of any

Meissner effect in the normal state because of the cancellation of the paramagnetic and diamagnetic current terms $\mathbf{J}_1$ and $\mathbf{J}_2$. As soon as $\Delta > 0$, however, the integral no longer gives the exact cancellation, and there is a finite $\lambda_L(T)$ leading to a Meissner effect if the sample is large enough. As T decreases and $\Delta(T)/kT$ increases further, the integral becomes less and less, eventually becoming exponentially small when $\Delta/kT \gg 1$. Thus, $\lambda_L(T)$ defined by (3.111) does reduce properly to $\lambda_L(0)$ as defined by (1.9) when $T \to 0$, justifying our notations. The general behavior is shown in Fig. 3.11. The physical origin of K_1 is that excitations on the trailing edge of the displaced momentum distribution in $\mathbf{k}$ space have lower energy and thus are more highly populated, so that the excited quasi-particles carry a net current in the reverse direction which partially cancels the current $\mathbf{J}_2$. It is conceptually important to recognize that the *total* response, including the (negative) quasi-particle contribution, is the supercurrent. The quasi-particle contribution does not die away because it represents a distribution giving minimum free energy.

In considering Fig. 3.11, we should keep in mind that $\lambda_L(T)$ will give the temperature dependence of the experimentally observed penetration depth only if the strength of the response to the self-consistent vector potential is adequately approximated by the response to the component with $\mathbf{q} = 0$. We now consider the q dependence of $K(q, T)$ to clarify this point.

3.10.2 Calculation of $K(q, 0)$

For simplicity, we shall examine the q dependence only at $T = 0$, although the calculation is readily extended by use of appropriate statistical techniques. At $T = 0$, the electrons must be in the BCS ground state $|\psi_G\rangle$. Then by ordinary first-order perturbation theory, the perturbed state $|\psi\rangle$ in the presence of $\mathscr{H}_1$ is

$$|\psi\rangle = |\psi_G\rangle - \sum_n \frac{\langle\psi_n|\mathscr{H}_1|\psi_G\rangle}{E_n}|\psi_n\rangle$$

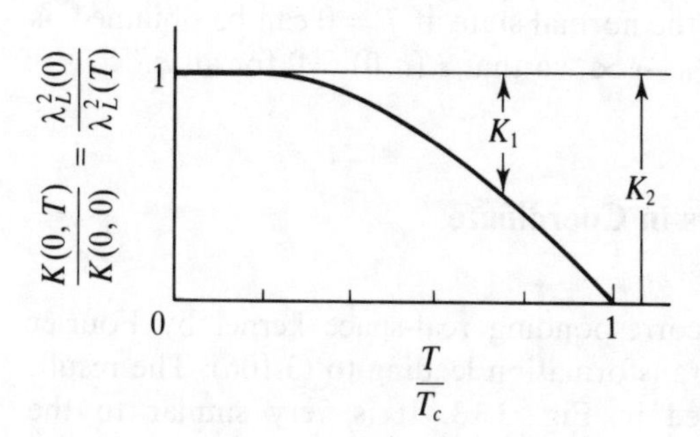

FIGURE 3.11
Temperature dependence of $K(0, T)/K(0, 0) = \lambda_L^2(0)/\lambda_L^2(T) = n_s(T)/n$ according to the BCS theory. Note that $K(0, T)$ is the result of the partial cancellation of the constant diamagnetic term K_2 by the temperature-dependent paramagnetic term K_1.

where the sum on n runs over the various excited states with excitation energy E_n. Taking the expectation value of $\mathbf{J}_1(\mathbf{q})$ over $|\psi\rangle$, we have

$$\langle\psi|\mathbf{J}_1(\mathbf{q})|\psi\rangle = \langle\psi_G|\mathbf{J}_1(\mathbf{q})|\psi_G\rangle - 2\,\mathrm{Re}\sum_n \frac{\langle\psi_G|\mathbf{J}_1(\mathbf{q})|\psi_n\rangle\langle\psi_n|\mathcal{H}_1|\psi_G\rangle}{E_n}$$

The first term is zero since the electrons all have paired momenta. In the second term, the only contribution comes from states ψ_n containing two quasi-particles generated by the term $\gamma^*_{\mathbf{k+q},0}\gamma^*_{\mathbf{k}1}$ in (3.98). Since the same coherence factor appears in the matrix elements of $\mathbf{J}_1(\mathbf{q})$ as in the elements of $\mathcal{H}_1$, we find

$$\langle\psi|\mathbf{J}_1(\mathbf{q})|\psi\rangle = \frac{2e^2\hbar^2}{m^2c}\sum_{\mathbf{k}} \frac{(v_{\mathbf{k}}u_{\mathbf{k+q}} - u_{\mathbf{k}}v_{\mathbf{k+q}})^2}{E_{\mathbf{k}} + E_{\mathbf{k+q}}}[\mathbf{k}\cdot\mathbf{a}(\mathbf{q})]\mathbf{k} \tag{3.112}$$

This term obviously gives a current parallel to $\mathbf{A}$, partially canceling the diamagnetic current $\mathbf{J}_2$, even at $T = 0$. The amount of cancellation depends on q, going to zero at $q = 0$, as noted earlier.

Using arguments similar to those used to reach (3.111), (3.112) leads to

$$K(q,0) = \lambda_L^{-2}(0)\left\{1 - \int_{-\infty}^{\infty} \frac{(v_{\mathbf{k}}u_{\mathbf{k+q}} - u_{\mathbf{k}}v_{\mathbf{k+q}})^2}{E_{\mathbf{k}} + E_{\mathbf{k+q}}}\,d\xi\right\} \tag{3.113}$$

for the response function K. Without going into details, one can see that this implies that as q first increases from zero, $K(q,\,0)$ decreases by a fractional amount of the order of $q^2\xi_0^2$, where, following BCS, ξ_0 is defined by

$$\xi_0 \equiv \frac{\hbar v_F}{\pi\Delta(0)} \tag{3.114}$$

[This follows since the coherence factor in the numerator can be written for small $\mathbf{q}$ as $(\xi_{\mathbf{k+q}} - \xi_{\mathbf{k}})\partial(u_{\mathbf{k}} - v_{\mathbf{k}})/\partial\xi_{\mathbf{k}}$; the first factor is of order $\hbar v_F q$, and the second is of order $1/\Delta(0)$ for the important part of the integration region.] On the other hand, if $q\xi_0 \gg 1$, to a first approximation the integral in (3.113) cancels the first term, leaving a residual value of

$$K(q,0) = K(0,0)\frac{3\pi}{4q\xi_0} \qquad q\xi_0 \gg 1 \tag{3.115}$$

as anticipated in (3.106c). The complete dependence of $K(q,0)$ on q is shown in Fig. 3.12. Note that the response of the normal state at $T = 0$ can be obtained as the limit $\Delta(0) \to 0$, which leads to $\xi_0 \to \infty$, so that $K(q,0) = 0$ for all q.

3.10.3 Nonlocal Electrodynamics in Coordinate Space

Given $K(q,T)$, one can find the corresponding real-space kernel by Fourier transformation, the inverse of the transformation leading to (3.106). The result, called $J(R,T)$ by BCS, is sketched in Fig. 3.13. It is very similar to the

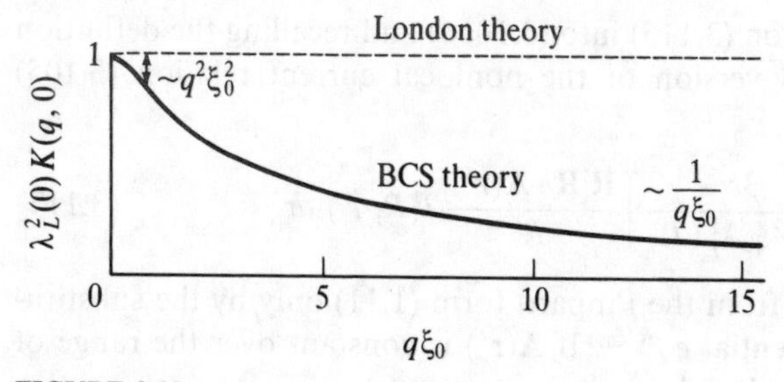

FIGURE 3.12
Comparison of q-dependent response of the nonlocal BCS theory with the q-independent response of the local London theory. In both cases, the curves are drawn for pure metals, with infinite mean free paths.

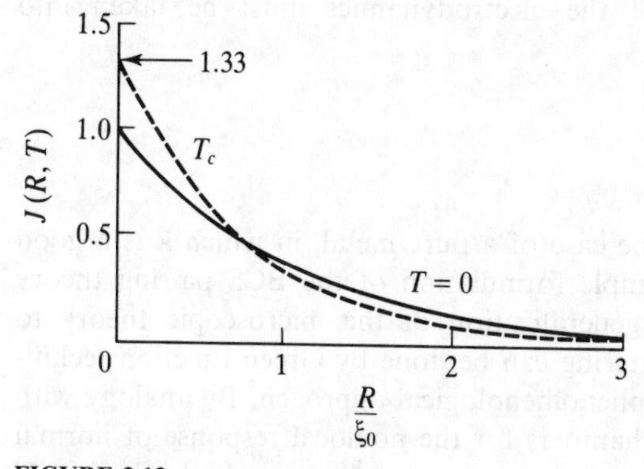

FIGURE 3.13
Schematic comparison of the BCS range function $J(R,T)$ at $T=0$ and T_c. Note that the range of nonlocality is reduced by a factor of about 0.75 on going from $T=0$ to T_c.

exponential form e^{-R/ξ_0} proposed by Pippard.[49] In fact, the $J(R,T)$ of BCS is normalized so that

$$\int_0^\infty J(R,T)\, dR = \xi_0 = \int_0^\infty e^{-r/\xi_0}\, dR \tag{3.116}$$

with ξ_0 as defined in (3.114). At $R=0$, the value of J ranges smoothly from 1 at $T=0$ to 1.33 at $T=T_c$, whereas in the Pippard approximation, the value of the exponential kernel at $R=0$ remains unity at all temperatures. Apart from this minor difference, the similarity is remarkably complete.

---

[49] A. B. Pippard, *Proc. Roy. Soc.* (London) **A216**, 547 (1953).

Inserting the normalization (3.115) into (3.106*a*) and recalling the definition (3.107), we see that the BCS version of the nonlocal current relation (3.105) becomes

$$\mathbf{J}(\mathbf{r}) = -\frac{3c}{16\pi^2\xi_0\lambda_L^2(T)}\int\frac{\mathbf{R}[\mathbf{R}\cdot\mathbf{A}(\mathbf{r}')]}{R^4}J(R,T)\,d\mathbf{r}' \tag{3.117}$$

where $\mathbf{R} = \mathbf{r} - \mathbf{r}'$. This differs from the Pippard form (1.11) only by the substitution of $J(R, T)$ for the exponential e^{-R/ξ_0}. If $\mathbf{A}(\mathbf{r}')$ is constant over the range of $J(R, T)$, (3.117) reduces to the London relation

$$\mathbf{J}(\mathbf{r}) = -\frac{c}{4\pi\lambda_L^2(T)}\mathbf{A}(\mathbf{r}) \tag{3.118}$$

However, if $\mathbf{A}(\mathbf{r})$ is varying significantly on the scale of ξ_0, this reduction cannot be made, and the nonlocality of the electrodynamics must be taken into account.

3.10.4 Effect of Impurities

Our results so far have been for the case of a pure metal, in which $\mathbf{k}$ is a good quantum number, allowing the simple formulation of the BCS pairing theory which we have given. Although generalization of the microscopic theory to include the effects of impurity scattering can be done by Green function techniques, we shall use instead a more phenomenological approach. By analogy with the corresponding expression of Chambers for the nonlocal response of normal electrons to an electric field, we expect a factor of $e^{-R/\ell}$ to multiply the kernel $J(R, T)$, where ℓ is the mean free path. This has the effect of making the electrodynamic response more local, and if ℓ is sufficiently small on the scale of spatial variation of $\mathbf{A}$, we can always simplify to a local response of the London type (3.118), but with a modified (reduced) coefficient characterized by a λ_{eff} ($>\lambda_L$). The appropriate value of λ_{eff} is determined by

$$\frac{\lambda_L^2(T)}{\lambda_{\text{eff}}^2(\ell,T)} = \frac{K(0,T,\ell)}{K(0,T,\infty)} = \frac{\int_0^\infty J(R,T)e^{-R/\ell}\,dR}{\xi_0} \tag{3.119}$$

where we have used (3.106*a*) and (3.116). [Note that we restrict use of the notation $\lambda_L(T)$ to characterize the response of *pure* metal at $q = 0$].

To avoid numerical evaluation of (3.119), we first consider the *extreme dirty limit*, $\ell \ll \xi_0$, in which the integral in the numerator reduces simply to $J(0, T)\ell$. Then

$$\lambda_{\text{eff}}(\ell,T) = \lambda_L(T)\left(\frac{\xi_0}{\ell}\right)^{1/2}[J(0,T)]^{-1/2} \tag{3.120}$$

Since $J(0, T)$ ranges only from 1 at $T = 0$ to 1.33 at T_c, the final factor introduces only a small correction.

Another convenient approximation is to use the Pippard exponential e^{-R/ξ_0} for $J(R,T)$. If this is done, the integral in (3.119) becomes simply ξ, where

$$\xi^{-1} = \xi_0^{-1} + \ell^{-1} \tag{3.121}$$

and then
$$\lambda_{\text{eff}}(\ell, T) = \lambda_L(T)\left(\frac{\xi_0}{\xi}\right)^{1/2} = \lambda_L(T)\left(1 + \frac{\xi_0}{\ell}\right)^{1/2} \tag{3.122}$$

This approximation is useful even when ℓ is not very small since it reduces to λ_L in the pure limit, and as a result, it has been commonly used. However, the approximation is inexact in that it is missing the factor $[J(0,T)]^{-1/2}$ of the microscopic theory in the dirty limit (3.120).

This defect can be remedied by a simple improvement. We retain the exponential approximation for the form of $J(R,T)$, but we replace e^{-R/ξ_0} by $J(0,T)$ $\exp -[J(0,T)R/\xi_0]$. This new form agrees with the microscopic $J(R,T)$ in its initial value at $R=0$ as well as in its integral over R. With this improvement, (3.122) is replaced by

$$\lambda_{\text{eff}}(\ell, T) = \lambda_L(T)\left(\frac{\xi_0'}{\xi'}\right)^{1/2} = \lambda_L(T)\left(1 + \frac{\xi_0'}{\ell}\right)^{1/2} \tag{3.123}$$

where the modified Pippard coherence lengths ξ' and ξ_0' are defined by the relation

$$\frac{1}{\xi'} \equiv \frac{1}{\xi_0'} + \frac{1}{\ell} \equiv \frac{J(0,T)}{\xi_0} + \frac{1}{\ell} \tag{3.123a}$$

Clearly, (3.123) reduces properly to (3.120), but also behaves well when ℓ is not small. Thus, (3.123) gives a very good approximation for the $q \approx 0$ response over the entire range of its arguments. It will also give a good approximation to the actual penetration depth if $\xi \ll \lambda_{\text{eff}}$, where ξ is given by (3.121). Note that (3.123) reduces to (3.122) at $T=0$ and to

$$\lambda_{\text{eff}}(\ell, T) = \lambda_L(T)\left(1 + 0.75\frac{\xi_0}{\ell}\right)^{1/2} \tag{3.123b}$$

for $T \approx T_c$. The latter form is often used in connection with the Ginzburg-Landau theory, which is valid near T_c.

3.10.5 Complex Conductivity

The discussion of the complex conductivity $\sigma(\omega)$ in a two-fluid approximation in Sec. 2.5.1 is restricted to frequencies well below the energy-gap frequency ω_g. In the present section, we take up the complementary regime that occurs when ω is of the same order as ω_g.

In the section on transition probabilities and absorption, we worked out an expression [(3.91) with the upper sign] for the ratio of the real part of the complex conductivity in the superconducting state to that of the normal state, σ_{1s}/σ_n, as a function of frequency. Now that we have worked out the low-frequency lossless response to a vector potential, we can embed the conductivity result in a more general picture. Since we can write

$$\mathbf{E} = -\frac{1}{c}\frac{\partial \mathbf{A}}{\partial t} = -\frac{i\omega \mathbf{A}}{c}$$

for a periodic electromagnetic field, we can define a complex conductivity proportional to $K(q, \omega, T)$. Equating two expressions for the current

$$\mathbf{J}(q, \omega, T) = \sigma(q, \omega, T)\mathbf{E}(q, \omega) = -\frac{c}{4\pi}K(q, \omega, T)\mathbf{A}(q, \omega)$$

we have

$$\sigma(q, \omega, T) = \frac{ic^2}{4\pi\omega}K(q, \omega, T) \tag{3.124}$$

Of course, we have worked out only some special cases of $K(q, \omega, T)$ [or $\sigma(q, \omega, T)$] in any detail, but these serve to outline much of the general behavior.

A simple analytic form can be given for the temperature variation of the low-frequency limit of σ_2/σ_n, namely,

$$\frac{\sigma_2}{\sigma_n} = \frac{\pi\Delta}{\hbar\omega}\tanh\frac{\Delta}{2kT} \qquad \hbar\omega \ll 2\Delta \tag{3.125}$$

which has the limiting forms

$$\frac{\sigma_2}{\sigma_n} \to \begin{cases} \dfrac{\pi\Delta}{\hbar\omega} & T \ll T_c \qquad (3.125a) \\[2ex] \dfrac{\pi}{2}\dfrac{\Delta^2}{kT\hbar\omega} & T \approx T_c \qquad (3.125b) \end{cases}$$

Relating these expressions to the London theory, in which

$$\sigma_{2L} = \frac{n_s e^2}{m\omega} \tag{3.126}$$

we see that these results correspond to $n_s \sim \Delta$ for $T \ll T_c$, but to $n_s \sim \Delta^2$ for $T \approx T_c$. Given Gor'kov's demonstration that near T_c the phenomenological ψ function of the Ginzburg-Landau (GL) theory is proportional to the gap Δ in the BCS theory, the latter result corresponds to the central identification of $n_s \sim |\psi|^2$ in the GL theory, which we discuss in the next chapter.

In terms of our present, more complete picture of a nonlocal or q-dependent response, we must ask: What is the region of validity of the expression for σ_{1s}/σ_n as given by (3.91) or (3.95)? It turns out that these expressions are valid when the response is determined by the value of the nonlocal response kernel for $R = 0$. This will be the case in the dirty limit, $\ell \ll \xi_0$, where the factor $e^{-R/\ell}$ cuts off before $J(R, \omega, T)$ changes very much. It will also be the case when $q\xi_0 \gg 1$, the *extreme anomalous limit* since in this case the appropriate Fourier transform (3.106*b*) gives weight mainly to a region extending only to about π/q from the origin. In both these limits, σ_n is real, so that $\sigma_n = \sigma_{1n}$. In the dirty limit, both σ_s and σ_n are proportional to ℓ; in the extreme anomalous limit, both are proportional to $1/q$. Thus, in both limits, the ratio σ_s/σ_n is dependent only on ω and T, permitting us to drop q and ℓ from the notation so long as we remember the limitation to $q\xi_0 \gg 1$ or $\ell \ll \xi_0$.

An explicit form for σ_2 at $T = 0$ analogous to (3.95) for σ_1 can be worked out, namely,

$$\frac{\sigma_{2s}}{\sigma_n} = \frac{1}{2}\left(1 + \frac{2\Delta}{\hbar\omega}\right) E(k') - \frac{1}{2}\left(1 - \frac{2\Delta}{\hbar\omega}\right) K(k') \tag{3.127}$$

where $k' = (1 - k^2)^{1/2}$ and $k = |(2\Delta - \hbar\omega)/(2\Delta + \hbar\omega)|$. The frequency dependence of this function is sketched in Fig. 3.14. Of particular importance is the proportionality of σ_2 to $1/\omega$ for $\hbar\omega \ll 2\Delta$. This corresponds to the fact that $K(q, \omega)$ is essentially independent of frequency there, as required for a frequency-independent penetration depth. More simply, this dependence is a consequence of the free-acceleration aspect of the supercurrent response as described by the London equation $\mathbf{E} = \partial(\Lambda \mathbf{J}_s)/\partial t$. For $\hbar\omega \gtrsim 2\Delta$, σ_2 falls to zero more rapidly than $1/\omega$, but at $\hbar\omega = 2\Delta$, $K(q, \omega)$ has decreased by only a factor of $2/\pi$ from its dc value. Thus, the superfluid response is essentially independent of frequency until microwave frequencies are reached.

The real and imaginary parts of σ both enter into the determination of the response of a superconducting system to time-dependent electromagnetic fields. For example, in the transmission experiments with very thin films, cited earlier in Sec. 3.9, the fractional transmissivity is readily shown to be

$$T = \left[\left(1 + \frac{\sigma_1\, dZ_0}{n+1}\right)^2 + \left(\frac{\sigma_2\, dZ_0}{n+1}\right)^2\right]^{-1} \tag{3.128}$$

where n is the index of refraction of the substrate, Z_0 (377 Ω per square in mks units) is the impedance of free space, and d is the film thickness. From (3.128) we see that $T_s \to 0$ as $\omega \to 0$ since $\sigma_{2s} \sim 1/\omega \to \infty$, leading to complete reflection. On the other hand, for $\hbar\omega \gg 2\Delta$, $\sigma_{2s} \to 0$ and $\sigma_{1s} \to \sigma_n$, so that $T_s \to T_n$. In between,

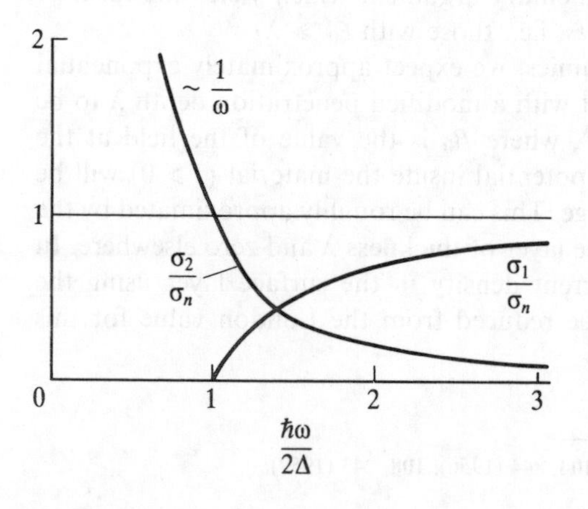

FIGURE 3.14
Complex conductivity of superconductors in extreme anomalous limit (or extreme dirty limit) at $T = 0$. The rise of σ_2 as $1/\omega$ below the gap describes the accelerative supercurrent response. Its coefficient is proportional to the "missing area" under the $\sigma_1(\omega)$ curve at finite frequencies (see the discussion in Sec. 3.9.3).

for $\hbar\omega \approx 2\Delta$, both σ_{1s} and σ_{2s} are smaller than σ_n, as shown in Fig. 3.14, and there is a peak in transmission at which $T_s > T_n$. Observation of these characteristic features by Glover and Tinkham[50] gave strong support to an energy-gap model of superconductivity and quantitative support to the BCS model in particular.

3.11 THE PENETRATION DEPTH

Having set up the general formalism in the preceding section for treating the electrodynamics of superconductors within the framework of the BCS theory, in this section we apply these results to calculating the predicted magnitude and temperature dependence of the experimentally observed penetration depth λ for a number of important cases. We define λ by the relation

$$\lambda \equiv h(0)^{-1} \int_0^\infty h(z)\, dz \tag{3.129}$$

for penetration into a bulk sample, whether or not the field penetration is exponential. This λ also serves, via (2.2), to determine the parameter n_s of the phenomenological London and GL theories.

To apply the formulas of Sec. 3.10, we need to choose the proper gauge for the vector potential. For the present simple case of a plane surface in a parallel magnetic field, the gauge choice will be uniquely determined by requiring that **A** be parallel to the surface but perpendicular to **h** and fall off to zero in the interior of the bulk superconductor. On the other hand, in dealing with a thin film in a parallel field of equal strength on both sides, symmetry requires that $|\mathbf{A}|$ pass through zero in the midplane. In other situations, other gauge choices may be required.

3.11.1 Preliminary Estimate of λ for Nonlocal Case

Before going into the rigorous solution for the penetration depth with nonlocal electrodynamics, we first give an elementary argument which yields the form of the results for *Pippard superconductors*, i.e., those with $\xi_0 \gg \lambda_L$.

Even with nonlocal electrodynamics, we expect approximately exponential penetration of the magnetic field, but with a modified penetration depth λ to be determined. If we write $h_y \approx B_0 e^{-z/\lambda}$, where B_0 is the value of the field at the surface, then the form of the vector potential inside the material $(z > 0)$ will be $A_x \approx \lambda B_0 e^{-z/\lambda}$ in the appropriate gauge. This can be roughly approximated by the constant value $\bar{A} = \lambda B_0$ over a surface layer of thickness λ and zero elsewhere. In calculating the resulting average current density in the surface layer using the nonlocal form (3.117), we get a value reduced from the London value for this

[50] R. E. Glover and M. Tinkham, *Phys. Rev.* **104**, 844 (1956); **108**, 243 (1957).

average $\bar{A}$ by a factor $\sim\lambda/\xi_0$ since this is the ratio of the volume ($\sim\xi_0^2\lambda$) in which $\bar{A}$ exists to the effective integration volume ($\sim\xi_0^3$). Thus, we expect

$$\bar{J} \approx -\frac{c}{4\pi\lambda_L^2}\frac{\lambda}{\xi_0}\bar{A} = \frac{c\lambda^2 B_0}{4\pi\lambda_L^2\xi_0}$$

Applying a Maxwell equation to the surface layer,

$$\frac{B_0}{\lambda} \approx |< \text{curl}\,\mathbf{h} >| = \frac{4\pi\bar{J}}{c} \approx \frac{\lambda^2 B_0}{\lambda_L^2\xi_0}$$

Solving, we find

$$\lambda \approx (\lambda_L^2\xi_0)^{1/3} \tag{3.130}$$

with neglect of numerical factors. Thus, when nonlocality is important, i.e., when $\xi_0 \gg \lambda_L$, the actual penetration depth λ will exceed λ_L by a modest factor of order $(\xi_0/\lambda_L)^{1/3}$. If $\xi_0 < \lambda_L$, of course, the preceding argument does not apply because the response is local, and $\lambda \approx \lambda_L$.

Note that (3.130) has a qualitative implication about the temperature dependence of λ in Pippard superconductors. Very near T_c, all superconductors become local since $\lambda_L(T) > \xi_0$, so that $\lambda(T) \approx \lambda_L(T) \sim (T_c - T)^{-1/2}$. At lower temperatures, when $\lambda_L(T) < \xi_0$, it follows from (3.130) that $\lambda(T) \approx (\lambda_L^2\xi_0)^{1/3} \sim (T_c - T)^{-1/3}$. This implies a crossover of the temperature dependence between the two expressions near the temperature at which $\lambda_L(T) \approx \xi_0$. Since this occurs at different values of T/T_c for different superconductors because of their different values of $\lambda_L(0)/\xi_0$, there cannot be a universal temperature dependence for $\lambda(T)$ in all superconductors. In particular, the famous Gorter-Casimir two-fluid dependence $\lambda(T) = \lambda(0)/[1-(T/T_c)^4]^{1/2}$ cannot be expected to apply to all materials equally well.

3.11.2 Solution by Fourier Analysis

A convenient technique for obtaining an exact solution to the field penetration problem is to apply Fourier analysis to **J** and **A** and to use (3.101) to obtain a self-consistent solution. The details of this procedure are given in the appendix. Some care is needed in handling the surface since our expressions from Sec. 3.10 for the response function $\mathbf{K}(q)$ are valid only in an infinite medium. This problem can be handled by the mathematical artifice of introducing externally supplied source currents in the interior of the infinite medium to simulate the field applied at a surface. Different procedures are required for the two limiting cases of completely diffuse and completely specular reflection of electrons at the surface. The results are found to agree for local superconductors and to differ at most by a factor of $\frac{8}{9}$ in the extreme nonlocal limit. This near agreement is fortunate since it is not clear which is the better approximation for the surfaces of real samples. Detailed numerical calculations of the penetration depth in pure and impure superconductors, with various ratios of $\lambda_L(0)/\xi_0$, have been made by Miller[51] using the exact BCS results for $K(q)$.

In order to avoid these numerical calculations, considerable attention has been given to two limiting cases in which analytic results can be obtained, even though the true situation usually lies in between.

The *local approximation* replaces $K(q)$ for all q by $K(0)$, a constant, thus reducing the problem to the London form, but with a modified penetration depth. Using the generalized Pippard approximation discussed in Sec. 3.10, one finds the results already quoted in (3.123), namely,

$$\lambda_{\text{eff}}(\ell, T) = \lambda_L(T)(1 + \xi_0'/\ell)^{1/2} \tag{3.131}$$

which reduces to $\lambda_{\text{eff}}(\ell, T) = \lambda_L(T)(1 + \xi_0/\ell)^{1/2}$ at $T = 0$ and to $\lambda_{\text{eff}}(\ell, T) = \lambda_L(T)(1 + 0.75\xi_0/\ell)^{1/2}$ near T_c. This approximation is reasonably well justified in dirty superconductors [if $\ell < \lambda(T)$], and even in pure superconductors very near T_c, where $\lambda(T) > \xi_0$. It should also apply to the high-temperature superconductors, provided the BCS model is relevant, because, in them, $\lambda(T) > \xi_0$ at all temperatures and purity levels.

The other approximation is the *extreme anomalous limit*, in which $K(q)$ is replaced at all q values by its asymptotic form for $q \to \infty$, where $K(q) \sim 1/q$. This should be a reasonable approximation for those pure superconductors with $\lambda_L(0) \ll \xi_0$. For the case of diffuse surface scattering, this calculation yields

$$\lambda_{\infty,\,\text{diff}} = 0.65(\lambda_L^2\xi_0')^{1/3} \tag{3.132}$$

where $\xi_0' = \xi_0/J(0, T)$ is the modified coherence length based on the BCS electrodynamics. Since $[J(0, T)]^{1/3}$ varies only from 1 to 1.10 as T ranges from zero to T_c, it is clear that our simple estimate (3.130) captured the essence of this more quantitative result (3.132). For specular surface scattering, the extreme anomalous limit yields the same value as (3.132) times a numerical factor of $\frac{8}{9}$.

Some perspective on the applicability of these results can be gained by consideration of numerical parameters for several pure metals well below T_c. Aluminum is well approximated by λ_∞ since $\lambda_L \approx 160$ Å, while $\xi_0 \approx 15{,}000$ Å, so that $\xi_0/\lambda_L \approx 100$. For tin, $\lambda_L \approx 350$ Å, while $\xi_0 \approx 3{,}000$ Å, so that λ_∞ is only a moderately good approximation. For lead, $\lambda_L \approx \xi_0/2$, and the London local approximation is actually better than λ_∞. In the high T_c materials, one typically has $\lambda_L \approx 1500$ Å and $\xi_0 \approx 15$ Å, so that the electrodynamics is completely local, and λ_∞ is irrelevant. The same is true of typical alloy superconductors, in which the short mean free path assures that $\xi \approx \ell \ll \lambda_L$.

From this survey we see that as a practical matter, the nonlocal electrodynamics plays a major role only in materials which have high Fermi velocity, low T_c, and a long mean free path, such as clean aluminum. Nonetheless, we have

---

[51] P. B. Miller, *Phys. Rev.* **113**, 1209 (1959).

devoted considerable attention to the subject because of its historical importance in working out the role of the coherence length ξ_0 in explaining why the classic pure superconductors all have measured penetration depths of ~ 500 Å, despite having values of λ_L which vary by a factor of 3 from aluminum to lead; this provides a quantitative confirmation of the BCS/Pippard theory.

3.11.3 Temperature Dependence of λ

As noted earlier, $\lambda(T)/\lambda(0)$ cannot have a universal temperature dependence on (T/T_c) in the BCS theory because of the variation of the ratio $\xi_0/\lambda_L(0)$ for different metals. Numerical calculations are required for each case. Yet, the calculated differences are not so great as to be inconsistent with the *approximately* universal "two-fluid" dependence observed experimentally in the classic superconductors, especially since the measurements are often not carried sufficiently near $T = 0$ because of sensitivity limitations. (For high T_c superconductors, the temperature dependence appears to be distinctly different.) To understand this better, let us compare the various predictions with the two-fluid dependence.

The simplest prediction is for a *pure* sample in the *local* limit, where $\lambda(T)/\lambda(0) = \lambda_L(T)/\lambda_L(0)$, whose dependence on T/T_c is given by (3.111). For a pure metal in the extreme *anomalous* limit, (3.132) implies that

$$\frac{\lambda_\infty(T)}{\lambda_\infty(0)} = \left[\frac{\lambda_L^2(T)}{\lambda_L^2(0)J(0,T)}\right]^{1/3} = \left[\frac{\Delta(T)}{\Delta(0)}\tanh\beta\Delta(T)/2\right]^{-1/3} \tag{3.133}$$

where the second form involves the relation (3.125) and provides a mathematical definition of $J(0, T)$ in terms of $\lambda_L(T)$ and $\Delta(T)$, quantities already defined in (3.111) and (3.53) and included in the numerical tabulation of Mühlschlegel.[52] Finally, in the *dirty local* limit, (3.131) can be written as

$$\frac{\lambda_{\text{eff}}(T)}{\lambda_{\text{eff}}(0)} = \frac{\lambda_L(T)}{\lambda_L(0)J^{1/2}(0,T)} = \left[\frac{\lambda_\infty(T)}{\lambda_\infty(0)}\right]^{3/2} \tag{3.134}$$

In Fig. 3.15, these various theoretical dependences are compared with the two-fluid approximation

$$\frac{\lambda(T)}{\lambda(0)} = \frac{1}{[1-(T/T_c)^4]^{1/2}} \tag{3.135}$$

Evidently, the temperature dependence of λ_∞ comes the closest of the theoretical forms to agreement with (3.135). This is satisfying since we expect λ_∞ to be the most appropriate approximation for the typical pure superconductors such as tin, on which the most careful early measurements were made. But, as mentioned previously, the approximation λ_∞ must break down near T_c, when $\lambda(T)$ exceeds

[52] B. Mühlschlegel, *Z. Phys.* **155**, 313 (1959).

ξ_0. In particular, the infinite slope of λ_∞^{-2} near T_c would be replaced by the finite slope of λ_L^{-2} in the exact calculation. Note that the difference in zero-temperature normalizations of λ_∞ and λ_L in Fig. 3.15 implies that the slope of λ_L^{-2} near T_c should be increased by a factor $\sim 0.4[\xi_0/\lambda_L(0)]^{2/3}$ before being compared with the slope of the curve for λ_∞^{-2}. Inserting appropriate values in this correction factor brings the slope close to that of the empirical form. In this way, we can see that the full microscopic theory gives a temperature dependence that is closer to the empirical two-fluid law than simply λ_∞ itself. It should also be noticed that Fig. 3.15 implies that a clean, local, weak-coupling BCS superconductor should show a temperature dependence of λ^{-2} which is generally closer to $(1 - t^2)$ than to $(1 - t^4)$, where $t = T/T_c$. Accordingly, the two-fluid temperature dependence should be used only with caution and not assumed to be universal.

3.11.4 Penetration Depth in Thin Films: λ_{eff} and $\lambda_\perp$

The theory of the penetration depth in thin films of nonlocal superconductors has received much attention for several reasons. For one, many of the earlier measurements of $\lambda(T)$ were made intentionally on thin films of thickness $d \sim \lambda$, so that there would be measurable changes in the magnetic moment with changes of T. Accordingly, one needed to ask whether such measured values of λ would be representative of bulk samples. Another reason for interest is that many experiments on superconductors in magnetic fields are performed on thin films, and we

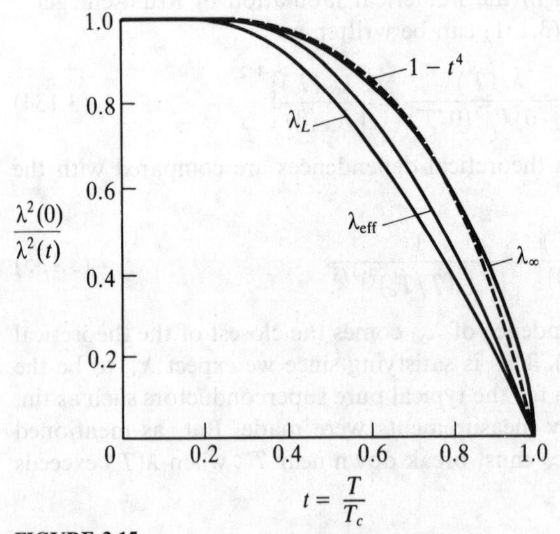

FIGURE 3.15
Comparison of the predicted temperature dependence for $1/\lambda^2$ in various limiting cases of the BCS theory. The dashed curve depicts the empirical approximation (3.135).

wish to understand the interaction between field and film as accurately as possible. Finally, analysis of the behavior of thin films illustrates a further consequence of the nonlocal electrodynamics which does not arise in the basic analysis of the field penetration at the surface of a bulk superconductor.

The end result of this analysis[53] is simple. For a sufficiently thin film in *parallel* magnetic field, where we can take the vector potential to have the form $A_y = H_z x$ characteristic of the unscreened field, the supercurrent response of a *nonlocal* superconductor is essentially equivalent to that of a *local* superconductor with

$$\lambda_{\text{eff}} \approx \lambda_L (\xi_0'/d)^{1/2} \qquad (\text{for } d \ll \xi_0') \tag{3.136}$$

with a numerical prefactor estimated to be $\frac{4}{3}$ for diffuse surface scattering and $(\frac{4}{3})^{1/2}$ for specular surface scattering. [Recall that $\xi_0' = \xi_0/J(0, T)$ was defined in (3.123*a*).] Note that this result is very similar to (3.131) for the penetration depth of a dirty superconductor with mean free path $\ell \approx d$, the film thickness, as would occur if ℓ were limited by surface scattering.

When a magnetic field is applied *perpendicular* to a thin film ($d \ll \lambda$), quite different considerations enter. In this configuration, the currents serve to control where the flux goes through the film. Accordingly, the vector potential and current density are essentially uniform through the thickness of the film, rather than reversing in direction about the midplane of the film as do the screening currents in the parallel-field case discussed in the previous paragraph. Because **A** is uniform through the thickness of the film, the effect of nonlocality on the response depends on the nature of the surface scattering. If it is diffuse, d is the effective thickness over which the uniform **A** is integrated in the nonlocal current expression, and λ_{eff} is again given by (3.136) apart from a numerical factor of order unity. On the other hand, if the surface scattering were perfectly specular, the boundary condition implies that a uniform **A** should be integrated over all space, and λ_{eff} would be the same as in bulk material of the same purity and mean free path. In practice, (3.136) is usually a reasonable approximation because of a combination of at least partial diffuse surface scattering and an internal mean free path ℓ which is comparable to d.

A much more important consequence of the perpendicular field orientation is that the screening distance λ_{scr} is not given by this λ_{eff} but, rather, by a thickness-dependent $\lambda_\perp \approx \lambda^2/d \gg \lambda$, as shown in a classic paper by Pearl.[54] This $\lambda_\perp$ reflects the effect of the two-dimensional geometry on solutions of Maxwell's equations; it does *not* reflect a change in the intrinsic n_s in the local constitutive relation between **J** and **A**, which is still described by the λ_{eff} of (3.136). To high-

[53] See, e.g., M. Tinkham, *Introduction to Superconductivity*, 1st ed., McGraw-Hill, New York (1975), pp. 81–86.

[54] J. Pearl, *Appl. Phys. Lett.* **5**, 65 (1964).

light the centrality of the two- vs. three-dimensional nature of the screening, we now give a heuristic argument which treats both cases in parallel.

We consider the response to an externally imposed filament of current I flowing along a straight line within the superconducting medium and work out the screening length λ_{scr}. Within this distance of the filament, the medium supplies an antiparallel screening supercurrent $-I$, so that at greater distances from the source filament than λ_{scr}, the magnetic field goes quickly to zero. Within the screening region of depth λ_{scr}, the magnetic field of the current will be of order $2I/c\lambda_{scr}$, which implies a vector potential **A** (parallel to the filament) of order $2I/c$ in the London gauge. Applying the London equations in terms of the vector potential, $\mathbf{J}_s = -(c/4\pi\lambda_{\text{eff}}^2)\mathbf{A}$, the screening current density near the source is $J_{scr} \approx -I/2\pi\lambda_{\text{eff}}^2$. In a three-dimensional bulk superconductor, the screening current flows in a circular cross section of radius λ_{eff}; the total screening current will be of order $\pi\lambda_{scr}^2 J_{scr}$. Equating this to the source current I, we find $\pi\lambda_{scr}^2 \approx 2\pi\lambda_{\text{eff}}^2$, or $\lambda_{scr} \approx \lambda_{\text{eff}}$, as expected. We now contrast this with the case of a current filament in (or immediately above) a thin superconducting film of thickness d. The screening current density $\mathbf{J}_{scr}$ found above is also correct in this case, but it will flow in a cross-sectional area of only $\sim 2\lambda_{scr} d$. Equating the total screening current to the source current, we find $2\lambda_{scr} d \approx 2\pi\lambda_{\text{eff}}^2$, or

$$\lambda_\perp = \lambda_{scr} \approx \lambda_{\text{eff}}^2/d \qquad (d \ll \lambda_{\text{eff}}) \tag{3.137}$$

within small numerical factors.

The most important application of this general result is to the current vortex associated with a quantum of flux passing through a thin film at normal incidence. As Pearl showed in his original paper, the scale at which the radial dependence of the circulating sheet current density changes from $1/r$ to $1/r^2$ is set by this $\lambda_\perp$. For comparison, the radial dependence of the current density in a similar vortex in *bulk* material with penetration depth λ changes from $1/r$ to an exponential cutoff at $r \sim \lambda$. Thus, one can say that the radius of a vortex expands from $\sim\lambda$ in bulk material to $\sim\lambda_\perp$ in a film of thickness $d < \lambda$. Moreover, in a thin film, Pearl's solution shows that the long-range falloff is only power law $\sim 1/r^2$, rather than exponential $\sim e^{-r/\lambda}$ as in bulk material.

3.11.5 Measurement of λ

Before leaving the subject of the penetration depth, we briefly review some experimental techniques which have been used to measure this quantity. The earliest experiments[55] involved use of a large number of colloidal particles or thin films with a small dimension d comparable to λ. Varying the temperature, and hence $\lambda(T)$, then caused substantial fractional changes in magnetic susceptibility, which

[55]For example, D. Shoenberg, *Proc. Roy. Soc.* (London) **A175**, 49 (1940); J. M. Lock, ibid., **A208**, 391 (1951).

could be measured. To the extent that the particle-size distribution of the sample was known, this allowed $\lambda(T)$ to be estimated, using (2.5), but evidently there were major quantitative uncertainties.

It was pointed out by Casimir[56] that an ac susceptibility technique should be sufficiently sensitive to allow the temperature-dependent change in field penetration $[\lambda(T) - \lambda(0)]$ at the surface of a single bulk sample to be measured. This experiment was first carried out successfully by Laurmann and Shoenberg[57] by using a mutual inductance bridge operating at 70 Hz, but their sensitivity did not allow very quantitative results to be obtained. A 10-fold increase in sensitivity, allowing changes of λ at the 2 Å level to be detected, was obtained by Pippard[58] by raising the frequency to $\sim 10^{10}$ Hz and by using microwave techniques to measure the temperature-dependent shift in the resonant frequency of a cavity caused by the changes in $\lambda(T)$. This technique greatly improved his sensitivity but did require significant corrections to be made for normal electron effects which become increasingly serious at the higher frequencies. By using the normal-state skin depth as a reference, he was able to obtain a measurement of the absolute value of $\lambda(T)$, not just its change with temperature, but this was of lesser accuracy.

After the BCS theory appeared, Schawlow and Devlin[59] remeasured $\lambda(T)$ in tin, working at about 10^5 Hz, where the high-frequency corrections needed by Pippard were negligible but without loss of sensitivity because digital frequency counters had become available to improve the accuracy with which shifts in the cavity frequency could be measured. While their results generally followed the two-fluid relation that $\lambda(T) \propto y = (1 - t^4)^{-1/2}$, a plot of $d\lambda/dy$ vs. y showed a rather sharp rise of this quantity below $y \approx 1.5$. Exactly this sort of behavior was predicted by BCS because of the exponential cutoff of excitations due to the gap, which controls the detailed temperature dependence of $\lambda_L(T)$ as given by (3.111). Although there was no complete quantitative agreement between theory and experiment, this probably can be attributed to the fact that the theory is for an idealized isotropic metal, whereas tin has a complex Fermi surface, with known anisotropy in the gap and in the penetration depth. These classic high-frequency measurements of $\lambda(T)$ have been reviewed by Waldram.[60]

It is important to remember that the most sensitive methods measure only *changes* in $\lambda(T)$ with temperature. As a result, quoted "experimental" values of $\lambda(0)$ are usually values inferred by fitting data to a theory of $\lambda(T)$. This is usually the old two-fluid dependence since with BCS *two* parameters, $\lambda(0)$ and ξ_0, must be fitted.

---

[56] H. B. G. Casimir, *Physica* **7**, 887 (1940).

[57] E. Laurmann and D. Shoenberg, *Nature* **160**, 747 (1947); *Proc. Roy. Soc.* (London) **A198**, 560 (1949).

[58] A. B. Pippard, *Proc. Roy. Soc.* (London) **A191**, 399 (1947); **203**, 98 (1950).

[59] A. L. Schawlow and G. E. Devlin, *Phys. Rev.* **113**, 120 (1959).

[60] J. R. Waldram, *Adv. Phys.* **13**, 1 (1964).

The advent of the high-temperature superconductors brought a revival of interest in the measurement of $\lambda(T)$ since its absolute value gives information about n_s, and its temperature dependence is predicted to be different for different proposed underlying microscopic theories. For example, for a local clean-limit BCS superconductor with a finite gap all over the Fermi surface, $[\lambda(T)/\lambda(0) - 1]$ should go to zero exponentially as $T^{-1/2}e^{-\Delta(0)/kT}$ as $T \rightarrow 0$,[61] as can be derived from (3.111). On the other hand, a superconductor with nodes in the gap has a finite density of low-lying states, and $[\lambda(T)/\lambda(0) - 1]$ should go to zero as a power law $(T/T_c)^n$, where n depends on whether the nodes are points or lines, etc.

Unfortunately, uncertainties about the surface quality of samples of these materials adds a major additional obstacle to those noted in connection with high-frequency measurements on the classic superconductors. Nonetheless, very sensitive measurements by Anlage et al.[62] on a resonant stripline seem to indicate that $\lambda^{-2}(T)$ in YBCO follows a dependence closer to $(1 - t^2)$ than the usual two-fluid form $(1 - t^4)$. In other words, the penetration depth is varying down to lower temperatures than in the classic superconductors. In general, this implies lower-lying quasi-particle states than in BCS. This could either reflect substantial gap anisotropy, with some regions of small gap, or it could reflect unconventional pairing, which entails nodes where the gap goes to zero.

A new experimental approach, capable of absolute measurement of $\lambda(T)$ and also insensitive to surface conditions, is the use of muon spin rotation measurements to determine the distribution of local magnetic fields in a superconductor in the mixed state.[63] The range of field-variation scales with $1/\lambda^2(T)$, which controls the strength of the local vortex currents that produce the local field variations, allowing $\lambda(T)$ to be inferred. Although the precision of this technique is not as great as the high-frequency methods, and there are complications in the interpretation stemming from the thermal motion of the flux tubes about their ideal lattice locations, this technique has been very useful in studying the high-temperature superconductors. These results will be discussed in Sec. 9.9.3, together with other types of measurements on these interesting materials.

3.12 CONCLUDING SUMMARY

In this chapter, we have set out the basic features of the BCS theory: the Cooper pairing due to a weak, phonon-mediated attraction between electrons, the nature of the superconducting ground state and its condensation energy relative to the

---

[61] J. Halbritter, *Z. Physik* **243**, 201 (1971).

[62] S. M. Anlage, B. W. Langley, G. Deutscher, J. Halbritter, and M. R. Beasley, *Phys. Rev.* **B44**, 9764 (1991).

[63] See, e.g., Y. J. Uemura et al., *Phys. Rev.* **B38**, 909 (1988); D. R. Harshman et al., *Phys. Rev.* **B39**, 851 (1989); Y. J. Uemura et al., *Phys. Rev. Lett.* **62**, 2317 (1989) and **65**, 2665 (1991); C. Niedermayer et al., *Phys. Rev. Lett.* **71**, 1764 (1993); J. Sonier et al., *Phys. Rev. Lett.* **72**, 744 (1994).

normal state, the excited quasi-particle states above the energy gap, and the temperature dependence of the gap and the thermodynamic quantities. We then showed how electron tunneling has been able to confirm the density of excited states above the gap in detail, and even to confirm the quantitative correctness of the electron-phonon mechanism for setting up the superconductive state in a number of the classic superconductors. Our analysis of the transition probabilities for quasi-particle excitation and scattering gave further testimony to the correctness of the theory since very different temperature dependences were predicted and observed for ultrasonic attenuation and nuclear relaxation processes, which would have the *same* temperature dependence in a *simple* two-fluid model without the coherence factors derived from the pairing model. Finally, we have treated the electrodynamics of superconductors, first computing the absorption due to quasi-particle processes, and then computing the lossless supercurrent response in the presence of a vector potential. The latter computation confirmed the correctness of the Pippard nonlocal generalization of the London electrodynamics, and was then applied to work out predictions for the dependence of $\lambda(T)$ on λ_L, ξ_0, and ℓ.

Having seen that the BCS theory provides a microscopic foundation for the relatively simple semimicroscopic or phenomenological London and Ginzburg-Landau theories of superconductivity, we now use the latter to explore numerous superconducting phenomena. This program will occupy most of the remaining chapters of this book.

CHAPTER 4

GINZBURG-LANDAU THEORY

The BCS microscopic theory, as described in Chap. 3, gives an excellent account of the data in those cases to which it is applicable, namely, those in which the energy gap Δ is constant in space. However, there are many situations in which the entire interest derives from the existence of spatial inhomogeneity. For example, in treating the intermediate state of type I superconductors in Sec. 2.3, we had to consider the interface where the superconducting state joined onto the normal state. This sort of spatial inhomogeneity becomes all pervasive in the mixed state of type II superconductors. In such situations, the fully microscopic theory becomes very difficult, and much reliance is placed on the more macroscopic Ginzburg-Landau[1] (GL) theory.

As originally proposed, this theory was a triumph of physical intutition, in which a pseudowavefunction $\psi(\mathbf{r})$ was introduced as a complex order parameter. $|\psi(\mathbf{r})|^2$ was to represent the local density of superconducting electrons, $n_s(\mathbf{r})$. The theory was developed by applying a variational method to an assumed expansion of the free-energy density in powers of $|\psi|^2$ and $|\nabla\psi|^2$, leading to a pair of coupled differential equations for $\psi(\mathbf{r})$ and the vector potential $\mathbf{A}(\mathbf{r})$, which we have quoted as (1.16) and (1.17). The result was a generalization of the London theory to deal with situations in which n_s varied in space, and also to deal with the nonlinear response to fields that are strong enough to change n_s. The local approximation of the London electrodynamics was retained, however.

[1] V. L. Ginzburg and L. D. Landau, *Zh. Eksperim. i. Teor. Fiz.* **20**, 1064 (1950).

Although quite successful in explaining intermediate-state phenomena, where the need for a theory capable of dealing with spatially inhomogeneous superconductivity was evident, this theory was initially given limited attention in the western literature because of its phenomenological foundation.

This situation changed in 1959 when Gor'kov[2] showed that the GL theory was, in fact, derivable as a rigorous limiting case of the microscopic theory, suitably reformulated in terms of Green functions to allow treating a spatially inhomogeneous regime. The conditions for validity of the GL theory were shown to be a restriction to temperatures sufficiently near T_c and to spatial variations of ψ and $\mathbf{A}$ which were not too rapid. In this reevaluation of the GL theory, $\psi(\mathbf{r})$ turned out to be proportional to the gap parameter $\Delta(\mathbf{r})$, both being in general complex quantities. At first it was thought that $|\Delta(\mathbf{r})|$, found from solving the newly interpreted GL equations, was simply a BCS gap which might vary in space or with applied magnetic fields, or both. This led to a period in which experiments were (incorrectly) interpreted in this overly simple way. It is now clear, however, that a solution to the GL equations for a given problem is only a useful first step toward understanding the spectral density of excitations. The key point is that fields, currents, and gradients act as "pairbreakers" which tend to blur out the sharp edge of the BCS gap as well as to reduce the value of Δ. Detailed discussion of these effects will be deferred until Chap. 10.

The greatest value of the theory remains in treating the macroscopic behavior or superconductors, in which the overall free energy is important instead of the detailed spectrum of excitations. Thus, it will be quite reliable in predicting critical fields and the spatial structure of $\psi(\mathbf{r})$ in nonuniform situations. It also provides the qualitative framework for understanding the dramatic supercurrent behavior as a consequence of quantum properties on a macroscopic scale.

Although one could in principle now give a *derivation* of the GL theory following Gor'kov, this would require techniques beyond the level of our presentation. Instead, we shall follow Ginzburg and Landau in phenomenologically postulating the form of the theory on grounds of plausibility, and then simply appealing to the results of microscopic theory (or experiment) to evaluate the few parameters of the theory by considering simple special cases.

4.1 THE GINZBURG-LANDAU FREE ENERGY

The basic postulate of GL is that if ψ is small and varies slowly in space, the free-energy density f can be expanded in a series of the form

$$f = f_{n0} + \alpha|\psi|^2 + \frac{\beta}{2}|\psi|^4 + \frac{1}{2m^*}\left|\left(\frac{\hbar}{i}\nabla - \frac{e^*}{c}\mathbf{A}\right)\psi\right|^2 + \frac{h^2}{8\pi} \tag{4.1}$$

[2] L. P. Gor'kov, *Zh. Eksperim. i. Teor. Fiz.* **36**, 1918 (1959) [*Soviet Phys.—JETP* **9**, 1364 (1959)].

Evidently, if $\psi = 0$, this reduces properly to the free energy of the normal state $f_{n0} + h^2/8\pi$, where $f_{n0}(T) = f_{n0}(0) - \frac{1}{2}\gamma T^2$. We now consider the remaining three terms describing the superconducting effects.

In the absence of fields and gradients, we have

$$f_s - f_n = \alpha|\psi|^2 + \tfrac{1}{2}\beta|\psi|^4 \tag{4.2}$$

which can be viewed as a series expansion in powers of $|\psi|^2$ or n_s, in which only the first two terms are retained.[3] These two terms should be adequate so long as one stays near the second-order phase transition at T_c, where the order parameter $|\psi|^2 \to 0$. Inspection of (4.2) shows that β must be positive if the theory is to be useful; otherwise the lowest free energy would occur for arbitrarily large values of $|\psi|^2$, where the expansion is surely inadequate.

As is illustrated in Fig. 4.1, two cases arise, depending on whether α is positive or negative. If α is positive, the minimum free energy occurs at $|\psi|^2 = 0$, corresponding to the normal state. On the other hand, if $\alpha < 0$, the minimum occurs when

$$|\psi|^2 = |\psi_\infty|^2 \equiv -\frac{\alpha}{\beta} \tag{4.3}$$

where the notation ψ_∞ is conventionally used because ψ approaches this value infinitely deep in the interior of the superconductor, where it is screened from any

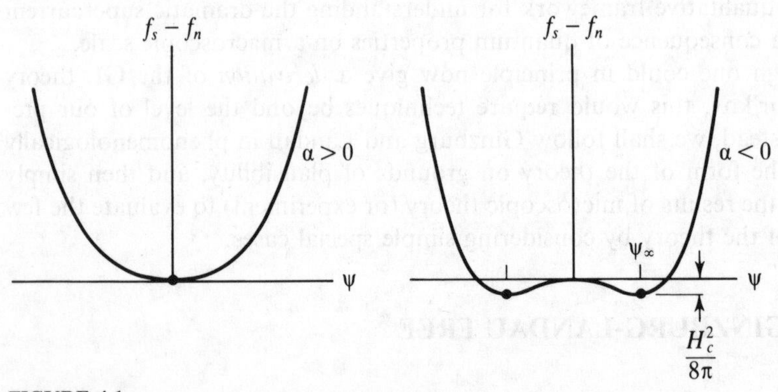

FIGURE 4.1
Ginzburg-Landau free-energy functions for $T > T_c$ ($\alpha > 0$) and for $T < T_c$ ($\alpha < 0$). Heavy dots indicate equilibrium positions. For simplicity, ψ has been taken to be real.

[3] Some considerations which restrict the choice to this form of expansion are the following: An expansion in powers of ψ itself is excluded since f must be real. This difficulty cannot be avoided by taking the real part of ψ since f should not depend on the absolute phase of ψ. Odd powers of $|\psi|$ are excluded because they are not analytic at $\psi = 0$.

surface fields or currents. When this value of ψ is substituted back into (4.2), one finds

$$f_s - f_n = \frac{-H_c^2}{8\pi} = \frac{-\alpha^2}{2\beta} \tag{4.4}$$

using the definition of the thermodynamic critical field H_c.

Evidently, $\alpha(T)$ must change from positive to negative at T_c, since by definition T_c is the highest temperature at which $|\psi|^2 \neq 0$ gives a lower free energy than $|\psi|^2 = 0$. Making a Taylor's series expansion of $\alpha(T)$ about T_c, and keeping only the leading term, we have

$$\alpha(t) = \alpha'(t-1) \qquad \alpha' > 0 \tag{4.5}$$

where $t = T/T_c$. Note that in view of (4.4), this assumption is consistent with the linear variation of H_c with $(1-t)$, if β is regular at T_c. Putting these temperature variations of α and β into (4.3), we see that

$$|\psi|^2 \propto (1-t) \tag{4.6}$$

for T near, but below, T_c. This is consistent with correlating $|\psi|^2$ with n_s, the density of superconducting electrons in the London theory since $n_s \propto \lambda^{-2} \propto (1-t)$ near T_c.

To make these considerations quantitative, we now consider the remaining term in the expansion (4.1), the term dealing with fields and gradients. If we write $\psi = |\psi|e^{i\varphi}$, it takes on the more transparent form

$$\frac{1}{2m^*}\left[\hbar^2(\nabla|\psi|)^2 + \left(\hbar\nabla\varphi - \frac{e^*\mathbf{A}}{c}\right)^2|\psi|^2\right] \tag{4.7}$$

The first term gives the extra energy associated with gradients in the *magnitude* of the order parameter, as in a domain wall. The second term gives the kinetic energy associated with supercurrents in a gauge-invariant form. In the London gauge, φ is constant, and this term is simply $e^{*2}A^2|\psi|^2/2m^*c^2$. Equating this to the kinetic-energy density for a London superconductor based on (1.8), namely, $A^2/8\pi\lambda_{\text{eff}}^2$, we find

$$\lambda_{\text{eff}}^2 = \frac{m^*c^2}{4\pi|\psi|^2 e^{*2}} \tag{4.8}$$

With the identification $n_s^* = |\psi|^2$, this agrees with the usual definition of the London penetration depth, except for the presence of the starred effective number, mass, and charge values. The kinetic-energy density term can then be written as $n_s^*(\frac{1}{2}m^*v_s^2)$, where the supercurrent velocity is given by

$$m^*\mathbf{v}_s = \mathbf{p}_s - \frac{e^*\mathbf{A}}{c} = \hbar\nabla\varphi - \frac{e^*\mathbf{A}}{c} \tag{4.9}$$

It should be noted that by writing the energy associated with the vector potential in the simple form (4.7), we have restricted the theory to the approximation of *local* electrodynamics.

Now let us deal with the starred effective parameters. In the original formulation of the theory, it was thought that e^* and m^* would be the normal electronic values. However, experimental data turned out to be fitted better if $e^* \approx 2e$. The microscopic pairing theory of superconductivity makes it unambiguous that $e^* = 2e$ exactly, the charge of a pair of electrons. In the free-electron approximation, it would then be natural to take $m^* = 2m$ and $n_s^* = \frac{1}{2}n_s$, where n_s is the number of single electrons in the condensate. With these conventions, $n_s^* e^{*2}/m^* = n_s e^2/m$, so the London penetration depth is unchanged by the pairing.

The situation is more complicated in real metals. Band structure and phonon "dressing" effects may lead to an effective mass for a single electron in the normal state which typically differs from the free-electron mass by 50 percent. Moreover, the most important class of applications of GL theory is to dirty superconductors, in which $\lambda_{\text{eff}}^2 \approx \lambda_L^2(\xi_0/\ell) \gg \lambda_L^2$. These increased penetration depths can be attributed formally to either an increase in m^* or a decrease in n_s. It might appear that an independent mass determination could be made by an experiment in which a superconductor is rotated at angular velocity ω and the resulting magnetic moment is measured. However, as shown by Alben,[4] using a general argument based on Larmor's theorem, the induced moment must be the same as that due to a magnetic field $H_L = (2mc/e)\omega$, with the free-electron value of e/m, as indeed Brickman[5] had found. A more recent experiment by Tate et al.,[6] using the Josephson effect to measure the rotation-induced flux with sufficient sensitivity to detect relativistic corrections,[7] confirms the free electron e/m to within 100 ppm (parts per million). These experiments indicate that rotation experiments do not add much to the information available from magnetic moment measurements. In contrast, cyclotron resonance has allowed carrier band-mass determination in semiconductors and metals, but historically it has not been used for superconductors because the necessary high magnetic field destroyed the superconductivity. In a pioneering far-infrared experiment on the high-temperature superconductor YBCO (which also has a high critical field), Karrai et al.[8] inferred a cyclotron mass of about $3m$ from magneto-optical activity stemming from an unresolved cyclotron resonance. This value is consistent with band calculations for this material, but its applicability is limited.

---

[4] R. Alben, *Phys. Lett.* **29A**, 477 (1969).

[5] N. F. Brickman, *Phys. Rev.* **184**, 460 (1969).

[6] J. Tate, S. B. Felch, and B. Cabrera, *Phys. Rev.* **B42**, 7885 (1990).

[7] A summary of the theory is given by B. Cabrera and M. Peskin, *Phys. Rev.* **B39**, 6425 (1989).

[8] K. Karrai, E. Choi, F. Dunmore, S. Liu, X. Ying, Qi Li, T. Venkatesan, H. D. Drew, and D. B. Fenner, *Phys. Rev. Lett.* **69**, 355 (1993).

In view of the experimental inaccessibility of m^*, it is convenient simply to assign it the value of twice the mass of the free electron. With m^* fixed, all variations of λ, whether due to temperature, band structure, phonons, impurities, or even nonlocal electrodynamics, are taken up by choosing an appropriate[9] value of $|\psi_\infty|^2 = n_s^* = n_s/2$. Even at $T = 0$, this n_s will not in general correspond to any obvious integral number of electrons per atom. Rather, our point of view is that n_s simply measures that part of the oscillator strength in the sum rule

$$\int_0^\infty \sigma_1(\omega)\, d\omega = \frac{\pi n e^2}{2m} \tag{4.10}$$

which is located in the superfluid response at $\omega = 0$ in the form of a term $(\pi n_s e^2/2m)\delta(\omega)$.

An upper bound on n_s even at low temperatures is set by the oscillator strength of the conduction electrons in the normal state, which is spread over frequencies up to the collision rate $1/\tau$ for $q = 0$, and up to qv_F for $q \neq 0$ even if $\tau \to \infty$. More specifically , for spatially uniform fields, we expect the usual Drude frequency dependence of σ_1

$$\sigma_1(\omega, 0) = \frac{ne^2\tau/m}{1 + \omega^2\tau^2} \tag{4.10a}$$

while if $\tau = \infty$, we have the Lindhard[10] result that

$$\sigma_1(\omega, q) = \frac{3\pi n e^2}{4mv_F q}\left(1 - \frac{\omega^2}{v_F^2 q^2}\right) \tag{4.10b}$$

for $\omega < qv_F$, and $\sigma_1 = 0$ for $\omega > qv_F$. Naturally, both these expressions satisfy the sum rule (4.10). Speaking qualitatively, the normal-state oscillator strength lying at frequencies below the gap frequency $\omega_g = 2\Delta/\hbar$ will be converted to n_s in the transition to the superconducting state, whereas that above the gap will be relatively unaffected. If we consider a superconductor with local electrodynamics, so that the approximation $q = 0$ can be used, we see from (4.10*a*) that almost all the oscillator strength will appear as n_s in a pure metal, where $\omega_g\tau > 1$; in this case, $n_s \approx n$ and $\lambda \approx \lambda_L$. On the other hand, n_s/n will be reduced to something of the order of $\omega_g\tau \approx \ell/\xi_0$ if the metal is dirty enough to have $\omega_g\tau < 1$. As a result, in dirty superconductors we have $\lambda/\lambda_L = (n/n_s)^{1/2} \approx (\xi_0/\ell)^{1/2}$, a result obtained more rigorously in (3.123). If we consider instead a pure, nonlocal superconductor, (4.10*b*) implies that $n_s(q)/n \approx \omega_g/qv_F \approx 1/q\xi_0$. This q-dependent superfluid fraction is a reflection of the fall of $K(q)$ as $1/q$ in Fig. 3.12. Taking a typical value

[9]In a famous remark, de Gennes pointed out that one could do the same if one chose m^* to be the mass of the *sun*, but the normalization of $|\psi_\infty|^2$ would be very different and n_s would be completely nonphysical. If the normalization of ψ is fixed by having it equal the energy gap in suitable units, as in more microscopic theories, then m^* is no longer a free parameter.

[10]J. Lindhard, *Kgl. Danske Videnskab. Selskab., Mat.-fys. Medd.* **28**, no. 8 (1954).

$q \approx 1/\lambda$ for the currents in the penetration layer, this implies that $\lambda_L^2/\lambda^2 = n_s/n \approx \lambda/\xi_0$, or $\lambda \approx (\lambda_L^2 \xi_0)^{1/3}$, as found more rigorously in (3.132). For use in the GL theory, which is local, we must take n_s to be such an average value appropriate to the actual penetration depth, not to λ_L. The survey in this paragraph reminds us of the power of the sum rule–energy gap argument in making simple physical estimates of the effective superfluid density in diverse situations.

Having noted that $e^* = 2e$, and taking the convention that $m^* = 2m$, we can now evaluate the parameters of the theory by solving (4.3), (4.4), and (4.8). The results are

$$|\psi_\infty|^2 \equiv n_s^* \equiv \frac{n_s}{2} = \frac{m^* c^2}{4\pi e^{*2} \lambda_{\text{eff}}^2} = \frac{mc^2}{8\pi e^2 \lambda_{\text{eff}}^2} \tag{4.11a}$$

$$\alpha(T) = -\frac{e^{*2}}{m^* c^2} H_c^2(T) \lambda_{\text{eff}}^2(T) = -\frac{2e^2}{mc^2} H_c^2(T) \lambda_{\text{eff}}^2(T) \tag{4.11b}$$

$$\beta(T) = \frac{4\pi e^{*4}}{m^{*2} c^4} H_c^2(T) \lambda_{\text{eff}}^4(T) = \frac{16\pi e^4}{m^2 c^4} H_c^2(T) \lambda_{\text{eff}}^4(T) \tag{4.11c}$$

where e and m are now the usual free-electron values and λ_{eff} and H_c are measured values, or those computed from the microscopic theory.

Since the electrodynamics of some superconductors are significantly nonlocal, it is evident that this prescription in terms of an effective London λ is straightforward only sufficiently near T_c so that $\lambda_L(T) > \xi_0$, or in samples dirty enough so that $\xi \approx \ell < \lambda(T)$, i.e., where the nonlocality is unimportant. It is only under these conditions that the GL theory is really exact. Fortunately, these are important cases, including the high-temperature superconductors. Moreover, the qualitative conclusions of the theory seem to have much wider validity; semiquantitative results can usually be obtained even when nonlocality is important by using a suitable λ_{eff}, such as the one we computed for films, (3.136). For pure bulk samples, as noted earlier, it is probably most appropriate to take $\lambda_{\text{eff}} = \lambda_{\text{exp}}$, the experimental value, if we attempt to apply the theory far enough below T_c so that the nonlocality of the electrodynamics makes $\lambda_{\text{exp}} > \lambda_L$.

It is worth noting that if we insert the empirical approximations $H_c \propto (1 - t^2)$ and $\lambda^{-2} \propto (1 - t^4)$ into (4.11), we find

$$|\psi_\infty|^2 \propto 1 - t^4 \approx 4(1 - t)$$

$$\alpha \propto \frac{1 - t^2}{1 + t^2} \approx 1 - t$$

$$\beta \propto \frac{1}{(1 + t^2)^2} \approx \text{const} \tag{4.12}$$

Since the theory is usually exactly valid only very near T_c, it is customary to carry only the leading dependence on temperature; i.e., $|\psi_\infty|^2$ and α are usually taken to vary as $(1 - t)$ and β is taken to be constant, as anticipated in our preliminary discussion. Still, the more complete forms in (4.12) give some idea of how the theory can be extended over a wider range of temperature, and they have a certain amount of experimental support.

Finally, we recall that although our discussion of (4.7) has centered on the kinetic energy of the supercurrent, this term also describes the energy associated with gradients in the magnitude of ψ. Moreover, no additional parameters are introduced since gauge invariance requires a particular combination of ∇ and $\mathbf{A}$ in (4.1). Thus, the coefficients in the theory are completely determined by the values of $\lambda_{\text{eff}}(T)$ and $H_c(T)$. Since we showed earlier how the microscopic theory determines these parameters, we have effectively shown how the GL theory is set up to serve as an extension of BCS to the case of gradients and strong fields, but with a restriction to $T \approx T_c$.

4.2 THE GINZBURG-LANDAU DIFFERENTIAL EQUATIONS

In the absence of boundary conditions which impose fields, currents, or gradients, the free energy is minimized by having $\psi = \psi_\infty$ everywhere. On the other hand, when fields, currents, or gradients are imposed, $\psi(\mathbf{r}) = |\psi(\mathbf{r})|e^{i\varphi(\mathbf{r})}$ adjusts itself to minimize the overall free energy, given by the volume integral of (4.1). This variational problem leads, by standard methods, to the celebrated GL differential equations

$$\alpha\psi + \beta|\psi|^2\psi + \frac{1}{2m^*}\left(\frac{\hbar}{i}\nabla - \frac{e^*}{c}\mathbf{A}\right)^2\psi = 0 \tag{4.13}$$

and

$$\mathbf{J} = \frac{c}{4\pi}\operatorname{curl}\mathbf{h} = \frac{e^*\hbar}{2m^*i}(\psi^*\nabla\psi - \psi\nabla\psi^*) - \frac{e^{*2}}{m^*c}\psi^*\psi\mathbf{A} \tag{4.14}$$

or

$$\mathbf{J} = \frac{e^*}{m^*}|\psi|^2\left(\hbar\nabla\varphi - \frac{e^*}{c}\mathbf{A}\right) = e^*|\psi|^2\mathbf{v}_s \tag{4.14a}$$

where in the last step we have repeated the identification (4.9). Note that the current expression (4.14) has exactly the form of the usual quantum-mechanical expression for particles of mass m^*, charge e^*, and wavefunction $\psi(\mathbf{r})$. Similarly, apart from the nonlinear term, the first equation has the form of Schrödinger's equation for such particles, with energy eigenvalue $-\alpha$. The nonlinear term acts like a repulsive potential of ψ on itself, tending to favor wavefunctions $\psi(\mathbf{r})$ which are spread out as uniformly as possible in space.

In carrying through the variational procedure, one must provide boundary conditions. A possible choice, which assures that no current passes through the surface, is

$$\left(\frac{\hbar}{i}\nabla - \frac{e^*}{c}\mathbf{A}\right)\psi\bigg|_n = 0 \tag{4.15}$$

This is the boundary condition used by GL, and it is appropriate at an insulating surface. Using the microscopic theory, de Gennes[11] has shown that for a metal-superconductor interface with no current, (4.15) must be generalized to

$$\left(\frac{\hbar}{i}\nabla - \frac{e^*}{c}\mathbf{A}\right)\psi\bigg|_n = \frac{i\hbar}{b}\psi \tag{4.15a}$$

where b is a real constant. As shown in Fig. 4.2, if $A_n = 0$, b is the extrapolation length to the point outside the boundary at which ψ would go to zero if it maintained the slope it had at the surface. The value of b will depend on the nature of the material to which contact is made, approaching zero for a magnetic material and infinity for an insulator, with normal metals lying in between.

4.2.1 The Ginzburg-Landau Coherence Length

To help get a feeling for the differential equation (4.13), we first consider a simplified case in which no fields are present. Then $\mathbf{A} = 0$, and we can take ψ to be real since the differential equation has only real coefficients. If we introduce a normalized wavefunction $f = \psi/\psi_\infty$, where $\psi_\infty^2 = -\alpha/\beta > 0$, the equation becomes (in one dimension)

$$\frac{\hbar^2}{2m^*|\alpha|}\frac{d^2 f}{dx^2} + f - f^3 = 0 \tag{4.16}$$

This makes it natural to define the characteristic length $\xi(T)$ for variation of ψ by

$$\xi^2(T) = \frac{\hbar^2}{2m^*|\alpha(T)|} \propto \frac{1}{1-t} \tag{4.17}$$

Note that this $\xi(T)$ is certainly not the same length as Pippard's ξ, which we used in our discussion of the nonlocal electrodynamics, since this $\xi(T)$ diverges at T_c, whereas the electrodynamic ξ is essentially constant. In fact, on the face of it, it is not clear why they should even be related. We retain this traditional notation, despite its considerable power to confuse because it is almost invariably used in

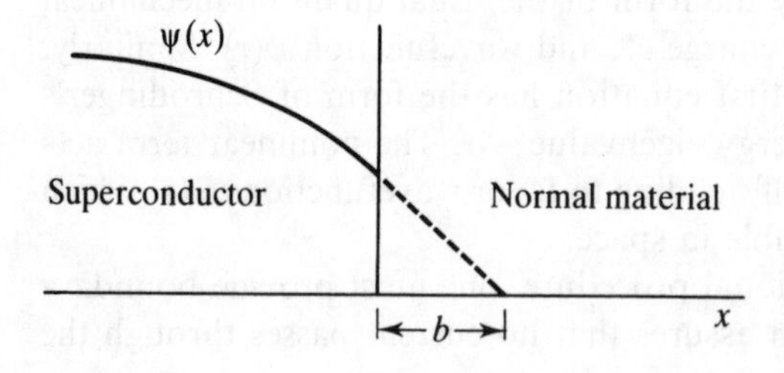

FIGURE 4.2
Schematic diagram illustrating the boundary condition (4.15a) at an interface characterized by an extrapolation length b.

[11]P. G. de Gennes, *Superconductivity of Metals and Alloys*, W. A. Benjamin, New York (1966), p. 227; reprinted by Addison-Wesley (1989).

the literature, and because it does turn out that $\xi(T) \approx \xi_0$ for pure materials well below T_c. In terms of $\xi(T)$, (4.16) becomes

$$\xi^2(T)\frac{d^2f}{dx^2}+f-f^3=0 \tag{4.18}$$

The significance of $\xi(T)$ as a characteristic length for variation of ψ (or f) can be made even more evident by considering a linearized form of (4.18), in which we set $f(x) = 1 + g(x)$, where $g(x) \ll 1$. Then we have, to first order in g,

$$\xi^2 g''(x) + (1+g) - (1+3g+\cdots) = 0$$

or

$$g'' = \left(\frac{2}{\xi^2}\right)g$$

so that

$$g(x) \sim e^{\pm\sqrt{2}x/\xi(T)} \tag{4.19}$$

which shows that a small disturbance of ψ from ψ_∞ will decay in a characteristic length of order $\xi(T)$.

Now that we have an idea of the significance of the length $\xi(T)$, let us see what its value is. Substituting the value of α from (4.11b) into the definition (4.17), we find

$$\xi(T) = \frac{\Phi_0}{2\sqrt{2}\pi H_c(T)\lambda_{\text{eff}}(T)} \tag{4.20}$$

where

$$\Phi_0 = \frac{hc}{e^*} = \frac{hc}{2e} \tag{4.21}$$

is the fluxoid quantum, which will play an important role in our future disussions. The fact that this $\xi(T)$ is at least related to the ξ_0 of Pippard and BCS is shown by the existence of the relation

$$\Phi_0 = \left(\frac{2}{3}\right)^{1/2}\pi^2\xi_0\lambda_L(0)H_c(0) \tag{4.22}$$

which follows readily from our earlier BCS results $\xi_0 = \hbar v_F/\pi\Delta(0)$ and $H_c^2(0)/8\pi = \frac{1}{2}N(0)\Delta^2(0)$, if we assume the free-electron relation between $N(0)$ and n. Combining (4.20) and (4.22), we can write

$$\frac{\xi(T)}{\xi_0} = \frac{\pi}{2\sqrt{3}}\frac{H_c(0)}{H_c(T)}\frac{\lambda_L(0)}{\lambda_{\text{eff}}(T)} \tag{4.23}$$

From this we can see that near T_c

$$\xi(T) = 0.74\frac{\xi_0}{(1-t)^{1/2}} \qquad \text{pure} \tag{4.24a}$$

$$\xi(T) = 0.855\frac{(\xi_0\ell)^{1/2}}{(1-t)^{1/2}} \qquad \text{dirty} \tag{4.24b}$$

The precise coefficients here were determined by using the exact results of BCS in the limit of $T \approx T_c$, namely,

$$H_c(t) = 1.73H_c(0)(1 - t) \tag{4.25}$$

$$\lambda_L(t) = \frac{\lambda_L(0)}{[2(1-t)]^{1/2}} \tag{4.26a}$$

$$\lambda_{\text{eff}}(t)\Bigg|_{\substack{\text{dirty}\\ \text{limit}}} = \lambda_L(t)\left(\frac{\xi_0}{1.33\ell}\right)^{1/2} \tag{4.26b}$$

The relation (4.24*a*), giving $\xi(T)$ for pure superconductors, has clear validity only in the extremely narrow temperature range near T_c in which the local electrodynamics are valid; outside this range, the appropriate effective value of ξ will be dependent on the sample configuration. Equation (4.24*b*) has a much broader range of validity for dirty superconductors because there the local approximation remains good.

It is also useful to introduce the famous dimensionless Ginzburg-Landau parameter κ, which is defined as the ratio of the two characteristic lengths

$$\kappa = \frac{\lambda_{\text{eff}}(T)}{\xi(T)} = \frac{2\sqrt{2}\pi H_c(T)\lambda_{\text{eff}}^2(T)}{\Phi_0} \tag{4.27}$$

With the empirical approximations $H_c \propto (1 - t^2)$ and $\lambda^{-2} \propto (1 - t^4)$, we see that κ should vary as $(1 + t^2)^{-1}$. Of course, this is only a rough approximation, but we can safely conclude that κ is regular at T_c, and varies only slowly with temperature. Using the preceding numerical results, we find the following results in the pure and dirty limits at T_c:

$$\kappa = 0.96\frac{\lambda_L(0)}{\xi_0} \qquad \text{pure} \tag{4.27a}$$

$$\kappa = 0.715\frac{\lambda_L(0)}{\ell} \qquad \text{dirty} \tag{4.27b}$$

In the classic pure superconductors, $\kappa \ll 1$, but in dirty superconductors or in the high-temperature superconductors, κ may be much greater than 1. As will be discussed later in more detail, the value $\kappa = 1/\sqrt{2}$ separates superconductors of types I and II.

4.3 CALCULATIONS OF THE DOMAIN-WALL ENERGY PARAMETER

We have now developed the methods required to compute the surface-energy parameter $\gamma = H_c^2\delta/8\pi$ that we used in our discussion of the intermediate state in Chap. 2. The one-dimensional variations of $\psi(x)$ and of $h(x)$ in the domain wall are sketched in Fig. 4.3, contrasting the cases with $\kappa \ll 1$ and $\kappa \gg 1$.

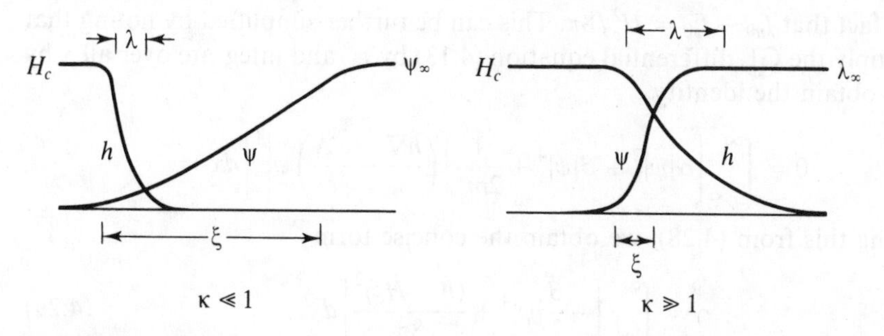

FIGURE 4.3
Schematic diagram of variation of h and ψ in a domain wall. The case $\kappa \ll 1$ refers to a type I superconductor (positive wall energy); the case $\kappa \gg 1$ refers to a type II superconductor (negative wall energy).

Qualitatively, we see that the surface energy is positive for $\kappa \ll 1$ since there is a region of thickness $\sim(\xi - \lambda)$ from which the magnetic field is held out (contributing to the positive diamagnetic energy) while not enjoying the full condensation energy associated with ψ_∞. The argument is reversed for $\kappa \gg 1$, leading to a negative surface energy. Let us now see how this argument is made quantitative by using the GL theory.

We seek solutions of the differential equations (4.13) and (4.14) subject to the boundary conditions

$$\begin{array}{llll} \psi = 0 & \text{and} & h = H_c & \text{as} \quad x \to -\infty \\ \psi = \psi_\infty & \text{and} & h = 0 & \text{as} \quad x \to +\infty \end{array}$$

Because the problem is one-dimensional, it is possible to take ψ real. Doing so simplifies the equations by eliminating cross terms of ∇ and $\mathbf{A}$ and also by reducing the second equation to the simple form $\mathbf{J} \propto |\psi(x)|^2\mathbf{A}$. Nonetheless, complete solutions can only be obtained numerically. We start by formally simplifying the expression for the surface energy.

First, we recognize that the appropriate quantity to calculate is the Gibbs free energy, since H is fixed at H_c, while B depends on the location of the domain wall. As indicated in the discussion leading to (2.16), the density of Gibbs free energy at H_c has the same value in superconducting or normal material, namely, f_{s0}, the Helmholtz free energy of the superconductor in the absence of fields or currents. Thus, the surface energy we seek is the excess of G over what it would be if its density were f_{s0} everywhere, i.e.,

$$\gamma = \int_{-\infty}^{\infty} (g_{sH} - f_{s0})dx = \int_{-\infty}^{\infty} \left(f_{sH} - \frac{hH_c}{4\pi} - f_{s0} \right) dx$$

$$= \int_{-\infty}^{\infty} \left[\alpha|\psi|^2 + \frac{\beta}{2}|\psi|^4 + \frac{1}{2m^*}\left| \left(\frac{\hbar\nabla}{i} - \frac{e^*\mathbf{A}}{c} \right)\psi \right|^2 + \frac{(h - H_c)^2}{8\pi} \right] dx \qquad (4.28)$$

using the fact that $f_{n0} - f_{s0} = H_c^2/8\pi$. This can be further simplified by noting that if we multiply the GL differential equation (4.13) by ψ^* and integrate over all x by parts, we obtain the identity

$$0 = \int_{-\infty}^{\infty} \left[\alpha|\psi|^2 + \beta|\psi|^4 + \frac{1}{2m^*} \left| \left(\frac{\hbar\nabla}{i} - \frac{e^*\mathbf{A}}{c} \right) \psi \right|^2 \right] dx$$

Subtracting this from (4.28), we obtain the concise form

$$\gamma = \int_{-\infty}^{\infty} \left[-\frac{\beta}{2}|\psi|^4 + \frac{(h - H_c)^2}{8\pi} \right] dx \tag{4.29}$$

which is to be equated to $(H_c^2/8\pi)\delta$. Finally, using (4.11), we can write this as

$$\delta = \int_{-\infty}^{\infty} \left[\left(1 - \frac{h}{H_c} \right)^2 - \left(\frac{\psi}{\psi_\infty} \right)^4 \right] dx \tag{4.30}$$

This form clearly displays how δ is determined by the balance between the positive diamagnetic energy and the negative condensation energy due to the superconductivity, as argued qualitatively in connection with Fig. 4.3.

Numerical solutions for $h(x)$ and $\psi(x)$ must be made in order to evaluate (4.30), except in limiting cases. For these cases, the following exact results have been obtained:

$$\delta = \frac{4\sqrt{2}\xi}{3} = 1.89\xi \quad \kappa \ll 1 \tag{4.31a}$$

$$\delta = \frac{-8(\sqrt{2} - 1)\lambda}{3} = -1.104\lambda \quad \kappa \gg 1 \tag{4.31b}$$

These results support our qualitative reasoning that δ should be of the order of $(\xi - \lambda)$.

Special consideration is required to show that the exact crossover from positive to negative surface energy occurs for $\kappa = 1/\sqrt{2}$. This was found by numerical integration by Ginzburg and Landau in their original paper, and they already anticipated that a conventional laminar intermediate state would only occur for lower values of κ. But until Abrikosov's path-breaking paper,[12] no one fully anticipated the radically different behavior that resulted from the negative surface energy at higher values of κ. In one stroke, his paper created the study of type II superconductivity, the name he gave to materials with $\kappa > 1/\sqrt{2}$. Since this is the subject of the next chapter, for the present we shall simply remark that the negative surface energy causes the flux-bearing (normal) regions to subdivide until a quantum limit is reached in which each quantum of flux $\Phi_0 = hc/2e$ passes through the sample as a distinct flux tube. These flux tubes form a regular array, and $\psi \to 0$ along the axis of each one.

---

[12] A. A. Abrikosov, *Zh. Eksperim. i. Teor. Fiz.* **32**, 1442 (1957) [*Soviet Phys.—JETP* **5**, 1174 (1957)].

Unlike the intermediate state of type I superconductors, this so-called *mixed state* of type II superconductors occurs over a substantial field range even if the sample demagnetizing factor is zero.

4.4 CRITICAL CURRENT OF A THIN WIRE OR FILM

Having taken a quick look at the calculation of the interface energy, in which we immediately find that numerical solutions are required, let us now step back and treat a number of important simpler examples in which exact analytic solutions are possible. In this way, we shall develop some familiarity with the GL theory before returning to more complex problems.

The very simplest applications are those in which the perturbing fields and currents are so weak that $|\psi| = \psi_\infty$ everywhere, and the GL theory reduces to the London theory.

A more interesting class of examples is that in which strong fields or currents change $|\psi|$ from ψ_∞, but in which $|\psi|$ has the same value everywhere. This will be the case if the sample is a thin wire or film so oriented with respect to any external field that any variation of $|\psi|$ would need to occur in a thickness $d \ll \xi(T)$. In that case, the term in the free-energy proportional to $(\nabla|\psi|)^2$ would give an excessively large contribution if any substantial variations occurred. As a result, they do not, and we can approximate $\psi(\mathbf{r})$ by $|\psi|e^{i\varphi(\mathbf{r})}$, where $|\psi|$ is constant. In this case, the expressions for the current and free-energy densities take on the simple forms

$$\mathbf{J}_s = \frac{2e}{m^*}|\psi|^2\left(\hbar\nabla\varphi - \frac{2e}{c}\mathbf{A}\right) = 2e|\psi|^2\mathbf{v}_s \tag{4.32}$$

$$f = f_{n0} + \alpha|\psi|^2 + \frac{\beta}{2}|\psi|^4 + |\psi|^2\frac{1}{2}m^* v_s^2 + \frac{h^2}{8\pi} \tag{4.33}$$

Although it is our standard convention to set $m^* = 2m$, we retain the more general formulas here to permit other normalizations of ψ to be used, if desired, and also as a reminder of the conventional nature of the parameter m^*.

Let us now apply these equations to treat the case of a uniform current density through a thin film or wire. Since the total energy due to the field term $h^2/8\pi$ is less than the kinetic energy of the current by a factor of the order of the ratio of the cross-sectional area of the conductor to λ^2, we can always neglect it for a sufficiently thin conductor. Then, for a given v_s, we can minimize (4.33) to find the optimum value of $|\psi|^2$. The result is

$$|\psi|^2 = \psi_\infty^2\left(1 - \frac{m^* v_s^2}{2|\alpha|}\right) = \psi_\infty^2\left[1 - \left(\frac{\xi m^* v_s}{\hbar}\right)^2\right] \tag{4.34}$$

where the second form is stated in terms of ξ and $m^* v_s$, quantities invariant under changes in conventions. The corresponding current is

$$J_s = 2e\psi_\infty^2 \left(1 - \frac{m^* v_s^2}{2|\alpha|}\right) v_s \tag{4.35}$$

As indicated in Fig. 4.4, this has a maximum value when $\partial J_s/\partial v_s = 0$, namely, when $\frac{1}{2} m^* v_s^2 = |\alpha|/3$ and $|\psi|^2/\psi_\infty^2 = 2/3$. We identify this maximum current with the critical current. Thus,

$$J_c = 2e\psi_\infty^2 \frac{2}{3}\left(\frac{2|\alpha|}{3m^*}\right)^{1/2} = \frac{cH_c(T)}{3\sqrt{6}\pi\lambda(T)} \propto (1-t)^{3/2} \tag{4.36}$$

where, again, the second form is entirely in terms of operationally significant quantities and the indicated proportionality to $(1-t)^{3/2}$ holds near T_c. The corresponding critical momentum is

$$p_c = m^* v_c = \frac{\hbar}{\sqrt{3}\xi(T)} \tag{4.37}$$

The critical velocity itself is poorly defined since it depends on the conventional choice of m^*.

It may be noted that we have taken v_s rather than $J_s \propto |\psi|^2 v_s$ as our independent variable. This was not a capricious choice; it was necessary since we are using the Helmholtz free energy, which is appropriate only if there are no induced emfs to effect energy interchanges with the source of current. This corresponds to specifying v_s, since an emf is needed to change that. If we wish to use the *current* as independent variable, then we must introduce a Legendre transformation on the free energy, as we did in (2.13) in dealing with magnetic energies. The appropriate term to subtract here to take into account work done by the generator is $m^* \mathbf{v}_s \cdot \mathbf{J}_s/2e$, so we could consider a Gibbs free-energy density

$$g = f - \frac{m^* v_s J_s}{2e} \tag{4.38a}$$

or

$$g = f_{n0} + \alpha|\psi|^2 + \frac{\beta}{2}|\psi|^4 + \frac{m^* J_s^2}{8e^2|\psi|^2} + \frac{h^2}{8\pi} \tag{4.38b}$$

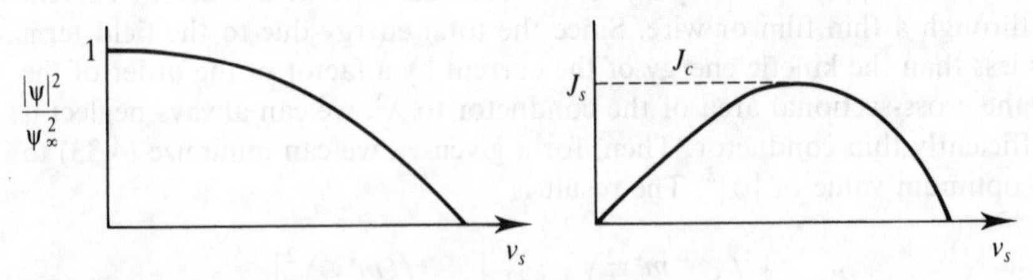

FIGURE 4.4
Variation of $|\psi|^2$ and of J_s with the superfluid velocity v_s.

where we have used (4.32) to eliminate v_s. Minimizing this with respect to $|\psi|^2$ for given J_s leads to a cubic equation in $|\psi|^2$. Although algebraically more awkward, this condition is consistent with what we found earlier. For example, we can write it in the form

$$\frac{m^* J_s^2}{8e^2} = -\alpha|\psi|^4 - \beta|\psi|^6 \tag{4.39}$$

whose maximum value occurs when $|\psi|^2 = -\frac{2}{3}(\alpha/\beta) = \frac{2}{3}\psi_\infty^2$, at which J_s has the critical value J_c found in (4.36).

It is of interest to compare this GL critical current with that of the London theory, where it is found by equating the density of kinetic energy to that of condensation energy

$$\frac{1}{2}n_s m v_c^2 = \frac{2\pi}{c^2}\lambda^2 J_c^2 = \frac{H_c^2}{8\pi}$$

so that

$$J_{c,\,\text{London}} = \frac{cH_c}{4\pi\lambda} \tag{4.40}$$

as stated in (2.7). This exceeds the more exact GL result (4.36) by a factor of $(3\sqrt{6}/4) = 1.84$ because it fails to take into account the decrease in $|\psi|^2$ with increasing current given by the nonlinear treatment.

It is also of interest to compare the GL result with that of the microscopic theory, where, of course, numerical computations must be made except in special cases. Bardeen has given a very useful review[13] of such calculations. Near T_c, the GL results are recovered, as expected. In the zero-temperature limit, the situation is quite different. In the presence of a uniform velocity $\mathbf{v}_s$, the quasi-particle energies are shifted by $\hbar\mathbf{k}\cdot\mathbf{v}_s$. [This may be seen from (3.108) by noting that a velocity $e\mathbf{a}(0)/mc$ is induced by a uniform vector potential $\mathbf{a}(0)$. It is discussed more thoroughly in Sec. 10.1.2.] Thus, the gap goes to zero for some states when

$$v_s = \frac{\Delta(0)}{\hbar k_F} = \frac{\hbar}{\pi m \xi_0} \tag{4.41}$$

Below this *depairing velocity*, all electrons contribute to the supercurrent and J_s is strictly proportional to v_s. Above this depairing velocity, some excitations occur at zero energy, the gap drops precipitously, and the maximum possible current is only 2 percent more than that at the velocity where depairing begins. The resulting $J_s(v_s)$ curve for $T = 0$ therefore shows a linear rise followed by a very steep drop to zero, in marked contrast with the GL result plotted in Fig. 4.4, appropriate near T_c.

---

[13] J. Bardeen, *Rev. Mod. Phys.* **34**, 667 (1962).

Experimental confirmation of these results is most straightforward if both transverse dimensions of the conductor can be made small compared to both λ and ξ. It is then safe to take both J_s and $|\psi|^2$ to be constant over the cross section, as is assumed in the theory. Some of the first careful experiments on samples of this sort were those of Hunt,[14] who worked on very narrow strips of a thin evaporated film. Later several groups worked with tin "whiskers" only about 1 μm (micrometer) in diameter, generally composed of a single crystal and having smooth surfaces, which are nearly ideal for the purpose. Since these experiments have concentrated on the fluctuation effects giving rise to resistance even at currents below the J_c computed earlier, we shall defer further discussion of them until a later chapter.

For reasons of experimental simplicity, many other measurements of critical currents have been made on thin-film samples which are *not* narrow on the scale of λ or ξ. With these, the measured J_c is usually much less than (4.36) for a number of reasons. First, it is somewhat difficult to make films of uniform thickness and structure. More seriously, the electrodynamic equations cause the supercurrent to pile up at the edges of the film because the external magnetic flux density is greatest there as the flux lines circle the film strip.[15] This effect makes the current density nonuniform, and also emphasizes the properties of the edges of the film, which generally are thinner and less perfect. This problem can be minimized in three ways: (1) One can simply make the strip narrower than $\lambda_\perp$, so that the product of the thickness d and the width w is less than λ^2; in this case, J_s will be nearly uniform even if $w > \lambda$. (2) One can use a ground-plane geometry, in which the film under study is deposited on a larger thick superconductor with only a thin insulating layer in between; in this geometry, the superconducting substrate forces the field lines to be parallel to the film, which in turn requires a uniform current density in the film. (3) One can use a cylindrical film, so that there are no edges, and symmetry guarantees a uniform current density provided a concentric current return is used. It is possible to reach critical currents within about 10 percent of the theoretical values by any of these techniques if enough care is taken.

A more microscopic test of the theory can be made by tunneling measurements, which allow a determination of the gap $\Delta(J)$ for $J < J_c$. According to (4.34), $|\psi|$, and hence Δ, should decrease as J^2 at first, going over to a more complicated variation at higher currents. Evidence for this decrease, based on an increase in the differential conductance at zero voltage, was first given by Levine[16]; more definitive results, based on measuring the full spectral density of states, were subsequently obtained by Mitescu.[17]

---

[14] T. K. Hunt, *Phys. Rev.* **151**, 325 (1966).

[15] See, e.g., W. J. Skocpol, *Phys. Rev.* **B14**, 1045 (1976).

[16] J. L. Levine, *Phys. Rev. Lett.* **15**, 154 (1965).

[17] C. D. Mitescu, thesis, California Institute of Technology, Pasadena (1966).

4.5 FLUXOID QUANTIZATION AND THE LITTLE-PARKS EXPERIMENT

An ingenious experiment in which $m^* v_s$ rather than the current is constrained by external conditions, and which clearly demonstrates that it is the fluxoid rather than the flux which is quantized, was performed by Little and Parks.[18] The experiment consists of measuring the resistive transition of a thin-walled superconducting cylinder in an axial magnetic field. From this one can infer shifts $\Delta T_c(H)$ in the critical temperature depending on the magnetic flux enclosed by the cylinder. In this section, we shall give an analysis[19] of this experiment in the framework of the Ginzburg-Landau theory.

4.5.1 The Fluxoid

In analyzing the state of a multiply connected superconductor in the presence of a magnetic field, F. London introduced the concept of the *fluxoid* Φ' associated with each hole (or normal region) passing through the superconductor. His definition was

$$\Phi' = \Phi + \left(\frac{4\pi}{c}\right)\oint \lambda^2 \mathbf{J}_s \cdot d\mathbf{s} = \Phi + \left(\frac{m^* c}{e^*}\right)\oint \mathbf{v}_s \cdot d\mathbf{s} \tag{4.42}$$

where

$$\Phi = \int \mathbf{h} \cdot d\mathbf{S} = \oint \mathbf{A} \cdot d\mathbf{s}$$

is the ordinary magnetic flux through the integration circuit. Since it is easily seen that $\Phi' = 0$ for any path which encloses no hole but only superconducting material, in which the London equation (1.4) holds, it follows that Φ' has the same value for *any* path around a given hole. Similarly, if there is a time variation of $\mathbf{h}$, the change in $\mathbf{J}_s$ induced according to the other London equation is just enough to hold Φ' constant (unless, of course, the change is so violent as to drive the superconductor into the normal state). These two conservation laws imply that Φ' has a unique constant value for all contours enclosing any given hole. In fact, London argued[20] that the values of Φ' should be restricted to a discrete set, integral multiples of a fluxoid quantum hc/e^*. This can be seen simply by applying the Bohr-Sommerfeld quantum condition to (4.42), as follows:

$$\begin{aligned}\Phi' &= \frac{c}{e^*}\oint\left(m^*\mathbf{v}_s + \frac{e^*\mathbf{A}}{c}\right)\cdot d\mathbf{s} = \frac{c}{e^*}\oint \mathbf{p}\cdot d\mathbf{s} \\ &= n\frac{hc}{e^*} = n\Phi_0\end{aligned} \tag{4.43}$$

[18] W. A. Little and R. D. Parks, *Phys. Rev. Lett.* **9**, 9 (1962); *Phys. Rev.* **133**, A97 (1964).

[19] M. Tinkham, *Phys. Rev.* **129**, 2413 (1963).

[20] F. London, *Superfluids*, vol. I., Wiley, New York (1950) p. 152.

Not anticipating the pairing theory of superconductivity, London presumed that e^* was e. We now know that $e^* = 2e$, so the fluxoid quantum has the value

$$\Phi_0 = \frac{hc}{2e} = 2.07 \times 10^{-7} \text{ G-cm}^2$$
$$= 2.07 \times 10^{-15} \text{ Wb (webers)} \qquad (4.44)$$

This has been shown experimentally by measurements[21] of the flux trapped in hollow cylinders with walls sufficiently thick so that the total flux and the fluxoid are indistinguishable because $\mathbf{v}_s \to 0$ in (4.42) once the skin depth is passed.

It should not be thought that the exactness of the concept of fluxoid quantization is limited by the preceding argument, with its use of the semiclassical Bohr-Sommerfeld language and the inexact London equations. From the viewpoint of GL theory it is based simply on the existence of a single-valued complex superconducting order parameter ψ. This requires that the phase φ must change by integral multiples of 2π in making a complete circuit; i.e.,

$$\oint \nabla\varphi \cdot d\mathbf{s} = 2\pi n \qquad (4.45)$$

which, with (4.9), leads directly to the results found earlier. We thus see that fluxoid quantization is the macroscopic analog of the quantization of angular momentum in an atomic system. Accordingly, it is not surprising that it provides a powerful tool for dealing with many problems of superconductors penetrated by magnetic flux.

4.5.2 The Little-Parks Experiment

Now let us apply this principle to the analysis of the Little-Parks experiment. Let R be the radius of the thin-walled cylinder and H be the applied field. [No distinction needs to be made between the applied field and the field inside the cylinder since we are seeking the shifted $T_c(H)$, at which point $|\psi|^2 \to 0$, so $J_s \to 0$ and the fields are the same.] Then, using (4.42) with $\Phi = \pi R^2 H$ and $\Phi' = n\Phi_0$, the supercurrent velocity is fixed by

$$v_s = \frac{\hbar}{m^* R}\left(n - \frac{\Phi}{\Phi_0}\right) \qquad (4.46)$$

For the value of Φ imposed by a given H, the energy of the currents in the cylinder will be least for that integer n for which v_s is a minimum, and accordingly that choice of n will allow the system to remain superconducting at the highest possible temperature. With n chosen in this way, v_s will be a periodic function of Φ/Φ_0, as

---

[21] B. S. Deaver and W. M. Fairbank, *Phys. Rev. Lett.* 7, 43 (1961); R. Doll and M. Näbauer, *Phys. Rev. Lett.* 7, 51 (1961).

indicated in Fig. 4.5. Since v_s is now specified, we may apply (4.34) to find the reduction in $|\psi|^2$. In particular, the transition occurs when $|\psi|^2 = 0$, i.e., when

$$\frac{1}{\xi^2} = \left(\frac{m^* v_s}{\hbar}\right)^2 = \frac{1}{R^2}\left(n - \frac{\Phi}{\Phi_0}\right)^2$$

Using the relations (4.24) to provide the proportionality constants between ξ^{-2} and $(1 - t)$, we find that the fractional depression of T_c is given by

$$\frac{\Delta T_c(H)}{T_c} = \begin{cases} 0.55 \dfrac{\xi_0^2}{R^2}\left(n - \dfrac{\Phi}{\Phi_0}\right)^2 & \text{clean} \\[2ex] 0.73 \dfrac{\xi_0 \ell}{R^2}\left(n - \dfrac{\Phi}{\Phi_0}\right)^2 & \text{dirty} \end{cases} \tag{4.47}$$

The maximum depression of T_c occurs when $n - \Phi/\Phi_0 = \frac{1}{2}$. At that point, $\Delta T_c/T_c$ reaches $0.14\xi_0^2/R^2$ and $0.18\xi_0\ell/R^2$ for the clean and dirty cases, respectively. The samples actually used were typically dirty tin films evaporated around organic filaments $\sim 1\,\mu$m in diameter. Taking typical values—$R = 7 \times 10^{-5}$ cm, $\xi_0 = 2 \times 10^{-5}$ cm, and $\ell = 10^{-6}$ cm—this leads to $\Delta T_c|_{\max} \sim 0.8 \times 10^{-3} T_c \approx 3 \times 10^{-3}$ K, which is readily measurable. For this diameter, the periodicity in field would be set by $\Phi_0/\pi R^2 = 14$ G.

Although we have idealized the problem to finding $T_c(H)$, the actual experiment measures the periodic variation of the resistance of the film with changing

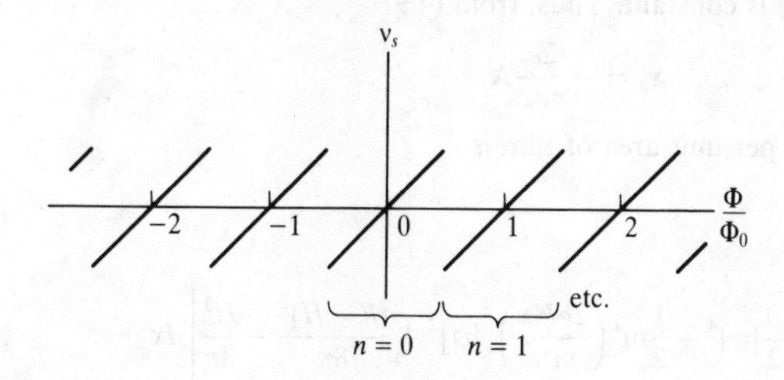

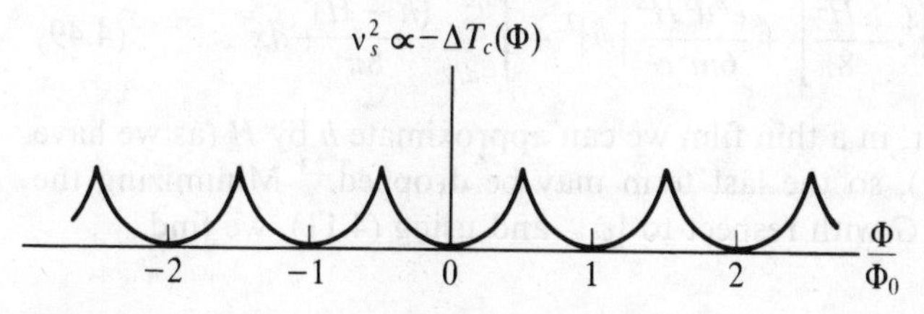

FIGURE 4.5
Variation of v_s and v_s^2 with flux threading the hollow cylinder in the Little-Parks experiment. The depression of T_c, and hence the increase in resistance in the actual experiment, is proportional to v_s^2 and thus displays the scalloped shape of the lower curve.

H. This takes advantage of the finite breadth of the resistive transition to simplify the measurements since $\Delta R(H)$ may be converted to $\Delta T_c(H)$ by using the measured dR/dT in the transition region. This is not entirely quantitative, however, since the shape of the resistive transition is observed to change with H, so that the inferred $\Delta T_c(H)$ depends to some extent on the choice of resistance level within the transition region. Another complication is an aperiodic quadratic shift of T_c with H, which is sometimes observed and may be due to misalignment. Fortunately, improvements of the experiment and analysis[22] cleared up most of these problems and led to a rather satisfactory agreement with the theory.

4.6 PARALLEL CRITICAL FIELD OF THIN FILMS

If we confine our attention initially to finding the critical field, and if we anticipate that thin films have second-order phase transitions, in which $|\psi|^2 \to 0$ and $\lambda_{\text{eff}}(H) \to \infty$, then we may neglect screening and write

$$A_y = \int_0^x h(x')\,dx' \approx Hx \tag{4.48}$$

where H is the applied field. The latter form will also be a good approximation for films of thickness $d < \lambda$, even far from the transition.

This A_y is an example of the *London gauge choice* mentioned in Sec. 1.2 and used to obtain λ_{eff} for a thin film in (3.36). With this gauge choice, the phase φ as well as the amplitude $|\psi|$ is constant. Thus, from (4.9)

$$\mathbf{v}_s = -\frac{2e}{m^*c}\mathbf{A}$$

so the Gibbs free energy per unit area of film is

$$\begin{aligned} G &= \int_{-d/2}^{d/2}\left(f - \frac{hH}{4\pi}\right)dx \\ &= \int_{-d/2}^{d/2}\left[f_{n0} + \alpha|\psi|^2 + \frac{\beta}{2}|\psi|^4 + \frac{1}{2}m^*\left(\frac{2eHx}{m^*c}\right)^2|\psi|^2 + \frac{(h-H)^2}{8\pi} - \frac{H^2}{8\pi}\right]dx \\ &= d\left[f_{n0} + \alpha|\psi|^2 + \frac{\beta}{2}|\psi|^4 - \frac{H^2}{8\pi}\right] + \frac{e^2d^3H^2}{6m^*c^2}|\psi|^2 + \int_{-d/2}^{d/2}\frac{(h-H)^2}{8\pi}dx \end{aligned} \tag{4.49}$$

Now, as mentioned earlier, in a thin film we can approximate h by H (as we have done in writing $A_y = Hx$), so the last term may be dropped.[23] Minimizing the remaining expression for G with respect to $|\psi|^2$ and using (4.11), we find

---

[22] R. P. Groff and R. D. Parks, *Phys. Rev.* **176**, 567 (1968).

[23] It is easy to show that the last term is down by a factor of order $d^2e^2|\psi|^2/mc^2 \ll 1$ compared to the term proportional to d^3, which we keep.

$$|\psi|^2 = \psi_\infty^2 \left(1 - \frac{d^2H^2}{24\lambda^2 H_c^2}\right) \tag{4.50}$$

where λ here refers to the appropriate λ_{eff} in zero field. Thus, the film becomes normal, i.e., $|\psi|^2 \to 0$, when $H = H_{c\|}$, given by

$$H_{c\|} = 2\sqrt{6}\frac{H_c \lambda}{d} \tag{4.51}$$

This parallel critical field can exceed the thermodynamic critical field H_c, by a large factor if d/λ is small enough. The physical reason for this is simply that the thin film, largely penetrated by the field, has little diamagnetic energy for a given applied field in comparison to an equal volume of a bulk superconductor.

Rewriting (4.50) in terms of $H_{c\|}$, we obtain

$$\frac{|\psi|^2}{\psi_\infty^2} = 1 - \frac{H^2}{H_{c\|}^2} \tag{4.52}$$

Recalling the proportionality of ψ and Δ, we see that this predicts that the energy gap in a thin film will be depressed continuously to zero by increasing a parallel magnetic field up to $H_{c\|}$. This behavior was confirmed qualitatively by early electron tunneling experiments.[24] However, later work[25] showed that the BCS excitation spectrum is progressively smeared out by the field until at $H \approx 0.95H_{c\|}$ the spectrum becomes gapless, although the film stays superconducting as measured by its resistance. Such *gapless* superconductivity, first proposed by Abrikosov and Gor'kov,[26] turns out to be quite characteristic of superconductors subjected to time-reversal noninvariant perturbations (typically magnetic), but we shall defer detailed discussion of this until Sec. 10.2.

4.6.1 Thicker Films

What if the film is not extremely thin? So long as it is thin enough to have a second-order phase transition, (4.51) remains *exactly*[27] correct because at $H_{c\|}$, $|\psi|^2 \to 0$, so $\lambda_{\text{eff}}(H) \to \infty$ and $d/\lambda(H) \to 0$. The question then is: How thin must the film be to show a second-order transition at $H_{c\|}$? To answer this question, the full expression for the free energy must be used, allowing for the fact that screening currents will make $h < H$ inside the film.

---

[24] R. Meservy and D. H. Douglass, Jr., *Phys. Rev.* **135**, A24 (1964).

[25] J. L. Levine, *Phys. Rev.* **155**, 373 (1967); J. Millstein and M. Tinkham, *Phys. Rev.* **158**, 325 (1967).

[26] A. A. Abrikosov and L. P. Gor'kov, *Zh. Eksperim. i. Teor. Fiz.* **39**, 1781 (1960) [*Soviet Phys.—JETP* **12**, 1243 (1961)].

[27] Of course, there will be small corrections if d/ξ is not very small because then $|\psi|$ will not be exactly constant across the film thickness, as has been assumed. A variational calculation shows that this correction increases $H_{c\|}$ above (4.51) by a fractional amount of order $d^2/100\xi^2$. This never exceeds 3 percent before the film is thick enough [$d \approx 1.8\xi(T)$] to have an entirely different solution in which $|\psi|$ is maximum near a surface. This case will be treated later in Sec. 4.10.2.

So long as $d \ll \xi$, we can retain the approximation that $|\psi|$ is constant over the film. Then h is governed by the simple London equations with a field-dependent λ_{eff} given by

$$\frac{1}{\lambda_{\text{eff}}^2(H)} = \frac{16\pi e^2|\psi|^2}{m^* c^2} \tag{4.53}$$

Thus, the usual symmetric combination of exponential solutions, satisfying the boundary conditions $h = H$ at both surfaces, will hold, so that

$$h = H \frac{\cosh x/\lambda_{\text{eff}}(H)}{\cosh d/2\lambda_{\text{eff}}(H)} \tag{4.54a}$$

The corresponding vector potential is

$$A_y = H\lambda_{\text{eff}}(H) \frac{\sinh x/\lambda_{\text{eff}}(H)}{\cosh d/2\lambda_{\text{eff}}(H)} \tag{4.54b}$$

When these are used in the expression for G, and the condition for a minimum is computed, the result is a somewhat awkward transcendental equation. After a little manipulation,[28] it emerges that the maximum thickness for which a second-order transition occurs is

$$d_{\text{max,2d-order}} = \sqrt{5}\lambda \tag{4.55}$$

where this λ is again the value of $\lambda_{\text{eff}}(H)$ at zero field. Below this thickness, $H_{c\|}$ is rigorously given by (4.51), but the decrease of $|\psi|^2$ to zero as H approaches $H_{c\|}$ is not in general as simple as (4.52). For $d > \sqrt{5}\lambda$, the transition becomes first order, with a discontinuous drop in $|\psi|^2$ to zero, as in the transition of a bulk sample in a magnetic field. However, if d is not much larger than $\sqrt{5}\lambda$, $|\psi|^2$ decreases substantially before the transition occurs. These results are shown schematically in Fig. 4.6.

4.7 THE LINEARIZED GL EQUATION

In the examples treated earlier, we have always restricted our consideration to thin films or wires in which $d \ll \xi(T)$, so that $|\psi|$ did not vary appreciably. Evidently, this restriction excludes most of the interesting cases, such as bulk samples and films in fields other than those parallel to the surface. We now wish to treat such cases. However, to avoid facing immediately the full complications of solving a pair of coupled nonlinear partial differential equations, we first study the solutions of the *linearized* GL equation, obtained by dropping the term $\beta|\psi|^2\psi$ in (4.13), which corresponds to dropping $\frac{1}{2}\beta|\psi|^4$ in (4.1). These omissions

[28]For details, see, e.g., P. G. de Gennes, *Superconductivity of Metals and Alloys*, W. A. Benjamin, New York (1966), p. 189. These results were obtained first in the original paper of Ginzburg and Landau.

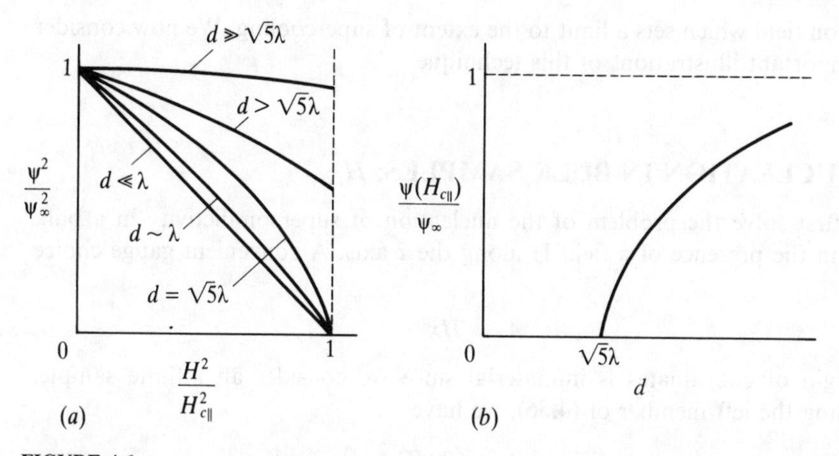

FIGURE 4.6
Dependence of ψ on the magnetic field for various film thicknesses. The size of the discontinuity of ψ at the first-order transition for thickness $d > \sqrt{5}\lambda$ is shown in (*b*). It is assumed that $d \ll \xi(T)$ throughout.

will be justified only if $|\psi|^2 \ll \psi_\infty^2 = -\alpha/\beta$ because when $\psi \approx \psi_\infty$, the term in β is of the same order of magnitude as the one in α which we retain. Thus, this linearized theory will be appropriate only when the magnetic field has reduced ψ to a value much smaller than ψ_∞. Using our definition (4.17) relating α to ξ, we can write the linearized form of the differential equation (4.13) as

$$\left(\frac{\nabla}{i} - \frac{2\pi \mathbf{A}}{\Phi_0}\right)^2 \psi = -\frac{2m^*\alpha}{\hbar^2}\psi \equiv \frac{\psi}{\xi^2(T)} \tag{4.56}$$

A further essential simplification arises from the fact that in (4.56) $\mathbf{A} = \mathbf{A}_{\text{ext}}$ since all screening effects due to supercurrents are proportional to $|\psi|^2$, and hence lead to higher-order terms which are dropped in the linearized approximation. Thus, in this approximation, the second GL equation (4.14), giving the current, is decoupled from the first, which governs ψ, leading to great mathematical simplification.

We note that (4.56) is identical with the Schrödinger equation for a free particle of mass m^* and charge $e^* = 2e$ in a magnetic field $\mathbf{h} = \text{curl}\,\mathbf{A}$, with $-\alpha = |\alpha|$ playing the role of the energy eigenvalue. This property allows various solutions and methods, familiar from quantum mechanics, to be applied directly to superconductivity. In particular, we may determine the fields at which solutions of the linearized GL equation exist, and hence at which solutions to the full nonlinear GL equations are possible with infinitesimal amplitude, by simply equating $1/\xi^2(T)$ with the field-dependent eigenvalues of the operator on the left in (4.56). The field values determined in this way correspond to the critical fields for second-order phase transitions, or if the transition is of first order, to the nucleation field which sets a limit to the extent of supercooling. We now consider some important illustrations of this technique.

4.8 NUCLEATION IN BULK SAMPLES: H_{c2}

Let us first solve the problem of the nucleation of superconductivity in a bulk sample in the presence of a field **H** along the z axis. A convenient gauge choice is

$$A_y = Hx$$

The origin of coordinates is immaterial since we consider an infinite sample. Expanding the left member of (4.56), we have

$$\left[-\nabla^2 + \frac{4\pi i}{\Phi_0} Hx \frac{\partial}{\partial y} + \left(\frac{2\pi H}{\Phi_0}\right)^2 x^2\right]\psi = \frac{1}{\xi^2}\psi \tag{4.57}$$

Since the effective potential depends only on x, it is reasonable to look for a solution of the form

$$\psi = e^{ik_y y} e^{ik_z z} f(x) \tag{4.58}$$

Substituting this into (4.57) and rearranging terms, we find

$$-f''(x) + \left(\frac{2\pi H}{\Phi_0}\right)^2 (x - x_0)^2 f = \left(\frac{1}{\xi^2} - k_z^2\right) f \tag{4.59}$$

where

$$x_0 = \frac{k_y \Phi_0}{2\pi H} \tag{4.59a}$$

Thus, inclusion of the factor $e^{ik_y y}$ only shifts the location of the minimum of the effective potential. This is unimportant for the present, but it will become important when we deal with superconductivity near surfaces of finite samples, and when we construct a space-filling solution rather than a localized one.

We can obtain the solutions to (4.59) immediately by noting that (after multiplying by $\hbar^2/2m^*$) it is the Schrödinger equation for a particle of mass m^* bound in a harmonic oscillator potential with force constant $(2\pi\hbar H/\Phi_0)^2/m^*$. This problem is formally the same as that of finding the quantized states of a normal charged particle in a magnetic field, which leads to the so-called *Landau levels*, separated by the cyclotron energy $\hbar\omega_c$. The resulting harmonic oscillator eigenvalues are

$$\epsilon_n = \left(n + \frac{1}{2}\right)\hbar\omega_c = \left(n + \frac{1}{2}\right)\hbar\left(\frac{2eH}{m^*c}\right)$$

In view of (4.59), these are to be equated to $(\hbar^2/2m^*)(\xi^{-2} - k_z^2)$. Thus,

$$H = \frac{\Phi_0}{2\pi(2n+1)}\left(\frac{1}{\xi^2} - k_z^2\right) \tag{4.60}$$

Evidently, this has its highest value if $k_z = 0$ and $n = 0$. The corresponding value, defined as H_{c2}, is

$$H_{c2} = \frac{\Phi_0}{2\pi\xi^2(T)}$$

This is the highest field at which superconductivity can nucleate in the interior of a large sample in a decreasing external field. As may be verified by substitution in (4.59), the corresponding eigenfunction is

$$f(x) = \exp\left[-\frac{(x - x_0)^2}{2\xi^2}\right] \tag{4.61}$$

The relation of H_{c2} to the thermodynamic critical field H_c is clarified if we reexpress H_{c2} in terms of H_c using (4.20) and (4.27). In this way, we arrive at the three equivalent expressions for H_{c2}

$$H_{c2} = \frac{\Phi_0}{2\pi\xi^2} = \frac{4\pi\lambda^2 H_c^2}{\Phi_0} = \sqrt{2}\kappa H_c \tag{4.62}$$

The third form makes it clear that the value $\kappa = 1/\sqrt{2}$ does indeed separate the materials for which $H_{c2} > H_c$ (type II superconductors) from those for which $H_{c2} < H_c$ (type I). Because of the significance of H_{c2} as a nucleation field, these inequalities imply that in a decreasing field, type II superconductors become superconducting in a second-order phase transition (with $|\psi|^2$ starting up continuously from zero) at $H_{c2} > H_c$. On the other hand, type I superconductors "supercool," remaining normal even below H_c, ideally until $H_{c2} < H_c$ is reached. At this point, nucleation occurs, followed by a discontinuous and irreversible jump of $|\psi|^2$ to ψ_∞^2. These features are illustrated in Fig. 4.7. In practice, nucleation at sample defects usually limits the amount of supercooling actually observed in type I superconductors to less than the theoretical limit set by H_{c2}. Nonetheless, rather ideal and tiny samples may remain normal all the way to the theoretical limit.[29] For example, Feder and McLachlan[30] found $\kappa = 0.062$ for indium by observing supercooling down to $H_{c2} \approx 0.09 H_c$.

4.9 NUCLEATION AT SURFACES: H_{c3}

Since real superconductors are finite in size, behavior near the surfaces must be considered. In fact, Saint-James and de Gennes[31] showed that superconductivity can nucleate at a metal-insulator interface in a parallel field H_{c3} higher by a factor

---

[29] To avoid nucleation at the surface at $H_{c3} > H_{c2}$, as treated in the next section, the sample surface must be plated with a normal metal.

[30] J. Feder and D. S. McLachlan, *Phys. Rev.* **177**, 763 (1969).

[31] D. Saint-James and P. G. de Gennes, *Phys. Lett.* 7, 306 (1963).

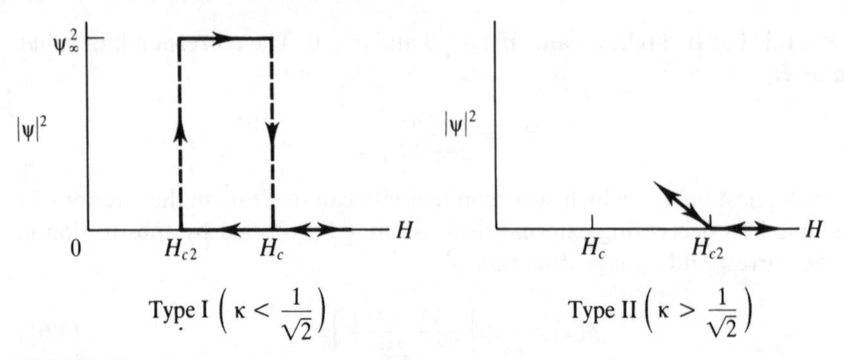

FIGURE 4.7
Contrast of behavior of order parameter at H_{c2} in type I and type II superconductors. Note hysteretic behavior with type I and reversible behavior with type II.

of 1.695 than H_{c2}. For field values between H_{c2} and H_{c3}, there is a superconducting surface sheath of thickness $\sim\xi(T)$, while $\psi \to 0$ in the interior. Let us see how this comes about.

At an insulating surface, the boundary condition on ψ is

$$\left(\frac{\nabla}{i} - \frac{2\pi \mathbf{A}}{\Phi_0}\right)\psi\bigg|_n = 0 \tag{4.63}$$

as stated in connection with (4.15*a*). For $\mathbf{H}\|\hat{\mathbf{z}}$ normal to the surface, this is satisfied by (4.58) for $k_z = 0$ (the case relevant for H_{c2}) since with our gauge choice $\mathbf{A}$ is along $\hat{\mathbf{y}}$, which is in the plane of the surface. Thus, H_{c2} as found above would correctly give the nucleation field near a surface *normal* to the field.

Now consider a surface *parallel* to the field, e.g., the *yz* plane. In our chosen gauge, $A_x = A_n = 0$, so the boundary condition becomes simply

$$\frac{\partial\psi}{\partial x}\bigg|_{\text{surface}} = \frac{df}{dx}\bigg|_{\text{surface}} = 0 \tag{4.64}$$

Our eigenfunction (4.61) satisfies this if $x_0 = 0$ or ∞, in both cases with the eigenvalue corresponding to H_{c2}. An eigenfunction with lower eigenvalue, still satisfying (4.64), can be constructed if k_y is chosen so as to put x_0, the minimum of the effective potential, inside the surface by a distance of the order of ξ. This can be seen to be the case by a qualitative argument: Imagine the potential well about x_0 to be extended by a mirror image outside the surface, as shown in Fig. 4.8*b*, thus forming a potential symmetric about the surface. Since the lowest eigenfunction of a symmetric potential is itself symmetric, it has $df/dx = 0$ at the surface, satisfying the boundary condition. [Of course, the half of $f(x)$ outside the sample surface has no physical significance and is discarded.] It is clear by inspection that this new *surface* eigenfunction must have a lower eigenvalue than the *interior* ones because it arises from a potential curve that is lower and broader than the simple

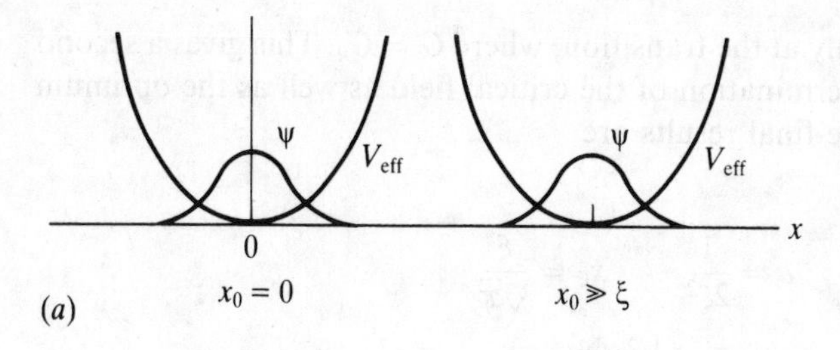

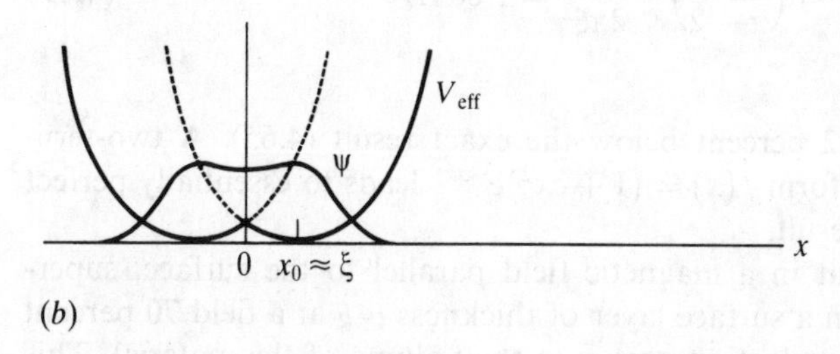

FIGURE 4.8
(*a*) Surface and interior nucleation at H_{c2}. (*b*) Surface nucleation at H_{c3}.

parabola about x_0. The exact solution shows that this eigenvalue is lower by a factor of 0.59, leading to

$$H_{c3} = \frac{1}{0.59} H_{c2} = 1.695 H_{c2} = 1.695(\sqrt{2}\kappa H_c) \tag{4.65}$$

This exact result was obtained by using the tabulated Weber functions. However, a very simple variational approach (suggested by C. Kittel) gives quite a good approximation and illustrates the usefulness of variational methods in working with the GL theory. We outline the calculation here, leaving the details as an exercise. Motivated by (4.61), we take our trial function to be

$$\psi = f(x) e^{ik_y y} = e^{-ax^2} e^{ik_y y} \tag{4.66}$$

With x measured from the sample surface, this function automatically satisfies the boundary condition (4.64). The parameters a and k_y are then determined variationally so as to minimize the Gibbs free energy per unit surface area. In the linearized approximation, this can written as

$$G - G_n = \frac{\hbar^2}{2m^*} \int_0^\infty \left[-\frac{1}{\xi^2} |\psi|^2 + \left| \left(\frac{\nabla}{i} - \frac{2\pi}{\Phi_0} \mathbf{A} \right) \psi \right|^2 \right] dx \tag{4.67}$$

Substituting (4.66) into this expression and then differentiating under the integral sign yields

$$x_0 = \frac{k_y \Phi_0}{2\pi H} = (2\pi a)^{-1/2} \tag{4.68}$$

Using this value of x_0, $G - G_n$ becomes a function of a, H, and $\xi(T)$. Minimizing this with respect to a determines a as a function of H for given ξ. However, the linearized theory is valid only at the transition, where $G = G_n$. This gives a second condition, allowing the determination of the critical field as well as the optimum value of a at that field. The final results are

$$a = \frac{1}{2\xi^2} \qquad x_0 = \frac{\xi}{\sqrt{\pi}}$$

$$H_{c3} \approx \left(\frac{\pi}{\pi - 2}\right)^{1/2} \frac{\Phi_0}{2\pi\xi^2} = 1.66\, H_{c2} \tag{4.69}$$

This value of H_{c3} is only 2 percent below the exact result (4.65). A two-term variational function of the form $f(x) = (1 + cx^2)e^{-ax^2}$ leads to essentially perfect agreement with the exact result.

Our conclusion is that in a magnetic field parallel to the surface, superconductivity will nucleate in a surface layer of thickness $\sim\xi$ at a field 70 percent higher than that at which nucleation occurs in the volume of the material. This means that a sample may be able to carry a surface supercurrent over a wide range of fields in which there is no volume superconductivity as measured, e.g., by magnetization. As is readily imagined, the theoretical discovery of surface superconductivity by Saint-James and de Gennes provided a rational explanation for great amounts of experimental data on the persistence of superconductivity at high fields which had previously been dismissed as due to inhomogeneous samples.

An interesting consequence of these results is that H_{c3}, not H_{c2}, should limit the range of supercooling of a type I superconductor since the surface sheath of superconductivity will serve to initiate the transformation of the interior. In the experiments of Feder and McLachlan, cited earlier, both H_{c3} and H_{c2} were measured as supercooling fields by the expedient of plating some of the samples with a normal metal. The plated samples supercooled to H_{c2} because the normal plating changes the boundary condition on ψ, so that (4.64) is replaced by the less favorable condition $d\psi/dx = -\psi/b$. In more physical terms, surface superconductivity is inhibited by the normal metal overlay because any pairs formed at the surface tend to diffuse into the normal metal and to be destroyed. In other words, the contact with normal metal serves as a pair-breaking mechanism which locally augments the effect of the magnetic field, thus suppressing the surface superconductivity.

For materials with κ between $1/1.695\sqrt{2}$ and $1/\sqrt{2}$, i.e., in the range $0.42 < \kappa < 0.707$, we have the inequality $H_{c2} < H_c < H_{c3}$. Thus, the type I superconductors with κ in this range should not show any supercooling at H_c, despite the fact that the volume of the sample makes a first-order transition there. Such superconductors are sometimes referred to as *type $1\frac{1}{2}$ superconductors*, a somewhat questionable classification since its validity depends on the nature of the boundary condition on ψ at the sample surface, not on the intrinsic properties of the bulk of the material.

4.10 NUCLEATION IN FILMS AND FOILS

In our discussion of the surface nucleation field H_{c3}, we tacitly assumed that the medium was semi-infinite, allowing us to ignore all surfaces except the one of direct concern. We now consider the nucleation of superconductivity in a film, where both surfaces must be considered. First we discuss the angular dependence of the critical field for a film thin enough $[d \ll \xi(T)]$ so that we may treat $|\psi|$ as constant through the thickness of the film. We then consider films of intermediate thickness $d \approx \xi$, not only for their intrinsic interest, but also because they provide a simple introduction to the vortex state of bulk samples.

4.10.1 Angular Dependence of the Critical Field of Thin Films

We noted earlier that H_{c2} would be the critical field for nucleation of superconductivity near a surface normal to the magnetic field since the corresponding eigenfunction has $\partial\psi/\partial z = 0$ everywhere, and hence automatically satisfies the boundary condition of zero normal derivative. For the same reason, H_{c2} will also give the value of $H_{c\perp}$, the perpendicular critical field of a thin film. It is only necessary to insert in (4.62) the appropriate effective value of κ, which will depend on the thickness d if the film is very thin. In fact, it is often more convenient to think in terms of the thickness dependence of λ_{eff}, as was analyzed in Sec. 3.11.4, and to use the intermediate form in (4.62).

The other limiting case, the parallel critical field, was worked out for thin films in Sec. 4.6. Since these two limiting values are very different in the case of thin films, $H_{c\parallel}$ being much greater than $H_{c\perp}$, it is of interest to work out the angular dependence which interpolates between these limits. In first treating this problem, Tinkham[32] used a simple physical argument based on fluxoid quantization to obtain the answer that $H_c(\theta)$ is determined implicitly by the relation

$$\left|\frac{H_c(\theta)\sin\theta}{H_{c\perp}}\right| + \left(\frac{H_c(\theta)\cos\theta}{H_{c\parallel}}\right)^2 = 1 \tag{4.70}$$

Note that this implies a cusp in $H_c(\theta)$ at $\theta = 0$, i.e., field parallel to film, since

$$\left|\frac{dH_c}{d\theta}\right|_{\theta=0} = \frac{H_{c\parallel}^2}{2H_{c\perp}} > 0 \tag{4.70a}$$

These results were subsequently[33] confirmed as valid thin-film limits by using the linearized GL equation. We now outline this calculation.

[32] M. Tinkham, *Phys. Rev.* **129**, 2413 (1963).

[33] M. Tinkham, *Conf. on the Phys. of Type-II Superconductivity*, Cleveland, OH (1964, unpublished); F. E. Harper and M. Tinkham, *Phys. Rev.* **172**, 441 (1968).

We choose a coordinate system in which x is measured normal to the film from its midplane. The magnetic field is chosen to lie in the xz plane at an angle θ from the plane of the film. For convenience, the vector potential is chosen to have only a y component, given by

$$A_y = H(x\cos\theta - z\sin\theta) \tag{4.71}$$

When this is inserted in (4.56), the resulting partial differential equation is difficult to solve with the appropriate boundary condition (4.63) because it does not separate in these coordinates. We can safely take ψ to be independent of y, however, since y does not enter the differential equation. Moreover, if $d \ll \xi$, we can take ψ to be independent of x, as we did in Sec. 4.6, which automatically satisfies the boundary condition of zero normal derivative at the surface. Thus, in the limit $d \to 0$, we seek a function of a single variable $\psi(z)$. This problem may be approached variationally, using (4.67) with suitably changed integration limits. Applying the calculus of variations to the resulting expression, we are led to the ordinary differential equation

$$-\frac{d^2\psi}{dz^2} + \left(\frac{2\pi H\sin\theta}{\Phi_0}\right)^2 z^2\psi = \left[\frac{1}{\xi^2} - \left(\frac{\pi H d\cos\theta}{\sqrt{3}\Phi_0}\right)^2\right]\psi \tag{4.72}$$

[This could have been obtained directly from the general form (4.56), with $\mathbf{A}$ given by (4.71), by taking $\partial\psi/\partial x = \partial\psi/\partial y = 0$ and making the plausible replacement of x and x^2 by the average values $\langle x\rangle = 0$ and $\langle x^2\rangle = d^2/12$.] Since (4.72) has the same structure as (4.59), we can take over the solution found there. Thus, the eigenfunction has the form

$$\psi \propto \left(\exp\left[\frac{-\pi H z^2\sin\theta}{\Phi_0}\right]\right) \tag{4.73}$$

and by equating the corresponding eigenvalue of the operator on the left-hand side of (4.72) with the coefficient on the right, we are led to (4.70). In this way, (4.70) is established as a rigorous result in the limit of a very thin film.

The quality of the approximation may be tested by considering a more general variational trial function, one allowing variation in the x direction. In this way, we may estimate that corrections to $H_{c\parallel}$ are of order $d^2/100\xi^2$, whereas corrections to $|dH_c(\theta)/d\theta|_{\theta=0}$ are of order $d^2/5\xi^2$. Thus, so long as $d/\xi \ll 1$, the simple results for the limiting values and for the interpolation for intermediate angles should be quite accurate. A somewhat more accurate form for the case when d/ξ is not too small has been given by Yamafuji et al.[34] Saint-James[35] has carried out exact computer solutions of the initial slope at $\theta = 0$. His results follow our simple analytic approximation (4.70*a*) quite well for $d < \xi$, but they

[34] K. Yamafuji, T. Kawashima, and F. Irie, *Phys. Lett.* **20**, 123 (1966).

[35] D. Saint-James, *Phys. Lett.* **16**, 218 (1965).

reveal a singular change in behavior at the critical thickness $d_c \approx 1.8\xi$, at which surface solutions become favored. We now consider nucleation in films of such intermediate thicknesses. For simplicity, we shall restrict our attention to the case of a film in a parallel magnetic field.

4.10.2 Nucleation in Films of Intermediate Thickness

We start with $d \ll \xi$ and consider what happens as d increases. First, we must relax the approximation that ψ is independent of x. As remarked earlier, one way to do this is to use a variational approach. For example, we might take an x dependence

$$1 + c\cos\frac{2\pi x}{d}$$

which satisfies the boundary condition $d\psi/dx = 0$ at $x = \pm d/2$ and choose c so as to minimize the free energy. The result of doing this is that $c > 0$; i.e., ψ tends to decrease near the surfaces of the film, where $\frac{1}{2}m^* v_s^2$ is large, as is obvious on qualitative grounds. As remarked earlier, this improved ψ leads to a slightly higher critical field, namely,

$$H_{c\parallel} = \frac{2\sqrt{6}H_c\lambda}{d}\left(1 + \frac{9d^2}{\pi^6\xi^2}\right) \tag{4.74}$$

This correction term never exceeds 3 percent because when $d > d_c \approx 1.8\xi$, this symmetrical solution is superseded by the surface solution of Saint-James and de Gennes, suitably modified by the presence of two surfaces in close proximity. Let us now see how this comes about.

In our analysis of nucleation at a single surface, we found it optimum to choose k_y (or x_0), so that the minimum of the effective potential was located a distance $\sim\xi$ behind the surface. Evidently, a qualitative change in the nature of the solutions at the two surfaces must be expected when they are close enough together to cause these two minima to come together, i.e., at some critical thickness $d_c \sim 2\xi$. For $d < d_c$, the lowest eigenvalue (highest critical field) is obtained for $k_y = x_0 = 0$, so that the minimum of the effective potential is in the midplane of the film. These are the symmetric solutions of the previous paragraph. However, for $d > d_c$, the optimum place for x_0 shifts away from the center so as to stay an optimum distance behind one of the surfaces. It is difficult to predict the details of the changeover at d_c without numerical computation. The results of such work[36] are that $x_0 = k_y = 0$ below $d_c = 1.81\xi$, and that above d_c they rise from zero with infinite initial slope.

---

[36] D. Saint-James, *Phys. Lett.* **16**, 218 (1965); H. J. Fink, *Phys. Rev.* **177**, 732 (1969); H. A. Schultens, *Z. Phys.* **232**, 430 (1970).

It is evident that when $x_0 \neq 0$, there must be two equivalent positions for the minimum, at $x = \pm|x_0|$ with $k_y = \pm|k_y|$, both having eigenfunctions of the linearized GL equation with exactly the same eigenvalue. These are shown schematically in Fig. 4.9. Either eigenfunction may be used for calculating the critical field, as was done by Saint-James and de Gennes for arbitrary values of d/ξ. However, as soon as one goes slightly below the nucleation field H_{c3} of the film, so that ψ becomes finite, then the nonlinear term of the full GL equations comes into play. This has the effect of resolving degeneracies in favor of extended solutions since the $\beta|\psi|^4$ term penalizes wavefunctions with ψ peaked up in a small volume. Thus, we must expect a linear combination of the two asymmetric solutions, with equal weights, such as

$$\begin{aligned}\psi = \psi_+ + \psi_- &= e^{ik_y y} f(x) + e^{-ik_y y} f(-x) \\ &= \cos k_y y [f(x) + f(-x)] + i \sin k_y y [f(x) - f(-x)]\end{aligned} \tag{4.75}$$

Because of the interference of the two solutions, there are nodes along the midplane ($x = 0$) whenever $\cos k_y y = 0$, i.e., at intervals

$$\Delta y = \frac{\pi}{k_y} = \frac{\Phi_0}{2x_0 H} \approx \frac{\Phi_0}{H(d - d_c)} \tag{4.76}$$

The approximate equality would hold if $|x_0|$ were given by $d/2 - d_c/2$, so that the potential minima stayed a constant distance behind the surfaces for $d > d_c$, as is approximately the case. If we work out the phase of (4.75), we find that it varies through 2π in a circuit around each of these nodes, corresponding to vortices of current as shown in Fig. 4.10. Thus, by superimposing two essentially one-dimensional solutions, we have generated a two-dimensional solution with currents circulating around nodes of $|\psi|$.

As the slab gets thicker, the two surface solutions pull farther apart, and the vortex spacing Δy decreases. According to the numerical calculations, after a certain point is reached, the vortex pattern becomes more complex. However, the overlap of the two surface waves $\psi_\pm$ also decreases. Eventually, the energetic

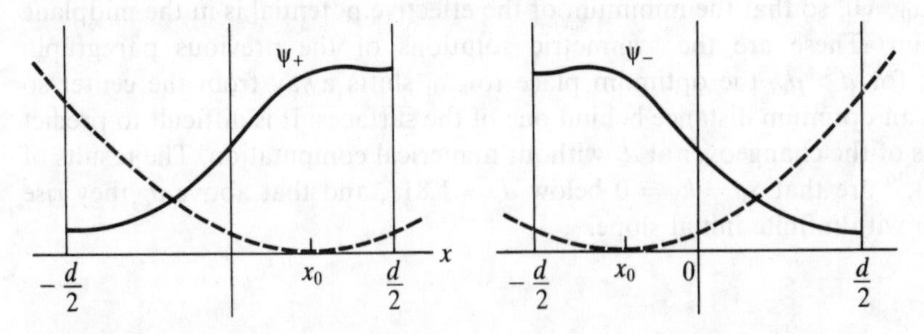

FIGURE 4.9
The two equivalent asymmetric solutions for nucleation in a film of intermediate thickness. The dashed curves centred at $\pm x_0$ represent the effective potential due to the magnetic field.

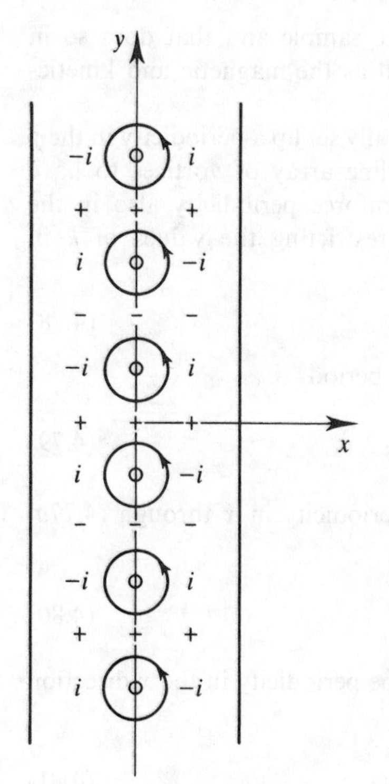

FIGURE 4.10
Vortex pattern in superconducting film of intermediate thickness set up by superposition of the two asymmetric solutions of Fig. 4.9. Notations $\pm$, $\pm i$ denote phase factor $e^{i\varphi}$ of ψ. Arrows indicate $\nabla\varphi$, to which J_s is proportional.

coupling of the two surface solutions becomes negligible compared to kT, and the two separate surface solutions become independent, as treated earlier.

4.11 THE ABRIKOSOV VORTEX STATE AT H_{c2}

Just as there are two degenerate surface solutions to the linearized GL equations at H_{c3} for a sufficiently thick film, for a bulk sample, there are an infinite number of interior solutions at H_{c2}, each of the form

$$\psi_k = e^{iky} f(x) = \exp(iky) \exp\left[-\frac{(x - x_k)^2}{2\xi^2}\right] \tag{4.77}$$

with k denoting k_y and

$$x_k = \frac{k\Phi_0}{2\pi H} \tag{4.77a}$$

as found above. Each of these describes a gaussian slice of superconductivity at the plane $x = x_k$. All ψ_k are orthogonal because of the different e^{iky} factors. Each solution is equally valid exactly at H_{c2}, and all of them give the same H_{c2}. Just as in the case of the thin film, however, as soon as H is reduced below H_{c2} by any finite amount, the minimum free-energy solution of the nonlinear equations will have to be one that fills the entire sample and that does so in such a way as to minimize the $\beta|\psi|^4$ term as well as the magnetic and kinetic-energy terms.

The two solutions for the film case automatically set up a periodicity in the y direction. Since qualitatively we expect a crystalline array of vortices to have lower energy than a random one, we want to enforce periodicity also in the interior solution. This can be done simply by restricting the values of k in (4.77) to a discrete set

$$k_n = nq \tag{4.78}$$

in which case there will be a periodicity in y with period

$$\Delta y = \frac{2\pi}{q} \tag{4.79}$$

This restriction also automatically facilitates a periodicity in x through (4.77a) since the gaussian solutions are located at

$$x_n = \frac{k_n \Phi_0}{2\pi H} = \frac{nq\Phi_0}{2\pi H} \tag{4.80}$$

Thus, if all ψ_n enter with equal weight, there will be periodicity in the x direction with period

$$\Delta x = \frac{q\Phi_0}{2\pi H} = \frac{\Phi_0}{H\Delta y} \tag{4.81}$$

From (4.81) we see at once that

$$H\Delta x \Delta y = \Phi_0 \tag{4.82}$$

so that each unit cell of the periodic array carries one quantum of flux. Although we have shown this only for the case when $H = H_{c2}$, it is, in fact, true also below H_{c2} if we replace H by B. This can be seen simply from the fluxoid quantization discussion since by symmetry, $\mathbf{J}_s = 0$ on the boundary between two identical cells of the vortex pattern.

More generally, we may consider a function

$$\psi_L = \sum_n C_n \psi_n = \sum_n C_n \exp(inqy) \exp\left[-\frac{(x - x_n)^2}{2\xi^2}\right] \tag{4.83}$$

This is a general solution to the *linearized* GL equation at H_{c2}, periodic in y by construction. It will also be periodic in x if the C_n are periodic functions of n, such that $C_{n+\nu} = C_n$, for some ν. For example, the square lattice of Abrikosov arises if all C_n are equal, so that $\nu = 1$. This is the case considered previously in the qualitative discussion leading to (4.81). On the other hand, a triangular lattice results if $\nu = 2$ and $C_1 = iC_0$.

All solutions of the form (4.83) are possible at H_{c2}. To determine which one should actually be observed, it is necessary to bring in the nonlinear terms and to do some numerical calculations. As noted by Abrikosov, the parameter determining the relative favorability of various possible solutions is

$$\beta_A \equiv \frac{\langle \psi_L^4 \rangle}{\langle \psi_L^2 \rangle^2} \tag{4.84}$$

This parameter is obviously independent of the normalization of ψ. It has the value unity if ψ is constant and becomes increasingly large for functions which are more and more peaked up and localized. For example, if ψ were approximately constant over a localized fraction f of the volume, and approximately zero everywhere else, $\beta_A \approx f^{-1} \gg 1$.

An indication of why this quantity enters can be obtained as follows. First, imagine that ψ is forced to vary in space to satisfy external conditions, but that the variation in space is so slow that gradient and current terms in the energy may be neglected. Then the free energy may be approximated by the two terms in (4.2). To allow the form and the amplitude of ψ to be adjusted separately, we write $\psi(\mathbf{r}) = c\chi(\mathbf{r})$. Inserting this in (4.2) and minimizing with respect to c^2, we find $c^2 = -(\alpha/\beta)\langle\chi^2\rangle/\langle\chi^4\rangle$. When this is put back into (4.2), we find

$$\langle f_s - f_n \rangle = -\frac{\alpha^2 \langle \chi^2 \rangle^2}{2\beta \langle \chi^4 \rangle} = -\frac{\alpha^2}{2\beta}\beta_A^{-1} \tag{4.85}$$

If χ is constant, so $\beta_A = 1$, this reduces to the usual condensation energy (4.4). If χ is not constant, $\beta_A > 1$, and the more β_A increases, the less favorable is the energy. Again taking the extreme example of a localized solution filling only a fraction f of the sample volume, for which $\beta_A \approx f^{-1}$, we see that the condensation energy is only a fraction f of what would be obtained with a space-filling solution, for which $\beta_A \approx 1$. This result is, of course, intuitively very plausible, and it shows that any form of space-filling solution of the linearized equation will be clearly favored over any localized one, when the quartic terms in the free energy are taken into consideration.

A more realistic version of (4.85) can be derived by including in the free energy the term in (4.7) arising from the gradient of $|\psi|$ but still excluding currents and vector potentials, which require more complex self-consistent solutions. If that is done, and the calculation previously outlined is repeated, an additional factor appears, so that (4.85) is replaced by

$$\langle f_s - f_n \rangle = -\left(\frac{\alpha^2}{2\beta}\right)\beta_A^{-1}\left[1 - \xi^2 \frac{\langle |\nabla\chi|^2 \rangle}{\langle \chi^2 \rangle}\right]^2 \tag{4.86}$$

This extra factor goes to zero at a second-order phase transition point, where χ satisfies the linearized GL equation. Thus, we see that β_A^{-1} still measures the effectiveness of a wavefunction with respect to minimizing the effect of the nonlinear terms, and the condensation energy still increases quadratically with temperature below the point of the second-order phase transition. That transition point is shifted, however, from T_c (where $\alpha \rightarrow 0$) to some lower temperature at which nucleation can occur in the presence of gradients. This temperature is determined by the vanishing of the final factor in (4.86), i.e., by

$$\xi^2(T) = \frac{\langle \chi^2 \rangle}{\langle |\nabla \chi|^2 \rangle} \tag{4.87}$$

Although we have not proved it here, these qualitiative features of (4.86) carry through to the cases of interest here, in which magnetic field and current terms in the energy play a critical role.

Returning now to the optimization of (4.83), we see that it is equivalent to finding the set of C_n for which β_A is the smallest since the final factor in (4.86) is the same for any linear combination of solutions with the same H_{c2}. Numerical calculations show that for the square lattice of Abrikosov, $\beta_A = 1.18$, whereas for the triangular lattice mentioned earlier (with $iC_{2n+1} = C_{2n} =$ const), $\beta_A = 1.16$.[37] Considering this small difference, it is understandable that a numerical error could have led Abrikosov originally to find that the square array was more stable. Later work by Kleiner, Roth, and Autler[38] rectified this error and showed that the triangular array had, in fact, the most favorable value of β_A of all possible periodic solutions.

It is interesting that this result agrees with that of a simple argument based on the fact that the triangular array is a "closed-packed" one, in which each vortex is surrounded by a hexagonal array of other vortices. (See Fig. 4.11.) In this array, the nearest neighbor distance is

$$a_\Delta = \left(\frac{4}{3}\right)^{1/4}\left(\frac{\Phi_0}{B}\right)^{1/2} = 1.075\left(\frac{\Phi_0}{B}\right)^{1/2} \tag{4.88a}$$

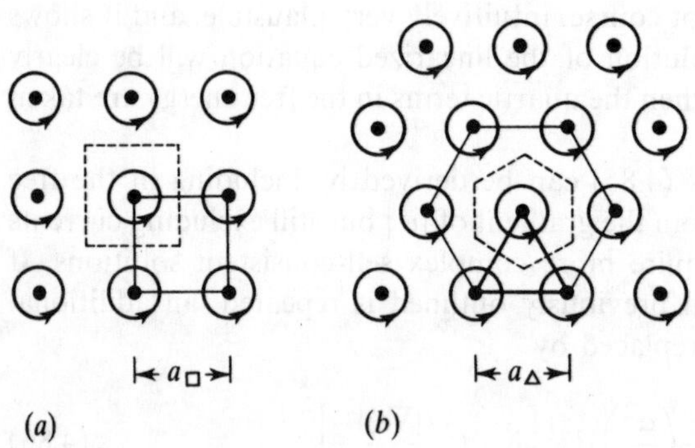

FIGURE 4.11
Schematic diagram of square and triangular vortex arrays. The dashed lines outline the basic unit cell.

---

[37]Some feeling for these numbers can be gained by noting that they correspond to cells in which 'normal cores' with $\psi = 0$ occupy some 15 percent of the cell area, with $\psi \approx$ const over the rest. In reality, ψ goes *smoothly* to zero at the center of each vortex.

[38]W. H. Kleiner, L. M. Roth, and S. H. Autler, *Phys. Rev.* **133**, A1226 (1964).

whereas for the four neighbors in a square array

$$a_{\square} = \left(\frac{\Phi_0}{B}\right)^{1/2} \tag{4.88b}$$

Thus, for a given flux density, $a_{\triangle} > a_{\square}$. Taking into account the mutual repulsion of the vortices, it is reasonable that the structure with the greatest separation of the nearest neighbors would be favored.

In general, experiments[39] confirm the triangular array, but in some materials, symmetries of the underlying crystal structure appear to dominate over the small theoretical energy difference for a structureless medium, leading to the observation of square or even rectangular arrays. Also, defects in the material may introduce sufficient inhomogeneity to destroy the regular array entirely, leading to observation of a glassy rather than a crystalline array of vortices.

Since the detailed nature of the vortex array is unimportant for most applications, we shall not carry through the details of the Abrikosov solution. Rather, in the next chapter, we shall outline the principal results and discuss their physical consequences largely with the simplification that the unit cell of the vortex array may be approximated by a circular one of the same area. This simplification is analogous to that of replacing the actual Wigner-Seitz cell by a spherical one of the same volume in the calculation of the electronic energy bands in solids. Typical numerical results such as β_A differ about as much between circular and hexagonal cells as between hexgonal and square cells, i.e., by a few percent. Of course, H_{c2} itself must be exactly the same for *any* solution of the linearized GL equation in the interior of an infinite medium.

[39]U. Essmann and H. Träuble, *Phys. Lett.* **24A**, 526 (1967); A. Seeger, *Comments Solid State Phys.* **3**, 97 (1970); U. Essmann, *Phys. Lett.* **41A**, 477 (1972).

CHAPTER 5

MAGNETIC PROPERTIES OF CLASSIC TYPE II SUPERCONDUCTORS

In the previous chapter, we found that superconductors with $\kappa > 1/\sqrt{2}$, i.e., type II superconductors, have solutions of the GL equations with $|\psi| > 0$ until fields $H_{c2} > H_c$ are reached. In particular, the Abrikosov solution (4.83) corresponded to a regular array of vortices of current surrounding nodal lines of ψ. Each unit cell of the array carried total flux equal to $\Phi_0 = hc/2e$.

It is the objective of the present chapter to investigate the behavior of type II superconductors over the entire field range from zero to H_{c2}. This will allow us to find H_{c1}, the field at which flux first penetrates (in an ideally long, thin sample of zero demagnetizing factor), and the entire magnetization curve which describes the increase of B from zero at H_{c1} to H at H_{c2}. We shall then consider the effect of the Lorentz force due to a transport current on the flux in a type II superconductor, and show how it can lead to electrical resistance associated with flux creep or flow. Finally, we examine the design conflict between thermal stability and low ac loss in practical magnets.

For simplicity, we assume throughout this chapter that the superconductor is isotropic, leaving for Chap. 9 the important modifications resulting from the extreme anisotropy of the high-temperature layered superconductors. Similarly, we restrict the discussion of fluctuation-induced resistance to the case of relatively weak fluctuations, again leaving the discussion of strong fluctuation effects, including thermally induced melting of the Abrikosov flux-line lattice, to Chap. 9.

5.1 BEHAVIOR NEAR H_{c1}: THE STRUCTURE OF AN ISOLATED VORTEX

When the first flux enters a type II superconductor, it is carried within an array of vortices sparsely distributed through the material. So long as the separation is large compared to λ, there will be negligible overlap or interaction of the vortices, so that each can be treated in isolation. Given the axial symmetry of the situation, the problem reduces to finding a self-consistent solution of the GL equations for $\psi(r)$ and $h(r)$. From this one can calculate the extra free energy ϵ_1 per unit length of the line. This determines H_{c1} in the following manner. By definition, when $H = H_{c1}$ the Gibbs free energy must have the same value whether the first vortex is in or out of the sample. Thus, at H_{c1}

$$G_s \Big|_{\text{no flux}} = G_s \Big|_{\text{first vortex}}$$

or since $G = F - (H/4\pi) \int h\, d\mathbf{r}$, $G_s = F_s$ in the absence of flux, and the condition becomes

$$F_s = F_s + \epsilon_1 L - \frac{H_{c1} \int h\, d\mathbf{r}}{4\pi}$$

$$= F_s + \epsilon_1 L - \frac{H_{c1} \Phi_0 L}{4\pi}$$

where L is the length of the vortex line in the sample. Thus,

$$H_{c1} = \frac{4\pi\epsilon_1}{\Phi_0} \tag{5.1}$$

The calculation of ψ, h, and ϵ_1 for arbitrary κ unfortunately requires a numerical solution of the GL equations. Thus, considerable attention has been given to the extreme type II limit, in which $\kappa = \lambda/\xi \gg 1$, because useful analytical results can be obtained. The simplification results because ψ can rise from zero to a limiting value (which will be ψ_∞ if we are dealing with an isolated vortex) within a *core* region of radius $\sim\xi$. Thus, over most of the vortex (of radius $\sim\lambda \gg \xi$) the superconductor will act like an ordinary London superconductor.

Before making this restrictive assumption, however, let us first see how far we can go in approaching the full solution of the nonlinear GL equations (4.13) and (4.14). It is convenient to introduce a vortex wavefunction of the form

$$\psi = \psi_\infty f(r) e^{i\theta} \tag{5.2}$$

which builds in the axial symmetry and the fact that the phase of ψ varies by 2π in making a complete circuit, corresponding to the existence of a single flux quantum associated with the vortex. This phase choice for ψ fixes the gauge choice[1] for **A** so that

$$\mathbf{A} = A(r)\hat{\boldsymbol{\theta}} \tag{5.3}$$

with

$$A(r) = \left(\frac{1}{r}\right)\int_0^r r'h(r')\,dr' \tag{5.3a}$$

Near the center of the vortex, this becomes

$$A(r) = \frac{h(0)r}{2} \tag{5.3b}$$

whereas far from the center of an isolated vortex, it becomes

$$A_\infty = \frac{\Phi_0}{2\pi r} \tag{5.3c}$$

since the total flux contained is $\oint \mathbf{A}\cdot d\mathbf{s} = 2\pi r A_\infty = \Phi_0$.

When (5.2) is substituted in the GL equation (4.13), we find, after simplifying, that f satisfies the equation

$$f - f^3 - \xi^2\left[\left(\frac{1}{r} - \frac{2\pi A}{\Phi_0}\right)^2 f - \frac{1}{r}\frac{d}{dr}\left(r\frac{df}{dr}\right)\right] = 0 \tag{5.4}$$

The current has only a θ component, and from (4.14) it is

$$J = -\frac{c}{4\pi}\frac{dh(r)}{dr} = -\frac{c}{4\pi}\frac{d}{dr}\left[\frac{1}{r}\frac{d}{dr}(rA)\right] = \frac{e^*\hbar}{m^*}\psi_\infty^2 f^2\left(\frac{1}{r} - \frac{2\pi A}{\Phi_0}\right) \tag{5.5}$$

The problem is now to find simultaneous solutions of these two nonlinear differential equations for $f(r)$ and $A(r)$. Since this requires numerical methods, in general, we shall examine certain limiting cases in which progress can be made analytically.

First, let us look right at the center of the vortex, as $r \to 0$. Using (5.3*b*), (5.4) becomes

$$f - f^3 - \xi^2\left[\left(\frac{1}{r} - \frac{\pi h(0) r}{\Phi_0}\right)^2 f - \frac{1}{r}\frac{d}{dr}\left(r\frac{df}{dr}\right)\right] = 0 \tag{5.6}$$

[1]Note that this gauge choice is quite different from the London gauge, in which ψ is real, so $\mathbf{J} \propto \mathbf{A}$, and **A** vanishes exponentially with distance from the vortex. In the London gauge, our $A(r)$ is replaced by

$$A'(r) = A(r) - A_\infty = A(r) - \frac{\Phi_0}{2\pi r}$$

Since curl $(\hat{\theta}/r) = 0$ (except along the line $r = 0$), A and A' give the same $h(r)$, $J(r)$, and $|\psi(r)|$. We shall use $A(r)$ because it is nonsingular, but equivalent results could be obtained with either choice.

Let us assume that for $r \approx 0$ the solution starts as

$$f = cr^n \qquad n \geq 0$$

Then (5.6) becomes

$$cr^n - c^3 r^{3n} - \xi^2 \left[\left(\frac{1}{r} - \frac{\pi h(0) r}{\Phi_0} \right)^2 cr^n - n^2 cr^{n-2} \right] = 0$$

As $r \to 0$, the leading term is proportional to $r^{n-2}(1 - n^2)$. For this to vanish, $n = 1$, and f must start out proportional to r at the origin. (It can readily be shown, in general, that if the vortex contains m quanta of flux, $\psi \sim e^{im\theta}$ and $f \sim r^m$, as $r \to 0$.) From the structure of (5.6) we can see that only odd powers of r now enter in the expansion of f. Working out the coefficient of the next term, we find

$$f \approx cr \left\{ 1 - \frac{r^2}{8\xi^2} \left[1 + \frac{h(0)}{H_{c2}} \right] \right\} \tag{5.7}$$

which shows that the rise of $f(r)$ starts to saturate at $r \approx 2\xi$, as might be expected. To get the normalization constant c, we must go further to bring the f^3 term into play. [It plays no role in the terms $\sim f$ which give (5.7).] However, it is clear that c must be $\sim 1/2\xi$ for isolated vortices, so that the series (5.7) will join on to the distant solution, where $f \to 1$. A reasonable approximation to f over the entire range is

$$f \approx \tanh \frac{\nu r}{\xi} \tag{5.8}$$

where ν is a constant ~ 1. This dependence is sketched in Fig. 5.1.

5.1.1 The High-κ Approximation

Because f rises almost to unity in a distance $\sim \xi$, we can make a very convenient approximation when $\lambda \gg \xi$, or $\kappa \gg 1$. Namely, over all except a core region of

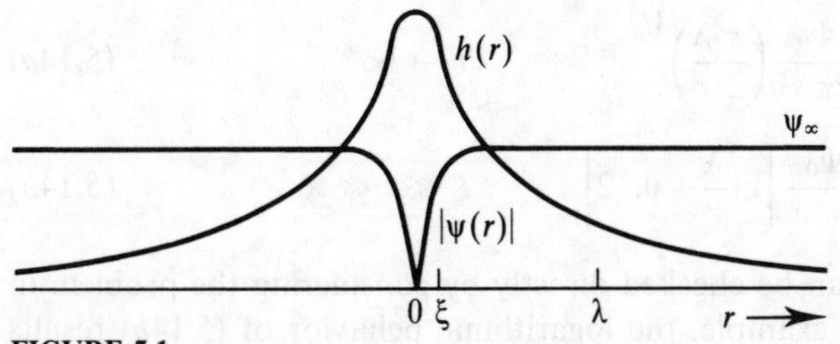

FIGURE 5.1
Structure of an isolated Abrikosov vortex in a material with $\kappa \approx 8$. The maximum value of $h(r)$ is approximately $2H_{c1}$.

radius $\sim\xi$, we can treat f as constant $\simeq 1$, in which case the London equations govern the fields and currents. Thus, outside the core

$$\frac{4\pi\lambda^2}{c}\,\mathrm{curl}\,\mathbf{J}_s + \mathbf{h} = 0 \tag{5.9}$$

If this relation held everywhere, the fluxoid for any path would be zero. We correct this by inserting a term to take into account the presence of the core, so that (5.9) becomes

$$\frac{4\pi\lambda^2}{c}\,\mathrm{curl}\,\mathbf{J}_s + \mathbf{h} = \hat{\mathbf{z}}\Phi_0\delta_2(\mathbf{r}) \tag{5.10}$$

where $\hat{\mathbf{z}}$ is a unit vector along the vortex and $\delta_2(\mathbf{r})$ is a two-dimensional δ function at the location of the core. Note that (5.10) can be obtained directly by taking the curl of (5.5). Combining (5.10) with the Maxwell equation

$$\mathrm{curl}\,\mathbf{h} = \frac{4\pi}{c}\,\mathbf{J} \tag{5.11}$$

we obtain

$$\lambda^2\,\mathrm{curl}\,\mathrm{curl}\,\mathbf{h} + \mathbf{h} = \hat{\mathbf{z}}\Phi_0\delta_2(\mathbf{r}) \tag{5.12}$$

Since div $\mathbf{h} = 0$, this can be written

$$\nabla^2\mathbf{h} - \frac{\mathbf{h}}{\lambda^2} = -\frac{\Phi_0}{\lambda^2}\,\hat{\mathbf{z}}\delta_2(\mathbf{r}) \tag{5.13}$$

This equation has the exact solution

$$h(r) = \frac{\Phi_0}{2\pi\lambda^2}\,K_0\left(\frac{r}{\lambda}\right) \tag{5.14}$$

where K_0 is a zeroth-order Hankel function of imaginary argument. Qualitatively, $K_0(r/\lambda)$ cuts off as $e^{-r/\lambda}$ at large distances and diverges logarithmically as ln (λ/r) as $r \to 0$. Of course, in reality this divergence is cut off at $r \sim \xi$, where $|\psi|^2$ starts dropping to zero. Thus, $h(r)$ is actually regular at the center of the vortex, as shown in Fig. 5.1. More precisely, the two limiting forms of (5.14) are known to be

$$h(r) \to \frac{\Phi_0}{2\pi\lambda^2}\left(\frac{\pi}{2}\frac{\lambda}{r}\right)^{1/2} e^{-r/\lambda} \qquad r \to \infty \tag{5.14a}$$

$$h(r) \approx \frac{\Phi_0}{2\pi\lambda^2}\left[\ln\frac{\lambda}{r} + 0.12\right] \qquad \xi \ll r \ll \lambda \tag{5.14b}$$

These two limiting forms can be checked directly by considering the problem in the appropriate limits. For example, the logarithmic behavior of (5.14*b*) results from $J \propto v_s = \hbar/m^*r$, which follows from fluxoid quantization so long as $r \ll \lambda$, so that the flux enclosed in a circle of radius r is much less than Φ_0.

5.1.2 Vortex-Line Energy

Now let us find the line tension, or free energy per unit length, ϵ_1. Neglecting the core, we have only the contributions from the field energy and the kinetic energy of the currents

$$\epsilon_1 = \frac{1}{8\pi}\int (h^2 + \lambda^2|\text{curl}\,\mathbf{h}|^2)\,dS \tag{5.15}$$

This can be transformed by a vector identity to

$$\begin{aligned}\epsilon_1 &= \frac{1}{8\pi}\int (\mathbf{h} + \lambda^2\,\text{curl}\,\text{curl}\,\mathbf{h}) \cdot \mathbf{h}\,dS + \frac{\lambda^2}{8\pi}\oint (\mathbf{h} \times \text{curl}\,\mathbf{h}) \cdot d\mathbf{s} \\ &= \frac{1}{8\pi}\int |\mathbf{h}|\Phi_0\delta_2(\mathbf{r})\,dS + \frac{\lambda^2}{8\pi}\oint (\mathbf{h} \times \text{curl}\,\mathbf{h}) \cdot d\mathbf{s}\end{aligned} \tag{5.15a}$$

where the line integrals are around the inner and outer perimeter of the integration area. Since the integration excludes the core, the first term contributes nothing. The second term goes to zero at infinity but gives a finite contribution in encircling the core, namely,

$$\epsilon_1 = \frac{\lambda^2}{8\pi}\left[h\frac{dh}{dr}\,2\pi r\right]_\xi$$

Using (5.14b), $dh/dr = \Phi_0/2\pi\lambda^2 r$, so this reduces to

$$\epsilon_1 = \frac{\Phi_0}{8\pi}\,h(\xi) \approx \frac{\Phi_0}{8\pi}\,h(0) \tag{5.16}$$

where $h(\xi) \approx h(0)$ because $f \to 0$, and hence $j_s \to 0$ in the core.[2] Using (5.14b) again, but dropping the 0.12 as not significant, in view of the approximation made in imposing a cutoff at ξ, we have

$$\epsilon_1 \approx \left(\frac{\Phi_0}{4\pi\lambda}\right)^2 \ln\kappa \tag{5.17}$$

Since this depends only logarithmically on the core size, the result should be quite reliable, despite the crude treatment of the core.

The magnitude can be reexpressed in more physical terms by using the relation (4.20), namely, $\Phi_0 = 2\sqrt{2}\pi\lambda\xi H_c$. Thus,

$$\epsilon_1 = \frac{H_c^2}{8\pi}\,4\pi\xi^2 \ln\kappa \tag{5.17a}$$

This shows that the line energy is of the same order of magnitude as the condensation energy lost in the core, but it is larger by a factor of order $4\ln\kappa$. Thus,

---

[2] It is interesting to note that (5.16) would also arise [but from the *first* term of (5.15a)] if the core were *not* excluded from the integration.

for $\kappa \gg 1$, errors in handling the core should be unimportant. Similarly, we can see that ϵ_1 is of the order of the total field energy $h^2/8\pi$ since $h \approx \Phi_0/\pi\lambda^2$, and this is integrated over an area $\sim \pi\lambda^2$, giving an energy $\Phi_0^2/8\pi^2\lambda^2$. This estimate differs from (5.17) only by the absence of a factor of $\frac{1}{2}\ln\kappa \approx 1$.

Now that we have evaluated the line tension ϵ_1, we can substitute back into (5.1) to get H_{c1}, the field at which flux first penetrates. This is

$$H_{c1} = \frac{4\pi}{\Phi_0}\epsilon_1 \approx \tfrac{1}{2}h(0) \approx \frac{\Phi_0}{4\pi\lambda^2}\ln\kappa = \frac{H_c}{\sqrt{2}\kappa}\ln\kappa \tag{5.18}$$

Thus, apart from the $\ln\kappa$ term, $H_c/H_{c1} = H_{c2}/H_c = \sqrt{2}\,\kappa$, so that H_c is approximately the geometric mean of H_{c1} and H_{c2}.

5.2 INTERACTION BETWEEN VORTEX LINES

If we continue to make the approximation $\kappa \gg 1$, it is easy to treat the interaction energy between two vortices, since in this approximation the medium is linear and we may use superposition. Thus, the field is given by

$$\begin{aligned}\mathbf{h}(\mathbf{r}) &= \mathbf{h}_1(\mathbf{r}) + \mathbf{h}_2(\mathbf{r})\\ &= [h(|\mathbf{r}-\mathbf{r}_1|) + h(|\mathbf{r}-\mathbf{r}_2|)]\hat{\mathbf{z}}\end{aligned} \tag{5.19}$$

where $\mathbf{r}_1$ and $\mathbf{r}_2$ specify the positions of the cores of the two vortex lines and $h(r)$ is given by (5.14). The energy may be calculated by substituting this into (5.15) and using vector transformations as we did in reaching (5.16). The result for the total increase in free energy per unit length can be written

$$\begin{aligned}\Delta F &= \frac{\Phi_0}{8\pi}[h_1(\mathbf{r}_1) + h_1(\mathbf{r}_2) + h_2(\mathbf{r}_1) + h_2(\mathbf{r}_2)]\\ &= 2\left[\frac{\Phi_0}{8\pi}h_1(\mathbf{r}_1)\right] + \frac{\Phi_0}{4\pi}h_1(\mathbf{r}_2)\end{aligned}$$

by symmetry. The first term is just the sum of the two individual line energies. The second term is the interaction energy that we were looking for

$$F_{12} = \frac{\Phi_0 h_1(\mathbf{r}_2)}{4\pi} = \frac{\Phi_0^2}{8\pi^2\lambda^2}K_0\left(\frac{r_{12}}{\lambda}\right) \tag{5.20}$$

As noted earlier, this falls off as $r_{12}^{-1/2}e^{-r_{12}/\lambda}$ at large distances and varies logarithmically at small distances. The interaction is *repulsive* for the usual case, in which the flux has the same sense in both vortices.

We may compute the *force* arising from this interaction by taking a derivative of F_{12}. For example, the force on line 2 in the x direction is

$$f_{2x} = -\frac{\partial F_{12}}{\partial x_2} = -\frac{\Phi_0}{4\pi}\frac{\partial h_1(\mathbf{r}_2)}{\partial x_2} = \frac{\Phi_0}{c}J_{1y}(\mathbf{r}_2) \tag{5.21}$$

using the Maxwell equation: curl $\mathbf{h} = 4\pi\mathbf{J}/c$. Putting this back into vector form, the force per unit length on vortex 2 is

$$\mathbf{f}_2 = \mathbf{J}_1(\mathbf{r}_2) \times \frac{\mathbf{\Phi}_0}{c} \tag{5.22}$$

where the direction of $\mathbf{\Phi}_0$ is parallel to the flux density. Making the obvious generalization to an arbitrary array, we obtain

$$\mathbf{f} = \mathbf{J}_s \times \frac{\mathbf{\Phi}_0}{c} \tag{5.23}$$

where now $\mathbf{J}_s$ represents the total supercurrent density due to all other vortices (and even including any net transport current) at the location of the core of the vortex in question.

An implication of (5.23) is that a vortex line can be in static equilibrium at any given position only if the superfluid velocity from all other sources is zero there. This can be accomplished if each vortex is surrounded by a symmetrical array, as in the square or triangular arrays discussed earlier. However, it turns out that the square array has only *unstable* equilibrium, so that small displacements tend to grow. On the other hand, the triangular array is stable,[3] as is reasonable since it has the lowest energy. Further, this result gives warning that even the triangular array will feel a force transverse to any transport current, so that the vortices will move unless *pinned* in place by inhomogeneities in the medium. Since flux motion causes energy dissipation and induces a longitudinal *resistive* voltage, this situation is crucial in determining the usefulness of type II superconductors in the construction of high-field superconducting solenoids, where strong currents and fields inevitably must coexist. We shall return to this point a little later.

5.3 MAGNETIZATION CURVES

Now that we have worked out the energy of a single vortex and the interaction energy between two vortices, we can work out the magnetization curve from the first flux penetration up to the vicinity of H_{c2}. We can distinguish three regimes between H_{c1} and H_{c2}:

1. Very near H_{c1}, $\Phi_0/B \gg \lambda^2$, and the vortices are separated by more than λ. In this case, only a few nearest neighbors are important.
2. For moderate values of B such that $\xi^2 \ll \Phi_0/B \ll \lambda^2$, many vortices are within interaction range of any given one, so that more elaborate summing procedures are required. However, it is still a good approximation to neglect details of the core.

[3]See, e.g., A. L. Fetter, P. C. Hohenberg, and P. Pincus, *Phys. Rev.* **147**, 140 (1966).

3. Near H_{c2}, $\xi^2 \approx \Phi_0/B$, so that the cores are almost overlapping. This requires more detailed treatment of the core. Our simple superposition technique is no longer accurate, but the Abrikosov solution (4.83) to the linearized GL equation at H_{c2} is a helpful approximation.

In the first two of these regimes, we can write the increase in Gibbs function per unit volume as

$$G - G_{s0} = \frac{B}{\Phi_0}\epsilon_1 + \sum_{i>j} F_{ij} - \frac{BH}{4\pi} \tag{5.24}$$

where we have used the fact that B/Φ_0 gives the number of vortices per unit area normal to the field. For fixed H, B should then take on the value which minimizes G. Since $\Sigma_{i>j} F_{ij}$ is positive and increases as B increases, we see at once that for $H < H_{c1} = 4\pi\epsilon_1/\Phi_0$, the minimum of G occurs when $B = 0$, in agreement with our earlier analysis. If H is slightly above H_{c1}, flux will enter until limited by the increase of the interaction energy between the vortices. Formally,

$$\frac{\partial G}{\partial B} = 0 = \frac{\epsilon_1}{\Phi_0} - \frac{H}{4\pi} + \frac{\partial}{\partial B}\sum_{i>j} F_{ij}$$

so B is determined by

$$\frac{\partial}{\partial B}\sum_{i>j} F_{ij} = \frac{H - H_{c1}}{4\pi} \tag{5.25}$$

5.3.1 Low Flux Density

To start, let us consider the first regime and assume a regular array of vortices such that each vortex has z nearest neighbors at a distance $a = c(\Phi_0/B)^{1/2}$. For example, from (4.88), $c_\square = 1$ for the square array ($z_\square = 4$) and $c_\triangle = 1.075$ for the triangular array ($z_\triangle = 6$). Given the exponential decrease of F_{ij}, we neglect all contributions except those of the nearest neighbors. Then, using (5.20), we obtain

$$\begin{aligned}\sum_{i>j} F_{ij} &= \left(\frac{B}{\Phi_0}\right)\frac{z}{2}\frac{\Phi_0^2}{8\pi^2\lambda^2}K_0\left(\frac{a}{\lambda}\right) \\ &\approx \frac{Bz\Phi_0}{16\pi^2\lambda^2}\left(\frac{\pi}{2}\frac{\lambda}{a}\right)^{1/2} e^{-a/\lambda}\end{aligned} \tag{5.26}$$

Because this varies exponentially with a but only linearly with z, we can see that the triangular array will surely be lower in energy if a is large enough (i.e., if B is small enough).

Given (5.26), it is straightforward to take the derivative with respect to B (including the change occurring via the change in a with B) and to insert the result in (5.25) to find $B(H)$. In view of the dominance of the exponential variation of F_{ij} with $a/\lambda = c(\Phi_0/B)^{1/2}/\lambda$, we might anticipate that $(H - H_{c1})$ would vary roughly as $e^{-a/\lambda}$, so that B would vary inversely as the square of the logarithm of $H - H_{c1}$. This turns out to be the case, the leading term at H_{c1} being

$$B = \frac{2\Phi_0}{\sqrt{3}\lambda^2}\left\{\ln\left[\frac{3\Phi_0}{4\pi\lambda^2(H - H_{c1})}\right]\right\}^{-2} \tag{5.27}$$

Note that B is continuous at H_{c1}, corresponding to a second-order phase transition, but it rises there with infinite initial slope. The physical reason for this steep rise is that once $H > H_{c1}$ so that the first vortex can enter, there is little to inhibit a large number of vortices from entering, until they get within a distance $\sim\lambda$ of each other. Since $H_{c1} \approx (\Phi_0/4\pi\lambda^2)\ln\kappa$, the mutual repulsion will not become strong until B has risen from zero to some significant fraction of H_{c1}. In fact, experimental data[4] show that in some low-κ systems, B rises discontinuously by a finite amount, exactly at H_{c1}, corresponding to a first-order transition.

The basic qualitative requirement for such unexpected behavior is that at some intermediate distance the vortices have a lower energy than at infinite separation, although they repel at still smaller separations. A possible cause is the reversal of the sign of a penetrating field predicted in the nonlocal electrodynamics of superconductors with low κ and demonstrated for a thin film by Drangeid and Sommerhalder.[5] This effect is completely missing in the standard GL theory, which uses local electrodynamics. In fact, detailed calculations by Eilenberger and Büttner[6] using the microscopic theory showed the existence of a damped oscillatory component in the spatial variation of $\Delta(\mathbf{r})$ and $A(\mathbf{r})$ if $\kappa \leq 1.7$. This behavior may provide a qualitative explanation for a locally attractive potential.

5.3.2 Intermediate Flux Densities

For flux densities in the range $\Phi_0/\lambda^2 \lesssim B \ll \Phi_0/\xi^2$, which corresponds to $H_{c1} \lesssim B \ll H_{c2}$, we can continue to use our modified London equation to treat the electrodynamics, but we must carefully include the interaction with many neighbors. This is most conveniently done by Fourier analysis of the local flux density $h(x, y)$ in a plane perpendicular to the field. Since the vortex array is periodic, a Fourier *series* is used:

$$h_z(\mathbf{r}) = \sum_{\mathbf{Q}} h_{\mathbf{Q}} e^{i\mathbf{Q}\cdot\mathbf{r}} \tag{5.28}$$

---

[4] U. Kumpf, *Phys. Stat. Sol.* **44**, 829 (1971); **52**, 653 (1972); J. Auer and H. Ullmaier, *Phys. Rev.* **B7**, 136 (1973).

[5] K. E. Drangeid and R. Sommerhalder, *Phys. Rev. Lett.* **8**, 467 (1962).

[6] G. Eilenberger and H. Büttner, *Z. Phys.* **224**, 335 (1969).

where the **Q**'s run over the two-dimensional reciprocal lattice of the array. For example, in a square array, the **Q**'s are of the form

$$\mathbf{Q}_{mn} = \frac{2\pi}{a_\square}(m\hat{\mathbf{x}} + n\hat{\mathbf{y}}) \tag{5.29}$$

In the triangular array, the primitive translations are not orthogonal, and the situation is a bit more complicated. If we take the translations in coordinate space to be

$$\mathbf{a}_1 = a_\triangle\,\hat{\mathbf{x}} \qquad \mathbf{a}_2 = \frac{a_\triangle}{2}(\hat{\mathbf{x}} + \sqrt{3}\,\hat{\mathbf{y}}) \tag{5.30}$$

then the **Q**'s are linear combinations of integral multiples of

$$\mathbf{Q}_1 = \frac{2\pi}{a_\triangle}\left(\hat{\mathbf{x}} - \frac{\hat{\mathbf{y}}}{\sqrt{3}}\right) \qquad \mathbf{Q}_2 = \frac{2\pi}{a_\triangle}\frac{2}{\sqrt{3}}\,\hat{\mathbf{y}} \tag{5.31}$$

which have the required property that $\mathbf{a}_i \cdot \mathbf{Q}_j = 2\pi\delta_{ij}$. For the moment, though, we retain the generality of an arbitrary periodic array.

We now determine the coefficients $h_\mathbf{Q}$ by requiring that $h(x, y)$ satisfy the modified London equation (5.12)

$$\mathbf{h} + \lambda^2 \operatorname{curl}\operatorname{curl}\mathbf{h} = \Phi_0\,\hat{\mathbf{z}}\sum_i \delta_2(\mathbf{r} - \mathbf{r}_i) \tag{5.32}$$

where the sum over $\mathbf{r}_i$ runs over the locations of the various vortices. Inserting the series (5.28), we see that this becomes

$$\sum_\mathbf{Q}(h_\mathbf{Q} + \lambda^2 Q^2 h_\mathbf{Q})e^{i\mathbf{Q}\cdot\mathbf{r}} = B\sum_\mathbf{Q} e^{i\mathbf{Q}\cdot\mathbf{r}} \tag{5.33}$$

In obtaining the right member, we have taken the origin of coordinates at the center of a vortex and used the fact that the area of the unit cell is Φ_0/B. Solving for $h_\mathbf{Q}$, we find

$$h_\mathbf{Q} = \frac{B}{1 + \lambda^2 Q^2} \tag{5.34}$$

so that

$$h_z(\mathbf{r}) = B\sum_\mathbf{Q} \frac{e^{i\mathbf{Q}\cdot\mathbf{r}}}{1 + \lambda^2 Q^2} \tag{5.35}$$

Given $h(\mathbf{r})$, we can calculate the increase in free energy per unit length (still neglecting core effects) by

$$F - F_{s0} = \frac{1}{8\pi}\int [h^2 + \lambda^2|\operatorname{curl}\mathbf{h}|^2]\,dS$$

Using the same vector transformations as were used to reach (5.16) or (5.20), we can reduce this to

$$F - F_{s0} = \frac{\Phi_0}{8\pi} \sum_i h(\mathbf{r}_i)$$

where the $h(\mathbf{r}_i)$ are the total fields at the cores of the vortices, i.e., self-field plus fields due to other vortices. Since all $h(\mathbf{r}_i) = h(0)$, and since there are B/Φ_0 vortices per unit area, we can write the increase in F per unit volume as

$$F - F_{s0} = \frac{Bh(0)}{8\pi} = \frac{B^2}{8\pi} \sum_{\mathbf{Q}} \frac{1}{1 + \lambda^2 Q^2} \tag{5.36}$$

using (5.35). As it stands, this sum is not convergent since the number of $\mathbf{Q}$ values between Q and $Q + \delta Q$ varies as $Q\,\delta Q$. Thus, a cutoff is required at high Q. It is reasonable to take this as $Q_{\max} \approx 1/\xi$ since Fourier components higher than $1/\xi$ come dominantly from the spurious logarithmic singularity of $h(r)$ as $r \to 0$, which results from use of (5.32) in the core region where $|\psi|^2$ is going to zero.

Without actually evaluating the sum in (5.36), which depends on the lattice structure of the vortices, let us use it formally to define the Gibbs free energy $G = F - BH/4\pi$. In this case, the equilibrium value of B is found by setting $\partial G/\partial B = 0$. This leads to the general relation

$$H = \frac{1}{2}\left[h(0) + B\frac{dh(0)}{dB}\right] \tag{5.37}$$

If $B = 0$, this reduces to $\frac{1}{2}h(0)$, which should be H_{c1}; comparing with (5.18), we see that this is correct. For arbitrary values of B, both terms play a role, and they may be computed by summing the series. Since the term with $\mathbf{Q} = 0$ will dominate when the vortices are highly overlapping and the field is almost uniform, it is useful to take this term out explicitly. When this is done, (5.37) can be written as

$$H = B\left\{1 + \frac{1}{2}\sum_{\mathbf{Q}}{}' [(1 + \lambda^2 Q^2)^{-1} + (1 + \lambda^2 Q^2)^{-2}]\right\} \tag{5.38}$$

where the summation is now over $\mathbf{Q} > 0$. Once $B \gg H_{c1}$, so that the vortices are highly overlapping, the second term under the summation sign becomes negligible compared to the first, and may be dropped. In the same spirit, we may make an integral approximation to the first sum, integrating from $Q_{\min}$ to $Q_{\max}$, with weighting factor $(\Phi_0/2\pi B)|\mathbf{Q}|d|\mathbf{Q}|$. In this, $Q_{\min}^2 \approx 4\pi B/\Phi_0$, to take into account the omitted term in which $Q = 0$ and $Q_{\max} \approx 1/\xi$. Proceeding in this way, and neglecting numerical factors, we can roughly approximate (5.38) by:

$$H \approx B + H_{c1}\frac{\ln(H_{c2}/B)}{\ln \kappa} \tag{5.39}$$

which gives a reasonable description of the data well above H_{c1}.

5.3.3 Regime Near H_{c2}

Near H_{c2}, the vortices are packed so tightly that the cores fill much of the volume. Thus, it becomes essential to abandon the simple approach of the modified London theory and to go back to the full GL equations. Fortunately, the small value of ψ/ψ_∞ as one approaches H_{c2} makes it possible to use an expansion scheme starting from the Abrikosov solution for the linearized problem (4.83). It turns out to be a good approximation to retain the form of the linearized solution, simply scaling the periodicity to correspond to $B(H)$ rather than to H_{c2}, in which case $\psi(\mathbf{r})$ can be characterized by a simple amplitude parameter, e.g., $\langle\psi^2\rangle$. Although we shall not go through the calculations here, Abrikosov was able to show that the local flux density was less than the applied field by an amount proportional to the local value of $|\psi(\mathbf{r})|^2$. As a result, $M = (B - H)/4\pi$ is proportional to $\langle\psi^2(\mathbf{r})\rangle$. Since $\langle\psi^2(\mathbf{r})\rangle$ goes to zero linearly with $(H_{c2} - H)$ in the second-order phase transition at H_{c2}, this leads to M vanishing in the same way. The explicit result is

$$B = H + 4\pi M = H - \frac{H_{c2} - H}{(2\kappa^2 - 1)\beta_A} \tag{5.40}$$

where $\beta_A = \langle\psi^4\rangle/\langle\psi^2\rangle^2 \geq 1$ is the parameter introduced in (4.84) to characterize the inconstancy of ψ over space. As noted there, the parameter β_A is independent of the amplitude of ψ, depending only on the configuration of the particular solution of the linearized GL equations that we are using. The most important case is the triangular lattice, for which $\beta_A = 1.16$; for the square lattice, it is 1.18. Note that our approximation (5.39) agrees with this result in the limit $B \approx H \approx H_{c2}$ if $2\kappa^2 \gg 1$ and if we take $\beta_A \approx 1$.

In view of the thermodynamic relation

$$\left(\frac{\partial G}{\partial H}\right)_T = -\frac{B}{4\pi} \tag{5.41}$$

the lattice with the lowest value of β_A is most stable. We can see this by integrating down from the normal state at H_{c2}, where $G_s(H_{c2}) = G_n(H_{c2})$ in all cases. Then

$$\begin{aligned} G_s(H) &= G_n(H_{c2}) + \frac{1}{4\pi}\int_H^{H_{c2}} B(H)\, dH \\ &= G_n(H_{c2}) + \frac{H_{c2}^2 - H^2}{8\pi} - \frac{1}{8\pi}\frac{(H_{c2} - H)^2}{(2\kappa^2 - 1)\beta_A} \end{aligned} \tag{5.42}$$

which makes it clear that the lower the value of β_A, the lower the value of $G_s(H < H_{c2})$, as anticipated in (4.86). Thus, the triangular lattice is more stable than the square lattice near H_{c2} as well as near H_{c1}.

An important feature of (5.40) is the presence of a factor $(2\kappa^2 - 1)$ in the denominator. This factor goes to zero at $\kappa = 1/\sqrt{2}$, which is exactly the criterion for distinguishing type I and II superconductors. This is reasonable since a type I superconductor undergoes a first-order transition at H_c, in which $4\pi|M|$ rises discontinuously from 0 to H_c. The qualitative change in the shape of the magnetization curve with the value of κ is sketched in Fig. 5.2. Despite these changes in shape, however, the area under the curve is in *all* cases given by the condensation energy $H_c^2/8\pi$. This may be shown by integrating (5.41), noting that $F = G$ when $H = 0$ and that $G_s = G_n$ when $H > H_{c2}$. Thus, whatever the value of κ, we can write

$$-\int M\,dH = F_n(0) - F_s(0) = \frac{H_c^2}{8\pi} \tag{5.43}$$

as noted in (2.6).

Given the various relations derived or stated earlier, κ values can be determined in several ways from the experimental magnetization curves. Very near T_c, where the simple GL equations can be rigorously justified by microscopic theory, we expect all these determinations to be consistent, but at lower temperatures there may be small discrepancies. Maki[7] first treated these deviations theoretically from a microscopic point of view; he introduced the notations κ_1, κ_2, and κ_3 for the values determined by fitting data to the three relations

$$H_{c2} = \sqrt{2}\,\kappa_1 H_c \tag{5.44a}$$

$$4\pi \left.\frac{dM}{dH}\right|_{H_{c2}} = (2\kappa_2^2 - 1)^{-1}\beta_A^{-1} \tag{5.44b}$$

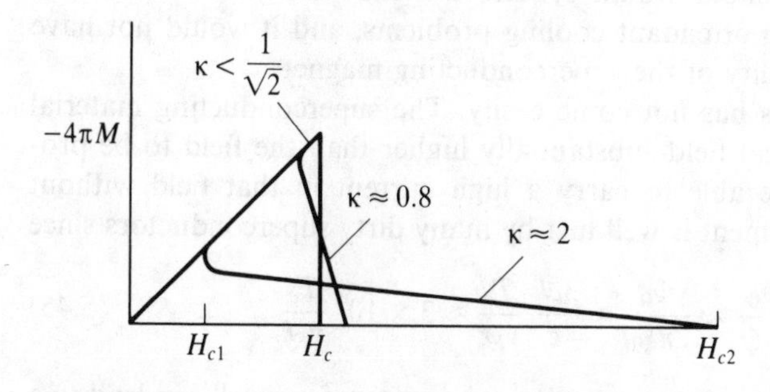

FIGURE 5.2
Comparison of magnetization curves for three superconductors with the same value of thermodynamic critical field H_c, but with different values of κ. For $\kappa < 1/\sqrt{2}$, the superconductor is of type I and exhibits a first-order transition at H_c. For $\kappa > 1/\sqrt{2}$, the superconductor is type II and shows second-order transitions at H_{c1} and H_{c2} (for clarity, marked only for the highest κ case). In all cases, the area under the curve is the condensation energy $H_c^2/8\pi$.

[7] K. Maki, *Physics* **1**, 21, 127 (1964).

$$H_{c1} = H_c \frac{\ln \kappa_3}{\sqrt{2}\,\kappa_3} \tag{5.44c}$$

More properly, (5.44*c*) should be replaced by the exact numerical relation between H_{c1}/H_c and κ, as first worked out by Harden and Arp,[8] to which (5.44*c*) is an analytic approximation, valid for large κ. This pioneering work of Maki was subsequently extended by Eilenberger,[9] whose work showed that the different temperature dependencies of the various values of κ_i should depend not only on l/ξ_0, but also on the degree of anisotropy in the impurity scattering. Roughly speaking, these differences among the κ_i reflect the different degrees of sensitivity of the various magnetic properties of a superconductor to the degree of nonlocality of the electrodynamics. Only as $T \to T_c$ does the true electrodynamics become fully local, and then all the κ_i approach a common limiting value, usually denoted simply κ, without subscript. Since the various κ_i in type II materials usually differ from κ by less than 20 percent, our results for a single, constant value retain semiquantitative validity despite all these complications. Hence, we shall not go further into these matters here.

5.4 FLUX PINNING, CREEP, AND FLOW

From the practical point of view, the most useful aspect of type II superconductivity to date has been the ability to make superconducting solenoids which can supply steady fields of over 100,000 G (gauss) without dissipation of energy because of the resistanceless persistent current. A comparable field produced by a water-cooled copper solenoid would require a steady dissipation of 2 MW (megawatts) of power, with attendant cooling problems, and it would not have the essentially infinite stability of the superconducting magnet.

Making such magnets has not come easily. The superconducting material must not only have a critical field substantially higher than the field to be produced, but it must also be able to carry a high current in that field without resistance. The first requirement is well met by many dirty superconductors since

$$H_{c2} = \frac{\Phi_0}{2\pi\xi^2} \approx \frac{\Phi_0}{2\pi\xi_0 l} \approx \frac{3ck}{e}\frac{T_c}{v_F\ell} \approx 3 \times 10^4 \frac{T_c}{v_F\ell} \tag{5.45}$$

Thus, given a high T_c and low Fermi velocity, a low mean free path can lead to a value of H_{c2} of up to ~250,000 Oe (oersteds) for such brittle materials as V_3Si and Nb_3Sn, and many materials with much more convenient mechanical properties have $H_{c2} \sim 100,000$ Oe. For the high-temperature superconductors, H_{c2} can be as high as 10^6 Oe or higher.

[8] J. L. Harden and V. Arp, *Cryogen.* **3**, 105 (1963).

[9] G. Eilenberger, *Phys. Rev.* **153**, 584 (1967).

The real problem lies in finding a material that is able to carry a usefully high current in the presence of this strong penetrating field without dissipation of energy. Any appreciable dissipation leads to heating, which degrades the performance further, leading to intolerable catastrophic flux jumps. The origin of the dissipation is the Lorentz force density

$$\mathbf{F} = \mathbf{J} \times \frac{\mathbf{B}}{c} \tag{5.46}$$

between the current in the superconductor and the flux threading through it. We derived this earlier (5.23) in the form

$$\mathbf{f} = \mathbf{J} \times \frac{\Phi_0}{c} \tag{5.46a}$$

which is the force on a single vortex. Because of this force, flux lines tend to move transverse to the current. If they do move, say, with velocity $\mathbf{v}$, they essentially *induce* an electric field of magnitude

$$\mathbf{E} = \mathbf{B} \times \frac{\mathbf{v}}{c} \tag{5.47}$$

which is parallel to $\mathbf{J}$. This acts like a resistive voltage, and power is dissipated.

Looking into this more carefully, we can write the Maxwell equation relating the current density to the curl of the field in two ways. The classic form is

$$\operatorname{curl} \mathbf{H} = 4\pi \frac{\mathbf{J}_{\text{ext}}}{c} \tag{5.48a}$$

in which $\mathbf{J}_{\text{ext}}$ represents only externally imposed currents, not the currents arising from the equilibrium response of the medium. For example, we saw that in the equilibrium intermediate state of a type I superconductor in a magnetic field, H was H_c everywhere and $\mathbf{J}_{\text{ext}}$ was zero. Alternatively, the Maxwell equation can be written microscopically, taking the medium as vacuum, so that it has the form

$$\operatorname{curl} \mathbf{h} = 4\pi \frac{\mathbf{J}}{c} \tag{5.48b}$$

where $\mathbf{J}$ is now the *total* current, including the equilibrium response of the medium. Again taking the case of the intermediate state, this equation describes the variation of $\mathbf{h}$ between $\mathbf{h}_n$ and 0 as a result of the microscopic screening currents in the London penetration depth of each superconducting lamina. Finally, $\mathbf{B}$ is the spatial average $\mathbf{h}$ over the laminar (or other) structure.

If we pass now to the case of an ideal type II superconductor in the mixed (i.e., vortex) state with no transport current imposed, $\mathbf{H}$ is again everywhere equal to the applied value since $\mathbf{J}_{\text{ext}} = 0$, while $\mathbf{B} = \mathbf{h}$ drops from $\mathbf{H}$ to $\mathbf{B}_{\text{eq}}(\mathbf{H})$ in a surface layer of depth $\sim\lambda$, where there is a microscopic surface current $\mathbf{J}$. Since this situation is, by definition, the equilibrium one, it is clear that there can be no net force on any vortices, even those which are exposed to this surface current. We conclude from this example that the current density which determines the net driving force on a vortex is not the total $\mathbf{J}$ but, rather, only the nonequilibrium part. By convention, we shall denote this by $\mathbf{J}_{ext}$, even though in some cases it may reflect only a nonequilibrium state of an isolated system. Thus, precisely speaking, the force density on the vortices which tends to make them move is

$$\alpha = \mathbf{J}_{ext} \times \frac{\mathbf{B}}{c} = (\text{curl } \mathbf{H}) \times \frac{\mathbf{B}}{4\pi} \tag{5.49}$$

This effective force density α differs from the force density (5.46) in leaving out equilibrium forces sustained by the medium, which do not tend to displace flux lines relative to the medium. Although conceptually it is important to make this careful distinction between **B** and **H** and between **J** and $\mathbf{J}_{ext}$, it should be recognized that in most applications of high-κ type II superconductors, $\mathbf{B} \approx \mathbf{H}$ because the microscopic screening currents associated with $4\pi M$ in equilibrium are much less than the useful transport currents. Thus, in the following discussion, we shall, for simplicity, sometimes ignore the difference between **B** and **H** and between **J** and $\mathbf{J}_{ext}$.

A convenient example is a hollow superconducting cylinder containing trapped flux held in by a current circulating in the wall. (See Fig. 5.3.) This model problem can be considered to be an idealization of a superconducting solenoid operating in the persistent current mode, i.e., with the leads connected by a superconducting short circuit. If the wall thickness d is small compared to the radius R, we may reduce the problem to a one-dimensional one by neglecting the curvature of the wall. In this case, the Maxwell equation given as (5.48a) can be written as

$$\frac{dH}{dx} = -4\pi \frac{J_{ext}}{c}$$

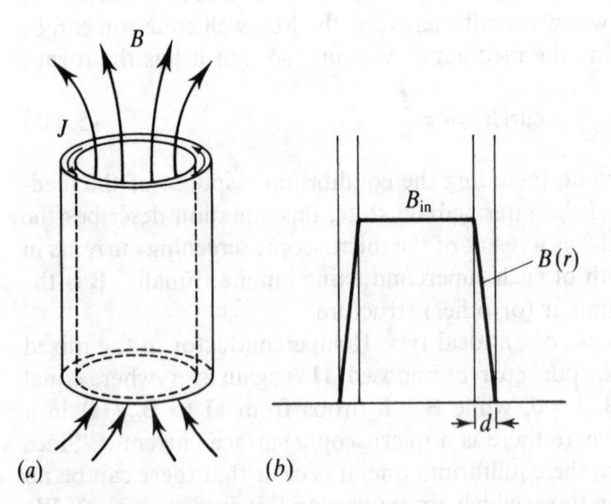

FIGURE 5.3
Flux trapped in a hollow cylinder of type II superconductor. (a) Sketch of overall geometry. (b) Local flux-density profile.

with H in the z direction and J_{ext} in the y direction. There is an outward force on each vortex of $f = (\Phi_0/4\pi)|dH/dx|$ per unit length. Summed over all the vortices, this gives a force per unit volume of

$$\alpha = \frac{B}{4\pi}\left|\frac{dH}{dx}\right| \tag{5.50}$$

To the approximation that $B \approx H$, this is equal to the gradient of the magnetic pressure $B^2/8\pi$, the difference being the part of the magnetic pressure taken up by the ideal medium rather than by the vortex pinning at material defects. Unless this force α is balanced by some other force, we must expect the vortices to move outward through the wall, each carrying one quantum of flux with it, reducing the trapped flux inside and resulting in a corresponding decrease in J. The decreasing flux implies an induced emf around the ring in the direction of the current and given by

$$\mathscr{E} = 2\pi RE = -\frac{1}{c}\frac{d\Phi}{dt} = \frac{1}{c}\,2\pi RBv \tag{5.51}$$

where E is the local electric field, and v is the outward velocity of the flux density B. This result is seen to be equivalent to (5.47). It could also have been arrived at by applying the Lorentz transformation for a change of coordinates from a frame moving with the vortices to one at rest in the laboratory.

Although derived by an induction argument for a transient situation (5.51) is also correct in steady-state situations in which an external energy source (such as a battery) maintains $\mathbf{B}$ and $\mathbf{J}$ steady on a macroscopic scale while vortices steadily move through the medium.[10] In either case, the electric field leads to energy dissipation at the rate $\mathbf{E} \cdot \mathbf{J}$. This corresponds to the decrease in magnetic-field energy in the hollow cylinder, or to the energy abstracted from the external source in the steady-state case.

Expression (5.51) for the emf $\mathscr{E}$ can be recast in a useful and significant way if we express the flux Φ in terms of the flux quantum so that $\Phi = n\Phi_0$. Then

$$\mathscr{E} = -\frac{1}{c}\frac{d\Phi}{dt} = \frac{-\Phi_0}{c}\frac{dn}{dt} = \frac{\hbar\omega}{2e} \tag{5.52}$$

where $\omega/2\pi$ is the frequency with which vortices leak out, or equivalently, the rate at which the fluxoid quantum number of the circuit is decreasing. Since the integrated phase change of $|\psi|e^{i\phi}$ around a path, $\Delta\varphi = \oint \nabla\varphi \cdot d\mathbf{s} = 2\pi n$, where n is the fluxoid quantum number, we can also write (5.52) in the form

$$2e\mathscr{E} = e^*\mathscr{E} = \hbar\frac{d\,\Delta\varphi}{dt} = \hbar\omega \tag{5.52a}$$

---

[10] An interesting analysis of the corresponding case of the driven motion of domains in the intermediate state has been given by P. R. Solomon and R. E. Harris, *Phys. Rev.* **B3**, 2969 (1971).

This is a form of the so-called *Josephson frequency relation*, which will be discussed at length in the next chapter.

From the preceding arguments we conclude that (above H_{c1}) a type II superconductor will show resistance and be unable to sustain a persistent current unless some mechanism exists which prevents the Lorentz force from moving the vortices. Such a mechanism is called a *pinning* force since it "pins" the vortices to fixed locations in the material. Pinning results from any spatial inhomogeneity of the material since local variations of ξ, λ, or H_c due to impurities, grain boundaries, voids, etc., will cause local variations of ϵ_1, the free energy per unit length of a flux line, causing some locations of the vortex to be favored over others. To be most effective, these inhomogeneities must be on a scale of the order of λ or ξ, i.e., $\sim 10^{-6}$ to 10^{-5} cm, rather than on the atomic scale where inhomogeneity causes electronic scattering which limits the mean free path ℓ. If the pinning is sufficiently strong, vortex motion can be made small enough so that the superconductor acts very much like a perfect conductor. However, for strong currents, there will always be thermally activated flux "creep," in which vortices hop from one pinning site to another, and in some cases, this will occur at a measurable rate. If the pinning is weak compared to the driving force, the vortices move in a rather steady motion, at a velocity limited by viscous drag. This regime is referred to as flux *flow* and usually gives a *flow resistivity* ρ_f, which is comparable with ρ_n, the resistivity of the material in the normal state. Hence, for practical applications, flux flow must be avoided and the creep rate held to a low level. The "melting" line in the *HT* plane, which will be discussed in Chap. 9, provides a rough guide to identify the onset of a regime of measurable resistance.

5.5 FLUX FLOW

Let us first consider the ideal case where there is *no* pinning and find what properties should be expected in the case of pure flux flow, in which vortex motion is retarded only by viscous damping. We can write down a simple phenomenology to start, simply assuming a viscous drag coefficient η such that the viscous force per unit length of a vortex line moving with velocity $\mathbf{v}_L$ is $-\eta \mathbf{v}_L$. Equating this to the driving force (5.46*a*), we find the magnitudes related by

$$J\frac{\Phi_0}{c} = \eta v_L$$

Combining this with (5.47), we find the space-averaged fields related by

$$\rho_f = \frac{E}{J} = B\frac{\Phi_0}{\eta c^2} \tag{5.53}$$

Thus, to the extent that η is independent of B, ρ_f should be proportional to B.

This analysis reduces the problem to that of finding η, which can be expressed in energetic terms by noting that the rate of energy dissipation per unit length of vortex is

$$W = -\mathbf{F} \cdot v_L = \eta v_L^2 \tag{5.54}$$

However, we still must find how the dissipation actually occurs due to a moving vortex. Two general approaches have been made to this problem: a relatively elementary model developed first by Bardeen and Stephen,[11] and somewhat later a more rigorous analysis by Schmid,[12] by Caroli and Maki,[13] and by Hu and Thompson,[14] using the time-dependent Ginzburg-Landau theory. For the present, we shall content ourselves with the former approach since it gives a clearer picture of the actual dissipation mechanism; the more sophisticated treatment is outlined in Chap. 10.

5.5.1 The Bardeen-Stephen Model

This model makes the approximation that the superconductor is local. Further, it assumes that there is a finite core of radius $\sim\xi$ which is fully normal, and that the dissipation occurs by ordinary resistive processes in this core. Since the idea of a fully normal core is central to this model, while clearly only a simplifying approximation, we must scrutinize the assumption to see how well justified it is.

The basis in microscopic theory for the concept of the normal core is the calculation of Caroli, de Gennes, and Matricon[15] of the quasi-particle excitation spectrum in a pure superconductor containing a vortex. They found that although $\psi(r)$ vanishes only at $r = 0$, rising roughly as $\psi_\infty(r/\xi)$ in the core region, there is a density of low-lying excitations localized in the vortex which is about the same as that of a normal cylinder of radius $\sim\xi$. This illustrates that $\Delta(r) \propto \psi(r)$ does not need to correspond to an actual local energy gap if ψ is varying in space. We shall return to this point in Chap. 10.

We can arrive at a similar estimate of the size of the normal core within the framework of the GL theory by finding the radius at which the vortex circulation velocity $v_s = \hbar/m^* r$ is large enough to reduce $|\psi|$ to zero according to the relation (4.34), namely,

$$\frac{|\psi|^2}{\psi_\infty^2} = 1 - \frac{m^{*2}\xi^2 v_s^2}{\hbar^2} = 1 - \left(\frac{\xi}{r}\right)^2$$

Thus, this model gives $r_{\text{core}} = \xi$. It should be emphasized, however, that this use of (4.34) is really unjustified since it was derived for the case of a velocity that is *uniform* in space. As we have noted, the actual solution of the GL equations for

---

[11] J. Bardeen and M. J. Stephen, *Phys. Rev.* **140**, A1197 (1965).

[12] A. Schmid, *Phys. Kondensierten Materie* **5**, 302 (1966).

[13] C. Caroli and K. Maki, *Phys. Rev.* **159**, 306, 316 (1967); **164**, 591 (1967); **169**, 381 (1968).

[14] C. R. Hu and R. S. Thompson, *Phys. Rev.* **B6**, 110 (1972).

[15] C. Caroli, P. G. de Gennes, and J. Matricon, *Phys. Lett.* **9**, 307 (1964). See also, P. G. de Gennes, *Superconductivity of Metals and Alloys*, W. A. Benjamin, New York (1966), p. 153.

the vortex case shows that ψ is nonzero except for r exactly equal to zero, contrary to the preceding relation.

An intermediate point of view can be obtained by referring to the shift of the excitation spectrum (by $\mathbf{v} \cdot \boldsymbol{p}_F$) used to obtain (4.41) and discussed further in Sec. 10.1. Again, this is really valid only for a uniform case, but if we apply it locally, we find that there will be gapless excitations inside a radius $\sim\xi$ if $\Delta/\Delta_\infty \approx \tanh r/\xi$ and $v_s = \hbar/m^* r$.

Finally, we note that since $H_{c2} = \Phi_0/2\pi\xi^2$, at H_{c2} normal cores of radius $\sim\xi$ would fill about half the sample volume, again a qualitatively reasonable situation since at H_{c2} there is a second-order transition to the fully normal state.

From the general agreement of all these approaches we conclude that the concept of a quasi-normal core of radius $\sim\xi$ should have considerable validity, particularly for excitations, even if it is not strictly justified. For definiteness and simplicity, we take an obviously oversimplified model in which there is a sharp discontinuity at radius $a \approx \xi$ between a fully normal core and fully superconducting material. We then treat the problem by using the London equations outside the core and Ohm's law inside.

We can find the microscopic electric field $\mathbf{e}$ outside the core by using the first London equation

$$\begin{aligned}\mathbf{e} &= \frac{\partial}{\partial t}(\Lambda \mathbf{J}_s) = \frac{\partial}{\partial t}\left(\frac{m^* \mathbf{v}_s}{e^*}\right) \\ &= -\mathbf{v}_L \cdot \nabla\left(\frac{m^* \mathbf{v}_s}{e^*}\right) = -\mathbf{v}_L \cdot \nabla\left(\frac{\hbar}{2e}\frac{\hat{\boldsymbol{\theta}}}{r}\right)\end{aligned} \tag{5.55}$$

For example, if $\mathbf{v}_L$ is along the x direction,

$$\mathbf{e} = -\left(\frac{v_{Lx}\Phi_0}{2\pi c}\right)\frac{\partial}{\partial x}\left(\frac{\hat{\boldsymbol{\theta}}}{r}\right) = \left(\frac{v_{Lx}\Phi_0}{2\pi c r^2}\right)(\cos\theta\,\hat{\boldsymbol{\theta}} - \sin\theta\,\hat{\mathbf{r}}) \tag{5.55a}$$

where θ is measured from the x direction. This field pattern, sketched in Fig. 5.4, has the form of the field of a line of electric dipoles, and averages to zero over the volume $r > a$ for which it applies. Thus, any overall average electric field will have

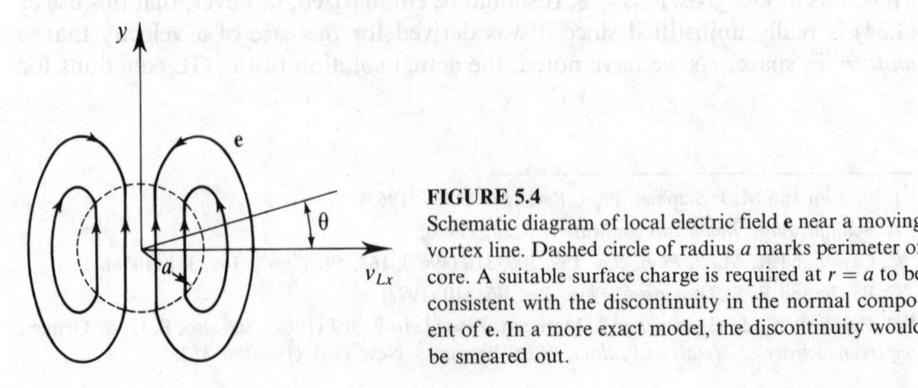

FIGURE 5.4
Schematic diagram of local electric field $\mathbf{e}$ near a moving vortex line. Dashed circle of radius a marks perimeter of core. A suitable surface charge is required at $r = a$ to be consistent with the discontinuity in the normal component of $\mathbf{e}$. In a more exact model, the discontinuity would be smeared out.

to come from the core. Requiring continuity of tangential **e** at $r = a$ with that given by (5.55a), we find a uniform field inside the core, namely,

$$\mathbf{e}_{\text{core}} = \frac{v_{Lx}\Phi_0}{2\pi a^2 c}\,\hat{\mathbf{y}} \tag{5.56}$$

Given $\mathbf{e}_{\text{core}}$, the dissipation of energy per unit length of core is

$$W_{\text{core}} = \pi a^2 \sigma_n e_{\text{core}}^2 = \frac{v_L^2 \Phi_0^2}{4\pi a^2 c^2 \rho_n} \tag{5.57}$$

According to Bardeen and Stephen, there is an additional equal amount of dissipation by normal currents in the transition region outside the core. This is readily verified near T_c since integration using (5.55a) shows that

$$\int_a^\infty \int_0^{2\pi} e^2(r)\, r\, dr\, d\theta = \frac{v_L^2 \Phi_0^2}{4\pi a^2 c^2}$$

Thus, if $\sigma_{1s} \approx \sigma_n$, as is the case near T_c, there will be exactly as much dissipation outside the core as inside. Away from T_c, the argument is less simple.

Other mechanisms of dissipation have been suggested. For example, even before the advent of the Bardeen-Stephen theory, Tinkham[16] had shown that dissipation comparable to that observed could be explained if the GL wavefunction could adjust to the time-varying field configurations of a moving vortex only in a finite relaxation time τ, such that $\tau^{-1} \approx \Delta(0)/\hbar$ at $T = 0$, decreasing as $(1 - t)$ near T_c. This qualitative conjecture was confirmed by the subsequent work of Schmid and others, cited earlier, in developing a time-dependent GL theory based on the microscopic theory. In fact, this approach now appears to offer the most rigorous treatment of the problem. Another approach has been given by Clem.[17] He has shown that dissipation results from the irreversible entropy flow from the trailing to leading edge of the vortex, corresponding to entropy increase at the leading edge where superconducting material is converted to normal, and the reverse at the trailing edge. It is not entirely clear to what extent all these various mechanisms are additive and to what extent they simply provide alternate ways of looking at the same thing.

To maintain the simplicity of our approach, we shall follow the Bardeen-Stephen results, which certainly give the main qualitative features, but it should be borne in mind that the model is rather oversimplified. Thus, we add a factor of 2 to (5.57) to allow for dissipation outside the core and then equate the result to (5.54) to obtain

$$\eta = \frac{\Phi_0^2}{2\pi a^2 c^2 \rho_n} \approx \frac{\Phi_0 H_{c2}}{\rho_n c^2} \tag{5.58}$$

[16] M. Tinkham, *Phys. Rev. Lett.* **13**, 804 (1964).

[17] J. R. Clem, *Phys. Rev. Lett.* **20**, 735 (1968).

Substituting this in (5.53), we obtain for the flow resistance ρ_f

$$\frac{\rho_f}{\rho_n} = \frac{2\pi a^2 B}{\Phi_0} = \left(\frac{a}{\xi}\right)^2 \frac{B}{H_{c2}} \approx \frac{B}{H_{c2}} \tag{5.59}$$

In writing the final approximate equalities in (5.58) and (5.59), we have used the fact that we expect the core radius a to be approximately equal to ξ. If we simply set $a = \xi$, ρ_r joins smoothly onto ρ_n at H_{c2}, which is reasonable since the transition there is of second order. Experimental data generally follow this simple result (5.59) reasonably well. By using short current pulses to minimize the heating effects of intense currents, Kunchur et al.[18] have recently demonstrated quantitative agreement with (5.59), provided the current is high enough to overwhelm pinning forces completely.

Note that this simple form would *not* result from a simple static distribution of normal cores, even if the fraction of normal metal were B/H_{c2}. In that case, the currents would simply avoid the normal cores and flow only through the superconducting matrix. The *motion* of the vortices is essential. Given this motion [with η given by (5.58)], it turns out that the (normal) current density in the core just equals the applied transport current driving the motion. Thus, the transport current flows right through the moving cores, producing dissipation. If the vortex velocity is not just right, e.g., because of the other contributions to η mentioned earlier or because of pinning forces, then these two current densities are not equal. This causes a *backflow* pattern of current to be superimposed on the uniform transport current, as indicated in Fig. 5.5. The sense of circulation is such as to give a core current density less than the transport current density J_T. Evidently, this leads to smaller dissipation in the core, i.e., to a lower electrical resistance. In the limit of a stationary core, there would be no resistance.

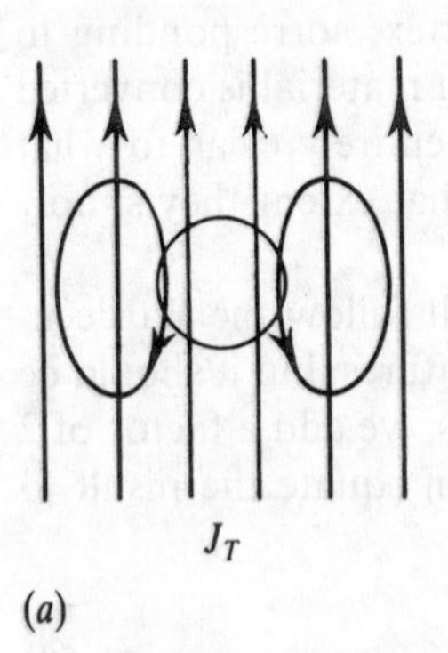

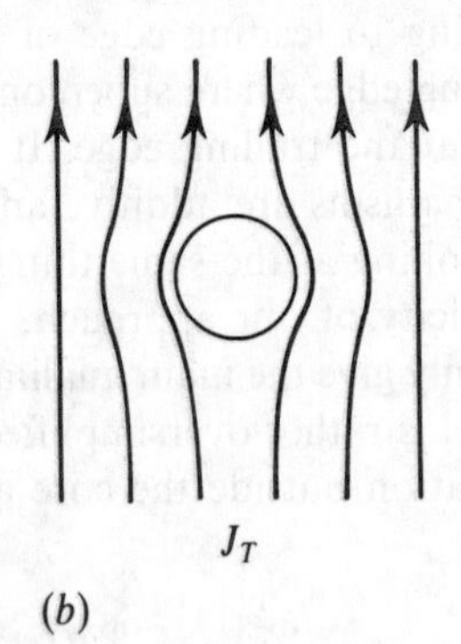

FIGURE 5.5
Backflow current pattern at pinned vortex. (*a*) Uniform transport current J_T and backflow current separately. (*b*) Superimposed current pattern, with zero current in core.

---

[18] M. N. Kunchur, D. K. Christen, and J. M. Phillips, *Phys. Rev. Lett.* **70**, 998 (1993).

5.5.2 Onset of Resistance in a Wire

Although not of much practical significance, it is of conceptual interest to apply these ideas to the case of an isolated wire of type II material carrying a current sufficient to cause a field at the surface greater than H_{c1}. Circumferential vortex rings will then form at the surface, shrink, and annihilate at the center, giving resistance. Let us see how the observed behavior should differ from that of type I superconductors, treated in Sec. 2.3, by using the London model of a static intermediate state.

First, the critical current for the appearance of resistance will be

$$I_{c1} = \tfrac{1}{2} H_{c1} ca \tag{5.60}$$

where a is now the radius of the wire. This I_{c1} will generally be quite small since $H_{c1} < H_c$. Once a vortex ring is created at the surface, how will it shrink? Equating the rate of decrease of vortex-ring energy to the energy dissipated, we have

$$\frac{d}{dt}(2\pi r \epsilon_1) = 2\pi r \eta \left(\frac{dr}{dt}\right)^2$$

Thus,

$$r\frac{dr}{dt} = \frac{\epsilon_1}{\eta} \tag{5.61}$$

and each vortex ring shrinks to annihilation in a time

$$T = \frac{\eta a^2}{2\epsilon_1} = \frac{H_{c2}}{H_{c1}} \frac{a^2}{\rho_n c^2} \tag{5.62}$$

where we have used the definitions of H_{c1}, H_{c2}, and η. For typical values, $T \approx 10^{-5}$ sec. A consequence of (5.61) is that the radial velocity $dr/dt \propto 1/r$; this implies that the time spent in any given radial increment δr is proportional to dt/dr or r. Thus, $B(r) \propto r$, just as was the case in the static intermediate-state structure of the London theory.

The preceding analysis is appropriate only very near I_{c1}, where interactions between vortex rings can be neglected since $B \approx 0$ in the wire. To derive more general relations, we equate the driving force to the viscous force:

$$\frac{J\Phi_0}{c} = \eta v$$

The inward radial velocity v of the vortex rings can be replaced by cE/B, where E is the induced longitudinal electric field and $\mathbf{B}(\mathbf{r})$ is, as usual, the local value of $\mathbf{h}(\mathbf{r})$ after averaging out the vortex structure. Also, in view of the discussion leading to (5.49), we may replace $\mathbf{J}$ by $(c/4\pi)\,\mathrm{curl}\,\mathbf{H}$. With these substitutions, we obtain

$$\frac{B}{r}\frac{\partial}{\partial r}(rH) = \frac{4\pi\eta c E}{\Phi_0} \tag{5.63}$$

where B and H are circumferential, while E is longitudinal.

Taking first the case of $I \approx I_{c1}$, we may take $H \approx H_{c1}$ throughout the wire since $dB/dH \to \infty$ at H_{c1}, causing $H \approx H_{c1}$ to be consistent with any small flux density. Then $\partial(rH)/\partial r$ will be simply H_{c1}, and (5.63) can be written

$$B(r) = \frac{4\pi\eta c E r}{\Phi_0 H_{c1}} \tag{5.64}$$

which is proportional to r, as anticipated previously. To find E, and hence the resistance, we equate the surface value of B to $B(H_s)$, where $H_s = 2I/ca$ is the surface value of H and $B(H_s)$ refers to the relationship derived in Sec. 5.3. If we also introduce the expression given in (5.58) for η, we can finally express E in terms of the corresponding normal-state value, so that

$$\frac{E}{E_n} = \frac{R}{R_n} = \frac{B(H_s)}{2H_{c2}} \qquad I \approx I_{c1} \tag{5.65}$$

In view of the form of $B(H)$ at H_{c1}, R/R_n rises with infinite slope at I_{c1}.

The other simple limit occurs at much higher currents (near I_{c2}), where it is reasonable to make the approximation $B = H$ over most of the radius. In this case, (5.63) can be integrated (after multiplying through by r^2), with the result

$$H^2 = B^2 = \frac{8\pi\eta c E r}{3\Phi_0} \tag{5.66}$$

Note that now B and H are proportional to $r^{1/2}$, whereas near I_{c1}, H was constant and B varied as r. In both limits, the product BH is proportional to r, with very nearly the same coefficient. Relating the electric field to that in the normal state at the same current, as above, we find

$$\frac{E}{E_n} = \frac{R}{R_n} = \frac{3}{4}\frac{H_s}{H_{c2}} = \frac{3}{4}\frac{I}{I_{c2}} \tag{5.67}$$

Finally, if $I > I_{c2}$, $H_s > H_{c2}$ and an outer shell of the conductor must be completely normal, while the inner core has the effective resistivity $\frac{3}{4}\rho_n$, according to (5.67). Combining these two conductances in parallel, we see that the effective average conductivity is $(1 + r_1^2/3a^2)\rho_n^{-1}$, where r_1 is the radius of the inner core. Requiring $H(r_1) = H_{c2}$, we find

$$\frac{r_1}{a} = \frac{2I}{I_{c2}}\left[1 - \left(1 - \frac{3I_{c2}^2}{4I^2}\right)^{1/2}\right]$$

$$\to \frac{3I_{c2}}{4I} \qquad I \gg I_{c2} \tag{5.68}$$

from which the current dependence of the resistance above I_{c2} may be readily computed. The limiting form at high currents is

$$\frac{R}{R_n} = 1 - \frac{3}{16}\left(\frac{I_{c2}}{I}\right)^2 \qquad I \gg I_{c2} \tag{5.69}$$

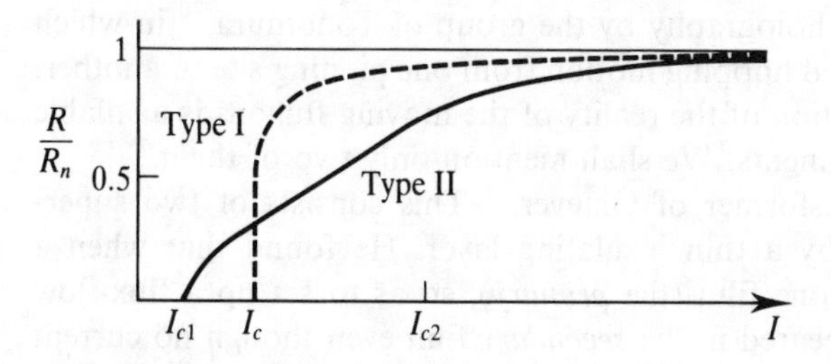

FIGURE 5.6
Onset of resistance in a wire of ideal type II superconductor with no pinning and $\kappa \approx 1.7$. For comparison, the dashed curve shows the corresponding behavior of a type I superconductor with the same H_c.

Thus, just as with the type I superconductor treated in Sec. 2.3, a partially superconducting core persists, which vanishes only as $I \to \infty$, if we neglect heating effects.

These various results are combined graphically in Fig. 5.6, where they are compared with R/R_n for a type I superconductor with the same H_c. For practical purposes, it is important only to note that R is of the same order of magnitude as R_n as soon as I appreciably exceeds I_{c1}, which is typically less than the I_c of a type I superconductor. Hence, a type II superconductor with ideal flux-flow characteristics would be of little value as a lossless current conductor. Pinning is essential for that purpose.

5.5.3 Experimental Verification of Flux Flow

Numerous experiments have been carried out to test the ideas described above. The basic ones were those of Kim and coworkers[19] which measured the resistance of strips of type II superconductors in perpendicular fields. These experimental results stimulated the development of the flux-flow theory in the first place. However, because of the deceptively simple form of the result $\rho/\rho_n \approx B/H_{c2}$, it was thought possible that the reasonable agreement with experiment might be fortuitous, and that another explanation (not involving a time-dependent flux pattern) might hold. With this motivation, other ingenious experiments were performed to test the idea of flux motion more specifically. In the intermediate state of type I superconductors, the domains are large enough so that dramatic motion pictures of their flow can be made by using magneto-optic techniques. Similar videotapes of the motion of single fluxons in type II superconductors have

[19] Y. B. Kim, C. F. Hempstead, and A. R. Strnad, *Phys. Rev. Lett.* **12**, 145 (1964); *Phys. Rev.* **139**, A1163 (1965).

been made by using electron holography by the group of Tonomura,[20] in which case the movement is a directed hopping motion from one pinning site to another. Further, less direct, confirmation of the reality of the moving fluxons is available from various electrical experiments. We shall mention only two of them.

The first is the dc transformer of Giaever.[21] This consists of two superconducting films separated by a thin insulating layer. He found that when a current was passed through one film (the *primary*), so as to set up a flux-flow voltage, an equal voltage appeared in the *secondary* film even though no current passed through it. This result is exactly what one would expect from the flux-flow picture since the moving flux spots in the primary would drag along similar flux spots in the secondary, leading to the same induced voltage. The transformer action fails if slippage occurs between the two vortex arrays because the coupling between the vortices is too weak due to too large a separation between the films, or because the flux pattern is modulated too weakly in space. Deltour and Tinkham[22] made an early study of the reduction of coupling strength with the increase of magnetic field, temperature, or bias current. Particularly careful later work on this topic was carried out by Ekin et al.[23] No plausible alternative explanation of this dc transformer action has been given in terms of a stationary resistive structure in the primary.

The other class of experiments we mention is the study of the frequency spectrum of the noise superimposed on the dc resistive voltage in superconductors. Such measurements were initiated by van Ooijen and van Gurp,[24] who reasoned that if vortices move across a conductor of width W at velocity $v_L = cE/B$, then the voltage should really be the sum of a large number of flat-topped pulses of duration $\tau = W/v_L$. If so, there should be a sort of "shot" noise spectrum, which cuts off above $\omega \approx 1/\tau$, and whose amplitude is a measure of the amount of flux moving in each independent, discrete entity. The measurements showed such noise, with a spectrum that was reasonably much as expected. However, except at the highest currents and fields, they found that the moving entities carried more than a single flux quantum Φ_0; typically as many as 1,000 such quanta appeared to move as a unit. Presumably, this result is due to pinning, which is expected to cause flux to move in *bundles* of vortices such that the total force on the bundle is sufficient to overcome a pinning barrier. In fact, the experiments give evidence that some vortices stay pinned, whereas others then must move faster than would be the case if all were moving. The conclusion is that the concept of flux motion is generally correct, but that defects in real material samples considerably complicate the idealized picture.

---

[20] J. E. Bonevich, K. Harada, T. Matsuda, H. Kasai, T. Yoshida, G. Pozzi, and A. Tonomura, *Phys. Rev. Lett.* **70**, 2952 (1993).

[21] I. Giaever, *Phys. Rev. Lett.* **15**, 825 (1965).

[22] R. Deltour and M. Tinkham, *Phys. Rev.* **174**, 478 (1968).

[23] J. W. Ekin, B. Serin, and J. R. Clem, *Phys. Rev.* **B9**, 912 (1974).

[24] D. J. van Ooijen and G. J. van Gurp, *Phys. Lett.* **17**, 230 (1965).

5.5.4 Concluding Remarks on Flux Flow

In the preceding discussion, we have tacitly assumed that the vortices would move in a purely transverse direction under the influence of the Lorentz force. This is in contrast with the behavior of vortices in a liquid, which, to a first approximation, drift along with the current. If the flux-bearing vortices move with a component of velocity along $\mathbf{J}_T$, this will lead to a transverse, or *Hall-effect*, voltage, given by the same $\mathbf{B} \times \mathbf{v}_L/c$ effect referred to in (5.47). We have ignored this possibility earlier because most experimental data indicate that the Hall angle is small. In fact, the Bardeen-Stephen model leads to a Hall effect of the same size as in the normal state for fields equal to those present in the core. Since most samples of type II materials studied are alloys or intermetallic compounds with rather short electronic mean free paths, the Hall angle is small and rather hard to measure with confidence. For example, any structural asymmetry left from the fabrication of the samples (such as a rolling direction) will tend to channel the vortices along a particular direction. This *guided motion* determines a spurious Hall angle having no intrinsic significance.

Another problem related to sample quality is caused by the pinning which is present in all real samples. This forces one to use a finite (and perhaps large) current before any flow of vortices occurs. Thus, one is not really able to measure the linear magnetoresistive behavior considered in the idealized theory. Efforts have been made to reduce pinning by careful annealing, and it appears that pinning effects can be minimized by superimposing ac currents on the dc current.[25] Still, one is never completely certain that intrinsic properties are accurately determined.

Despite these difficulties, some progress has been made, especially since the discovery of the high-temperature superconductors has spurred renewed interest. Although the simple Bardeen-Stephen model predicts that the Hall effect stems from the quasi-normal core and hence has the same sign as in the normal state, recent experiments show that the sign of the Hall effect can actually *reverse* as the magnetic field is swept down through the region of H_{c2}. Taken at face value, this sign reversal implies that the vortex starts drifting *upstream*, contrary to the universal behavior of vortices in ordinary fluids. The reality and significance of these observations, and the questions they raise about the forces acting on vortices, have been emphasized in papers by Hagen et al.[26] Attempts to account for this surprising behavior within the basic Bardeen-Stephen model include theories based on pinning[27] and quasi-particle backflow.[28] Another school of thought[29]

---

[25]J. A. Cape and I. F. Silvera, *Phys. Rev. Lett.* **20**, 326 (1968).

[26]S. J. Hagen, C. J. Lobb, R. L. Greene, M. G. Forrester, and J. H. Kang, *Phys. Rev.* **B41**, 11630 (1990); S. J. Hagen et al., *Phys. Rev.* **B47**, 1064 (1993).

[27]Z. I. Wang and C. S. Ting, *Phys. Rev. Lett.* **67**, 3618 (1991).

[28]R. A. Ferrell, *Phys. Rev. Lett.* **68**, 252 (1992).

[29]V. B. Geshkenbein and A. I. Larkin, *Phys. Rev. Lett.* **73**, 609 (1994); J. M. Harris, N. P. Ong, and Y. F. Yan, *Phys. Rev. Lett.* **73**, 610 (1994).

proposes that the effect is specific to high-temperature superconductors and can be described in terms of anisotropic scaling in the time-dependent GL equations. No further discussion will be given here since consensus on these phenomena is lacking at the time of this writing.

Another aspect of flux flow is that the moving vortex transports *entropy* as well as flux. A crude estimate of this transported entropy can be made for isolated vortices by simply taking the excess entropy of the quasi-normal core above that of a similar volume of superconducting material. Of course, more rigorous means are needed near H_{c2} when the vortices overlap strongly. The existence of entropy transport leads to thermomagnetic effects such as the Ettingshausen effect. That is, for steady state in a thermally isolated conductor, a transverse temperature gradient must develop parallel to the vortex flow to give a reverse flow of entropy equal to that carried by the vortices. These effects have been studied particularly extensively by Solomon and Otter,[30] and by Vidal.[31] A reasonably satisfactory theoretical understanding of the quantitative results seems to exist.

As a final remark, we note that rather ideal flux-flow resistance data can be obtained in the presence of pinning by making measurements at microwave frequencies. The point is that the pinning force $-k\,\delta x$ is proportional to the amplitude of the displacement of a vortex from its equilibrium position, while the viscous drag force is $\eta v = \eta\omega\,\delta x$. Thus, above a characteristic frequency $\omega_0 = k/\eta$, the pinning force becomes negligible. Gittleman and Rosenblum[32] were able to observe this changeover from a pinned regime showing no resistance to a flow regime with a well-defined ρ_f. It occurred typically near 10^7 Hz in their experiments. Thus, measurements of surface resistance of type II superconductors at microwave frequencies ($\sim 10^{10}$ Hz) generally are not complicated by pinning effects.

5.6 THE CRITICAL-STATE MODEL

We now move to the other extreme, in which pinning is strong enough to prevent any substantial vortex motion and associated electrical resistance. Although pinning forces and driving forces presumably act on individual vortices, it is appropriate to adopt a more macroscopic view because the motion of individual vortices is largely prevented by their mutual repulsion, which as we have seen tends to impose a crystalline structure on the array. Thus, we must expect flux to move in bundles when the driving force per unit volume exceeds the pinning force available in the same volume. This is in agreement with the

[30] P. R. Solomon and F. A. Otter, Jr., *Phys. Rev.* **164**, 608 (1967); P. R. Solomon, *Phys. Rev.* **179**, 475 (1969).

[31] F. Vidal, *Phys. Rev.* **B8**, 1982 (1973).

[32] J. I. Gittleman and B. Rosenblum, *Phys. Rev. Lett.* **16**, 734 (1968).

noise measurements of van Ooijen and van Gurp cited earlier. Since the force per unit volume is

$$\alpha \equiv \mathbf{J}_{\text{ext}} \times \frac{\mathbf{B}}{c} \tag{5.70}$$

the condition for zero dissipation is that α never exceed the maximum available pinning force per unit volume, α_c.

The implications of this concept are made clear by returning to consideration of the hollow superconducting cylinder, introduced in Sec. 5.4. Assume that we start with a large field B_0 inside the cylinder (perhaps established by a solenoid in the bore, which is then turned off). So long as $B_0 > H_{c1}$, flux-bearing vortices will immediately start to enter the wall, and the decrease of the flux left in the bore will induce a current in the wall. Since this current density is $J_\theta = (c/4\pi)\, dH/dr$, there will be a very strong current if $H(r)$ drops abruptly from B_0 to 0 in the wall. When put into (5.70) the combination of high field and high current will lead to $\alpha > \alpha_c$. In this case, vortices will penetrate farther into the wall, tending to reduce the gradient term. This process will continue until

$$\alpha \equiv |\alpha| = \frac{J_{\text{ext}} B}{c} = \frac{1}{4\pi} B \frac{dH}{dr} \approx \frac{d}{dr}\left(\frac{B^2}{8\pi}\right) \leq \alpha_c \tag{5.71}$$

everywhere. This situation is called the *critical state*.

Depending on the relation between B_0 and the wall thickness, the critical state will be reached in one of two ways, as illustrated in Fig. 5.7: (*a*) α may drop below α_c before the flux penetrates all the way to the outside, so that all the initial flux is still in the bore or in the wall, or (*b*) if the wall thickness is inadequate to confine the initial amount of flux, flux will leak out through the wall until that which remains can be retained with $\alpha \leq \alpha_c$ everywhere. Evidently, the maximum

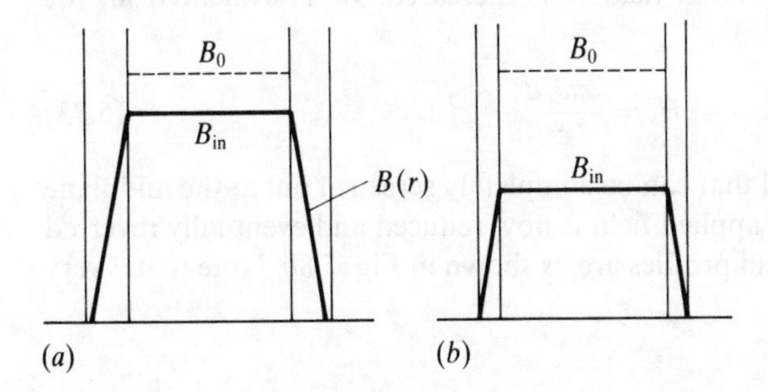

FIGURE 5.7
The critical state in a hollow superconducting cylinder. In (*a*), the wall thickness is sufficient to trap all the initial flux. In (*b*), the walls are too thin to do so. For simplicity, field profiles have been drawn using the Bean model, in which $J_c \propto dB/dr$ is constant. The same value of J_c has been taken in both (*a*) and (*b*).

B inside that can be retained (with $B = 0$ outside) is given by integrating the approximate form of (5.71), namely,

$$\frac{B_{max}^2}{8\pi} = \int_{R_{in}}^{R_{out}} \alpha_c[B(r)]\, dr \tag{5.72}$$

where we have allowed for the possibility that α_c is a function of the local flux density (as well as temperature, etc.).

If we treat α_c as constant, (5.72) shows that the maximum B that can be held in a hollow cylinder should increase as the square root of the wall thickness. This has an important implication for the design of superconducting solenoids; namely, the winding thickness must increase roughly as the square of the field in the bore. [Since (5.72) assumes complete freedom of the local current density to adjust as a function of r, the actual performance of a solenoid will generally be less favorable than (5.72) implies, even if multiple windings are used to give some control over $J(r)$.] Then since the winding thickness is typically large compared to the bore radius, the mass of superconducting material required rises roughly as the square of the wall thickness, or as B^4 (for constant length). Thus, even before one approaches H_{c2}, where α_c must be expected to go to zero, the size (and hence the cost) of a superconducting solenoid should increase very rapidly with the field capability in the bore. This is, in fact, the case in practice. For example, a series of commercial Nb_3Sn magnets rated at 110, 125, and 130 kG had design weights of 48, 110, and 130 lb, respectively, not far from the dependence on B^4 derived earlier.

It is clear that α_c cannot be constant all the way down to $B = 0$ since this would imply an infinite critical current in zero field. This difficulty is avoided in an alternative simple approximation due to Bean,[33] namely, $J_c =$ constant,[34] or $\alpha_c \propto B$. In fact, Bean suggested this model of the critical state before the more extensive work of Kim and coworkers. In the Bean model, the flux-density profiles are simply straight lines of slope $4\pi J_c/c$, which simplifies qualitative discussion. For example, Fig. 5.8*a* illustrates the profiles for flux penetrating into a thick slab of thickness d as the external field H is increased. As is evident from the figure,

$$H_s = \frac{2\pi J_c d}{c} \tag{5.73}$$

is the maximum external field that can be completely screened out at the midplane of the superconductor. If the applied field is now reduced and eventually reversed in direction, the successive field profiles are as shown in Fig. 5.8*b*. Note that a very

---

[33] C. P. Bean, *Phys. Rev. Lett.* **8**, 250 (1962).

[34] The implications within the Bean model of the *anisotropic* critical currents of the layered high-temperature superconductors have been explored by E. M. Gyorgy, R. B. van Dover, K. A. Jackson, L. F. Schneemeyer, and J. V. Waszczak, *Appl. Phys. Lett.* **55**, 283 (1989).

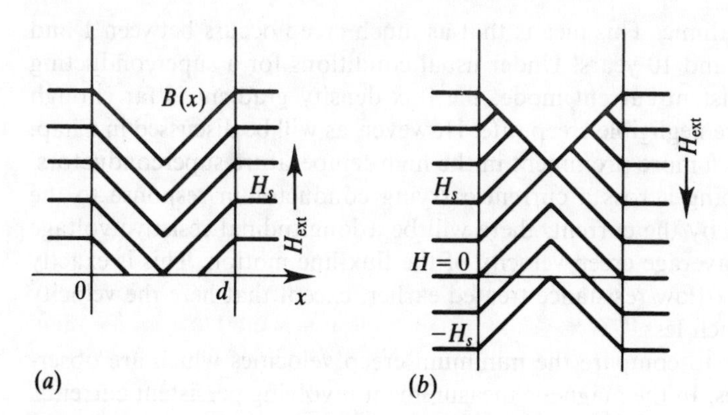

FIGURE 5.8
Internal flux-density profiles in a slab subjected to (*a*) increasing and (*b*) decreasing external field. H_s is the maximum applied field that can be screened at the midplane. Note the occurrence of canceling flux densities in (*b*) when $H_{ext} = -\frac{1}{2}H_s$.

substantial amount of flux may be left trapped in the slab after the external field is reduced to zero. One may even have flux densities which change sign in the interior of the slab. In that case, one would expect some annihilation of opposing vortices to occur, but if the pinning is strong, this would affect only a thin layer at the crossover of B from one sense to the other. As is evident from this figure, there is much hysteresis, and associated irreversibility, in the cycling of these "hard" superconductors. For example, if the external field is cycled through a maximum field $H_m < H_s$, one can see that the area inside the hysteresis loop $\oint B\,dH$ (and hence the energy Q dissipated as heat per cycle) will increase as H_m^3. On the other hand, if $H_m \gg H_s$, then $Q \propto H_m$. These hysteresis losses limit the value of type II superconductors for ac applications, as will be discussed more completely in Sec. 5.8.

5.7 THERMALLY ACTIVATED FLUX CREEP

At finite temperatures, thermal energy may allow flux lines to jump from one pinning point to another in response to the driving force of the current and flux-density gradient. The resulting flux creep is revealed in two ways: (1) It leads to slow changes in trapped magnetic fields and (2) it leads to measurable resistive voltages. Before going into details, we briefly compare these two manifestations.

If flux is trapped in a hollow cylinder, or in a superconducting solenoid in the persistent-current mode, there may be an observable decrease of this trapped field with time. This is in contrast with type I superconductors, in which no such change has ever been detected in macroscopic samples. Actually, even in classic type II superconductors, this creep is unobservably slow unless the flux-density gradient is very near the critical value. Since any creep that occurs will relieve the gradient, the creep gets slower and slower. In fact, we shall show that the time dependence is logarithmic. This means that as much creep occurs between 1 and 10 sec as between 1 and 10 years! Under usual conditions for a superconducting solenoid in the persistent-current mode, the flux-density gradient is far enough from critical to assure negligible creep rate. However, as will be discussed in Chap. 9, flux motion is *much* more prominent in the high-temperature superconductors.

If flux is creeping across a current-carrying conductor in response to the driving force caused by the current, there will be a longitudinal resistive voltage proportional to the average creep velocity of the flux-line motion. This is exactly analogous to the flux-flow resistance treated earlier, except that here the velocity of flux motion is much less.

It is of interest to compare the minimum creep velocities which are observable in the two cases. In the magnetic measurement involving persistent currents, one can easily detect motion in which the flux creeps the width of a conductor ($\sim$0.1 mm) in, say, a day; hence $v_{\min} \approx 10^{-7}$ cm/sec. In the reistance measurements with conventional (i.e. nonsuperconducting) electronics, one is typically limited to detecting voltage gradients of $\gtrsim 10^{-7}$ V/cm. With $B \approx 10^4$ G, this corresponds to a minimum detectable velocity of about 10^{-3} cm/sec. Thus, generally speaking, persistent-current measurements are the more sensitive.

5.7.1 Anderson-Kim Flux-Creep Theory

This theory[35] assumes that flux creep occurs by bundles of flux lines jumping between adjacent pinning points. Bundles are assumed to jump as a unit because the range λ of the repulsive interaction between flux lines is typically large compared to the distance between lines; this encourages cooperative motion. The jump rate is presumed to be of the usual form

$$R = \omega_0 e^{-F_0/kT} \tag{5.74}$$

where ω_0 is some characteristic frequency of flux-line vibration, unknown in detail, but assumed to lie in the range from 10^5 to 10^{11} sec^{-1}, and F_0 is the activation free energy, or barrier energy, i.e., the increase in system-free energy when the flux bundle is at the saddle point between two positions where the free energy is at a local minimum. In the absence of any flux-density gradient, jumps are as likely to occur in one direction as the other, and no net creep velocity exists.

Since various lengths, such as λ, ξ_0, ℓ, distance between pinning centres, width of pinning centers, etc., enter the theory in ways which are imperfectly known and depend on the details of the model, we shall introduce a single microscopic length $L \approx 10^{-5}$ cm for all these.[36] This will keep dimensional arguments

---

[35] P. W. Anderson, *Phys. Rev. Lett.* **9**, 309 (1962); P. W. Anderson and Y. B. Kim, *Rev. Mod. Phys.* **36**, 39 (1964).

[36] A detailed theoretical analysis of systematic experimental data on flux creep, which allows a more precise characterization of these parameters than we use here, has been given by M. R. Beasley, R. Labusch, and W. W. Webb, *Phys. Rev.* **181**, 682 (1969).

straight and allow adequate estimates of orders of magnitude, thus permitting the form of the theory to be developed with a maximum of simplicity.

We now introduce a transport current or flux-density gradient, which tilts the spatial energy dependence as indicated schematically in Fig. 5.9, making a jump easier in the direction of decreasing flux density than in the opposite direction. The shift in barrier heights is equal to the work done by the driving force in going over the barrier. Since the force density is α, the force on a flux bundle of cross section L^2 and of length L is αL^3, so the work done in moving it a distance L is $\Delta F = \alpha L^4$. This will lead to a *net* jump rate in the direction of the force α of

$$R = R_+ - R_- = \omega_0 e^{-F_0/kT}(e^{\Delta F/kT} - e^{-\Delta F/kT}) \tag{5.75}$$

This amounts to a net creep velocity of

$$v = 2v_0 e^{-F_0/kT} \sinh \frac{\alpha L^4}{kT} = 2V_0 \sinh \frac{\alpha L^4}{kT} \tag{5.76}$$

where $v_0 = \omega_0 L \approx 10^{3\pm3}$ cm/sec would be the creep velocity if there were no barrier.

To proceed further, we must make at least a rough estimate of the barrier energy F_0. It is useful to write this as

$$F_0 = p \frac{H_c^2}{8\pi} L^3 \tag{5.77}$$

with p giving the fractional modulation of the condensation energy in volume L^3 available as a pinning energy. Because any strong pinning centers (such as voids) will typically fill only a small fraction of the volume, while any extended pinning centers (such as strains) will cause only small fractional changes in superconducting properties, p will be small. For an order of magnitude estimate, we take

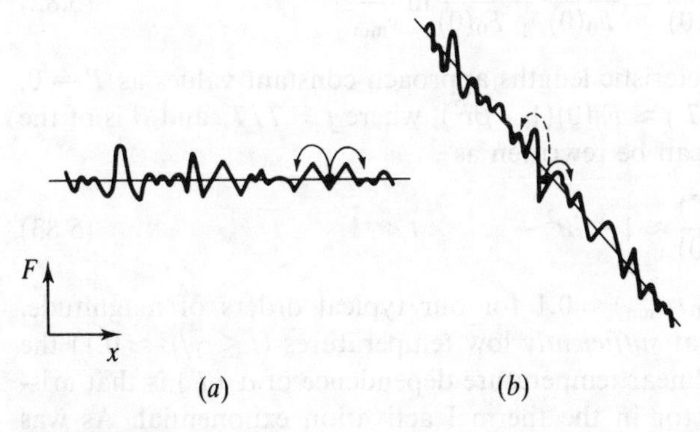

FIGURE 5.9
Schematic representation of flux bundles jumping over barriers to adjacent pinning sites. The ordinate represents the relative value of the total free energy as a function of the position of the center of the flux bundle. (*a*) Zero driving force. (*b*) Driving force due to current (or dB/dx) favoring jumps in a "downhill" direction.

$p \approx 10^{-3}$, and $H_c \approx 2{,}000$ Oe; then $F_0/k \approx 1200$ K, so F_0 is certainly very large compared to kT. Taking this estimate, we have

$$V_0 = \nu_0 e^{-F_0/kT} \approx 10^{3\pm3} e^{-1200/T} \approx 10^{3\pm3-500/T} \tag{5.78}$$

From this we see that even large uncertainties in $\nu_0 = \omega_0 L$ are completely overshadowed by small changes in the pinning energy, and we may essentially forget about all factors except the exponentials $e^{-F/kT}$. The preceding estimate shows that in the relevant temperature range of 1 to 10 K, $V_0 \approx 10^{-500}$ to 10^{-50} cm/sec. Thus, the net creep velocity $v = 2V_0 \sinh(\alpha L^4/kT)$ will be unobservably small unless the driving term $\sinh(\alpha L^4/kT)$ is huge. However, if this is the case, then $2\sinh(\alpha L^4/kT) \approx e^{\alpha L^4/kT}$, so that in the useful range we have simply

$$v = V_0(T) e^{\alpha L^4/kT} = \nu_0 e^{-F_0/kT} e^{\alpha L^4/kT} \tag{5.79}$$

Consider the situation at $T = 0$, where $v = 0$ unless $\alpha L^4 \gtrsim F_0$. This defines

$$\alpha_c(0) = \frac{F_0(0)}{L^4} \tag{5.80}$$

as the critical force-density parameter (at $T = 0$) used in our earlier discussion of the critical state. We can use this relation to eliminate L^4 from (5.79), obtaining

$$v = \nu_0 \exp\left[-\frac{F_0(T) - F_0(0)\dfrac{\alpha}{\alpha_c(0)}}{kT}\right] \tag{5.81}$$

Since some minimum velocity $v_{\min}$ is required to give detectable creep effects, we can define an effective $\alpha_c(T)$ as that value of α leading to this minimum detectable creep velocity. From the preceding expression, we see that this will be

$$\frac{\alpha_c(T)}{\alpha_c(0)} = \frac{F_0(T)}{F_0(0)} - \frac{kT}{F_0(0)} \ln\frac{\nu_0}{v_{\min}} \tag{5.82}$$

Now since H_c and the characteristic lengths approach constant values as $T \to 0$, for $T \ll T_c$ we can write $F_0(T) \approx F_0(0)(1 - \beta t^2)$, where $t = T/T_c$ and β is of the order of unity. Thus, (5.82) can be rewritten as

$$\frac{\alpha(T)}{\alpha_c(0)} \approx 1 - \beta t^2 - \gamma t \qquad t \ll 1 \tag{5.83}$$

where $\gamma = [kT_c/F_0(0)]$ ln $(\nu_0/v_{\min}) \approx 0.1$ for our typical orders of magnitude. From this form we see that at *sufficiently* low temperatures ($t \leq \gamma/\beta \approx 0.1$) the linear term dominates. This linear temperature dependence of $\alpha_c(T)$ is that arising from the explicit kT factor in the thermal activation exponential. As was pointed out by Kim and Anderson, this mechanism seems almost unique in being able to account for such a strong variation of α_c at very low temperatures, where H_c, ξ, and λ approach limiting values. Moreover, the order of magnitude of $d\alpha/dT$ obtained in this way was in reasonable agreement with the experimental results of Kim et al.[37] On the other hand, it is clear from (5.83) that for all except the lowest temperatures (typically $\lesssim 1$ K), the temperature dependence of α_c should be dominated by the change in pinning strength (reflected in βt^2), not by the change in creep rate (reflected in γt), if our numerical estimates are approximately correct. This agrees with the conclusion reached by Campbell et al.[38] from comparisons of the temperature dependences of J_c and of the reversible magnetization curves of various materials.

Next, we consider the time dependence of the creep phenomenon. Because of the mathematical intricacy of the full treatment, let us first show in a simple way that a logarithmic dependence on the time should be expected. The key idea is that the driving force is roughly proportional to the value of B that is still trapped in the hollow cylinder of our example. Thus, in view of the exponential dependence of the creep rate on the driving force, we expect

$$\frac{dB}{dt} \approx -Ce^{B/B_0} \tag{5.84}$$

where t now denotes time, which has the solution

$$B = \text{const} - B_0 \ln t \tag{5.85}$$

Now let us give a more careful treatment. We again neglect the difference between B and H, denoting the local value of both by B. For simplicity, we consider a one-dimensional problem: $\mathbf{B}$ is along $\hat{\mathbf{z}}$; ∇B_z and the flux-line velocity $\mathbf{v}$ are along $\hat{\mathbf{x}}$. The conservation of flux lines then requires that

$$\frac{\partial B}{\partial t} = -\frac{\partial(Bv)}{\partial x} \tag{5.86}$$

The velocity v is related to the force-density parameter α by (5.81), which we write as

$$v = V_0 e^{\alpha/\alpha_1} \tag{5.87}$$

where, as in (5.71),

$$\alpha = -\frac{\partial(B^2/8\pi)}{\partial x} \tag{5.88}$$

and, following (5.81),

$$\alpha_1 = \frac{kT\alpha_c(0)}{F_0(0)} \approx \frac{\alpha_c(0)}{300} \tag{5.89}$$

[37] Y. B. Kim, C. F. Hempstead, and A. R. Strnad, *Phys. Rev.* **131**, 2486 (1963).

[38] A. M. Campbell, J. E. Evetts, and D. Dew-Hughes, *Phil. Mag.* **18**, 313 (1968).

using our estimated value of $F_0/k = 1200\,\mathrm{K}$ and assuming $T \approx 4\,\mathrm{K}$. Taking the time derivative of (5.88), and interchanging the order of time and space differentiation, we have

$$\frac{\partial \alpha}{\partial t} = -\frac{\partial}{\partial x}\frac{\partial}{\partial t}\frac{B^2}{8\pi} = -\frac{\partial}{\partial x}\frac{B}{4\pi}\frac{\partial B}{\partial t} = \frac{\partial}{\partial x}\frac{B}{4\pi}\frac{\partial (Bv)}{\partial x}$$
$$\approx \frac{B^2}{4\pi}\frac{\partial^2 v}{\partial x^2} = \frac{B^2}{4\pi}V_0\frac{\partial^2}{\partial x^2}e^{\alpha/\alpha_1} \tag{5.90}$$

The justification for dropping spatial derivatives of B compared to those of v is that the exponential dependence of v on derivatives of B is of the order of $\alpha_c(0)/\alpha_1$, or about 300 times, as fast as the direct dependence. In the same spirit, we can locally replace the factor $B^2V_0/4\pi$ by a constant C. The form of the resulting equation suggests a trial solution of the form

$$Ce^{\alpha/\alpha_1} = (ax^2 + bx + c)g(t) \tag{5.91}$$

This satisfies the differential equation (5.90) if $g(t)$ obeys the equation

$$\frac{dg}{dt} = \frac{2ag^2}{\alpha_1}$$

Integrating, we find

$$g(t) = -\frac{\alpha_1}{2at}$$

(To avoid carrying a constant of integration, we have chosen to measure time t from an initial point at which $g = \infty$, corresponding to infinitely rapid initial creep.) Inserting this $g(t)$ in (5.91), and taking the logarithm, we have

$$\alpha = F(x) - \alpha_1 \ln t \tag{5.92}$$

where $F(x)$ is a function only of x. In fact, since creep is unobservably slow unless α is quite near α_c, it must be that $F(x) \approx \alpha_c$ in the relevant region of space. Thus, (5.92) can be replaced approximately by

$$\alpha = \alpha_c - \alpha_1 \ln t \tag{5.92a}$$

Applying this expression to the case of flux trapped in a hollow superconducting cylinder with wall thickness d, where $\alpha \approx B_{\text{in}}^2/8\pi d$, we see that

$$B_{\text{in}} \approx B_c\left(1 - \frac{\alpha_1}{2\alpha_c}\ln t\right) \tag{5.93}$$

where B_c is approximately the value of B_{in} giving the critical-state condition. With the estimate (5.89), $\alpha_1/\alpha_c \approx 1/300$. In using (5.93), we should keep in mind that the infinity at $t = 0$ is inaccessible because of the choice of origin of time. Also, this simple form will hold only for moderately small fractional changes in B; thus, no significance should be attached to the fact that (5.93) changes sign after extremely long times.

This characteristic and unusual logarithmic time dependence has been well verified experimentally. For example, in the pioneering experiment of Kim, Hempstead, and Strnad[39] on flux trapped in NbZr tubes, this logarithmic decay was followed repeatedly over a period of time from 10 to 5,000 sec after a sudden adjustment of the external field. In the typical example shown in Fig. 5.10, the trapped field B_{in} of some 4,000 G was observed to fall at a rate of about 5 G per decade of time. Comparing this with (5.93), we see that it corresponds to $\alpha_1/\alpha_c = (400 \ln 10)^{-1} \approx \frac{1}{900}$. Considering the extreme approximations and crude estimates that are involved, the order of magnitude agreement with our estimated value of $\frac{1}{300}$ for this ratio is satisfactory. On the other hand, extensive studies of the rapid flux creep in high-temperature superconductors (see Chap. 9) show time dependences which may be better described by a *stretched exponential* than by the simple logarithmic form.

It is interesting to extrapolate the logarithmic decay of the trapped field in order to estimate how long the circulating currents will take to die out. Taking (5.92*a*), with $\alpha = B_{\text{in}}^2/8\pi d$, we see that the crossover would occur when $t = e^{\alpha_c/\alpha_1} \approx e^{900} \approx 10^{400}$ sec $\approx 10^{390}$ years! Of course, the logarithmic dependence does not hold over such large changes in B. For example, when B drops below H_{c1}, flux leakage should effectively stop since flux lines cannot exist in the volume of the superconductor in equilibrium. Still, this estimate gives an idea of the time scale. The circulating currents are persistent for practical purposes, although in principle they would die out over eons of time. In practice, a superconducting magnet in the persistent current mode is operated at a low enough value of α to ensure that the decay in current over a period of days is negligible. The state of the superconductor is then roughly equivalent to that of one which started near α_c but for which a very long decay time has already elapsed, so that the "next decade" will be an extremely long period of time.

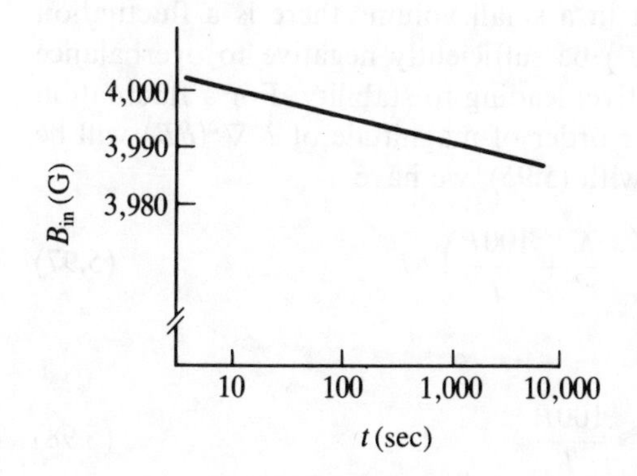

FIGURE 5.10
Evidence for logarithmic decay of "persistent" current in a hollow cylinder of type II superconductor. [*After Kim, Hempstead, and Strnad,* Phys. Rev. Lett. **9**, *306 (1962).*]

[39] Y. B. Kim, C. F. Hempstead, and A. R. Strnad, *Phys. Rev. Lett.* **9**, 306 (1962).

5.7.2 Thermal Instability

The dissipation of energy associated with flux creep can lead to disastrous consequences if it leads to a thermal runaway, in which the material rapidly heats up and the entire energy stored in the magnet is suddenly converted into thermal energy. To prevent this, the system must be thermally stable; i.e., if some region of the material acquires a temperature increment δT, stability requires that δT decay to zero, not continue to grow. The condition for stability is that for given δT, the increased outflow of heat to the surrounding material should be greater than the increased rate of dissipation because of more rapid creep.

We start by considering the effect of δT on P, the power dissipated per unit volume. P can be written simply as

$$P = \alpha v = \alpha v_0 e^{-(F_0 - \alpha L^4)/kT} \tag{5.94}$$

since α is the force per unit volume, force times velocity is the rate of doing work, and here all the work goes into heating the system because of dissipative processes in the cores of the flux lines. From this expression we obtain

$$\frac{T}{P}\frac{\partial P}{\partial T} = \frac{F_0}{kT} - \frac{\alpha L^4}{kT} - \frac{\partial F_0}{\partial(kT)} = -\ln\left(\frac{v}{v_0}\right) - \frac{\partial F_0}{\partial(kT)} \approx 100 \tag{5.95}$$

for the typical values used earlier. Thus, a small increase in temperature will lead to a very much larger increase in heating power, and instability will result unless cooling is very efficient.

To study this balance, we consider the heat-flow equation

$$C\frac{\partial T}{\partial t} = K\nabla^2 T + P \tag{5.96}$$

where C is the specific heat per unit volume, K is the thermal conductivity, and P is the power input per unit volume. In steady state, $T(\mathbf{r})$ is the solution of $K\nabla^2 T + P = 0$. We now imagine that in a small volume there is a fluctuation $\delta T > 0$. The question is: Will $K\nabla^2(\delta T)$ be sufficiently negative to overbalance $(\partial P/\partial T)\,\delta T$, so that $\partial(\delta T)/\partial t$ is negative, leading to stability. For a fluctuation localized in a volume of radius $\sim r$, the order of magnitude of $K\nabla^2(\delta T)$ will be $-(K/r^2)\,\delta T$. Combining this estimate with (5.95), we have

$$C\frac{\partial(\delta T)}{\partial t} = \left(-\frac{K}{r^2} + \frac{100P}{T}\right)\delta T \tag{5.97}$$

For stability, then,

$$\frac{K}{r^2} > \frac{100P}{T} \tag{5.98}$$

From this we see that the criterion is most demanding when r is taken to be as large as possible, i.e., characteristic of the magnet as a whole if there is no internal cooling. A useful way to estimate the value of (P/T) is from the steady-state temperature rise ΔT above the bath temperature. From $K\nabla^2 T + P = 0$, we have $-K(\Delta T)/r^2 + P = 0$. Using this relation to simplify (5.98), we obtain

$$\frac{\Delta T}{T} \lesssim \frac{1}{100} \tag{5.99}$$

Thus, to avoid thermal runaway, the magnet must always be operated under conditions such that the steady-state dissipation due to flux creep is so small that the resulting temperature rise is less than ~1 percent. In fact, magnets are normally operated under conditions in which thermally activated flux creep is negligible, and other considerations actually set the limits on performance. These are discussed further in the next section.

The preceding analysis indicates that good design requires good thermal conductivity and good contact with the helium bath. Magnet materials with high H_{c2} usually have short electronic mean free paths and hence poor thermal conductivities, making them intrinsically unstable. This tendency is almost invariably reduced in practice by cladding the superconductor with a metal like copper. This copper layer is an electrical insulator compared to a superconductor, but it provides excellent thermal conduction, at the expense of wasting some winding volume.

5.8 SUPERCONDUCTING MAGNETS FOR TIME-VARYING FIELDS

In addition to providing thermal stabilization, the addition of a copper coating will also cause eddy-current damping of magnetic-field changes. This helps to stabilize the magnet against fluctuations, but it also limits the rate at which *desired* field changes can be made without undue power dissipation. In view of the potential importance of superconducting magnets in pulsed-field applications (in particle accelerators) and in ac applications (at power frequencies), considerable effort has gone into finding ways to make magnets which are both stable and capable of rapid field sweep. The most successful approach involves the use of composite conductors consisting of twisted arrays of fine superconducting filaments—typically ~30 μm (micrometers) in diameter—embedded in a matrix of normal metal. An extensive analysis of this scheme has been given by Wilson, Walters, Lewin, and Smith.[40] In this section, we review some relevant considerations.

A fundamentally important comparison is the rate of movement of heat in relation to that of magnetic flux through the magnet material. From the heat-flow equation (5.96) we may define the thermal diffusivity constant

$$D_T = \frac{K}{C} \tag{5.100}$$

[40] M. N. Wilson, C. R. Walters, J. D. Lewin, and P. F. Smith, *J. Physics* **D3**, 1518 (1970).

The physical significance of this parameter is that the time τ required for heat to diffuse a distance L to relieve a temperature gradient is of the order of

$$\tau_T = \frac{L^2}{D_T} \tag{5.100a}$$

By applying Maxwell's equations, a similar magnetic-diffusion equation can be derived, in which the diffusion constant is

$$D_M = c^2 \frac{\rho}{4\pi} = 10^9 \frac{\rho}{4\pi} \tag{5.101}$$

where the two forms hold if ρ is in cgs units and in ohm-cm, respectively. The time required for magnetic flux to diffuse a distance L is then of the order of

$$\tau_M = \frac{L^2}{D_M} \tag{5.101a}$$

Typical numerical values for pure metals at 4 K are $D_T = 10^3\ \text{cm}^2/\text{sec}$ and $D_M = 1\ \text{cm}^2/\text{sec}$, so heat moves much faster than magnetic flux. On the other hand, in alloys, such as the superconducting material in the normal state or normal metal alloys, these numerical values are roughly interchanged, so that magnetic flux moves much more rapidly than heat. Note that in a composite filamentary conductor, $D_M(\text{Cu}) \approx D_T(\text{core}) \approx 1\ \text{cm}^2/\text{sec}$. Taking into account the factor L^2 in the respective times, we see that this implies that heat can escape from the filamentary cores faster than flux can diffuse through the pure copper matrix from one filament to another. This efficient local cooling of the filaments helps to make such composite materials highly stable.

5.8.1 Flux Jumps

We now examine the control of flux jumps, replacing the flux-creep analysis of the previous section by a more elementary one in which the material is described by the Bean model of the critical state, with a critical current density J_c below which there is no resistance and above which the material is normal. Since creep in classic superconductors is important only in a small range of temperature or field, this simplification is appropriate over the important practical range of conditions.

For analytical simplicity, we take a one-dimensional model, with superconducting layers of thickness d parallel to the field, and x normal to the plane, measured from the midplane of a typical conductor. We restrict attention here to the usual case of a conductor sufficiently thin so that the field penetrates throughout: i.e., we assume $H > H_s = 2\pi J_c d/c$, where H_s is the screening field defined in (5.73), and we consider a single layer. In this case, the internal field is

$$B(x) = H - \frac{4\pi J_c}{c}\left(\frac{1}{2}d - |x|\right) \tag{5.102}$$

In writing this symmetrical form, with H the same at both surfaces, we have made the approximation of ignoring the self-field of the transport current carried by the conductor under consideration.

Now assume there is a small upward fluctuation in temperature δT, causing a decrease $\delta J_c = (dJ_c/dT)\,\delta T$ in the critical current. This changes $B(x)$, inducing an electric field E. As has been shown by various authors, the power dissipation $\dot{Q}$ associated with the nonequilibrium current J_c can be written as the product of J_c and the electric field induced by the changing field. Averaged over unit volume, this heat is

$$\delta Q = \frac{\pi}{3c^2}\, d^2 J_c\, \delta J_c \tag{5.103}$$

The resulting temperature rise is $\delta T' = \delta Q/C$, where C is the specific heat per unit volume. So long as $\delta T' < \delta T$, the process converges to a finite multiple of the initial δT, and the system is stable against such fluctuations. As can be seen from the previous argument, this will be the case if

$$d^2 < \frac{3c^2 C}{\pi J_c}\left(\frac{-dJ_c}{dT}\right)^{-1} \tag{5.104}$$

This is the condition for *adiabatic* stability since we have assumed no heat leaves the volume in question during the duration of the fluctuation. As such, it is a conservative criterion since removal of heat by the copper matrix will tend to reduce the size of the fluctuation. With typical values of NbTi filaments, this criterion requires filament diameters $d \lesssim 0.01$ cm for stability; experimental results are consistent with this conclusion. In practice, to reduce losses in swept-field applications, d is usually chosen to be even smaller than is required by this stability criterion, so there is a substantial margin of safety.

It is of interest to work out the total heat released if J_c goes all the way down to zero, so that the flux jump goes to completion. Integrating (5.103), we have

$$Q = \frac{\pi}{6c^2}\, d^2 J_c^2 \tag{5.105}$$

Assuming a constant specific heat C, the temperature rise will be $\Delta T = Q/C \approx 3 \times 10^4 d^2$(K), for typical numerical values with d in centimeters. For any d large enough to be unstable according to the criterion (5.104), ΔT will be comparable with T itself. For such large temperature rsies, C ($\propto T^3$) will increase significantly, making it possible that (5.104) will become satisfied and the growth of the fluctuation stopped before J_c is reduced all the way to zero. Such *partial flux jumps* allow some relaxation of the internal shielding currents without necessarily interrupting the flow of transport current.

A *dynamic* stability criterion may be derived in the same general way, but taking into account the time required for the flux to change and the rate of diffusion of the heat into the copper. The result is similar to (5.104), apart from a factor of the order of the ratio of the diffusivities $D_T(\text{core})/D_M(\text{Cu})$, which is typically near unity, as noted earlier. In deriving this result, it was assumed that the copper temperature did not rise. This assumption breaks down for an edge-cooled tape-wound magnet since the heat must diffuse a long way to reach the bath, and other parameters of the system enter into the stability criterion in that case, as is discussed by Wilson et al.

5.8.2 Twisted Composite Conductors

The preceding arguments have shown that a composite conductor containing tiny superconducting filaments in a copper matrix should have excellent stability against flux jumps. We now turn to the case of time-varying fields, in which case the copper will carry electric current as well as heat current. There will then be two types of losses: eddy-current losses in the copper, which depend on sweep rate, and hysteresis losses in the superconductor, which are intrinsically independent of sweep rate. As will be shown in detail, normal eddy-current losses can be kept small by using fine wire with the turns insulated from one another. Thus, hysteresis losses typically dominate. We now compute how large they are.

Returning to our model with thin superconducting sheets, in which the internal field is given by (5.102), we see that $\dot{B}(x) = \dot{H}$, so the induced electric field is $E_y = \dot{H}x/c$, by Faraday's law. Setting P, the local rate of energy dissipation per unit volume, equal to $J_c E$ as before, and averaging over the thickness of the superconducting sheet, we have

$$P = \dot{Q} = \frac{J_c \dot{H} d}{4c} \quad \text{per unit volume} \tag{5.106}$$

Thus, the heat released for any given change ΔH is independent of the rate of change, as expected for a hysteresis loss, and we can write the dissipation for a complete cycle in which the field changes by $\pm\Delta H$ about an operating point as $Q = J_c d\, \Delta H/c$. This result shows that by making the superconducting filaments thin enough, this loss can be made very small.

In arriving at (5.106), we made the assumption that the changing field penetrated all the way through the filament. This assumption will be justified in applications in which the magnetic field is swept over a wide range, such as in a magnet pulsed between $H = 0$ and H_{max}. It will *not* be justified, however, in applications which are basically dc, with an ac ripple of small amplitude $\Delta H < H_s$. For these, it is necessary to take into account the fact that the changing field does not penetrate uniformly. The resulting more accurate expressions for loss per cycle of $\pm\Delta H$ about an operating point are

$$Q = \frac{J_c d\, \Delta H}{c} - \frac{4\pi J_c^2 d^2}{3c^2} \qquad \Delta H > H_s = \frac{2\pi d J_c}{c} \tag{5.107a}$$

and

$$Q = \frac{(\Delta H)^3 c}{12\pi^2 d J_c} \qquad \Delta H < H_s \tag{5.107b}$$

Our approximation (5.106) is equivalent to keeping only the leading term in (5.107a). The correction term in (5.107a) reduces this leading term by a factor of 3 when $\Delta H = H_s$, the lowest value of the ac field for which complete penetration occurs. At this point, a continuous transition is made to (5.107*b*), and the loss thereafter decreases as $(\Delta H)^3$. Thus, the true loss can be much lower than would be estimated by using (5.106).

Going further, (5.107) indicates that $Q \to 0$ in either limit $d \to 0$ or $d \to \infty$. Dissipation is greatest for intermediate values. This can be made more clear by introducing a dimensionless thickness variable

$$\beta = \frac{2\pi J_c d}{c\,\Delta H} = \frac{d}{d_s} = \frac{H_s}{\Delta H}$$

which is the ratio of d to the slab thickness d_s penetrated by a field change ΔH. In terms of β, (5.107) becomes

$$Q = \frac{(\Delta H)^2}{6\pi}(3\beta - 2\beta^2) \qquad \beta < 1 \tag{5.108a}$$

$$Q = \frac{(\Delta H)^2}{6\pi}\frac{1}{\beta} \qquad \beta > 1 \tag{5.108b}$$

These expressions have a maximum at $\beta = \frac{3}{4}$, where $Q = 3(\Delta H)^2/16\pi$, and they fall to zero as $\beta \to 0$ or $\beta \to \infty$. To minimize Q, there are two strategies: make β either large or small, but not near unity. With typical material parameters, the filament diameter d must be kept below $\sim 200\ \mu m$ for thermal stability, and it is hard to fabricate them smaller than $\sim 10\ \mu m$. For typical critical-current densities and this range of d, it then turns out that the larger filaments give lower losses for $\Delta H < 500$ Oe, whereas the smaller filaments are superior for larger ac amplitudes, such as in pulsed magnets.

An important qualification on the applicability of (5.107) is that each filament must act independently, so that the origin of x can be taken in the center of each one. This would certainly be the case if each filament were electrically insulated from its neighbors, but in practice the copper matrix provides a good electrical contact between all the filaments in one conductor. This profoundly modifies the current patterns at high field sweep rates because the induced voltages can drive current from one filament to another through the copper. In this case, the filaments are said to be *coupled*, and the effective diameter d is no longer that of one filament; rather, it can be as large as that of the entire bundle of 10 to 1,000 filaments in a conductor, and the effectiveness of the multiplicity of fine filaments is lost. As first pointed out clearly by P. F. Smith,[41] this degradation in performance at high sweep rates can be greatly reduced by twisting the conductors with a pitch less than a characteristic length l_c, which depends on sweep rate and

[41]P. F. Smith, M. N. Wilson, C. R. Walters, and J. D. Lewin, in A. D. McInturff (ed.), *Proc. 1968 Brookhaven Summer Study on Superconducting Devices and Accelerators*, Brookhaven National Lab. Publ. No. BNL-50155, p. 913. Also see ref. 40.

other parameters. We now analyze a simplified model which demonstrates these properties and leads to a derivation of the characteristic length.

Consider two thin superconducting sheets of thickness d, width w, and finite length $2l$, separated by a layer of normal metal of resistivity ρ and thickness d', where for algebraic simplicity we take $l \gg d' \gg d$. The geometry is illustrated in Fig. 5.11. Further, we assume there is an applied field parallel to w which is changing at a constant rate $\dot{H}$. We now apply Faraday's law to various circuits to compute the induced emf and the resulting currents.

First, let us compute the eddy-current loss in the normal metal if no superconductor were present. If we measure x from the midplane of the normal metal layer, $E_y \approx \dot{H}x/c$ and $E_x \ll E_y$. The local rate of power dissipation is then $P = JE = E^2/\rho$. Averaging over x this gives

$$P = \frac{\dot{H}^2 d'^2}{12\rho c^2} \tag{5.109}$$

as the average dissipation per unit volume. If we take d' equal to the overall diameter of the composite conductor, this will give an estimate of the eddy-current loss with no superconductors present. To get some feeling for the size of this loss, consider an example of a magnet wound with a 0.05-cm-diameter wire with $\rho = 0.01\rho_{\mathrm{Cu}(300\,\mathrm{K})}$. Then $P \approx 10^{-12}\dot{H}^2$ watts/cm^3 where $\dot{H}^2$ refers to an average over the volume of the magnet winding. Even at a sweep rate as high as 10^3 G/sec, this loss would be only 1 mW for a magnet with 10^3 cm^3 of winding. On the other hand, in an ac application, where $\dot{H} = \omega H_{\max}$, this loss can become very large at useful power frequencies. For example, at 60 Hz, with $H_{\max} = 10^4$, P would be 10^4 W for a magnet winding of 10^3 cm^3, a prohibitive value for cooling at 4 K. To reduce this figure to a more manageable one, we could use somewhat finer wire and replace the copper matrix by an alloy with much higher low-temperature resistance. However, it would be difficult to reduce the loss by more than a factor of 10^3 in this way. Our general conclusion is that normal-metal eddy-

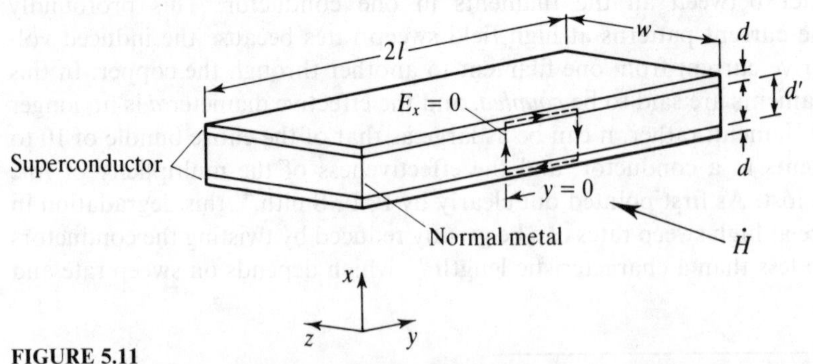

FIGURE 5.11
Laminar model of multifilamentary composite conductor. As discussed in the text, application of Faraday's law to circuits such as the one shown in dashed lines enables one to make approximate determinations of current and emf patterns for various cases.

current losses can be made negligible at slow-to-moderate sweep rates, but that above about 10^4 G/sec, they will be increasingly hard to overcome.[42]

Now let us take into account the presence of the superconducting layers. We apply Faraday's law to a circuit passing through both superconducting sheets, as indicated in Fig. 5.11. So long as the current density in the superconducting sheets is less than J_c, the entire induced voltage is dropped across the normal metal, leading to a current density between the sheets of $J_x = \dot{H}y/c\rho$. Integrating out from $y = 0$, we find the total current from one sheet to the other equals the critical current of a sheet when $y = l_c$, where

$$l_c^2 = \frac{2cJ_c d\rho}{\dot{H}} \tag{5.110}$$

Thus, if $l < l_c$, our assumption that no voltage was dropped in the superconductor is self-consistent, and there is no dissipation in the superconductor due to these circulating currents. However, the eddy-current loss in the normal metal is greatly increased because the induced emf is dropped over a much shorter distance. As a result, the factor d'^2 in (5.109) is replaced by $(2l)^2$.

On the other hand, if $l > l_c$, the current density in the center of the superconductor would apparently have to exceed J_c, indicating that our assumption of no dissipation in the superconductor is no longer self-consistent. The actual physical situation is that in the central part of the layers, farther than l_c from the ends, the supercurrent density is constant at J_c. Consequently, no current enters the normal metal from the superconductor, and accordingly, none of the emf is dropped there. Thus, in the central part, all the emf is dropped in the superconductor, giving an electric field $\dot{H}d'/2c$. The resulting dissipation is $J_c\dot{H}d'/2c$ per unit volume of superconductor, which is just as high as if the normal-metal layer were replaced by solid superconductor. If one averages the enhanced eddy-current loss in the copper over the length l_c at each end, where J_x in the normal metal rises as $(\dot{H}/c\rho)[y - (l - l_c)]$, one finds that it is two thirds as great (per unit length) as is the dissipation in the superconductor in the central part. Thus, essentially all the benefit of having the superconductor in thin layers is lost if $l > l_c$.

For sweep rates of 10 to 10^4 G/sec, this length l_c is typically in the range from 10 to 100 cm, so it is clear that practical magnet windings are much longer than l_c. However, as pointed out by Smith et al., it is possible to decouple the filaments by twisting the wire with a pitch which is much shorter than l_c. This effectively reverses the sign of the induced emf each time the filaments are

---

[42]One might wonder why these losses are not an equal problem in the windings of ordinary power transformers. The basic difference is that even for $B_{max} \approx 10^4$ G in the iron core, B_{max} in the copper winding may be only 10 G. Thus, eddy-current loss is primarily a problem in the core. This is reduced by lamination to a thickness of about 0.04 cm, and, of course, ρ of silicon steel is about 500 times greater than that of copper at 4 K. Taken together, these factors reduce the dissipation per unit volume by a factor of about 1,000 to a level which is tolerable for cooling at room temperatures.

interchanged in position, allowing the induced current in the superconductor to be kept always below J_c. In this case, the dissipation is given by the enhanced eddy-current loss in the normal metal, referred to earlier.

Specific assumptions about the geometry are required to scale our results for two flat layers up to a wire containing many layers of filaments. Nevertheless, we can estimate a new length l_c' by equating the total induced circulating current with the total critical current of the filaments on one side of the center of the wire. If d' is now taken to be the diameter of the bundle of filaments in the wire, and if a filling factor of $\frac{1}{2}$ is assumed, we find

$$l_c'^2 \approx \frac{cJ_cd'\rho}{2\dot{H}} = \frac{d'}{4d}\, l_c^2 \tag{5.111}$$

as an extension of (5.110). Thus, we expect the excess loss to be reduced by a factor of about $l^2/l_c'^2$ from the value for a solid superconductor of diameter d'. Note that with our definitions of l_c and l_c', the pitch for a twist by 2π is denoted $4l$.

If we now add this extra dissipation due to coupling to the value (5.106) for completely decoupled filaments, the total dissipation per unit volume of superconductor can be conveniently written as

$$P = \frac{J_c\dot{H}d_{\text{eff}}}{4c} \tag{5.112}$$

where $$d_{\text{eff}} \approx d' \qquad l > l_c' \tag{5.112a}$$

and $$d_{\text{eff}} \approx d\left(1 + \frac{4l^2}{l_c^2}\right) = d + \frac{2l^2\dot{H}}{J_c\rho c} \qquad l < l_c' \tag{5.112b}$$

[Note that in (5.112), we have reverted to using only the leading term of (5.107*a*), an approximation appropriate to small d and large ΔH.] As expected, P is proportional to $\dot{H}d$ for small $\dot{H}$, where the filaments are decoupled, and to $\dot{H}d'$ for large $\dot{H}$, when they are fully coupled. In the transition region between these regimes, P increases as $\dot{H}^2$ since $l_c^{-2} \propto \dot{H}$. There will be another transition to an $\dot{H}^2$ dependence when the normal eddy-current term (5.109) dominates. This will occur when $\dot{H} \gtrsim \rho cJ_c/d' \gtrsim 10^7$ G/sec. Thus, in the presently practical range of $\dot{H}$, losses are dominated by (5.112).

For typical parameter values, this dissipated power will fall in the range from 10^{-6} to $10^{-5}\ \dot{H}$ W/cm^3, depending on the filament size and the degree of decoupling. Noting that only a fraction of the volume is filled with superconductor and that the average of $\dot{H}$ over the winding is typically less than half the value in the bore, we see that the loss can be as small as $10^{-7}\ \dot{H}$ W/cm^3 when referred to toal winding volume and the field in the bore. Such a performance was demonstrated by Spurway, Lewin, and Smith.[43]

[43]A. H. Spurway, J. D. Lewin, and P. F. Smith, *J. Physics* **D3**, 1572 (1970).

As a final figure of merit, let us compute the ratio of the energy stored in the magnetic field to the energy dissipated in the process of establishing and removing the field. For simplicity, consider a solenoid of length L, radius R, and winding thickness D, with $D \ll R \ll L$. Then the stored energy will be $E \approx \pi R^2 L(H^2/8\pi)$. Assuming complete decoupling of the filaments, we can use (5.106) to compute the dissipation. Canceling the factor of 2 for the loss on removing the field as well as setting it up against the factor $\frac{1}{2}$, which relates the average field in the winding to the field in the bore, the total dissipated energy per cycle is $\Delta E = 2\pi RLD(J_c Hd/4c)$. Taking the ratio, and noting that $H = 4\pi JD/c$, we have

$$\frac{E}{\Delta E} = \frac{JR}{J_c d} \tag{5.113}$$

Thus, with $J \approx J_c$, one should be able to achieve energy storage ratios of the order of 1,000 with presently available materials and $R \approx 2$ cm. In fact, ratios of over 100 were obtained in early experiments by Dahl et al.,[44] without reaching J_c. For comparison, with a normal resistive coil, this ratio would be given by

$$\frac{E}{\Delta E} \approx \frac{\tau}{\Delta t} \tag{5.114}$$

where $\tau = L/R$ is the time constant of the magnet and Δt is the duration of the pulse. Noting that $\tau \approx 2\pi DR/c^2\rho$, which is of the order of 10 sec for typical dimensions and resistivities of pure metals at low temperatures, we see that the superconductive coil should be more efficient than even cryogenic normal coils for pulse durations greater than about 0.01 to 0.1 sec. It is for this reason that superconductive coils have been developed to generate the intense transient fields required in particle accelerators.

[44]P. F. Dahl, G. H. Morgan, and W. B. Sampson, *J. Appl. Phys.* **40**, 2083 (1969).

CHAPTER 6

JOSEPHSON EFFECT I: BASIC PHENOMENA AND APPLICATIONS

6.1 INTRODUCTION

In 1962, Josephson[1] made the remarkable prediction that a zero voltage supercurrent

$$I_s = I_c \sin \Delta\varphi \tag{6.1}$$

should flow between two superconducting electrodes separated by a thin insulating barrier. Here $\Delta\varphi$ is the difference in the phase of the Ginzburg-Landau wavefunction in the two electrodes, and the *critical current* I_c is the maximum supercurrent that the junction can support. He further predicted that if a voltage difference V were maintained across the junction, the phase difference $\Delta\varphi$ would evolve according to

$$d(\Delta\varphi)/dt = 2eV/\hbar \tag{6.2}$$

[1]B. D. Josephson, *Phys. Lett.* **1**, 251 (1962); *Adv. Phys.* **14**, 419 (1965).

so that the current would be an alternating current of amplitude I_c and frequency $\nu = 2eV/h$. Thus, the quantum energy $h\nu$ equals the energy change of a Cooper pair transferred across the junction. These predicted effects, known as the dc and ac Josephson effects, respectively, have been fully confirmed by an immense body of experiments.

Although his prediction was based on a microscopic theoretical analysis of the quantum mechanical tunneling of electrons through the barrier layer, it is now clear that the effects are much more general, and occur whenever two strongly superconducting electrodes are connected by a "weak link." As illustrated in Fig. 6.1, the weak link can be an insulating layer as Josephson originally proposed, or a normal metal layer made weakly superconductive by the so-called *proximity effect* (in which Cooper pairs from a superconducting metal in close proximity diffuse into the normal metal), or simply a short, narrow constriction in otherwise continuous superconducting material. These three typical cases are often referred to as *S-I-S*, *S-N-S*, or *S-c-S* junctions, respectively, where the *S*, *I*, *N*, and *c* denote superconductor, insulator, normal metal, and constriction, respectively.[2]

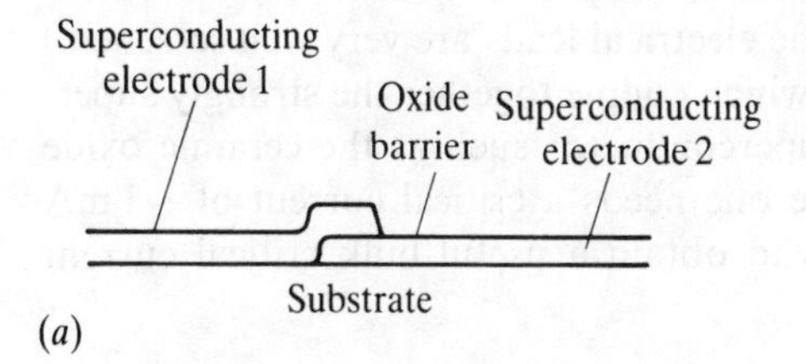

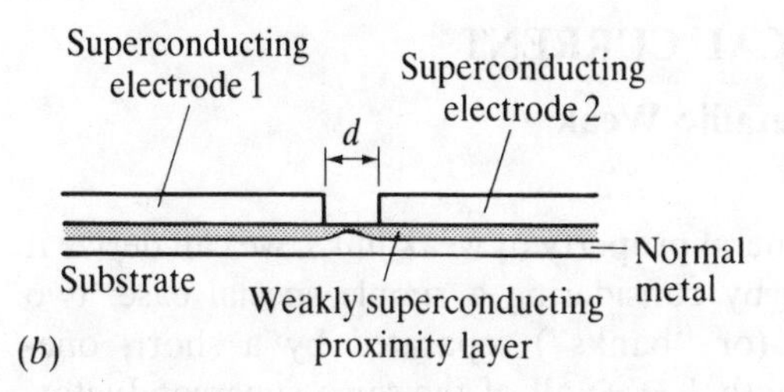

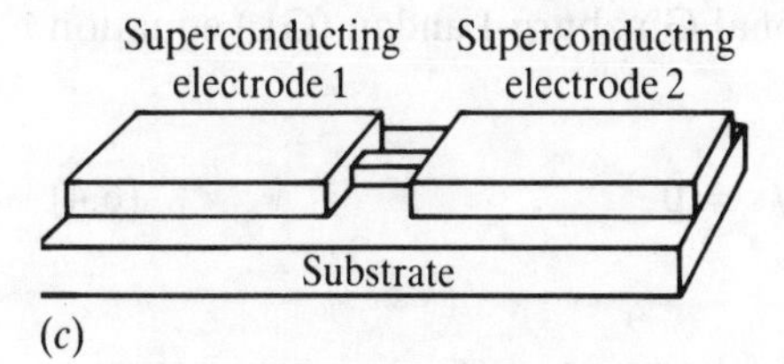

FIGURE 6.1
Three types of Josephson junction: (*a*) *S-I-S*, (*b*) *S-N-S*, and (*c*) *S-c-S*.

[2]In the high-temperature superconductors, a simple grain boundary can serve as a weak link because these materials have such short coherence lengths. Depending on details, it may be closer to *S-N-S* or *S-I-S* in character.

Given the two relations (6.1) and (6.2), one can derive the coupling free energy stored in the junction by integrating the electrical work $\int I_s V dt = \int I_s(\hbar/2e)\, d(\Delta\varphi)$ done by a current source in changing the phase. In this way, we find

$$F = \text{const.} - E_J \cos \Delta\varphi \qquad \text{where } E_J \equiv (\hbar I_c/2e) \tag{6.3}$$

Clearly, the energy is a minimum when the two phases are equal, so that $\Delta\varphi = 0$. This corresponds to the energy minimum in the absence of phase *gradients* in a bulk superconductor. The critical current is a measure of how strongly the phases of the two superconducting electrodes are coupled through the weak link. This depends on how thin and of what material the barrier is, or, in the case of constriction weak links, on the cross-sectional area and length of the neck.

In most applications I_c lies in the range of a microampere to a few milliamperes. The lower limit stems from the requirement that the coupling energy E_J exceed the thermal energy kT, i.e., $I_c > 2ekT/\hbar$; otherwise thermal fluctuations in $\Delta\varphi$ will wash out the phase-dependent supercurrent.[3] (Although the numerical factor $2ek/\hbar$ is only $\sim 0.04 \mu\text{A K}^{-1}$, I_c must be sufficient to overcome an effective noise temperature which may approach room temperature rather than the temperature of the superconductor, unless the electrical leads are very well screened.) At the other extreme are the weak links which couple together the strongly superconducting grains in a granular bulk superconductor such as the ceramic oxide high-temperature superconductors. Here one needs a critical current of ~ 1 mA between adjacent 1-μm diameter grains to obtain a useful bulk critical current density of the order of $10^5\ \text{A cm}^{-2}$.

6.2 THE JOSEPHSON CRITICAL CURRENT

6.2.1 Short One-Dimensional Metallic Weak Links

Insofar as the Josephson effect is a general property of weak links, we can derive it from the Ginzburg-Laundau theory by considering a simple special case: two massive superconducting electrodes (or "banks") separated by a short, one-dimensional link (or "bridge") of length $L \ll \xi$, all of the same superconductor. From (4.18) we know that the one-dimensional Ginzburg-Landau (GL) equation describing the bridge can be written as

$$\xi^2 \frac{d^2 f}{dx^2} + f - f^3 = 0 \tag{6.4}$$

---

[3]To see how this works, write (6.1) as $I_s = I_c \sin(\Delta\varphi + \delta\varphi)$, where $\Delta\varphi$ is the average value and $\delta\varphi$ is the fluctation. Expand the sine by a trigonometric identity and average over the fluctuations. Since $\sin \delta\varphi$ averages to zero, one has $\langle I_s \rangle = [I_c \langle \cos \delta\varphi \rangle] \sin \Delta\varphi$, so the apparent critical current is reduced by a factor $\langle \cos \delta\varphi \rangle$. For random fluctuations this can be approximated by a factor $\exp -\langle (\delta\varphi)^2 \rangle /2$, where $\langle (\delta\varphi)^2 \rangle$ is the mean-square fluctuation about the average phase difference $\Delta\varphi$. At low temperatures this mean-square deviation is $\sim kT/E_J$ by the equipartition theorem.

where $f = \psi/\psi_\infty$. We can assume the massive electrodes are in equilibrium, so that $|f| = 1$ in both of them (by definition of ψ_∞), but the phases of the order parameters may be different. Since *absolute* phase is undefined, without loss of generality we can take the phases to be 0 and $\Delta\varphi$. The appropriate solution of (6.4) in the bridge then is that which matches the boundary conditions $f = 1$ at $x = 0$ and $f = e^{i\Delta\varphi}$ at $x = L$. As pointed out by Aslamazov and Larkin[4] in 1968, so long as $L \ll \xi$, the first term in (6.4) dominates because it is larger than the other two terms by a factor that scales with $(\xi/L)^2$ for any nonzero $\Delta\varphi$. In the limit, this reduces the problem to Laplace's equation $d^2f/dx^2 = 0$, for which the most general solution (in one dimension) is $f = a + bx$. Applying the boundary conditions at the two ends of the bridge, we obtain the solution

$$f = (1 - x/L) + (x/L)e^{i\Delta\varphi} \tag{6.5}$$

It is suggestive to view (6.5) as the superposition of two partial solutions: the first representing the spread of the order parameter from the left bank with phase zero, and the second representing the spread of the order parameter from the right bank, with phase $\Delta\varphi$. If we insert $f(x)$ from (6.5) into the GL current expression, we obtain

$$I_s = I_c \sin \Delta\varphi \tag{6.6}$$

with I_c given by

$$I_c = (2e\hbar\psi_\infty^2/m^*)(\mathcal{A}/L) \tag{6.7}$$

where $\mathcal{A}$ is the cross-sectional area of the bridge. If instead we had computed the GL free energy of this $f(x)$ by integrating over the length of the bridge, we would have obtained

$$\Delta F = (\hbar/2e)I_c(1 - \cos \Delta\varphi) \tag{6.8}$$

with the same expression (6.7) for I_c, as expected from the general argument which led to (6.3).

If we compare (6.7) with the critical current, as given by (4.36), of a *long* superconducting bridge with the same $\mathcal{A}$ and other properties, we find that the critical current of the bridge differs by a factor $(3^{3/2}/2)(\xi/L)$. Since the approximation leading to (6.7) is only valid if $(\xi/L) \gg 1$, the bridge critical current is always greater than that of a long wire of the same cross section. The physical reason for this is that the strongly superconducting banks help to "support" ψ in the bridge, allowing it to sustain a higher phase gradient ($\sim 1/L$) than the limiting value $\sim 1/\xi$ in the long wire.

[4] L. G. Aslamazov and A. I. Larkin, *Zh. Eksp. Teor. Fiz. Pis. Red.* **9**, 150 (1968); transl. *JETP Lett.* **9**, 87 (1969).

By combining the two terms in (6.5), one finds the spatial dependence of the magnitude of the order parameter to be

$$|f|^2 = 1 - 4(x/L)(1 - x/L)\sin^2(\Delta\varphi/2) \tag{6.9}$$

which reduces to $|f|^2 = \cos^2(\Delta\varphi/2)$ at the center of the bridge. Thus, if $V > 0$, so that $\Delta\varphi$ increases steadily with time, $|f|^2$ oscillates up and down, going to zero at the center of the bridge once during each cycle when $\Delta\varphi = \pi$. One should recognize that as $\Delta\varphi$ increases without limit, the phase of f does not "wind up" in some sort of tightening helix as it does in a long wire. Rather, the entire $f(x, t)$ repeats exactly each time $\Delta\varphi(t)$ increases by 2π. It is often said that a "phase slip" by 2π occurs in this process each time $|f|^2$ touches zero, making the local phase at that point undefined, but, in fact, the time evolution of $f(x)$ is completely regular and continuous throughout the cycle.

6.2.2 Other Weak Links

GENERAL CONSTRICTION WEAK LINKS. From the proportionality of I_c to (A/L) in (6.7) it is clear that I_c scales with dimensions of the bridge exactly as the inverse of its resistance R_n in the normal state; thus, I_cR_n has an *invariant* value $I_cR_n = (2e\hbar\rho_n\psi_\infty^2/m^*)$, which depends *only* on the material and the temperature, and not on bridge dimensions. In fact, this invariance holds for more general geometries, such as two- and three-dimensional constrictions, so long as the linear dimensions are smaller than ξ. The argument is very similar to that above, except that Laplace's equation now occurs in the higher dimension. It does not need to be *solved* for the exact geometry, however, since the invariance of I_cR_n is based simply on the fact that the *same* solution of Laplace's equation appears in both I_c and $1/R_n$ and cancels out. Accordingly, the I_cR_n product is frequently used as a measure of how closely real Josephson junctions approach the theoretical limit.

AMBEGAOKAR-BARATOFF FORMULA FOR TUNNEL JUNCTIONS. By applying the microscopic theory to a tunnel junction geometry, as had Josephson in his original derivation, Ambegaokar and Baratoff[5] worked out an exact result for the full temperature dependence of I_c for that system, namely,

$$I_cR_n = (\pi\Delta/2e)\tanh(\Delta/2kT) \tag{6.10}$$

By using our BCS results for dirty superconductors to evaluate the GL parameter ψ_∞, we find that the I_cR_n product for a metallic weak link given in the previous paragraph reduces to this exact same form, at least near T_c, where the GL theory is valid. Thus, (6.10) is an important *general* result. At $T = 0$ it reduces to

[5] V. Ambegaokar and A. Baratoff, *Phys. Rev. Lett.* **10**, 486 (1963); erratum, **11**, 104 (1963).

$I_cR_n = \pi\Delta(0)/2e$. It is also convenient to note that by using the BCS $\Delta(T)$, we see that (6.10) varies linearly with T near T_c and can be approximated there by

$$I_cR_n = (2.34\pi k/e)(T_c - T) \approx (T_c - T) \times 635\,\mu\text{V/K} \qquad (6.10a)$$

KULIK AND OMEL'YANCHUK REFINEMENTS FOR METALLIC LINKS. Although (6.10) holds for tunnel junctions, and also, near T_c, for a sufficiently short and narrow constriction weak link in metal with an electronic mean free path ℓ that is sufficiently short to justify the diffusive approximation underlying the GL theory, it it not completely general. Kulik and Omel'yanchuk[6] have used the more sophisticated theories of Usadel[7] and of Eilenberger,[8] which are valid all the way to $T = 0$, to work out $I_c(T)$ for short metallic constrictions. The results reduce to (6.10) as $T \to T_c$ but deviate at lower temperatures. For the same slope of I_cR_n at T_c, they reach values at $T = 0$ which, for the dirty and clean limits, respectively, are 1.32 and 2 times greater than would be given by (6.10). These temperature dependences are compared with (6.10) in Fig. 6.2, which is taken from the excellent review by Likharev.[9]

***S-N-S* JUNCTIONS.** If the weak link is an *S-N-S* (superconducting-normal metal-superconducting) junction, the normal-metal coherence length ξ_n sets the length scale. In the clean limit, $\xi_n \equiv \xi_n^0 = \hbar v_F/2\pi k_B T$, but if the electronic mean free path

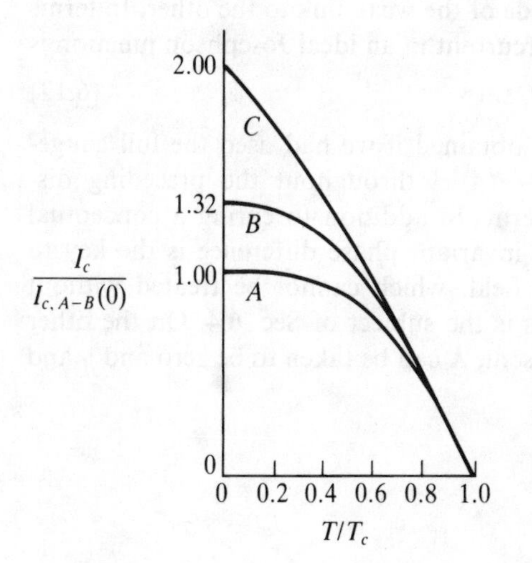

FIGURE 6.2
(*A*) Temperature dependence of $I_c(T)$ for the Ambegaokar-Baratoff theory, normalized to its value $\pi\Delta(0)/2eR_n$ at $T = 0$. (*B*) and (*C*) show the results of the Kulik-Omel'yanchuk theories for dirty and clean limits, respectively, with the same normalization. (*After Likharev review article.*)

[6] I. O. Kulik and A. N. Omel'yanchuk, *Zh. Eskp. Teor. Fiz. Pis. Red.* **21**, 216 (1975); transl. *JETP Lett.* **21**, 96 (1975); *Fiz. Nizk. Temp.* **3**, 945 (1977); transl. *Sov. J. Low Temp. Phys.* **3**, 459 (1978).

[7] K. D. Usadel, *Phys. Rev. Lett.* **25**, 507 (1970).

[8] G. Eilenberger, *Z. Phys.* **214**, 195 (1968).

[9] K. K. Likharev, *Rev. Mod. Phys.* **51**, 101 (1979).

$\ell \ll \xi_n^0$ so that the electronic motion is diffusive, then the appropriate expression for ξ_n is the geometric mean of $\ell/3$ and ξ_n^0, as expected for a random walk process. I_c falls with the thickness of the normal layer approximately as e^{-d/ξ_n}, i.e., with the value of ψ^2 in the middle of the normal region. Since R_n increases only *linearly* with d, the $I_c R_n$ product can fall exponentially far below the ideal value given by (6.10) if the thickness d of the normal barrier is *not* thin compared to ξ_n. The temperature dependence of I_c is also quite different from (6.10) if $d > \xi_n(T)$: In addition to going to zero at T_c, $I_c \sim e^{-d/\xi_n}$ rises exponentially with decreasing temperature as $e^{-\alpha T}$ or $e^{-\beta\sqrt{T}}$, depending on whether ξ_n is in the clean or dirty limit.

6.2.3 Gauge-Invariant Phase

For simplicity, we have carried out the foregoing discussion in terms of the phase difference $\Delta\varphi$. Since $\Delta\varphi$ is not a gauge-invariant quantity, it does not have a unique value for a given physical situation; hence, it cannot in general determine the current I_s, which is a well-defined gauge-invariant physical quantity. This difficulty is cured by replacing $\Delta\varphi$ throughout by the *gauge-invariant phase difference* γ, defined by

$$\gamma \equiv \Delta\varphi - (2\pi/\Phi_0)\int \mathbf{A}\cdot d\mathbf{s} \tag{6.11}$$

where the integration is from one electrode of the weak link to the other. In terms of γ, the general expression for the supercurrent in an ideal Josephson junction is

$$I_s = I_c \sin\gamma \tag{6.12}$$

This is the result which we would have obtained if we had used the full gauge-invariant gradient expression $[(\hbar\nabla/i) - e^*\mathbf{A}/c]$ throughout the preceding discussion instead of using only its first term. In addition to curing a conceptual problem, the introduction of the gauge-invariant phase difference is the key to working out the effects of a magnetic field, which cannot be treated without introducing the vector potential $\mathbf{A}$. This is the subject of Sec. 6.4. On the other hand, so long as no magnetic field is present, $\mathbf{A}$ can be taken to be zero and γ and $\Delta\varphi$ can be used interchangably.

6.3 THE RCSJ MODEL

6.3.1 Definition of the Model

Although (6.12) and $I_c(T)$ suffice to characterize the zero voltage dc properties of a weak link, for finite voltage situations involving the ac Josephson effect, a more complete description is required. This is usually provided by the RCSJ (resistively and capacitively shunted junction) model, in which we model the physical Josephson junction by an ideal one described by (6.12), shunted by a resistance R and a capacitance C, as sketched in Fig. 6.3*a*. The resistance R builds in

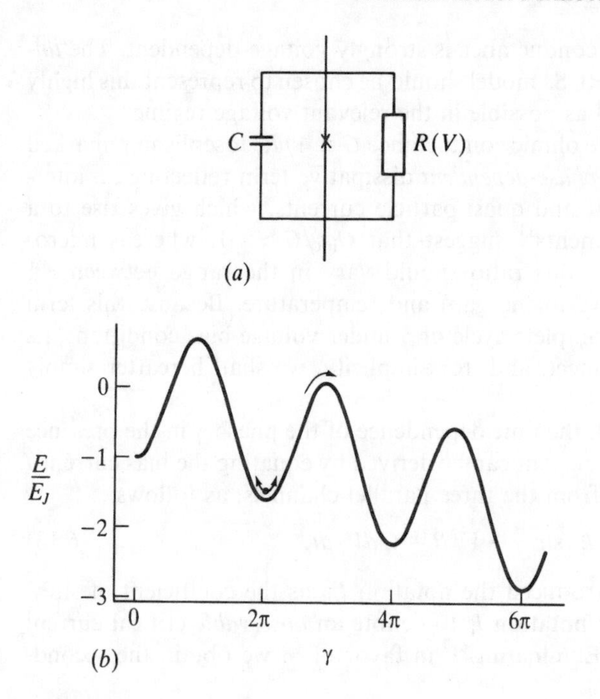

FIGURE 6.3
(*a*) Equivalent circuit of RCSJ model. (*b*) The "tilted-washboard" potential representation of the RCSJ model.

dissipation in the finite voltage regime, without affecting the lossless dc regime, while C reflects the geometric shunting capacitance *between the two electrodes*, not the capacitance of the electrodes to "ground."[10] The appropriate magnitude of R will be of order R_n in an *S-c-S* junction, an *S-N-S* junction, or an *S-I-S* junction near T_c. At lower temperatures, however, the appropriate R for a high-quality *S-I-S* junction rises approximately as $R_n e^{\Delta/k_B T}$ (for $V < V_g \equiv 2\Delta/e$). This expression takes into account the dominant exponential temperature dependence arising from the freeze-out of quasi-particles at low temperatures but not the weaker effect of the singular density of states at the gap edge in BCS theory which enters in (3.82). In this connection, it is useful to review the dc *I-V* curve (Fig. 3.7), which results from quasi-particle tunneling alone, if the Josephson supercurrent is suppressed by a strong magnetic field or by thermal fluctuations. For $V < V_g$, only thermally excited quasi-particles can tunnel, giving a high resistance; for $V > V_g$, the applied voltage can break Cooper pairs, giving a tunnel resistance comparable to R_n.

---

[10] The interelectrode capacitance dominates because it scales with A/d, where A is the junction area and d is the very small barrier thickness (~20 Å) separating the electrodes, whereas the self-capacitance to ground scales with the linear dimension L of the electrodes.

Obviously, the quasi-particle conductance is strongly voltage-dependent. The *linear* damping resistor R in the RCSJ model should be chosen to represent this highly *nonlinear* conductance as well as possible in the relevant voltage regime.

In addition to this simple ohmic conductance $G \equiv 1/R$, Josephson remarked that there should be another *phase-dependent* dissipative term reflecting an interference term between the pair and quasi-particle currents, which gives rise to a current $(G_{\text{int}} \cos\gamma)V$. Experiments[11] suggest that $G_{\text{int}}/G \approx -1$, whereas microscopic theory[12] indicates that this ratio should vary in the range between ± 1 depending on voltage (relative to the gap) and temperature. Because this term averages out to zero over a complete cycle of γ under voltage bias conditions, its effects are rather subtle to detect, and, for simplicity, we shall hereafter simply drop the $\cos\gamma$ term.

Within the RCSJ model, the time dependence of the phase γ in the presence of an externally supplied bias current can be derived by equating the bias current I to the total junction current from the three parallel channels, as follows:

$$I = I_{c0} \sin\gamma + V/R + C\,dV/dt. \tag{6.13}$$

In this equation, we have introduced the notation I_{c0} as the coefficient of $\sin\gamma$, anticipating future use of the notation I_c to denote an *observable* critical current which may be less than I_{c0}. Eliminating V in favor of γ, we obtain the second-order differential equation

$$d^2\gamma/d\tau^2 + Q^{-1}\,d\gamma/d\tau + \sin\gamma = I/I_{c0} \tag{6.14}$$

in which we have introduced a dimensionless time variable $\tau = \omega_p t$, with

$$\omega_p = (2eI_{c0}/\hbar C)^{1/2} \tag{6.14a}$$

being the so-called *plasma frequency* of the junction, and the "quality factor" Q is defined by

$$Q = \omega_p RC \tag{6.14b}$$

This Q is identical with $\beta_c^{1/2}$, where β_c is a frequently used damping parameter that was introduced by Stewart and McCumber.[13]

THE TILTED-WASHBOARD MODEL. Qualitative insight into the junction dynamics described by solutions to the differential equation (6.14) can be obtained from the so-called "tilted washboard" model. This describes a mechanical analog based on the fact that the equation of motion (6.14) is the

---

[11] N. F. Pedersen, T. F. Finnegan, and D. N. Langenberg, *Phys. Rev.* **B6**, 4151 (1972); C. M. Falco, W. H. Parker, and S. E. Trullinger, *Phys. Rev. Lett.* **31**, 933 (1973); D. A. Vincent and B. S. Deaver, Jr., *Phys. Rev. Lett.* **32**, 212 (1974).

[12] R. E. Harris, *Phys. Rev.* **B10**, 84 (1974).

[13] W. C. Stewart, *Appl. Phys. Lett.* **12**, 277 (1968); D. E. McCumber, *J. Appl. Phys.* **39**, 3113 (1968).

same as that of a particle of mass $(\hbar/2e)^2 C$ moving along the γ axis in an effective potential

$$U(\gamma) = -E_J \cos\gamma - (\hbar I/2e)\gamma \tag{6.14c}$$

sketched in Fig. 6.3*b* and subjected to a viscous drag force $(\hbar/2e)^2(1/R)\, d\gamma/dt$. Clearly, the characteristic energy scale in (6.14*c*) is the Josephson coupling energy $E_J = (\hbar/2e)I_{c0}$. Again, in anticipation of treating thermal fluctuations, we have introduced the notation I_{c0} to denote explicitly the *fluctuation-free* intrinsic critical current of the RCSJ model. The geometric significance of I_{c0} is that when $I = I_{c0}$, the local minima of the tilted cosine become only horizontal inflection points in the otherwise continually downward slope, so that for $I \gtrsim I_{c0}$ no stable equilibrium point exists. For physical intuition, we visualize a mass point moving in a gravitational field along a track having the contour (6.14*c*).

For more quantitative thinking, we use a different construction, in which the vertical coordinate of the particle indicates its *total* energy as it moves along its trajectory, so that the height of the trajectory *above* the cosine curve represents the instantaneous kinetic energy $\frac{1}{2}m(d\gamma/dt)^2 = \frac{1}{2}CV^2$. In the absence of energy loss from damping, such a trajectory is horizontal. Noise from coupling to a thermal heat bath randomly shifts the system energy up and down by an amount of order k_BT on a time scale set by the relaxation time RC. The probability of a large excursion by an amount ΔE falls as $e^{-\Delta E/k_BT}$. As we discuss in detail in Sec. 6.3.3, these fluctuations allow the system to escape from the shallow energy minima which exist for I slightly below I_{c0}, giving a measured critical current $I_c < I_{c0}$.

6.3.2 *I-V* Characteristics at $T = 0$

So long as $I < I_{c0}$, a static solution of (6.14) with $\gamma = \sin^{-1}(I/I_{c0})$ and $V = 0$ is allowed. However, if $I > I_{c0}$, only time-dependent solutions exist, and they determine even the dc *I-V* curve of the junction. In this section, we discuss the nature of the predicted *I-V* curves in the absence of thermal fluctuation effects, which are treated in the following section.

OVERDAMPED JUNCTIONS. If C is small so that $Q \ll 1$, (6.14) reduces to a first-order differential equation, which can be written as

$$\frac{d\gamma}{dt} = \frac{2eI_{c0}R}{\hbar}\left(\frac{I}{I_{c0}} - \sin\gamma\right)$$

From the form of this equation one sees that, for $I > I_{c0}$, $d\gamma/dt$ is always positive but varies periodically with $\sin\gamma$. The phase advances more slowly (corresponding to a low instantaneous voltage) when $\sin\gamma$ is positive (so that the supercurrent is in the forward direction), and vice versa when $\sin\gamma$ is negative and the supercurrent opposes the dc bias current. One finds the time average voltage by integrating this equation to determine the period of time T required for γ to advance by 2π, and then by using the Josephson frequency relation $2eV/\hbar = 2\pi/T$. The result is the simple form

$$V = R(I^2 - I_{c0}^2)^{1/2} \tag{6.15}$$

which smoothly interpolates between $V = 0$ for $I < I_{c0}$, and Ohm's law $V = IR$ for $I \gg I_{c0}$.

Referring back to the argument given in the previous paragraph, when I only slightly exceeds I_{c0}, the dc voltage is actually the time average of a periodic series of *pulses*. These occur with frequency $f = 2eV_{dc}/h$, when γ passes through $3\pi/2 \bmod 2\pi (\sin\gamma \approx -1)$, each pulse having $V_{max} \approx 2I_{c0}R$ and an area $\int V\,dt = h/2e = \Phi_0/c$, so that the Josephson frequency relation (6.2) implies a total phase advance by 2π per pulse. In between the pulses the phase advances slowly as γ passes through the vicinity of $\pi/2$, where the supercurrent is maximum ($\sim I_{c0}$, but still less than I). This concept that a phase slip by 2π is the elementary resistive process in a Josephson junction also carries over to dissipation in other superconducting configurations such as long filaments.

In the tilted-washboard model, the heavy damping ($Q \ll 1$) implies that viscous drag dominates inertia, so that the instantaneous velocity of the mass is proportional to the instantaneous force, i.e., proportional to the local slope of the washboard. When I barely exceeds I_{c0}, the particle spends most of its time slowly sliding down the gentle incline near the almost horizontal inflection points, interrupted by quick drops over the "waterfall" down to the next nearly horizontal level. These quick drops correspond to the voltage pulses in the Josephson junction case.

UNDERDAMPED JUNCTIONS. When C is large enough so that $Q > 1$, the I-V curve becomes hysteretic. In the absence of thermally activated processes, upon increasing I from zero, $V = 0$ until I_{c0}, at which point V jumps discontinously up to a finite voltage V, corresponding to a "running state" in which the phase difference γ increases at the rate $2eV/\hbar$. (In the washboard analog, this corresponds to the mass point sliding steadily down the inclined washboard.) In the simple RCSJ model, this $V \approx I_{c0}R$, but in an ideal tunnel junction at $T \ll T_c$, the voltage jumps up to near the energy gap voltage $V_g = 2\Delta/e$ (see Fig. 6.4). If I is now reduced below I_{c0}, V does *not* drop back to zero until a "retrapping current" $I_{r0} \approx 4I_{c0}/\pi Q$ is reached. (In the analog, this hysteresis reflects the effect of the inertia of the moving mass, which with light damping can carry it up and over a barrier which would have stopped it if damping were heavy.) The physical reason for the 1/Q dependence of I_{r0} is that the energy fed in by the bias current as the phase advances by 2π from one barrier to the next must be dissipated in the same time to allow the system to retrap back into the zero voltage state. We shall derive this explicit expression for I_{r0} in the next section to provide a physical basis for understanding how I_r will be affected by thermal fluctuations.

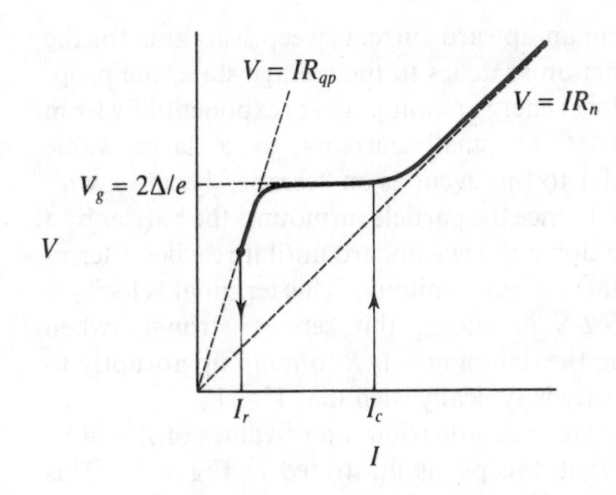

FIGURE 6.4
Hysteretic I-V curve of an underdamped Josephson junction. (Schematic.)

6.3.3 Effects of Thermal Fluctuations

When thermally activated processes are taken into account, the I-V curves of Josephson junctions can be strongly modified from the simple picture described earlier. This can be shown mathematically by adding a Johnson noise current term to the bias current in (6.14), or, more physically by considering the effect of thermal energies in the tilted-washboard model, introduced above. As noted in the introduction to this chapter, fluctuation effects are important unless the barrier height $2E_J$ is much larger than k_BT. Well below T_c, E_J is nearly constant and it follows from (6.10) that the ratio $2E_J/k_BT = 1.76(R_Q/R_n)(T_c/T)$, where the quantum resistance $R_Q = h/4e^2 = 6,453\Omega$. Thus, if $R_n > R_Q$ thermal fluctuation effects are strong down to $T/T_c \sim R_Q/R_n$, while if $R_n < R_Q$, they are important only fairly near T_c, where $E_J(T) \ll E_J(0)$. We now consider specific examples.

REDUCTION OF I_c IN UNDERDAMPED JUNCTIONS BY PREMATURE SWITCHING. When $k_BT \ll E_J$, thermally activated escapes from one potential minimum over the barrier to the next in an underdamped junction have a small probability $\sim e^{-\Delta U(I)/k_BT}$ at each attempt. The attempt frequency $\omega_A/2\pi$ is $\sim\omega_p/2\pi$, as the representative phase point oscillates back and forth in the well at the characteristic frequency of (6.14). More precisely, in the presence of a bias current, the attempt frequency is the frequency of small oscillations at the minimum of the *tilted* washboard, for which ω_p can be shown to be replaced by

$$\omega_A = \omega_p[1 - (I/I_{c0})^2]^{1/4} \tag{6.16}$$

Unless I is very near I_{c0}, this distinction between ω_A and ω_p is unimportant.

The exact current-dependent barrier height from (6.14*c*) is given by an algebraically awkward expression, but it can be approximated quite well by

$$\Delta U(I) \approx 2E_J(1 - I/I_{c0})^{3/2} \tag{6.17}$$

which goes to zero as $I \to I_{c0}$. Thus, in an upward current sweep searching for the "critical current" I_c at which the junction switches to the voltage state, the probability *per unit time* of escape from a local energy minimum rises exponentially from a very low value $\sim(\omega_p/2\pi)e^{-2E_J/k_BT}$ at small currents to a large value $\sim\omega_p/2\pi \sim 10^{10}\ \text{sec}^{-1}$ near I_{c0}. We refer to this event as an "escape" because, with the low damping associated with $Q > 1$, once the particle surmounts the barrier by a thermal fluctuation, it will accelerate down the washboard until it reaches a terminal velocity and never retrap in another energy minimum. This terminal velocity is determined by the damping. In *S-I-S* junctions, this sets in strongly when $V = (\hbar/2e)\, d\varphi/dt$ reaches V_g, causing the damping $\sim 1/R$ to jump up abruptly to $1/R_n$. Accordingly, the terminal velocity is typically such that $V \approx V_g$.

Since this escape is a stochastic process, a distribution of values of I_c will be measured on successive upward current sweeps, as illustrated in Fig. 6.5*b*. This distribution is characterized by its width δI_c and by its mean depression below the "unfluctuated" critical current I_{c0}. Although fully quantitative results require numerical computations (see Fulton and Dunkleberger[14]), one can show that the mean depression of I_c can be approximated by the formula

$$< I_c >= I_{c0}\{1 - [(k_BT/2E_J)\ln(\omega_p\Delta t/2\pi)]^{2/3}\} \tag{6.18}$$

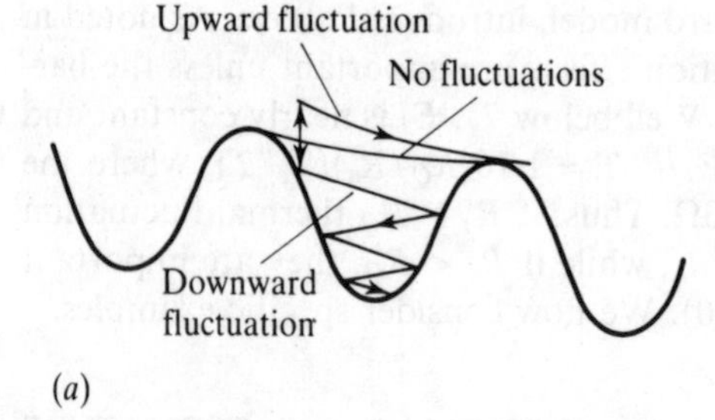

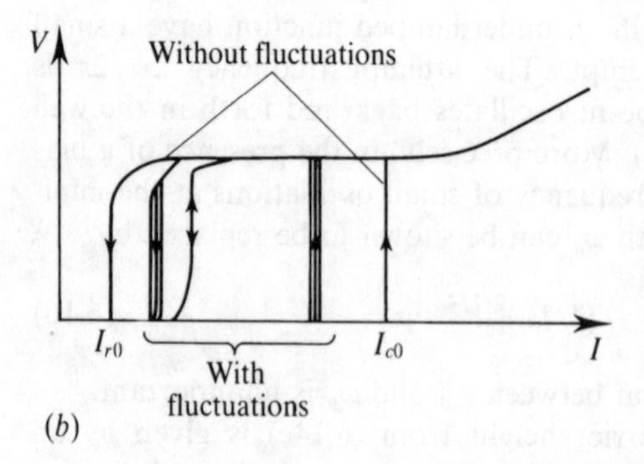

FIGURE 6.5
(*a*) Schematic of the effect of fluctuations on the retrapping process. (Total system energy is plotted vs. phase.) (*b*) Sketch of the effect of thermal fluctuations in reducing and eventually eliminating hysteresis in an underdamped Josephson junction.

[14] T. Fulton and L. N. Dunkleberger, *Phys. Rev.* **B9**, 4760 (1974).

where Δt is the time spent sweeping the bias current through the dense part of the distribution of observed I_c values. Since this Δt is of the order of 1 sec in a typical experiment, the logarithm typically will be of the order of $\ln 10^{10} \approx 23 \gg 1$ and very weakly sensitive to the current sweep rate. Because this logarithm is so large, fluctuation effects cause a major reduction in I_c as soon as $k_B T$ is as large as 5 percent of E_J. One can further show that the width δI_c of the switching distribution is approximately given by the mean depression of I_c divided by this same logarithmic factor.

THE RETRAPPING CURRENT I_r IN UNDERDAMPED JUNCTIONS. Although more attention has traditionally been focused on the critical current I_c, it is now recognized that the retrapping current I_r, at which the junction drops back into the zero voltage state upon reducing the bias current, provides a uniquely direct probe of the *damping* experienced by the junction. If there were *no* damping, the mass point sliding down from a maximum of the potential in Fig. 6.3*b* would not retrap as the tilt ($\sim I$) was reduced until the tilt reached zero, corresponding to I_r being zero. With finite damping, I_{r0} is fixed by the current (or washboard tilt) at which the energy dissipated in advancing γ from one maximum of the washboard to the next exactly equals the work done by the drive current during this same motion. Using this criterion, we can easily derive the formula $I_{r0} \approx 4I_{c0}/\pi Q$, which is valid for $Q \gg 1$ in the absence of fluctuations. We use the convention in which the height of the representative point above the washboard is its kinetic energy in the analog, or $\frac{1}{2}CV^2$ in the junction. If the representative point starts at the top of one maximum of the (slightly) tilted washboard, as shown in Fig. 6.5*a*, where it has zero velocity, I_{r0} is defined as the current at which the representative point just exactly reaches the next maximum, again with zero velocity. As noted above, with low damping ($Q \gg 1$), the representative point moves in a nearly horizontal trajectory from peak to peak. The kinetic energy is then $\frac{1}{2}CV^2 = E_J(1 + \cos\gamma)$, which implies a γ-dependent rate of dissipation V^2/R. If this dissipation is integrated over time as γ increases from $-\pi$ to $+\pi$, using the Josephson relation $d\gamma/dt = 2eV/\hbar$ to convert the time integral to an integral over γ, and the result is equated to the energy $Ih/2e$ fed in by the current during the same motion, one obtains $I_{r0} = 4I_{c0}/\pi Q$. Thus, I_{r0} is directly proportional to the damping $\sim 1/Q$, as stated earlier.

When one takes into account thermal fluctuations, one finds the perhaps surprising result that they *increase* I_r above the unfluctuated I_{r0}, whereas they *decreased* I_c below I_{c0}. The reason for this can be understood by the following physical argument. Assume $I = I_{r0}$, and consider a trajectory which just grazes successive maxima, as described above and as shown in Fig. 6.5*a*. A single downward fluctuation in energy leads to a trajectory that (in the absence of any subsequent fluctuation) spirals down to rest at the local minimum because there is dissipation but no net energy input. On the other hand, a single upward fluctuation relaxes back toward the initial trajectory because the increased energy dissipation $\sim V^2$ outweighs the increased energy input $\sim V$. Thus, $I = I_{r0}$ now leads to *certain* retrapping rather than being marginal, and the actual mean retrapping current I_r must be *greater* than I_{r0}. Of course, retrapping will also occur at a distribution of values on successive current downsweeps because of the stochastic nature of the process. To estimate this effect, Ben-Jacob et al.[15] have made an analysis resembling that leading to (6.18), despite the fact that here one is dealing with a driven *non*equilibrium regime. In this case, the characteristic energy is that of the (downward) energy fluctuation needed to initiate trapping, rather than the upward fluctuation needed to initiate escape.

In Fig. 6.5*b*, we schematically summarize the effect of thermal fluctuations on the *I-V* curve of an underdamped Josephson junction. In a low-resistance junction ($R \ll R_Q$) or at low temperatures, fluctuation effects are unimportant, and I_c and I_r are observed to occur at sharply defined values near the unfluctuated values I_{c0} and I_{r0}. In higher-resistance junctions (or higher temperatures), I_c and I_r each acquires a distribution of values, the means of which converge toward one another as the effect of fluctuations becomes more prominent. Finally, when $k_B T \gtrsim E_J$ and fluctuations dominate, the system is rapidly jumping back and forth between the trapped and running states, leading, on a laboratory time scale, to a broadened *nonhysteretic* resistive transition at a current value that is intermediate between I_{c0} and I_{r0}, at which the two states occur with equal likelihood.

AMBEGAOKAR-HALPERIN THEORY FOR OVERDAMPED JUNCTIONS. When a thermal noise current is included as an additional drive term in (6.14), Ambegaokar and Halperin[16] showed that the simple *I-V* relation for an overdamped junction, $V = R(I^2 - I_c^2)^{1/2}$ for $I > I_c$, quoted as (6.15), is fundamentally modified. In particular, as shown in Fig. 6.6, they found that there is always a *finite* resistance, even below I_c, because of thermally activated phase-slip processes. In terms of the tilted-washboard model, the representative phase point in an overdamped junction *diffuses* over the barriers in a continuing process, rather than making a single "escape" over the barrier. This difference occurs because the heavy damping brings the phase point back into an equilibrium distribution in the next minimum before it can diffuse to the next barrier, so that it has no chance to "run away" as it does in the lightly damped case. This activated resistance is nonlinear, increasing as $I \to I_c$, but it has a nonzero limiting value R_0 as $I \to 0$, which is related to the normal-state resistance by

$$R_0/R_n = [I_0(u/2)]^{-2} \propto ue^{-u} \tag{6.19}$$

Here $u = \hbar I_{c0}/ek_B T$ is the normalized activation energy, I_0 is the modified Bessel function, and the exponential dependence holds when $u \gg 1$. This exponentially small linear resistance at low currents is the single-junction analog of the so-called *TAFF* (thermally activated flux-flow) model of the resistance in bulk

[15] E. Ben-Jacob, D. J. Bergman, B. J. Matkowsky, and Z. Schuss, *Phys. Rev.* **A26**, 2805 (1982).
[16] V. Ambegaokar and B. I. Halperin, *Phys. Rev. Lett.* **22**, 1364 (1969).

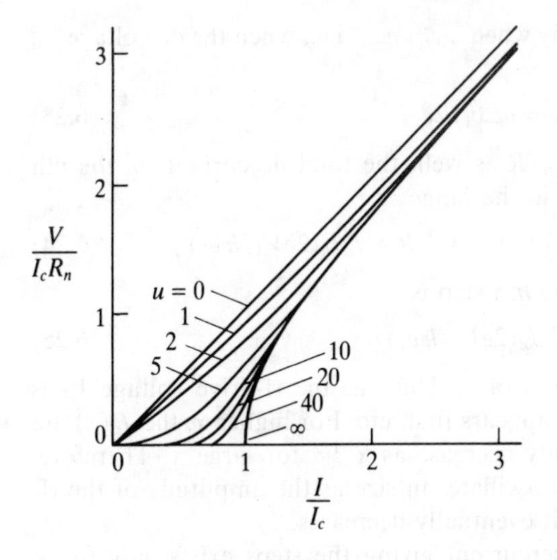

FIGURE 6.6
I-V characteristics of overdamped junction in the presence of thermal activation, as found by Ambegaokar and Halperin; here $u = \hbar I_c/ekT$.

superconductors, in which the activation energy is thought to be the energy to move a fluxon rather than the energy for a phase slip in a single junction.[17]

6.3.4 rf-Driven Junctions

When a Josephson junction is irradiated with radiation of angular frequency ω_1, the response of the *supercurrent* gives rise to constant-voltage *Shapiro steps*[18] in the dc *I-V* curve at voltages $V_n = n\hbar\omega_1/2e$, whereas the *photon-assisted tunneling* response of the *quasi-particles* gives rise to shifted images of the energy gap structure in the *I-V* curve, displaced in voltage by $\Delta V_n = n\hbar\omega_1/e$. In this section, we review each of these phenomena in turn.

SHAPIRO STEPS. For simplicity, we start by treating a junction with an ideal voltage bias of

$$V = V_0 + V_1 \cos \omega_1 t \tag{6.20}$$

Using γ to represent the phase difference across the junction, and integrating (6.20), we obtain

$$\gamma(t) = \gamma_0 + \omega_0 t + (2eV_1/\hbar\omega_1) \sin \omega_1 t \tag{6.21}$$

where γ_0 is a constant of integration and $\omega_0 \equiv 2eV_0/\hbar$. Inserting this $\gamma(t)$ into $I_s = I_c \sin \gamma$, and using the standard mathematical expansion of the sine of a sine in terms of Bessel functions, we have

$$I_s = I_c \sum (-1)^n J_n(2eV_1/\hbar\omega_1) \sin[\gamma_0 + \omega_0 t - n\omega_1 t] \tag{6.22}$$

[17] M. Tinkham, *Phys. Rev. Lett.* **61**, 1658 (1988).

[18] S. Shapiro, *Phys. Rev. Lett.* **11**, 80 (1963).

This contributes a dc component only when $\omega_0 = n\omega_1$, i.e., when the dc voltage V_0 has one of the Shapiro step values

$$V_n = n\hbar\omega_1/2e \tag{6.23}$$

If we include the normal current V_n/R as well, the total dc current on the nth Shapiro step can take on any value in the range

$$V_n/R - I_c J_n(2eV_1/\hbar\omega_1) \leq I \leq V_n/R + I_c J_n(2eV_1/\hbar\omega_1) \tag{6.24}$$

In other words, the half-width of the nth step is

$$I_n = I_c J_n(2eV_1/\hbar\omega_1) \tag{6.25}$$

The $J_n(x)$ vary as x^n for small values of x. Thus, as the rf-drive voltage V_1 is increased from zero, the lowest step appears first, etc. For higher x, the $J_n(x)$ are oscillatory functions which eventually decrease as $x^{-1/2}$ for large x. Therefore, once a given step appears, its width oscillates in size as the amplitude of the rf-drive voltage is increased and then it eventually decreases.

Note that the dc average supercurrent giving the steps exists *only* for V exactly equal to one of the V_n for this simple voltage-biased case. At all voltages between the V_n there would be no dc effect of the supercurrent. Of course, the same would be true of the dc *I-V* curve of the *un*irradiated junction, if it could be voltage-biased. The *I-V* curves that we have been discussing were all based on the more realistic assumption of dc-*current* bias, on the basis that the current of a zero resistance device is fixed by the external resistance in the circuit. Similarly, in the rf case, the drive is never an ideal voltage source, and in most cases, it is closer to a current drive for rf as well as for dc. Numerical calculations are then required to obtain the step widths, since an analytic solution of the sort leading to (6.25) is available only for voltage bias. For example, the numerical results of Russer[19] for ideal current bias show that there is a time-average supercurrent even for $V \neq V_n$, and the step widths oscillate in "quasi-Bessel function" form, showing a qualitative resemblance to the voltage-biased case.

PHOTON-ASSISTED TUNNELING. In the preceding discussion of the Shapiro steps, we approximated the parallel quasi-particle channel by an RCSJ model ohmic resistance R. In fact, in a superconducting tunnel junction with high-quality oxide barrier, the quasi-particle current is a highly nonlinear function of voltage, with a sharp onset of current at the gap voltage $V_g = 2\Delta/e$. Thus, a better approximation for the total current than that given above in (6.24) would be obtained by replacing the term V/R by the actual quasi-particle tunneling current $I_{qp}(V)$. However, this $I_{qp}(V)$ must take into account the effect of the radiation on the quasi-particle tunneling.

[19] P. Russer, *J. Appl. Phys.* **43**, 2008 (1972).

This can be done by the method introduced by Tien and Gordon.[20] It is based on the observation that an rf voltage applied *between* the two electrodes has no effect on their internal energy levels except to shift those in one electrode up or down relative to those in the other electrode by $eV_1 \cos \omega_1 t$. Thus, a quasi-particle energy $E_{\mathbf{k}}$ becomes $E_{\mathbf{k}} + eV_1 \cos \omega_1 t$, so that the quantum-mechanical frequency factor $e^{-iE_{\mathbf{k}}t/\hbar}$ becomes frequency-modulated and is replaced by $e^{-i\int(E_{\mathbf{k}}+eV_1 \cos \omega_1 t)\,dt/\hbar} = e^{-iE_{\mathbf{k}}t/\hbar} e^{-i(eV_1/\hbar\omega_1)\sin \omega_1 t}$. Using a Bessel function identity similar to that used to derive (6.22), we can write the factor involving V_1 as a sum of terms of the form $J_n(eV_1/\hbar\omega_1)e^{-in\omega_1 t}$. This result can be interpreted physically as saying that the quasi-particle level at $E_{\mathbf{k}}$ effectively splits into many levels at $E_{\mathbf{k}} \pm n\hbar\omega_1$, with probabilities given by the square of the amplitude coefficient, i.e., $[J_n(eV_1/\hbar\omega_1)]^2$. When this modified density of states is inserted into the semiconductor model, it has the effect that the quasi-particle tunneling *I-V* curve $I_{qp}^0(V)$ without radiation is replaced by

$$I_{qp}(V) = \sum J_n^2(eV_1/\hbar\omega_1) I_{qp}^0(V + n\hbar\omega_1/e) \tag{6.26}$$

where n runs over all positive and negative integers including zero. The main effect of (6.26) is to take the jump in current at the gap voltage and to break it up into a series of smaller current jumps at $V_g \pm n\hbar\omega_1/e$.

In their appearance on an *I-V* curve, these jumps qualitatively resemble the Shapiro current steps which occur at voltages $V_n = n\hbar\omega_1/2e$. However, they differ in many ways: The voltage separation is twice as great (because e, not $2e$, appears in the denominator). They are not truly flat, but only reflect the sharpness of the energy gap feature; and their strength varies as the *square* of the Bessel function of *half* the argument relative to the Shapiro step of the same order. Fig. 6.7 shows an *I-V* curve of a junction irradiated by far-infrared radiation, in which a superposition of both types of steps is visible. As we describe later in Sec. 6.7, these highly nonlinear *I-V* curves are the basis for practical detectors and mixers for use in astrophysical measurements.

6.4 JOSEPHSON EFFECT IN PRESENCE OF MAGNETIC FLUX

6.4.1 The Basic Principle of Quantum Interference

It is very important for understanding, as well as for such applications as the SQUID magnetometer, to work out the effect of an applied magnetic field on the Josephson effect. For simplicity, we shall ignore the effect of the field on the superconducting electrodes themselves and focus, instead, solely on its effect on the gauge-invariant phase difference via the $\mathbf{A}$ term in (6.11). To avoid sensitivity

[20]P. K. Tien and J. P. Gordon, *Phys. Rev.* **129**, 647 (1963).

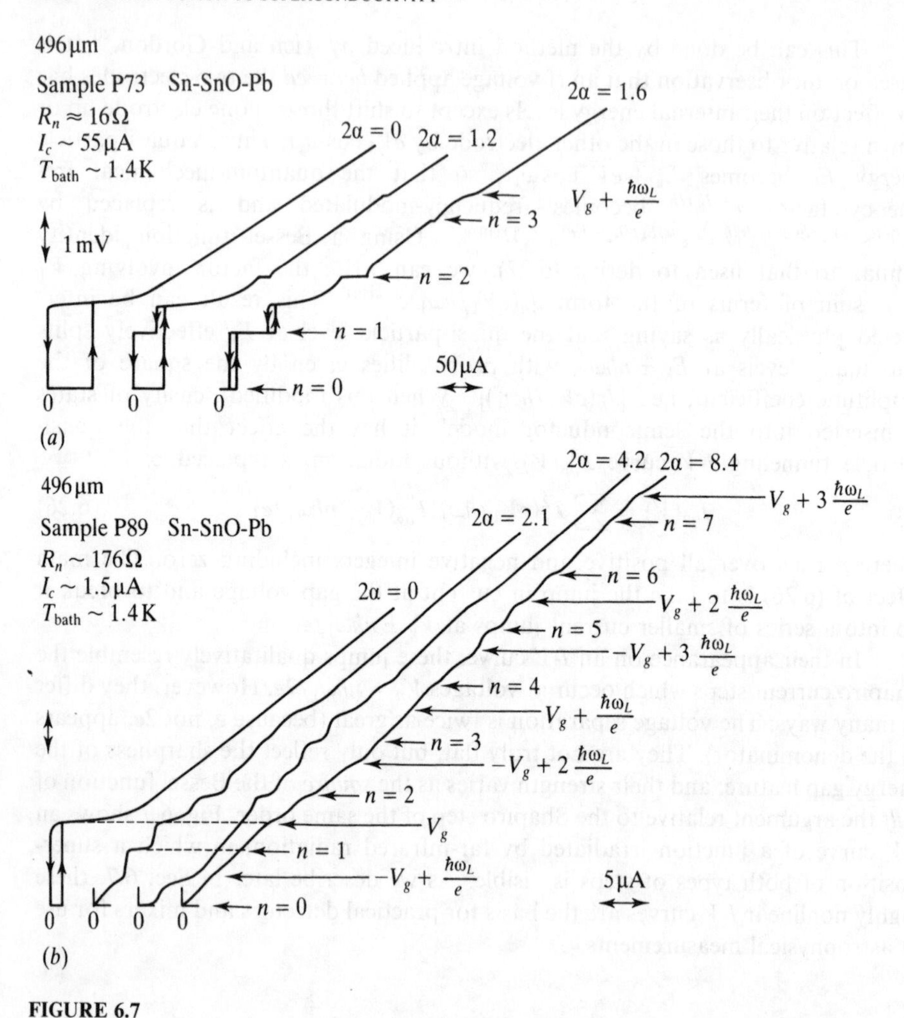

FIGURE 6.7
I-V curves of junctions irradiated by far-infrared radiation, showing both Shapiro steps and photon-assisted tunneling steps. The parameter $\alpha = eV_1/\hbar\omega_1$ is a measure of the amplitude of the radiation field. [*After Danchi et al.*, Appl. Phys. Lett. **41**, *883 (1982).*]

to the arbitrary gauge choice in **A**, we shall obtain results in terms of only the magnetic flux Φ through a specified contour, which is a gauge-invariant quantity.

The basic element of the analysis, from which we shall build solutions in more general cases, is the *phase-sensitive* summation of the supercurrent passed by two Josephson contacts (denoted 1, 2) between a pair of strong superconducting electrodes (denoted *A*, *B*) in Fig. 6.8*a*. Since **B** = curl **A**, the line integral of **A** around a contour passing through both links and the electrodes gives the enclosed flux Φ. If the electrodes are assumed to be thicker than λ, we can take the integration contour in the interior, where $\mathbf{v}_s$ is *zero*. Then the relation

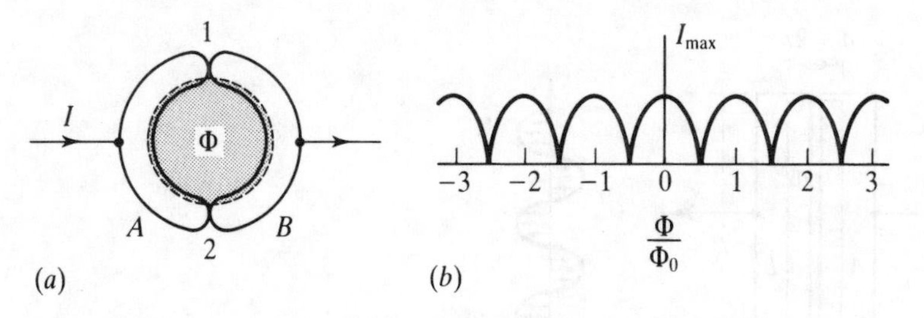

FIGURE 6.8
(*a*) Schematic diagram showing geometry for quantum interference of Josephson tunneling. (*b*) Resulting two-slit interference pattern.

$m^*\mathbf{v}_s = \hbar(\nabla\varphi - 2\pi\mathbf{A}/\Phi_0)$ [which is equivalent to (4.9)] implies that $\mathbf{A} = (\Phi_0/2\pi)\nabla\varphi$ in the electrodes, so

$$\Phi = \oint \mathbf{A}\cdot d\mathbf{s} = (\Phi_0/2\pi)\int_{\text{electrodes}} \nabla\varphi\cdot d\mathbf{s} + \int_{\text{links}} \mathbf{A}\cdot d\mathbf{s}. \tag{6.27}$$

Because the phase φ must be single-valued, $\int_{\text{electrodes}} \nabla\varphi\cdot d\mathbf{s}$ plus the sum of the finite phase differences $\Delta\varphi_i$ across the links must be zero, modulo 2π. This allows us to replace $\int_{\text{electrodes}} \nabla\varphi\cdot d\mathbf{s}$ by $(-\Sigma\Delta\varphi_i \bmod 2\pi)$. Combining this term with $\int_{\text{links}} \mathbf{A}\cdot d\mathbf{s}$, we find that the sum of the gauge-invariant phase differences γ_i (both taken in the same direction around the contour) is $2\pi\Phi/\Phi_0$. For calculating the total transport current from electrode *A* to electrode *B*, it is more convenient to define both γ_i in the sense from *A* to *B* as well, in which case we have

$$\gamma_1 - \gamma_2 = 2\pi\Phi/\Phi_0 (\text{mod } 2\pi) \tag{6.28}$$

This relation implies that γ_1 and γ_2 cannot simultaneously have the value $\pi/2$, as would be required to give the greatest total supercurrent, unless Φ is an integral multiple of Φ_0; otherwise the requirement of phase-coherent addition implies that the maximum supercurrent of the parallel combination must, in general, be less than $I_{c1} + I_{c2}$. In fact, if $I_{c1} = I_{c2} = I_c$, then the maximum supercurrent of the combination can be shown by a simple trigonometric argument to be

$$I_m = 2I_c|\cos(\pi\Phi/\Phi_0)| \tag{6.29}$$

This relation, plotted in Fig. 6.8*b*, has the same form as the analogous two-slit interference pattern of optics. It is the basis of the dc-SQUID magnetometer, which permits Φ to be measured to an extremely small fraction of Φ_0, as discussed in more detail in Sec. 6.5.

6.4.2 Extended Junctions

The same principle can be applied to determine the maximum supercurrent of an extended junction penetrated by flux Φ in the plane of the junction, as sketched in Fig. 6.9*a*. If the junction is rectangular and homogeneous with the field applied

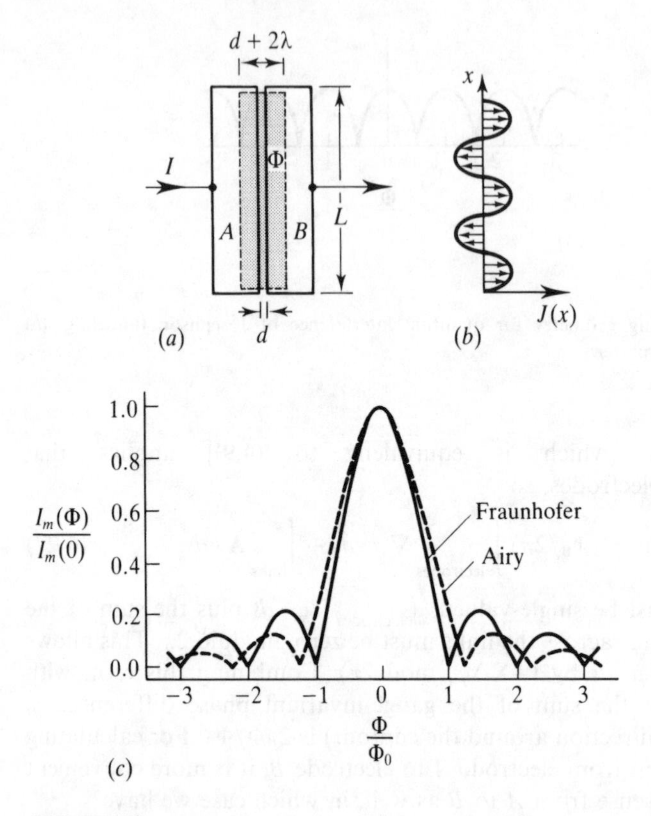

FIGURE 6.9
(*a*) Schematic diagram of extended junction geometry. (*b*) Position dependence of Josephson current density for case $\Phi \equiv (2\lambda + d)LH = \frac{5}{2}\Phi_0$ (*c*) Dependence of maximum supercurrent on enclosed flux for rectangular (solid curve) and circular junctions (dashed curve), showing Fraunhofer and Airy diffraction patterns, respectively. (For the circular case the diameter $2R$ of the junction replaces L in defining the enclosed flux Φ.)

parallel to one edge of the rectangle, and if the Josephson critical-current density J_c is small enough to allow neglect of field screening, the local value of the Josephson current $J_s = J_c \sin\gamma(x, y)$ oscillates sinusoidally with position as required by (6.28). Figure 6.9*b* shows an example, with $\Phi = \frac{5}{2}\Phi_0$. The net Josephson current cancels to zero over each complete cycle, so that in the example shown, the net current is that from one odd half-cycle. The complete cycles are sometimes referred to as "Josephson vortices" because they carry no net transport current. For an arbitrary flux Φ, the sinusoidal alternation in current density causes the maximum net supercurrent in one direction to be reduced from its zero field value by a factor

$$I_m(H)/I_m(0) = |\sin(\pi\Phi/\Phi_0)/(\pi\Phi/\Phi_0)| \tag{6.30}$$

a form often referred to as a *Fraunhofer diffraction pattern* by analogy with the case of light passing through a narrow rectangular slit. This function is depicted in Fig. 6.9*c*.

If the local critical supercurrent density $J_c(x)$ is *not* uniform, or if the junction area is *not* rectangular, but J_c is small enough to allow the neglect of screening, this simple result is replaced by a more general Fourier transform. Since it provides useful insights, we now develop this more general solution. Subsequently, we shall relax the condition that screening should be negligible.

JUNCTIONS WITH NEGLIGIBLE SCREENING. We choose a coordinate system such that the electrode surfaces are parallel to the xy plane, **H** lies along y, and the tunnel current is along z. If the electrodes are thick compared to λ and separated by an insulating barrier of thickness d, then the flux $\Delta\Phi$, enclosed by a contour including two tunneling paths separated by a distance Δx perpendicular to the field direction, is $\Delta\Phi \approx h(2\lambda + d)\Delta x$, where $h(x)$ is the local flux density in the barrier. Combining this relation with (6.28), we see that $\gamma(x, y)$ obeys the differential equations

$$\partial\gamma/\partial x = 2\pi(2\lambda + d)h/\Phi_0 \quad \text{and} \quad \partial\gamma/\partial y = 0 \tag{6.31}$$

If screening by the Josephson current can be neglected, $h = H$, the ambient magnetic field, and (6.31) can be integrated to yield

$$\gamma(x) = \gamma_0 + kx \quad \text{where} \quad k \equiv 2\pi H(2\lambda + d)/\Phi_0 \tag{6.31a}$$

The supercurrent through the junction is then given by

$$I_s = \iint J_c(x, y) \sin\gamma(x)\, dx\, dy \tag{6.32}$$

where the integration is over the junction area. After integration on y, we can rewrite (6.32) as

$$I_s = \int \mathcal{J}_c(x) \sin(\gamma_0 + kx)\, dx \tag{6.33}$$

where $\mathcal{J}_c(x) \equiv \int J_c(x, y)\, dy$ is the critical-current density *per unit length* along x. For example, in a rectangular junction with dimensions $X \times Y$ and uniform J_c, $\mathcal{J}_c(x) = YJ_c$, and the integration of (6.33) yields the familiar Fraunhofer diffraction pattern (6.30). On the other hand, for a uniform circular junction of radius R,

$$\mathcal{J}_c(x) = 2J_c(R^2 - x^2)^{1/2} \quad (\text{for } |x| \le R) \tag{6.34}$$

and the Fourier transform (6.33) yields the so-called Airy pattern (which also describes the diffraction of light through a circular aperture)

$$I_m(H)/I_m(0) = |J_1(kR)/(kR/2)| \tag{6.35}$$

where $kR \sim \Phi/\Phi_0 \propto H$. This function is also shown in Fig. 6.9*b*. If we ignore the periodic oscillation, this expression (6.35) falls asymptotically as $H^{-3/2}$, whereas the rectangular case (6.30) falls only as H^{-1}.

If the local $J_c(x,y)$ varies over the junction area in a known way, the Fourier transform (6.33) can be used to compute the field-dependent maximum supercurrent. Conversely, the field-dependent I_m can be used to infer $\mathcal{J}_c(x)$ from experimental data. For example, if $I_m(H)$ follows the Fraunhofer form, the $\mathcal{J}_c(x)$ must be uniform. Many illustrations of this process are given by Barone and Paterno.[21] An interesting example is that in which $J_c(x,y)$ contains a component $\delta J_c(x,y)$ which varies randomly on a short length scale ℓ because of structural defects in the material. As shown by Yanson,[22] this leads to a nonzero constant residual value of I_m at high fields, which only washes out when H reaches a value sufficiently high so that $k\ell \sim 1$. This persisting I_m can be viewed as a consequence of the Josephson vortices being "pinned" by random structural inhomogeneities, analogous to the pinning of vortices on defects in three-dimensional superconductors. Also, it may account for the residual J_c at high fields found[23] in some ceramic high-temperature superconductors. More explicitly, the overall Josephson coupling energy has a local minimum for some particular value of γ_0 in (6.31*a*), which "pins" the Josephson vortex pattern at that value of γ_0. The Fraunhofer pattern of $I_m(H)$ in Fig. 6.9*c* then reflects the fact that the depth of this pinning minimum is greatest when an odd number of half wavelengths of γ fit into the junction width. In this example of a rectangular junction, the pinning is not caused by random defects but, rather, by the discontinuous drop of $\mathcal{J}_c(x)$ at the edges of the junction. From this perspective one can interpret the more rapid fall of $I_m(H)$ in (6.35) for the round junction relative to (6.30) for the rectangular junction as reflecting the weaker "edge pinning" of the short-wavelength Josephson vortices by a round perimeter than by a straight edge aligned with the field direction in the rectangular case.

JUNCTIONS WITH SCREENING. Thus far we have ignored the magnetic field produced by the weak Josephson current, so that the field in the tunnel barrier is equal to the applied field H. Since H is a constant, the Josephson vortices were found to have a simple sinusoidal current variation. We now relax the requirement of negligible screening and take into account the Josephson currents. In weak applied fields, these currents will tend to screen the field from the interior of the junction, just as the London currents screen the field from a bulk sample. Similarly, above some critical field, flux will completely penetrate the junction, but the Josephson vortices will no longer be sinusoidal. Instead, they will acquire a soliton character.

---

[21]A. Barone and G. Paterno, *Physics and Applications of the Josephson Effect*, Wiley, New York (1982), pp. 70–95.

[22]I. K. Yanson, *Sov. Phys. JETP* **31**, 800 (1970); also *Tr. Fiz. Tekh. Inst. Niz. Temp. N.* **8**, 19 (1970).

[23]K. Heine, J. Tenbrink, and M. Thöner, *Appl. Phys. Lett.* **55**, 2443 (1989).

To work this out in detail, for simplicity we reinstate the requirement of a spatially uniform J_c. Taking into account the Josephson current, in (6.31) we can no longer assume that $h = H$. Rather, we use the Maxwell equation to write

$$\frac{\partial h}{\partial x} = \frac{4\pi J_z}{c} = \frac{4\pi J_c}{c} \sin\gamma \tag{6.36}$$

Upon differentiating (6.31) with respect to x and inserting (6.36) into the right-hand side, we obtain the second-order differential equation

$$\frac{d^2\gamma}{dx^2} = \frac{1}{\lambda_J^2} \sin\gamma \tag{6.37}$$

which defines the Josephson penetration depth

$$\lambda_J = \left[\frac{c\Phi_0}{8\pi^2 J_c(2\lambda + d)}\right]^{1/2} \tag{6.38}$$

For typical parameter values, λ_J is of the order of 1 mm. Clearly, λ_J plays the role of a penetration depth if $\gamma \ll 1$ since then (6.37) reduces to

$$\frac{d^2\gamma}{dx^2} = \frac{\gamma}{\lambda_J^2} \tag{6.39}$$

This has exponential solutions of the form $\gamma \sim e^{\pm x/\lambda_J}$, where x is measured from an outside edge of the junction. So long as the external field at the edge of the junction is much less than $4\pi J_c \lambda_J / c$, such "Meissner-effect" solutions are possible, with γ, h, and J_z all going exponentially to zero in the interior. On the other hand, if J_c is sufficiently small so that λ_J exceeds the lateral size of the junction, then screening effects are minimal and the treatment in the previous section is valid.

Insight into the nature of the solutions when γ is *not* restricted to be small can be gained by noting that (6.37) has the same form as the differential equation for a pendulum, if we make the transcription $x \to t$, $\gamma \to \theta$, and $\lambda_J^{-2} \to \omega_0^2 = g/L$, where θ is the angle of the pendulum from the *top* of its circular orbit and ω_0 is its natural frequency. In terms of this transcription, the solution for γ found in (6.31) by neglecting the effect of the Josephson current on the field corresponds to the motion of the pendulum whirling around and around with so much kinetic energy that gravitational acceleration is negligible. In (6.37), this corresponds to $\lambda_J \to \infty$, so that $d^2\gamma/dx^2 = 0$, $d\gamma/dx$ is constant, and there is a sinusoidal variation of J_z along x (the "Josephson vortices"). The finite junction width X implies that the pendulum-analog solution is relevant only for a finite time interval T, in which the pendulum makes as many revolutions as there are oscillations of $\gamma(x)$.

If one now considers a pendulum moving with less energy, but still with enough energy to have nonzero kinetic energy at the top of the circle, the motion $\theta(t)$ [and hence the variation $\gamma(x)$] will be periodic, but anharmonic. This leads to a nonsinusoidal, periodically reversing current distribution $J_z(x)$, of which each cycle contains one quantum of flux. Unlike the sinusoidal case discussed earlier, these Josephson vortices are actually localized entities, insofar as they are spaced with a separation which exceeds the length λ_J of the current distribution.

Finally, the Meissner-effect limit for a junction of length L corresponds to a pendulum moving with barely enough energy to go over the top. In this case, starting up with an initial angular velocity $(d\theta/dt)_0$ from an initial angle $-\theta_0$ (at a time t corresponding to $-L/2$), it decelerates nearly exponentially as it rises, moves very slowly for a long time (corresponding to the interior length of the junction) while going over the top where θ is small, and then exponentially accelerates down the other side, recovering the initial angular velocity at θ_0 (at a time corresponding to $+L/2$).

If the angular velocity at the top is negligible compared to the initial value, then θ_0 and $(d\theta/dt)_0$ are connected by the conservation of energy and are not independent. Translating back to the junction problem, we see that the corresponding initial condition is

$$\left(\frac{2\pi H}{\Phi_0}\right)^2 (2\lambda + d)^2 = \left(\frac{d\gamma}{dx}\right)_0^2 = \frac{2}{\lambda_J^2}(1 - \cos\gamma_0)$$

Solving, we have

$$\cos\gamma_0 = 1 - \frac{1}{2}\left(\frac{cH}{4\pi J_c \lambda_J}\right)^2 \tag{6.40}$$

Thus, for weak fields, γ_0 (the value of $|\gamma|$ at the edges of the junction) is given by

$$\gamma_0 = \frac{cH}{4\pi J_c \lambda_J}$$

On the other hand, the strongest field which can be screened is that corresponding to the pendulum starting at the bottom, namely, $\gamma_0 = \pi$, which yields

$$H_1 = \frac{8\pi J_c \lambda_J}{c} \tag{6.41}$$

which is typically of the order of 1 Oe. This field H_1 is the highest field for which a Meissner solution is possible and corresponds roughly to H_{c1} in a type II superconductor. [Actually, the screening at H_1 is only metastable. The maximum value of H for which screening is thermodynamically stable was shown by Josephson to be $H_{c1} = (2/\pi)H_1$.] Note that for fields approaching H_1 the screening is no longer exponential near the edge but, rather, becomes so in the interior when γ (or θ) have become small.

Continuing the analogy with a bulk superconductor, the maximum supercurrent that a large area junction can carry is essentially limited to J_c over a surface layer of thickness $\sim\lambda_J$ around the periphery of the junction. Any higher current would lead to magnetic fields that are greater than H_1, and hence to the establishment of a periodic "vortex" state, which would be dissipative in the presence of the transport current.

6.4.3 Time-Dependent Solutions

Now let us set up the differential equations governing the general variation in time and space of fields and currents in the junction. As before, we let the junction lie in the xy plane, with static magnetic field along y and phase variation along x.

NEGLIGIBLE SCREENING. For simplicity, we first return to the case of negligible screening currents, so that throughout the junction $h = H$ and $V = V_0$, the applied bias voltage. Then the time dependence is the same everywhere, given by the Josephson frequency relation $\partial\gamma/\partial t = 2eV_0/\hbar \equiv \omega_0$, whereas the spatial variation is governed by (6.31a). These equations are satisfied by a gauge-invariant phase difference given by

$$\gamma(x, t) = \gamma_0 + \omega_0 t + kx \tag{6.42}$$

This γ gives a periodic Josephson current distribution of exactly the same form as the "Josephson vortices" in the dc case, except that in the presence of a voltage it is moving with a velocity

$$v_0 = \frac{\omega_0}{k} = \frac{cV_0}{(2\lambda + d)H} \tag{6.43}$$

This motion is directly analogous to the flux-flow regime in the resistive state of bulk type II superconductors.

GENERAL CASE AND SINE-GORDON EQUATION. Now let us resume our development of equations to take into account the effect of the Josephson currents on the time-dependent electromagnetic fields in the barrier. Let $\mathbf{e}^0$ and $\mathbf{h}^0$ be the values in the barrier of the electric and magnetic fields $\mathbf{e}$ and $\mathbf{h}$. A key distinction is that $\mathbf{h}$ will extend a distance λ into the superconductors on either side, but $\mathbf{e}$ will be negligible except in the barrier; this asymmetry makes the junction a "slow-wave" structure. We can show this by integrating out the spatial dependence in the z direction, obtaining equations governing variations of $\mathbf{e}^0$ and $\mathbf{h}^0$ in the xy plane. Integrating the Maxwell equation curl $\mathbf{e} = -(1/c)\partial\mathbf{h}/\partial t$ over the narrow area indicated by the dashed contour in Fig. 6.10, we obtain

$$\frac{\partial e_z^0}{\partial x} = \frac{2\lambda + d}{d}\frac{1}{c}\frac{\partial h_y^0}{\partial t} \tag{6.44a}$$

Including the displacement current term, the Maxwell equation for curl $\mathbf{h}$ reduces to

$$\frac{\partial h_y^0}{\partial x} = \frac{4\pi}{c}J_z + \frac{\varepsilon}{c}\frac{\partial e_z^0}{\partial t} \tag{6.44b}$$

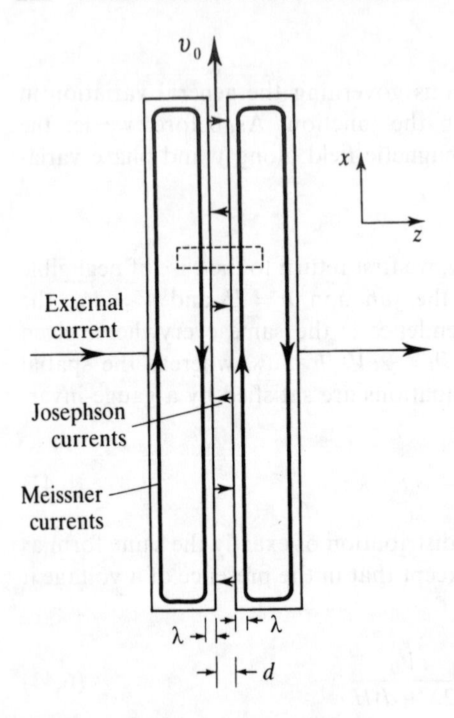

FIGURE 6.10
Diagram of an extended Josephson junction. The periodically reversing Josephson current pattern moves with velocity v_0 in the x direction if a voltage exists across the junction. The dashed contour is used to obtain (6.44*a*) using Faraday's law.

where J_z is the Josephson current and ε is the dielectric constant of the barrier. Combining the time derivative of (6.44*b*) with the space derivative of (6.44*a*), we obtain

$$\left(\frac{\partial^2}{\partial x^2} - \frac{1}{\bar{c}^2}\frac{\partial^2}{\partial t^2}\right)V = \frac{4\pi}{c^2}(2\lambda + d)\frac{\partial J_z}{\partial t} \tag{6.44c}$$

where $V = e_z^0 d$ is the voltage across the barrier, and

$$\bar{c}^2 = \frac{c^2}{\varepsilon(1 + 2\lambda/d)} \ll c^2 \tag{6.44d}$$

Since $\lambda \approx 500$ Å, while $d \approx 15$ Å, we see that $\bar{c} \approx c/20$ for a typical value $\varepsilon \approx 6$. As a consequence of the slow-wave nature of this junction, wavelengths are greatly reduced relative to free space. For example, microwaves of frequency $\nu = 10^{10}$ Hz, which have a free-space wavelength of 3 cm would have a wavelength in the junction of only ~1 mm. This disparity in wave velocities makes it difficult to couple electromagnetic energy in and out of the junction region.

It is convenient to transform (6.44*c*) into a form involving only the phase γ as the dependent variable. This can be done by using the Josephson current and frequency relations to eliminate J_z and V, followed by an integration with respect to time. To allow for a general orientation of $\mathbf{h}$ in the xy plane, we also include a term in $\partial^2/\partial y^2$. The result is

$$\left(\frac{\partial^2}{\partial x^2}+\frac{\partial^2}{\partial y^2}-\frac{1}{\bar{c}^2}\frac{\partial^2}{\partial t^2}\right)\gamma=\frac{\sin\gamma}{\lambda_J^2} \tag{6.45}$$

which is often referred to as the *sine-Gordon equation.*

If there is no time dependence, (6.45) reduces to (6.37), which, for $\gamma \ll 1$, shows exponential screening over a distance λ_J. A junction is called *small* if its width and length are small compared to λ_J, *large* if its length and width are large compared to λ_J, and *long* if its length $L \gg \lambda_J$ while its width $W \ll \lambda_J$.

Let us consider some more limiting cases: In the small junction limit, one can neglect the x, y variation, so that (6.45) reduces to

$$d^2\gamma/dt^2+\omega_p^2\sin\gamma=0 \tag{6.46}$$

with $\omega_p^2 \equiv \bar{c}^2/\lambda_J^2$. This equation reproduces (6.14), which was found from the RCSJ approximation for no damping ($1/Q=0$) and no bias current I, since the definition $\omega_p^2=2eI_c/\hbar C$ given there is equivalent to the ω_p^2 value found here (noting that the capacitance per unit area of the junction is $\varepsilon/4\pi d$). Of course, the sine-Gordon equation, like the RCSJ model, can be generalized by the inclusion of a damping term that is proportional to $\partial\gamma/\partial t$, as a perturbation,[24] but we shall not pursue that here.

Finally, if λ_J is very large, the solutions of (6.45) are simply plane waves with velocity $\bar{c}$. These electromagnetic waves will couple strongly to the (weak) Josephson currents if this velocity matches v_0, the velocity of the vortex pattern given by (6.43). This occurs when the applied voltage $V_0=[d(2\lambda+d)/\varepsilon]^{1/2}H$. A current peak in *I-V* curves was noted by Eck et al.[25] when this condition was satisfied, thus confirming these concepts. A related effect is the observation of steps in the *I-V* characteristic by Fiske.[26] These occur when the frequency of the Josephson currents match the frequencies of the various electromagnetic cavity modes of the slow wave in the junction.

SOLITONS IN LONG JUNCTIONS. Returning to the general equation (6.45), we see by its structure that it is invariant under a Lorentz transformation in which $\bar{c}$ plays the role of the light velocity. An interesting type of solution for the (infinitely) long junction geometry is the *soliton* or *fluxon* solution. In dimensionless units, with lengths x in units of λ_J and velocities u in units of $\bar{c}$, this has the form

$$\gamma(x,t)=4\tan^{-1}\left(\exp\left[\pm\frac{x-x_0-ut}{\sqrt{1-u^2}}\right]\right) \tag{6.47}$$

[24] D. W. McLaughlin and A. C. Scott, *Phys. Rev.* **A18**, 1652 (1978).

[25] R. E. Eck, D. J. Scalapino, and B. N. Taylor, *Phys. Rev. Lett.* **13**, 15 (1964).

[26] M. D. Fiske, *Rev. Mod. Phys.* **36**, 221 (1964); D. D. Coon and M. D. Fiske, *Phys. Rev.* **138**, A744 (1965).

This function maintains the value $\gamma = \pi$ at the moving point $x = x_0 + ut$ and goes from 0 to 2π as $[x - (x_0 + ut)]$ goes from $-\infty$ to $+\infty$ for the upper sign (\,fluxon) or vice versa for the lower sign (antifluxon). In other words, there is a positive Josephson current localized to a distance $\sim\lambda_J$ to the left of $(x_0 + ut)$ and a similar negative current to the right $(x_0 + ut)$. This is the localized nonlinear solution for a Josephson vortex which has evolved from the sinusoidal Josephson vortices discussed earlier for the case of negligible screening. Other exact solutions exist for the infinitely long lossless junction, representing fluxon-fluxon collisions, fluxon-antifluxon collisions, bound states, plasma waves, etc., and many have been observed experimentally.[27]

To consider a simple example, when a propagating fluxon reaches an open end of the junction transmission line, it is reflected back as an antifluxon. In a junction of length L, a full period of motion back and forth takes time $T = 2L/u\bar{c}$ and accomplishes an overall phase advance of 4π (since the passage of a fluxon and the return passage of an antifluxon both change γ by 2π). Thus, in the relativistic limit ($u \approx 1$) reached at large currents, the dc voltage across the junction will be $V = h\bar{c}/2eL$. If n fluxons are present, the voltage will be n times as great. These constant voltage values are referred to as *zero field steps* because they are based on fluxons trapped in the junction, which may be metastable even in the absence of an external magnetic field.

6.5 SQUID DEVICES

SQUIDs (*S*uperconducting *QU*antum *I*nterference *D*evices) are based on the principle of quantum interference developed in Sec. 6.4.1. In fact, (6.29) is the basic equation governing the two-junction dc SQUID, showing that its effective critical current is modulated by the flux enclosed, in units of the flux quantum Φ_0. The other popular alternative is the single-junction rf SQUID. Since the dc SQUID was the first to be developed, and presently offers the highest sensitivity, we shall concentrate on it and give only a summary of the operation of the rf version.

Because the state of the art of superconducting devices has been summarized in a recent book,[28] and continues to advance steadily, practical details will be treated only superficially. Nonetheless, it is hoped that this introductory discussion will serve to prepare the reader for more detailed technical analyses, as well as to give an illustration of the many applications in which superconducting junctions can be used.

---

[27]See, e.g., T. A. Fulton and R. C. Dynes, *Solid State Comm.* **12**, 57 (1973); B. Dueholm, O. A. Levring, J. Mygind, N. F. Pedersen, O. H. Soerensen, and M. Cirillo, *Phys. Rev. Lett.* **46**, 1299 (1981); K. Nakajima, H. Mizusawa, Y. Sawada, H. Akoh, and S. Takada, *Phys. Rev. Lett.* **65**, 1667 (1990).

[28]S. T. Ruggiero and D. A. Rudman (eds.), *Superconducting Devices*, Academic Press, San Diego (1990); see especially Chap. 2 by J. Clarke, "SQUIDS: Principles, Noise and Applications."

6.5.1 The dc SQUID

The modulation of the maximum supercurrent through two parallel junctions in a superconducting loop which encloses magnetic flux has been discussed in Sec. 6.4.1. Practical devices, however, are always biased above this current value and operate in a finite voltage regime. Insofar as the parallel junction combination acts as a single junction with flux-dependent critical current, we can apply our earlier *I-V* curve results to find the flux-dependent voltage. To avoid the complications and additional noise associated with hysteresis at the critical current, the junctions are usually operated in a slightly overdamped condition. (If the junctions are intrinsically underdamped *S-I-S* junctions, sufficient additional damping is supplied by fabricating the device with a metallic film strip as a shunt.) Drawing on (6.15), we can approximate the dc average voltage across the device in the dissipative regime by

$$V = (R/2)\{I^2 - [2I_c \cos(\pi\Phi/\Phi_0)]^2\}^{1/2} \tag{6.48}$$

where $R/2$ is the resistance of the two resistively shunted junctions in parallel, and the flux-modulated maximum supercurrent of the two junctions in parallel found in (6.29) has been inserted. (We retain the simplifying assumption that the two junctions have the same critical current I_c and resistance R.) Taking into account the trigonometric identity $\cos^2 x = (1 + \cos 2x)/2$, (6.48) demonstrates explicitly that V is periodic in Φ with period Φ_0.

Equation (6.48) shows that the SQUID is a flux-to-voltage transducer and can be used for that purpose in more complex devices. It is easily seen that the largest voltage swing is obtained if the device is biased at $I = 2I_c$, in which case V ranges from zero for integral numbers of flux quanta to $I_c R$ for half-integral flux values. It is important to note that because of the shunting needed to eliminate hysteresis, the value of R in this expression is much less than the normal resistance R_n of the tunnel junctions. Accordingly, the value of $I_c R$ is typically only a few microvolts, whereas the $I_c R_n$ product is of the order of the gap voltage, typically $\sim$1 mV. If the Q (or β_c) of the junctions could be lowered sufficiently to eliminate hysteresis by using small area, high current density junctions to reduce the capacitance (instead of by lowering the shunting resistance), these higher voltage swings should be in principle obtainable.

SCREENING EFFECTS. In this introductory discussion, we have implicitly neglected the difference between the externally supplied flux Φ_x and the flux Φ actually enclosed by the loop. These can differ for two reasons: the flux produced by the screening current I_s circulating in the loop and the flux coupled into the loop by the bias current. For simplicity of analysis, we shall assume that the current leads are attached symmetrically to the loop, so that the mutual inductance between the bias current and the loop is zero; in this case, the latter effect is zero. The flux produced by the screening current is not in general negligible, but it only distorts the functional form of the flux-voltage transfer function, not its periodicity. Let us look into this a bit more closely.

The transport supercurrent through the SQUID is the sum of the currents passed by junctions 1 and 2, namely,

$$I = I_c(\sin\gamma_1 + \sin\gamma_2) \tag{6.49a}$$

which is the basis for the flux-modulated maximum supercurrent given by (6.29). On the other hand, the circulating supercurrent which produces the screening flux is

$$I_s = (I_c/2)(\sin\gamma_2 - \sin\gamma_1) \tag{6.49b}$$

Both (6.49*a*) and (6.49*b*) are constrained from (6.28) by

$$\gamma_1 - \gamma_2 = 2\pi\Phi/\Phi_0 \ (\mathrm{mod}\ 2\pi) \tag{6.49c}$$

In this relation, Φ denotes the sum of Φ_x and Φ_s, where $\Phi_s = LI_s$ is the screening component of the flux and L is the inductance of the loop. Given I and Φ, one can solve for the two phase differences γ_i and then compute Φ_s and hence Φ_x. For example, if $I \approx 0$, then $\gamma_1 \approx -\gamma_2$, and

$$\Phi_x = \Phi + LI_c \sin(\pi\Phi/\Phi_0) \tag{6.49d}$$

This relationship can be inverted to obtain Φ as a function of Φ_x, as shown in Fig. 6.11.

An important special case is $\Phi = n\Phi_0$, for which $\gamma_1 = \gamma_2 \ (\mathrm{mod}\ 2\pi)$, so that $I_s = 0$ and $\Phi = \Phi_x$; thus, the periodicity of the SQUID response to Φ_x in integer multiples of Φ_0 is not affected by the screening. However, for convenient operation, one also requires that the relation between Φ and Φ_x be single-valued and nonhysteretic. To obtain a condition for this, we note that the maximum possible value of $\Phi_s = LI_s = LI_c$, according to (6.49*b*). Roughly speaking, if $|\Phi_s| < \Phi_0/2$, the screening is insufficient to give a multivalued relationship between Φ and Φ_x, given the period Φ_0. It is customary to define a screening parameter $\beta_m \equiv 2LI_c/\Phi_0$; more detailed analysis then shows that magnetic hysteresis is avoided, even in the absence of rounding by thermal fluctuations, if $\beta_m \lesssim 2/\pi \sim 1$.

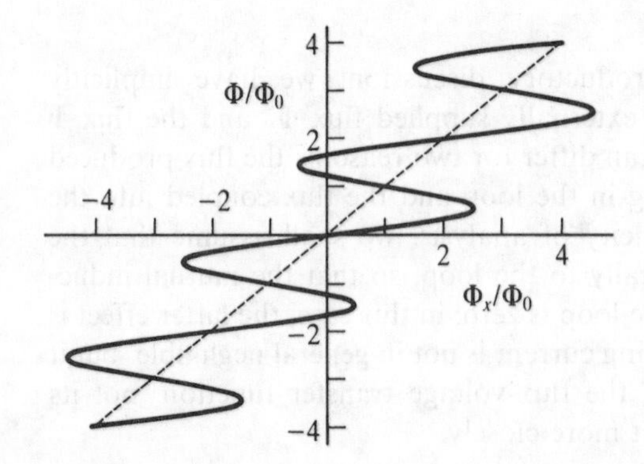

FIGURE 6.11
Total flux Φ vs. externally applied flux Φ_x as given by (6.49*d*), showing the effect of screening by circulating currents. The example plotted is for a dc SQUID with $LI_c/\Phi_0 = 2$ and operating at small bias current.

CHOICE OF PARAMETERS AND LIMITS ON SENSITIVITY. SQUID parameters must satisfy four constraints, two set by the need to avoid hysteresis and two set by thermal fluctuations. The two constraints which derive from the need to avoid hysteresis are $\beta_m \lesssim 1$ to avoid magnetic hysteresis and $\beta_c = Q^2 \lesssim 1$ to avoid hysteresis in the *I-V* curve. Restated explicitly in terms of parameter values, these can be written as

$$\beta_m \equiv 2LI_c/\Phi_0 \lesssim 1 \tag{6.50a}$$

$$\beta_c \equiv 2\pi I_c R^2 C/\Phi_0 \lesssim 1 \tag{6.50b}$$

The first of the two constraints based on thermal fluctuations is that phase fluctuations within each junction should not overcome the coupling energy E_J, causing loss of phase coherence. Roughly, this requires that $E_J > kT$, but because of the effects summarized by (6.18), one requires a stronger inequality. Computer simulations by Clarke and Koch[29] yield the more quantitative constraint

$$E_J = \hbar I_c/2e \gtrsim 5kT \tag{6.50c}$$

The second thermal requirement is set by flux fluctuations. By the equipartition theorem of statistical mechanics, the root-mean-square noise fluctuation of the flux in a loop of inductance L is given by $< \Phi_N^2 >^{1/2} = (kTL)^{1/2}$. If this exceeds $\sim\Phi_0/2$, the periodic response will be washed out by the noise. This imposes the constraint

$$L \lesssim \frac{\Phi_0^2}{4kT} \tag{6.50d}$$

Evaluated at 4.2 K, e.g., this implies that L must be less than $\sim$18 nH.

These four constraints are not all independent since (6.50*d*) is satisfied automatically if (6.50*a*) and (6.50*c*) are both satisfied. Thus, all four can be simultaneously satisfied by the choice of the three parameters I_c, L, and R^2C. We now turn to the question of how the actual limiting sensitivity set by thermal noise depends on parameter choices that are consistent with the constraints.

Viewed as a flux-to-voltage transducer, a SQUID can be characterized by the transfer coefficient $V_\Phi \equiv |(\partial V/\partial \Phi)_I|$. Evaluated at a point of maximum slope, (6.48) shows that $V_\Phi \approx 2I_c R/\Phi_0$. Given the constraint (6.50*a*), this is limited by $V_\Phi \lesssim R/L$. No matter how large this coefficient is, however, the minimum detectable flux change is determined by the competing flux noise level. Since present-day high-sensitivity SQUIDs operating at 4 K actually perform close to the ideal noise level set by thermal noise, a useful limiting case estimate can be made on that basis, as follows.

The flux resolution of the SQUID can be characterized by the equivalent flux noise $\Phi_N(t)$ with a spectral density $S_\Phi(f)$ at frequency f. Using the flux-to-

---

[29] J. Clarke and R. H. Koch, *Science* **242**, 217 (1988).

voltage transfer coefficient V_Φ, this can be related to the spectral density of voltage noise S_V, as

$$S_\Phi(f) = S_V(f)/(V_\Phi)^2 \approx 4kTL^2/R \tag{6.51}$$

In obtaining the last form, we have inserted the low-frequency Johnson noise expression $S_V(f) = 4kTR$ and used the limiting value $V_\Phi = R/L$. For the representative values $L = 1$ nH, $R = 1\,\Omega$, and $T = 4$ K, this corresponds to $\sim 10^{-5}\Phi_0$ in a bandwidth of 1 Hz.

Clarke and coworkers have shown that a more convenient figure of merit for characterizing the noise level in a SQUID is the flux noise spectral *energy* density

$$\varepsilon(f) = S_\Phi(f)/2L \approx 2kTL/R \tag{6.52}$$

The final form was obtained by using the same approximation as in (6.51); numerical simulations by Clarke's group indicate that a more realistic numerical prefactor is ~9 instead of 2. Because the option of lowering $\varepsilon(f)$ by increasing R is constrained by the need to keep $\beta_c < 1$ as given in (6.50*b*), it is useful to rewrite (6.52) eliminating R in favor of β_c, while continuing to assume that $\beta_m \approx 1$. Using the numerical prefactor from the simulations, we obtain

$$\varepsilon(f) \approx 16kT(LC/\beta_c)^{1/2} \tag{6.53}$$

with $\beta_c < 1$. This result has been found to give a good account of the noise energies measured in SQUIDs with a wide range of parameters. It clearly points to the possibility of increasing SQUID sensitivity by going to lower temperatures and to smaller structures to reduce L and C.

Evidently, (6.53) cannot hold all the way to $T = 0$ because quantum limits will prevent the sensitivity from becoming infinite. To make this approach to a quantum limit explicit, we customarily restate the noise energy $\varepsilon(f)$ as a multiple of $\hbar$, which has the same dimension (energy/frequency) as $\varepsilon(f)$. For a typical high-quality SQUID at 4 K, ε will be hundreds of times $\hbar$. The temperature dependence of (6.53) was tested by Wellstood et al.[30] by cooling to lower temperatures with a dilution refrigerator. They found that ε fell linearly with T down to ~150 mK, below which a special configuration was required to reduce hot electron effects. Using this, they found an effective temperature of ~50 mK at the lowest accessible bath temperature ~20 mK, at which point ε had been reduced to $\sim 5\hbar$.

PRACTICAL dc SQUIDs. Because of the constraint requiring that the SQUID loop itself have sufficiently low inductance to give good sensitivity, practical applications require an efficient inductive coupling into this small loop. Most SQUIDs use the spiral input coil in a square washer configuration, devised by Ketchen and

[30] F. C. Wellstood, C. Urbina, and J. Clarke, *Appl. Phys. Lett.* **50**, 772 (1987).

Jaycox.[31] A typical input coil described by Clarke has 50 turns and an inductance $\sim$800 nH, whereas the SQUID loop has $L \sim 0.4$ nH. The mutual inductance of $\sim$16 nH correponds to a coupling coefficient $k^2 = M^2/L_1L_2 \sim 0.75$, close to the theoretical limit of unity.

For convenience, SQUIDs are typically operated within a circuit called a *flux-locked loop*. In this arrangement, an additional coil is used to couple a small amplitude ($\leq \Phi_0/2$) ac ($\sim$100-kHz) modulation flux into the SQUID loop. This causes a modulation of the voltage across the SQUID, which depends on the difference between Φ and the nearest integer multiple of Φ_0. This signal is amplified and used to produce a feedback flux which is coupled into the loop to drive it back to the desired operating point Φ_{op}. Any external flux coupled in through the input circuit when this feedback loop is in operation then results in the generation of a canceling flux by the feedback circuit. The feedback current, which is directly proportional to the applied flux signal, is passed through a resistor to provide a voltage output of the SQUID circuit. In this way, one can measure changes in flux ranging from a small fraction of a flux quantum up to large numbers of quanta. Of course, the bandwidth of the device is limited to frequencies below the modulation frequency used in the feedback circuit.

6.5.2 The rf SQUID

Despite the fact that the dc SQUID offers higher sensitivity, because of its early commercial availability, the rf SQUID remains a commonly used form of the device. Accordingly, we shall give a brief sketch of the principles of its operation. For more details and a discussion of design optimization, the reader should consult the original literature.[32]

The rf SQUID loop is interrupted by only a single Josephson link instead of the two used in a dc SQUID. Since the Josephson element is shorted at dc by the superconducting loop, such a single-junction SQUID must be monitored at a radio frequency (typically $\sim$20 to 30 MHz). This is accomplished by coupling an rf current into the SQUID loop from a coil which forms part of a tank circuit resonant at the drive frequency ω_1, supplied by a constant rf-drive current I_1, as sketched in Fig. 6.12*a*. As shown below, the magnitude of the rf loss from the resonant circuit due to its coupling to the loop depends periodically on the dc flux enclosed by the loop, with period Φ_0. This flux-modulated rf loss modulates the Q of the tank circuit, and hence the magnitude of the rf voltage V_T across it as shown in Fig. 6.12*b*. This rf voltage is detected to provide the flux-dependent readout from the device. From this point on the device can be used in the same

[31] M. B. Ketchen and J. M. Jaycox, *Appl. Phys. Lett.* **40**, 736 (1982).

[32] See, e.g., A. H. Silver and J. E. Zimmerman, *Phys. Rev.* **157**, 317 (1967); J. E. Zimmerman, P. Thiene, and J. T. Harding, *J. Appl. Phys.* **41**, 1572 (1970); R. P. Giffard, R. A. Webb, and J. C. Wheatley, *J. Low Temp. Phys.* **6**, 533 (1972); W. W. Webb, *IEEE Trans. MAG*-**8**, 51 (1972).

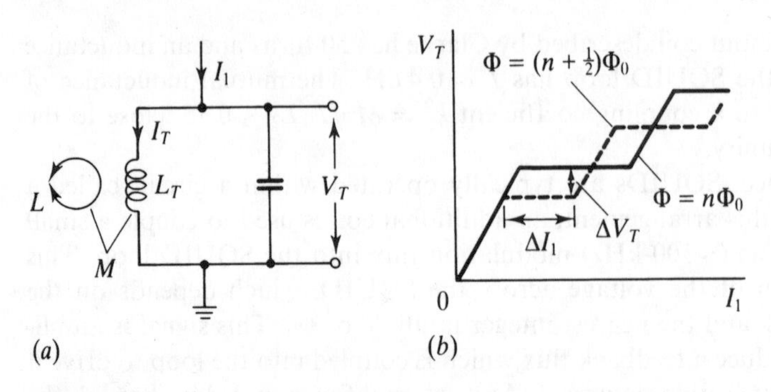

FIGURE 6.12
Operation of rf SQUID. (*a*) Schematic diagram of the circuit. A constant rf current I_1 is supplied, and the tank voltage V_T is monitored. (*b*) Relation between V_T and I_1 for the cases of integral and half-integral numbers of flux quanta. For intermediate values of flux; the voltage steps occur at intermediate values of V_T.

way as a dc SQUID; e.g., normally field modulation would be included so that it could be operated in a flux-locked loop.

Now let us look more closely at how the flux dependence of the rf loss arises. At the relatively low-operating frequency of the device, we can make a quasi-static approximation. The phase across the weak link is given by a variant of (6.28) for the single-junction case, namely,

$$\gamma = 2\pi\Phi/\Phi_0 \pmod{2\pi} \tag{6.54}$$

so that the circulating current is $I_s = I_c \sin(2\pi\Phi/\Phi_0)$, and the screened flux Φ in the loop satisfies

$$\Phi_x = \Phi + LI_c \sin(2\pi\Phi/\Phi_0) \tag{6.55}$$

This expression can be used to determine Φ as a function of Φ_x by a simple graphical construction similar to that used to obtain Fig. 6.11. If $2\pi LI_c \geq \Phi_0$, this relation is multivalued for at least some range of Φ_x, so that hysteresis will occur if Φ_x is cycled over a sufficient amplitude. It is the hysteretic jumps in Φ from one branch to another of (6.55), as Φ_x is swept, which provide the dominant dissipative process, on which the operation of the rf SQUID is based.

The energy ΔW dissipated in a complete drive cycle is given by the area of the hysteresis loop

$$\Delta W = \frac{1}{L}\oint \Phi_x d\Phi \tag{6.56}$$

For strong screening ($LI_c/\Phi_0 \gg 1$), the flux Φ is constrained by (6.55) to be near an integral flux value $n\Phi_0$, so that the flux change in each jump is typically $\sim\pm\Phi_0$. Thus, for a single-jump hysteresis loop, ΔW will be approximately given by (Φ_0/L) times the difference in the values of Φ_x at which the upward and downward jumps of Φ occur. On increasing Φ_x from zero, a quantum of flux will jump in when Φ_x first exceeds the maximum screening field LI_c. On the subsequent downward sweep, Φ_x exceeds the screening limit when $\Phi_x = \Phi_0 - LI_c$. The difference is $(2LI_c - \Phi_0)$, so the energy dissipated is

$$\Delta W \approx 2I_c\Phi_0 - \Phi_0^2/L \tag{6.57}$$

If the amplitude Φ_1 of the rf-drive flux is too small, no flux jumps and hysteresis loss can occur. Jumps and hysteresis loss start to occur when $\Phi_x \equiv \Phi_{dc} + \Phi_1 \pmod{\Phi_0}$ exceeds the maximum screening flux LI_c. For the two extreme cases, in which Φ_{dc} is an integral or half-integral multiple of Φ_0, this occurs at $\Phi_1 = LI_c$ and $LI_c - \Phi_0/2$, respectively. To relate this result to an observable quantity, we note that $\Phi_1 = MI_T = MV_T/\omega_1 L_T$, where M is the mutual inductance coupling of the SQUID loop to the tank coil, which has inductance L_T, across which an rf voltage V_T exists, and through which an rf current I_T flows. Thus, if one monitors V_T as the drive current I_1 to the tank circuit is increased from zero, V_T rises linearly with I_1 until V_T reaches $(\omega_1 LI_c/M)$ for the integer flux case, or the same factor times $(1 - \Phi_0/2LI_c)$ for the half-integer case.

If I_1 is increased further, V_T no longer continues its linear rise because the hysteretic loss effectively lowers the Q, and hence the impedance, of the tank circuit. Examining the behavior cycle by cycle on a time-resolved basis, one finds that each time a hystereic energy loss ΔW occurs, it reduces the stored energy in the tank circuit, lowering the circulating current so that Φ_1 becomes too small to trigger hysteretic loss on the next rf cycle. The continued input of energy from the source at a rate $I_1 V_T$ builds up the stored energy (and hence Φ_1) until the threshold is reached for another hysteretic loss to occur. With an increase of I_1, this buildup is hastened; but as soon as the threshold value is reached, the cycle repeats. This pins the observed value of V_T at the value corresponding to the threshold until I_1 reaches the value at which the buildup to threshold occurs after a *single* cycle of the rf drive. With a further increase of I_1, V_T will then again increase linearly until the threshold is reached for *two* hysteretic breakdowns to occur in a half-cycle; at this value of V_T a second voltage plateau occurs. With a further increase in I_1, a series of plateaus of V_T occurs, as sketched in Fig. 6.12*b*. A more quantitative discussion, taking into account the probabilistic nature of the switching transition, explains the fact that the plateaus actually have a nonzero slope.

From the above discussion and sketch we see that V_T will be modulated by $\Delta V_T = \omega\Phi_0 L_T/2M$ between integer and half-integer values of Φ_{dc}, if the SQUID is operated with I_1 fixed at a value for which a step in V_T exists at both the integer and half-integer flux values (and hence at all those values in between). On the face of it, it appears that the signal ΔV_T could be made arbitrarily large by making M smaller, but then the step widths would become too small for the device to work as described. In fact, one can show that the optimum value of the coefficient of coupling $k^2 = M^2/L_T L$ is $\sim 1/Q$. This corresponds to the smallest value of M giving overlapping step widths.

6.5.3 SQUID Applications

SQUIDs, either dc or rf, can be used for a wide range of measurement purposes, in all of which an auxiliary element provides a signal current to the input coupling coil. In this section, we give a brief survey to illustrate some of the range of applications which have already been demonstrated.

MAGNETOMETERS AND GRADIOMETERS. In a simple magnetometer, we connect a superconducting pickup loop to the terminals of the input coupling coil, thus forming a *flux transformer*. Because the entire input circuit is superconducting, fluxoid quantization holds for the circuit. Moreover, the kinetic terms are negligible compared to the electromagnetic ones if the conductors are macroscopic, so that flux itself is conserved in the circuit. This allows us to write

$$\Delta\Phi_p + (L_p + L_i)\,\Delta I = 0 \tag{6.58}$$

where $\Delta\Phi_p$ is the change in flux in the pickup loop, ΔI is the change in current in the circuit, and L_p and L_i, respectively, are the inductances of the pickup loop and of the input circuit of the SQUID. The flux coupled into the SQUID is then given by

$$\Delta\Phi = M_i\,\Delta I = M_i\,\Delta\Phi_p/(L_p + L_i) \tag{6.59}$$

where M_i is the mutual inductance of the input circuit to the SQUID loop. For choosing the dimensions of the pickup coil to optimize the design of such a magnetometer, we note that by making the coil bigger, we increase both $\Delta\Phi_p$ and L_p. It is plausible from (6.59) that this increases $\Delta\Phi$ until $L_p \approx L_i$, after which $\Delta\Phi$ decreases again because L_p increases faster with coil size or the number of turns than does $\Delta\Phi_p$. Thus, in an optimized case, $\Delta\Phi \approx M_i\,\Delta\Phi_p/2L_i$. This enables us to relate the flux energy in the pickup coil to that coupled into the SQUID by

$$\frac{(\Delta\Phi_p)^2}{2L_p} = \frac{4}{k^2}\frac{(\Delta\Phi)^2}{2L} \tag{6.60}$$

where $k^2 = M_i^2/L_iL$ is the coupling coefficient of the SQUID input to the SQUID loop. This relation allows us to determine the sensitivity of the magnetometer given the flux noise spectral energy density $\varepsilon(f)$ of the basic SQUID element, as estimated in (6.53) for a dc SQUID. In practice, magnetic-field sensitivities (at frequencies above the $1/f$ noise regime) of $5 \times 10^{-11}\ \mathrm{G/Hz^{-1/2}}$ have been obtained[33] with a pickup loop of a few millimeters diameter. This is a higher sensitivity than can be obtained by any other type of instrument.

SQUID magnetometers are a promising application for the high-temperature superconductors, which permit operation at liquid nitrogen temperatures,

[33] F. C. Wellstood, C. Heiden, and J. Clarke, *Rev. Sci. Inst.* **55**, 952 (1984).

77 K. Sensitivities comparable with commercial low-temperature SQUIDs have been demonstrated in laboratory devices, and progress continues.[34]

The *gradiometer* is an important generalization of the SQUID magnetometer; it does not measure the flux density B itself, but, rather, some spatial derivative of B. This is easily accomplished by replacing the single pickup loop by two or more loops, with areas and orientations arranged so that no net flux is coupled into the combination by a spatially uniform magnetic field. The only response of the SQUID is then to spatial variations of B. For example, if one uses two loops lying in the xy plane, each of area A, separated in the x direction by a distance X, and connected in series opposition, the net flux picked up will be $\Phi_p \approx XA(\partial B_z/\partial x)$, assuming that higher spatial derivatives of the field can be neglected. In an obvious extension, if one uses three coils in series, one can arrange to cancel both the uniform field and the first derivative, leaving Φ_p proportional to some chosen second derivative of B.

Gradiometers are particularly important for detecting weak magnetic signals which are generated locally in the presence of interfering noise signals from more remote sources. An outstanding example is in neuromagnetic studies of the functioning of the human brain, carried on amidst urban magnetic noise. If the distance of the pickup coils from the source is comparable with the separation of the coils in the gradiometer, there is very little cancellation of the desired signal, whereas magnetic noise from electric power lines, moving vehicles, elevators, etc., being spatially more uniform, is largely canceled out.

An important commercial application of the gradiometer principle is to create a *susceptometer*. If we set up a first-derivative gradiometer in a uniform magnetic field B, no flux is coupled. If we then mechanically introduce a sample into *one* of the loops of the gradiometer, a flux signal is generated which is proportional to its magnetic moment, or to χB, insofar as the susceptibility χ is constant. If the SQUID is connected in a flux-locked loop, this can be used to produce a direct readout of the susceptibility of the sample with very high sensitivity.

VOLTMETERS AND AMPLIFIERS. As early as 1966, Clarke[35] demonstrated the use of a SQUID to create a very sensitive voltmeter. The basic idea is to connect the source of voltage V to the input of the SQUID through a known resistance R. Then the current $I = V/R$ through the input coil produces a flux in the SQUID, which is read out in the usual way. In a better design, the output of the flux-locked loop is fed back to the input resistor in such a way as to null out the voltage source, effectively using the SQUID as a null indicator in a self-balancing measurement. In principle, this gives a very high effective input impedance, at least at

---

[34]J. Clarke, "High-T_c Josephson Junctions, SQUIDs and Magnetometers", in K. S. Bedell, M. Inui, D. Meltzer, J. R. Schrieffer, and S. Doniach (eds.), *Phenomenology and Applications of High-Temperature Superconductors*, Addison-Wesley, Reading, MA (1992), pp. 131–163.

[35]J. Clarke, *Phil. Mag.* **13**, 115 (1966).

dc. Nonetheless, a SQUID voltmeter is only effective when the source is of low impedance because of the resolution limit set by Johnson noise in the input circuit. For example, at 4 K this varies from $\sim 10^{-15}\,\text{V/Hz}^{-1/2}$ for $10^{-8}\,\Omega$ to $\sim 10^{-10}\,\text{V/Hz}^{-1/2}$ for $100\,\Omega$. Thus, the device can serve as a "femtovoltmeter" with a 1-sec response time only for measurements from sources with resistance below $\sim 10^{-8}\,\Omega$.

Although the SQUID voltmeter can be used to measure sufficiently low-frequency ac signals as well as dc voltages, substantial circuit modifications are required for operation as an amplifier at radio frequencies. Typically, a dc SQUID is used, but the flux-locked loop is eliminated. The input coupling is modified to accommodate high-frequency currents, perhaps by making it resonant at the desired frequency. For example, an amplifier built by Hilbert and Clarke,[36] operating at 4 K and 93 MHz, showed a gain of ~18 dB and a noise temperature of ~1.7 K, close to predicted values.

As T approaches absolute zero, the thermal noise reduces to the zero-point quantum noise. In an ideal amplifier with only this quantum noise, the noise temperature of a tuned amplifier should be given by $T_N \approx \hbar\omega/k \ln 2$, but such quantum-limited performance in a SQUID amplifier has not yet been demonstrated.

6.6 ARRAYS OF JOSEPHSON JUNCTIONS

In this section, we generalize the ideas that were developed in connection with the two-junction dc SQUID in a magnetic field to circuits containing much larger numbers of Josephson junctions connected by superconducting leads. Large planar arrays of Josephson junctions form a two-dimensional system with properties of interest in their own right, and when generalized to three dimensions as described in Chap. 9, they also form a model system for granular superconductors, such as the ceramic high T_c materials.

To develop the principles in the simplest way, we shall consider only *square* arrays of small superconducting islands with lattice spacing a, each of which is coupled to its four nearest neighbors by a Josephson weak link with coupling energy E_J. (See Fig. 6.13.) The four junctions forming a minimal square closed circuit, such as the one highlighted in the figure, comprise a *plaquette*. The sum of the four gauge-invariant phase differences going around the plaquette in a definite sense of circulation is constrained by a straightforward generalization of (6.28), namely,

$$\sum_{\text{plaquette}} \gamma_i = 2\pi\Phi/\Phi_0\,(\text{mod}\,2\pi) = 2\pi f\,(\text{mod}\,2\pi) = 2\pi(f-n) \tag{6.61}$$

[36] C. Hilbert and J. Clarke, *J. Low Temp. Phys.* **61**, 263 (1985).

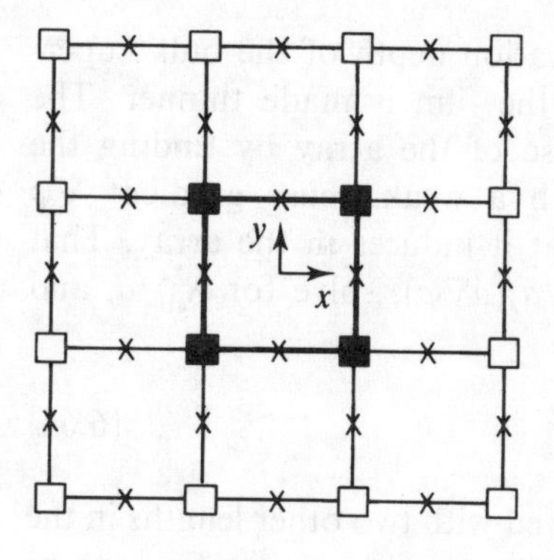

FIGURE 6.13
Schematic diagram of a section of a square array of Josephson junctions, with one plaquette highlighted. The boxes represent superconducting islands, and the lines with crosses denote the Josephson junctions joining them.

where Φ is the flux enclosed in the plaquette, f is the fraction of a flux quantum per unit cell, also referred to as the *frustration index*,[37] and n is an integer. We may form a further useful generalization by summing (6.61) over all the plaquettes contained within a larger closed contour. In adding up the sums $\Sigma\gamma_i$ for all the plaquettes inside the contour, we see that all internal links cancel in the sum because of entering twice with opposite signs from adjacent plaquettes, leaving only the sum over the exterior links running along the contour. Thus, the sum $\Sigma\gamma_i$ around the contour is given by

$$\sum_{\text{contour}} \gamma_i = 2\pi \sum_{\substack{\text{enclosed}\\ \text{cells}}} (f_j - n_j) \tag{6.61a}$$

To illuminate the main qualitative results in a simple way, we shall restrict our treatment to arrays in which the Josephson coupling is weak enough so that we can neglect macroscopic field-screening effects. Then $B = H$ everywhere, so that

$$f_j = f = Ha^2/\Phi_0 \tag{6.62}$$

where H is the field applied normal to the array, yielding a *uniformly frustrated array*. In this case, the energy of the array can be written as a sum over all the junctions of the single-junction energy:

$$E = E_J \sum_{\text{array}} (1 - \cos\gamma_i) \tag{6.63}$$

Because the neglect of screening is such an important approximation, let us examine the conditions under which it is justified. As discussed in Sec. 3.11.4, a two-dimensional superconducting system screens perpendicular fields in a characteristic length $\lambda_\perp$. For a thin film, this was shown in (3.137) to be $\lambda_\perp = \lambda_{\text{eff}}^2/d$,

[37]"Frustration" refers to the fact that the presence of magnetic flux in a closed ring of junctions frustrates the desire of each junction to minimize its own Josephson energy by having its $\gamma_i = 0$.

where λ_{eff} describes the three-dimensional penetration depth of the bulk superconducting material, so $\lambda_\perp$ diverges as $1/d$ as the film is made thinner. The arguments used there can be applied to the case of the array by finding the value of λ_{eff}^2/d for an equivalent film in which a weak phase gradient $\nabla\varphi$ would induce the same sheet current density that it induces in the array. That is, we equate $J_s d = (c/4\pi\lambda_{\text{eff}}^2)d(\Phi_0/2\pi)\nabla\varphi = (I_c/a)(a\nabla\varphi)$, solve for λ_{eff}^2/d, and set the result equal to $\lambda_\perp$. In this way, we obtain

$$\lambda_\perp = \frac{c\Phi_0}{8\pi^2 I_c} \tag{6.64}$$

which diverges as $I_c \to 0$. This $\lambda_\perp$ must be compared with two other lengths in the problem, the lattice spacing a and the radius R ($\gg a$) of the array. So long as $\lambda_\perp \gg R$, screening effects can be safely ignored, whereas they are very important if $\lambda_\perp \lesssim a$. In the intermediate regime, screening effects must be taken into account, but the qualitative picture presented here remains useful. The quantitative impact of screening and magnetic coupling between vortices in arrays has been addressed in computer simulations by Phillips et al.[38]

6.6.1 Arrays in Zero Magnetic Field

Let us start by considering the zero field case for which $f = 0$. Since the absolute minimum of (6.63) occurs for all $\gamma_i = 0$, which is consistent with (6.61) for $H = f = n = 0$, this represents the ground state of the system.

A simple excited state can be constructed by applying the constraint that one vortex is trapped in the array. Such a state can be specified by choosing a particular plaquette to be the site of the vortex "core," i.e., the plaquette for which $n = \pm 1$, with all the others retaining the value $n = 0$. By symmetry, we expect all four junctions on the core plaquette to have $|\gamma_i| = 2\pi/4 = \pi/2$ in the minimum energy configuration, whereas there is zero net circulation in all other plaquettes. Minimization of the system energy, subject to these boundary conditions, fixes all other γ_i, and hence the system energy, which can be computed numerically by an iterative procedure.

We can gain more physical insight, however, by the following approximate procedure: Since the sum of the γ_i around any contour enclosing the core must be 2π, it follows that the values of γ_i must fall off as $\sim 2\pi a/2\pi r = a/r$, where r is the "radius" of the path. For small γ_i, the energy per junction from (6.63) is $E_J\gamma_i^2/2 \approx E_J a^2/2r^2$, so that the $\sim 2\pi r/a$ junctions in the contour contribute $\sim \pi E_J a/r$ to the total energy. If we now approximate the discrete sum of the contributions of the concentric rings by an integral, we find that the total energy is

$$E = \pi E_J \ln(R/a) \tag{6.65}$$

[38] J. R. Phillips, H. S. J. van der Zant, J. White, and T. P. Orlando, *Phys. Rev.* **B47**, 5219 (1993).

where R is the outer limit of the integration, i.e., roughly the radius of the array. Apart from an additive constant to account for the errors made in handling the core, this result is correct. Note that this energy scales with E_J but goes to infinity logarithmically with array size. In practice, however, this infinite energy is usually cut off well before the limit set by the size of a (large) array. This cutoff can occur because of the formation of vortex-antivortex pairs in zero field, because of the screening length $\lambda_\perp$, or because of the presence of a weak background field. We discuss the first of these here, and return to the third in the next section.

THE KOSTERLITZ-THOULESS TRANSITION. Consider two vortices of opposite sign separated by a distance R_{12} in an unbounded array. Following (6.61*a*), we see that around any contour enclosing *both* cores the sum $\Sigma\gamma_i = 0$, whereas any contour enclosing only *one* of the cores will yield a sum of $\pm 2\pi$, depending on which core is enclosed. It follows that the sum over rings which led to (6.65) will be effectively cut off at $\sim R_{12}$ since there is no net circulation beyond that point, so that the energy of the pair of vortices will be given approximately by

$$E_{12} = 2\pi E_J \ln(R_{12}/a) \tag{6.66}$$

Note that E_{12} increases with increasing R_{12}, implying an *attractive* force of $\sim 2\pi E_J/R_{12}$ between vortices of the opposite sign.

This force law has led to the elaboration[39] of a two-dimensional Coulomb gas model as an analog which may be easier to think about than the vorticity. Applying Gauss's law in a two-dimensional world, one finds that the electric field of a charge Q is given by $\mathcal{E} = 4\pi Q/2\pi r = 2Q/r$, so that the attractive force on an equal and opposite charge at distance R_{12} will be $F = 2Q^2/R_{12}$. The corresponding energy relative to that at the closest approach distance a is $E = 2Q^2 \ln R_{12}/a$. Clearly, the energetics of vortices maps onto that of a two-dimensional Coulomb gas of charges of magnitude $Q = (\pi E_J)^{1/2}$.

At $T = 0$ the system seeks its lowest energy state, which, in zero field, will contain no vortices. At a finite temperature pairs of antiparallel vortices will be thermally generated, creating an equal population of vortices of both signs. At a sufficiently low temperature these will remain in tightly bound vortex-antivortex pairs. We can estimate the number of such bound pairs in an $N \times N$ array, with N^2 cells, as $N^2 e^{-\Delta E/kT}$, where $\Delta E \sim 2\pi E_J$. In minimizing the *free energy* $F = E - TS$ at higher temperatures, however, we note that the higher *entropy* of *dissociated* pairs must be balanced against their higher *energy*. The increase in entropy results from the increase in multiplicity of accessible states when the two cores *independently* occupy random sites in the array. In more detail, from (6.65) and (6.66) we estimate an energy increase of $\Delta E \approx 2\pi E_J \ln N$, whereas the increase in multiplicity from N^2 to $(N^2)^2$ increases the entropy by $\Delta S \approx 2k \ln N$.

[39]P. Minnhagen, *Phys. Rev.* **B24**, 6758 (1981); *Revs. Mod. Phys.* **59**, 1001 (1987).

Setting $\Delta F = \Delta E - T\,\Delta S = 0$, we *estimate* $kT \sim \pi E_J$ for the temperature at which vortices start to unbind.

As shown by the renormalization group analysis of Kosterlitz and Thouless[40] for the continuum superfluid ^{4}He film case, there is a *sharply defined* temperature, which we denote T_{KT}, at which vortex pairs start to unbind because of this entropy consideration. Beasley, Mooij, and Orlando[41] pointed out that the KT work was applicable to *superconducting* films sufficiently thin so that $\lambda_\perp$ exceeded the lateral dimension of the sample. A similar limit governs the applicability of the model to an array of Josephson junctions. The magnitude of this T_{KT} for the array case can be estimated in another way by considering a single vortex-antivortex pair in an infinitely large array, calculating their mean-square separation $\langle R_{12}^2 \rangle$, and finding the temperature at which this diverges.[42] Measuring R_{12} in units of a, we have

$$\langle R_{12}^2 \rangle \propto \int_1^\infty R^2 e^{-E(R)/kT} 2\pi R \; dR \propto \int_1^\infty R^3 e^{-2\pi E_J \ln R/kT} \, dR = \int_1^\infty R^3 R^{-2\pi E_J/kT} \, dR$$

which is infinite if $T \geq T_{KT}$, where T_{KT} is defined implicitly by

$$kT_{KT} = (\pi/2) E_J(T_{KT}) \tag{6.67a}$$

A rigorous renormalization group calculation taking into account the screening effects of other vortices at finite temperatures introduces a small correction factor (often referred to as the "dielectric constant" in the Coulomb gas transcription); but otherwise this result is exact. The key point is that $kT_{KT} \approx E_J$ in an array, with the exact coefficient depending on the number of nearest neighbors in the array geometry. Below this temperature, all vortices are in bound pairs of zero net vorticity, whereas, above it, thermally excited *free* vortices are created.

For completeness, we also quote here a number of equivalent implicit expressions for kT_{KT} which play the role of (6.67*a*) for thin superconducting films,[43] namely,

$$kT_{KT} = \frac{\pi \hbar^2 n_s^*(T_{KT}) d}{2m^*} = \frac{\Phi_0^2}{32\pi^2} \frac{d}{\lambda_{\text{eff}}^2(T_{KT})} = \frac{\Phi_0^2}{32\pi^2 \lambda_\perp(T_{KT})} = \frac{\Phi_0^2}{8\pi c^2 L_{K\square}(T_{KT})} \tag{6.67b}$$

In the last expression, $L_{K\square}$ is the directly measurable kinetic inductance per square of the film.

The Kosterlitz-Thouless transition temperature can be determined experimentally by monitoring the *I-V* characteristic of the sample as a function of temperature. The key point is that in order to see flux-flow dissipation, there

---

[40] J. M. Kosterlitz and D. J. Thouless, *J. Phys.* **C6**, 1181 (1973).

[41] M. R. Beasley, J. E. Mooij, and T. P. Orlando, *Phys. Rev. Lett.* **42**, 1165 (1979).

[42] Although this estimate happens to give the correct answer, a different numerical factor would be found in (6.67*a*) if one used the criterion that some other power of R_{12} has a divergent average.

[43] This equivalence is discussed by C. J. Lobb, D. W. Abraham, and M. Tinkham, *Phys. Rev.* **B27**, 150 (1983).

must be free vortices since no *net* Lorentz force is exerted on bound vortex pairs by a transport current through the array. Above T_{KT}, there is an equilibrium population of unbound vortices, which gives a *linear* resistance. This resistance can be shown to vary as $\sim e^{-const./\sqrt{T-T_{KT}}}$, which has a sharply defined onset at T_{KT}. Below T_{KT}, there are no free vortices at $I = 0$, and hence no linear resistance. The only free vortices are those generated by the current, which tends to dissociate a pair by exerting an equal and opposite Lorentz force on the two vortices. Just below T_{KT}, the number of vortices increases as I^2, so that the voltage rises as $V \sim I^{a(T)}$, with $a(T_c) \approx 3$. The jump from the linear dependence $V \sim I$ above T_{KT} to $V \sim I^3$ just below T_{KT} is a hallmark of the KT transition, which was confirmed experimentally by Resnick et al.[44] in measurements on a proximity-coupled array.

6.6.2 Arrays in Uniform Magnetic Field

If we apply a uniform magnetic field to a large array but continue to neglect screening, the lowest energy state will be one in which enough vortices enter so that there is no net macroscopic circulation around the perimeter, which would give rise to additional energy terms which are sample-size dependent. The circulation around a macroscopic contour is measured by the sum $\Sigma\,\gamma_i$ around the contour. From (6.61a) we see that this circulation will be zero if the average value of n equals f, which normally implies that the integer n values of the cells of the array will be one of the two integers which bracket f. For example, if $0 < f < 1$, a fraction f of the cells have $n = 1$ and the rest have $n = 0$.

An important special case is $f \ll 1$, so that the net density of vortices is small, and they are essentially independent excitations. If we think of the array as partitioned into zones, each of which is centered on the core cell of one of the vortices, application of the argument of the preceding paragraph to each zone would show no net circulation around its perimeter. (In the Coulomb gas analogy, the field produces a continuous background charge which cancels the vortex charge within each zone, leaving a neutral object.) Thus, in computing the energy of a single vortex as in (6.65), we find that the sum leading to the logarithm is cut off at $R \approx (\Phi_0/\pi H)^{1/2}$, and that we can write the energy per unit area as

$$E/\text{area} \approx (\pi E_J/a^2) f \ln(\pi f)^{-1/2} \qquad (f \ll 1) \tag{6.68}$$

independent of the overall array size. Since the circulation vanishes at the perimeter of each zone, the density of kinetic energy is small there, and hence this result is not sensitive to the exact "fitting together" of the zones. Apart from the weak logarithmic dependence on f, (6.68) gives an energy linear in the number of vortices and the system acts much like a collection of independent vortices. The

[44] D. J. Resnick, J. C. Garland, J. T. Boyd, S. Shoemaker, and R. S. Newrock, *Phys. Rev. Lett.* **47**, 1542 (1981). See also D. W. Abraham, et al., *Phys. Rev.* **B26**, 5268 (1982).

validity of this simple picture will be greatest for T sufficiently below T_{KT} so that there are few thermally excited vortex pairs, and this is the experimental regime in which these ideas are best tested.

Experimentally, an array with $f \ll 1$ does show (Rzchowski et al.[45]) a resistance which is linear in f, as expected for a collection of essentially independent vortices, or "fluxons." At low temperatures resistance appears above a critical current which is about 10 percent of the ideal value for the array of parallel Josephson junctions with no fluxons present. At higher temperatures a linear resistance appears for small currents, which is found to be thermally activated, with an activation energy of order E_J. The small critical current and the activation energy are attributed to the "intrinsic pinning" associated with the *discreteness* of the array, plus any extrinsic pinning caused by defects which destroy the exact periodicity of the array.

The intrinsic pinning energy was first worked out by Lobb et al.[46] They used an iterative numerical procedure to calculate the difference in the total energy of a large array containing a single vortex for two internal boundary conditions: (1) the vortex centered in one cell, giving a minimum energy as discussed in the previous section, and (2) the vortex centered on a link (for which $\gamma_i = \pi$) separating two cells, giving a saddle-point energy. In this way, they obtained a barrier height of only $0.199E_J$ for the square lattice, which accounts for the ~10-fold critical-current reduction, and a still smaller value $\sim 0.043E_J$ for a triangular lattice. This calculation was generalized by Rzchowski et al. to obtain a complete vortex energy surface for the square lattice case, as shown in Fig. 6.14. The experimental data (Rzchowski et al.[45]) on thermally activated resistance in square arrays of S-N-S junctions confirm the predicted proportionality to E_J but suggest a somewhat larger coefficient ~0.3 to 0.5. Measurements on other types of arrays have shown similar results. A possible theoretical explanation for this numerical discrepancy is offered by the numerical study by Phillips et al.[47] of self-field, or screening, effects in arrays in which mutual inductance between plaquettes is not negligible compared to the Josephson inductance.

Alternatively, this modest quantitative discrepancy may stem from *extrinsic* pinning caused by the inevitable variation of parameter values from cell to cell in a real fabricated array. The four junctions of the core plaquette contribute $4E_J$ to the vortex energy, which is half or more of the total vortex energy for any $f \gtrsim 0.001$. Thus, it is a variation of this core energy from cell to cell which dominates extrinsic pinning. Even a 10 percent fluctuation in the local average E_J from one cell to the next would give a pinning energy of $0.4E_J$, which is twice the calculated intrinsic barrier between cells, and of the correct order of magnitude to account for the excess of the measured activation energy over the

---

[45] M. S. Rzchowski, S. P. Benz, M. Tinkham, and C. J. Lobb, *Phys. Rev.* **B42**, 2041 (1990).

[46] C. J. Lobb, D. W. Abraham, and M. Tinkham, *Phys. Rev.* **B27**, 150 (1983).

[47] J. R. Phillips, H. S. J. van der Zant, J. White, and T. P. Orlando, *Phys. Rev.* **B47**, 5219 (1993).

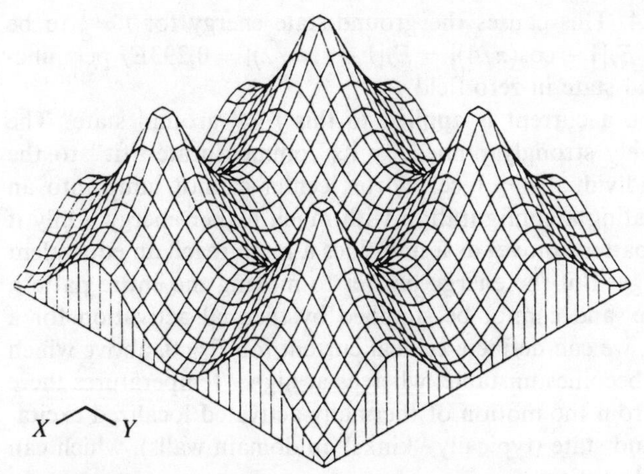

FIGURE 6.14
Effective potential energy surface for a vortex moving in a square array. The energy barrier between the minima is $\sim 0.2E_J$, giving rise to a critical current of $\sim 0.1I_c$, where I_c is the critical current of the array with no fluxon present. (*After Rzchowski et al.*)

theoretical intrinsic value. (These higher pinning energies do not carry over directly to higher critical currents since the critical current is limited by the *most weakly* pinned vortex, while the thermal activation measures a thermally weighted *average*.)

STRONGLY COMMENSURATE FIELDS. Novel phenomena arise when the frustration index takes on a rational value $f = p/q$, with p and q being *small* integers; such field values are called "strongly commensurate." For this case, the ground state or the lowest energy vortex configuration is typically a $q \times q$ superlattice relative to the underlying array lattice. For the important example of $f = \frac{1}{2}$, the superlattice is a simple checkerboard pattern, with cells with $n = 0$ and 1 represented by the black and white squares, respectively. (See Fig. 6.15.) Including the term $f = \frac{1}{2}$ as well as these values of n in (6.61), we see that the plaquette sums $\Sigma\gamma_i$ for the two types of cells are equal to $\pm\pi$. By symmetry, all $|\gamma_i|$ must be equal, and

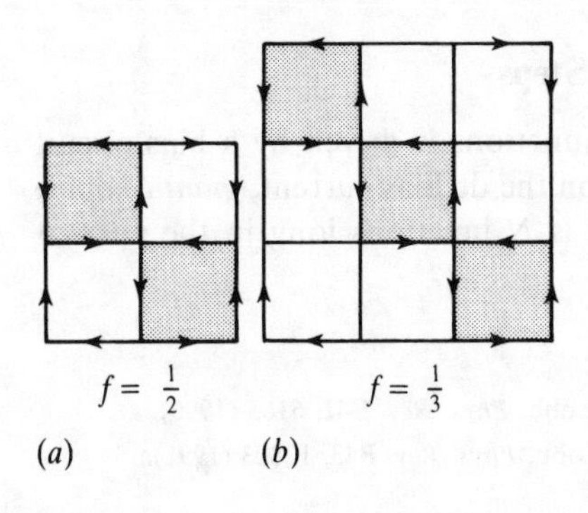

FIGURE 6.15
Schematic diagram of superlattice configurations in ground states for $f = \frac{1}{2}$ and $f = \frac{1}{3}$ and zero transport current. Arrows indicate the direction of current flow. Shaded cells have positive vorticity. (*a*) $f = \frac{1}{2}$, $q = 2$; currents equal $i_c \sin(\pi/4)$. (*b*) $f = \frac{1}{3}$, $q = 3$; currents equal $i_c \sin(\pi/3)$.

hence have the value $\pi/4$. This causes the ground-state energy for $f=\frac{1}{2}$ to be increased by an amount $E_J[1-\cos(\pi/4)] = E_J[1-(1/\sqrt{2})] = 0.293E_J$ per junction relative to the ground state in zero field.

What happens when a current is applied to this $f=\frac{1}{2}$ ground state? The vortex pattern is effectively strongly pinned by its commensurate "fit" to the underlying lattice. An individual $n=1$ cell vortex center cannot jump into an adjacent cell without creating a configuration with much higher energy. Only if the whole checkerboard pattern moves as a unit does a jump reach an equivalent state of equally low energy. But the energy barrier to moving the entire pattern scales with the array size, and cannot be supplied by thermal activation for a macroscopic array. Thus, we can define a critical current for $T=0$, above which the entire fluxon pattern becomes unstable, whereas at higher temperatures there will be linear resistance from the motion of thermally activated localized excitations from the $f=\frac{1}{2}$ ground state (typically "kinks" in domain walls), which can be viewed as "generalized vortices."

The $T=0$ critical current for $f=\frac{1}{2}$ is found by considering a periodically repeated 2×2 superlattice cell with a set of three distinct γ_i which deviate from their zero current values $(\pi/4)$ in such a way as to carry a net transport current in the desired direction. The maximum current (parallel to a unit cell edge) for which such a static solution exists is found[48] to be $(\sqrt{2}-1)I_c = 0.414I_c$ per junction, or $0.828I_c$ per 2×2 unit cell. By a generalization of this method to a 3×3 superlattice cell, the corresponding critical current for $f=\frac{1}{3}$ is found to be $0.268I_c$, or $0.804I_c$ per 3×3 unit cell. This same analysis produces the current-phase relationship for a superlattice cell treated as a renormalized weak link; it is found to resemble the $\sin\varphi$ of a single junction, but with a small admixture of odd harmonics. It should be noted that these critical currents ($0.414I_c$ and $0.268I_c$) from collective commensurate pinning are considerably larger than the corresponding value ($\sim 0.1I_c$) for the intrinsic pinning of an *isolated* vortex in an ideal lattice. However, if the lattice is deliberately made inhomogeneous by deleting links to provide extrinsic pinning, simulations[49] show that the critical current for depinning a *single* vortex can be raised to $\sim 0.6I_c$. This large increase is consistent with the large modulation in vortex energy if a link is missing from the core plaquette.

6.6.3 Arrays in rf Fields: Giant Shapiro Steps

When a uniform square array of Josephson junctions is driven by a bias circuit that superposes an ac current of frequency ν on the dc bias current, *giant Shapiro steps* are seen on the *I-V* curve. If the array is N junctions long in the current

[48] S. P. Benz, M. S. Rzchowski, M. Tinkham, and C. J. Lobb, *Phys. Rev.* **B42**, 6165 (1990).

[49] M. B. Cohn, M. S. Rzchowski, S. P. Benz, and C. J. Lobb, *Phys. Rev.* **B43**, 12823 (1991).

direction and M junctions wide, these giant Shapiro steps appear at dc voltages which are just N times the value for a single junction, namely, at

$$V_n = n\left(\frac{Nh\nu}{2e}\right)$$

This is what might be expected if all the junctions were exactly equivalent, and driven by equal bias drive, so that all of them would change the step number n at the same dc bias current. The fact that it is also observed in imperfect actual arrays implies that some region of stability of this collective time dependence is enforced by coupling between the junctions of the array. Similar arrays have been shown by Benz and Burroughs[50] to show promise for applications as a *voltage-tunable* microwave *source*, in which a dc bias causes generation of a microwave current at the Josephson frequency corresponding to the bias voltage.

More surprisingly, *fractional* giant Shapiro steps were observed in arrays which were irradiated while placed in a strongly commensurate magnetic field. If the field corresponded to frustration index $f = p/q$, with $q > 1$, steps were induced at integer multiples of a voltage unit which was $1/q$ times smaller than the giant step size $(Nh\nu/2e)$. Given the Josephson relation between the voltage and the rate of change of phase, this factor $1/q$ implies that q cycles of the rf field are required before the initial configuration of phases is repeated. The fractional steps can then be accounted for by noting that a $q \times q$ superlattice will require q rf cycles to repeat its starting configuration if one assumes that each cycle shifts the vortex pattern by a single cell of the underlying array. Although this argument appears to be essentially correct for the classic square array with current parallel to the cell edge direction, it must be refined[51] to account for other features determined by details of the array geometry, such as the observed *absence* of fractional steps if the bias current is run through the array along the cell diagonal direction.

6.7 *S-I-S* DETECTORS AND MIXERS

Although the Josephson effect is highly nonlinear and persists to high frequencies, it has not proved useful in practice as the basis for detectors of high-frequency radiation. In contrast, the highly nonlinear *quasi-particle* tunneling characteristic of *S-I-S* junctions provides the basis for highly sensitive detectors in the millimeter wave spectral region, which are in routine use in radio telescopes. In this section, we treat the basis of these devices. Two types of receiver will be discussed: the direct or square-law detector and the heterodyne detector, which uses a mixer. We start with the direct detector, which is most useful for broad bandwidths at high frequencies, and then take up the heterodyne mixer system, which is more useful for narrow bandwidths at lower frequencies. Our discussion begins with a brief

---

[50]S. P. Benz and C. J. Burroughs, *Appl. Phys. Lett.* **58**, 2162 (1991).

[51]L. L. Sohn, M. S. Rzchowski, J. U. Free, S. P. Benz, and M. Tinkham, *Phys. Rev.* **B44**, 925 (1991).

review of classical detector theory, and is necessarily rather superficial; the reader is directed to available reviews for a more comprehensive treatment.[52]

6.7.1 *S-I-S* Detectors

CLASSICAL THEORY. In the classical theory, one assumes that the response of a nonlinear element to an applied voltage can be expressed in a power-law expansion about an operating point. For example, one describes a voltage-driven device by writing

$$I(V) = I(V_0) + I'|_{V_0}\delta V + \tfrac{1}{2}I''|_{V_0}(\delta V)^2 + \cdots \tag{6.69}$$

where V_0 is the dc bias voltage at the operating point, δV is the departure from V_0, and I' and I'' are derivatives evaluated at V_0. The response $I(V)$ is assumed to be instantaneous, so that the coefficients are real and frequency independent. We now assume that

$$\delta V = V_1 \cos\omega_1 t \tag{6.70}$$

and compute the time-average current. Because the cosine averages to zero but has $\frac{1}{2}$ as the average of its square, we obtain

$$\bar{I} = I(V_0) + \tfrac{1}{4}I''V_1^2 \tag{6.71}$$

Such a device is called a "square-law detector" since it gives a response δI, which is proportional to the square of the input signal V_1. Clearly, a large curvature I'' is desirable, but to get a more useful figure of merit, we should normalize to the *power* input δP, not to the voltage. We can write $\delta P = \overline{I_1(t)V_1(t)} = I'V_1^2/2$, so that the *responsivity* $\mathcal{R}$ is given by

$$\mathcal{R} \equiv \delta I/\delta P = I''/2I' \tag{6.72}$$

which has the units of amperes per watt or V^{-1}.

To illustrate how this works, consider a classical semiconductor thermal diode in which

$$I = I_0(e^{eV/kT} - 1) \tag{6.73}$$

so that $I' = (e/kT)I_0e^{eV/kT}$ and $I'' = (e/kT)^2I_0e^{eV/kT}$. In this case, the responsivity is given by

$$\mathcal{R}_{\rm cl} = e/2kT \tag{6.74}$$

independent of details.

If we now try to apply this theory to an ideal *S-I-S* tunnel junction, we run into a problem. The obvious operating point for maximum nonlinearity is at V_g,

[52]See, e.g. Q. Hu and P. L. Richards, "Quasiparticle Mixers and Detectors," in S. T. Ruggiero and D. A. Rudman (eds.), *Superconducting Devices*, Academic Press, San Diego (1990), and references cited therein.

where the quasi-particle tunnel current rises discontinuously from a small current to $I \sim V_g/R_n$, as shown in Fig. 6.4. Insofar as it is an ideal junction, with no rounding at the energy gap, this "corner" in the *I-V* curve represents a point where the curvature I'' is *infinite*! According to (6.72), this would imply an infinite responsivity, which is surely nonphysical. The resolution of this problem is provided by a quantum version of this analysis, worked out by Tucker.

QUANTUM THEORY. Tucker's theory[53] takes into account the quantum nature of the radiation, according to which the minimum energy transfer at frequency ω is $\hbar\omega$. This implies that even for high-frequency fields of arbitrarily small amplitude V_1, the minimum energy transfer corresponds to a fixed finite voltage $V_\omega \equiv \hbar\omega/e$. A consequence of this in a simplified version of the Tucker theory is that we can use the classical responsivity formula, but with the derivatives replaced by finite-difference quantum generalizations. That is, we use

$$I' = (e/2\hbar\omega)[I^0(V + \hbar\omega/e) - I^0(V - \hbar\omega/e)] \tag{6.75a}$$

$$I'' = (e/\hbar\omega)^2[I^0(V + \hbar\omega/e) - 2I^0(V) + I^0(V - \hbar\omega/e)] \tag{6.75b}$$

where $I^0(V)$ denotes the dc *I-V* characteristic of the junction in the absence of any radiation. If we approximate the ideal tunneling *I-V* curve by $I = 0$ for $V < V_g$, and $I = V_g/R_n$ for V just above V_g, and choose an operating point V just below V_g, then both $I^0(V - \hbar\omega/e)$ and $I^0(V)$ are zero. Using (6.72), we find the finite responsivity

$$\mathcal{R}_{qm} = I''/2I' = e/\hbar\omega \tag{6.76}$$

despite the infinite classical curvature at this operating point. Obviously, (6.76) implies the transfer of one additional electron per photon, as in a photoelectric detector for optical photons where the quantum limit has long been familiar. In the present case, however, thermal dark current background is usually a greater problem than in photocells. On comparing (6.76) with the classical result (6.74), we see that they are equal when $kT = \hbar\omega/2$, the zero point energy associated with an oscillator at the frequency ω.

One can gain insight into the origin of (6.75*b*) by comparing it with the small amplitude limit of the Tien-Gordon formula (6.26) for photon-assisted tunneling. Noting that the leading terms in the power series expansions of $J_0(x)$ and $J_1(x)$ are $(1 - x^2/4)$ and $x/2$, respectively, (6.26) reproduces (6.71) if I'' is taken to have the form (6.75*b*).

---

[53] J. R. Tucker, IEEE *J. Quantum Electron.* **15**, 1234 (1979); see also J. R. Tucker and M. J. Feldman, *Rev. Mod. Phys.* **57**, 1055 (1985) for an excellent review of this field.

6.7.2 *S-I-S* Mixers

CLASSICAL THEORY. As in the discussion of direct detectors, our starting point is (6.69), the assumed power-law expansion of the *I-V* curve of the device about an operating point V_0. Now, however, we assume that *two* rf signals at different frequencies are applied, so that

$$\delta V = V_1 \cos \omega_1 t + V_2 \cos \omega_2 t \tag{6.77}$$

In the case relevant for mixer action, we take V_1 to be the relatively strong local oscillator voltage, whereas $V_2 \ll V_1$ is the signal voltage at a nearby frequency which is to be detected. When (6.77) is inserted in (6.69), the resulting current has components at many frequencies, including 0, ω_1, ω_2, $2\omega_1$, $2\omega_2$, and $(\omega_1 \pm \omega_2)$. Of these, the only one of interest here is the one at $\omega_{if} = (\omega_1 - \omega_2)$, called the "if" (intermediate frequency) or "beat frequency." The amplitude of this current is found from the expansion to be

$$I(\omega_{if}) = (I''/2)V_1 V_2 \tag{6.78}$$

This current is fed into an amplifier tuned to ω_{if}, so all the other current components can be ignored. In a typical configuration, ω_{if} is held constant, and the receiver is tuned by sweeping the local oscillator frequency ω_1.

The advantage of the heterodyne system for detection of narrow band signals is clear from the form of (6.78), namely, $I(\omega_{if})$ varies with the *first* power of the (weak) signal voltage V_2, rather than with its square, as does δI in the direct detector. On the face of it, (6.78) suggests that $I(\omega_{if})$ could be made as large as desired simply by increasing the local oscillator voltage V_1. This is, of course, not the case since the power series expansion (6.69) on which our analysis is based becomes nonconvergent unless $V_1 \lesssim 2I'/I''$. With that constraint, $I(\omega_{if}) < I'V_2$.

A more useful measure of the performance of a mixer is its conversion efficiency or conversion gain G. By this we mean the ratio of the *if* power out to the *rf* power in. This is

$$G = \frac{P_{if}}{P_{rf}} = \frac{I^2(\omega_{if})R_{if}}{V_2^2/R_{\omega_2}} \approx \left(\frac{I''V_1}{2I'}\right)^2 \leq 1 \tag{6.79}$$

where R_{if} is the input impedance of the *if* amplifier, $1/R_{\omega 2}$ is the input admittance of the mixer, and both R_{if} and $R_{\omega 2}$ are assumed to match the differential resistance $1/I'$ of the mixer for optimum performance. We have also used (6.78) and the constraint that $V_1 \leq 2I'/I''$. A more rigorous analysis yields the slightly more conservative result that $G \leq 1/2$.

QUANTUM THEORY. Perhaps the most startling result of the Tucker quantum theory of mixer action is the prediction that conversion *gain* should be possible in certain regimes, contrary to the standard prediction (6.79) for classical resistive junctions. Unfortunately, the theory is complicated, and it seems difficult to extract a simple explanation of why gain is possible in the quantum regime. It is clear, however, that just as in our discussion of the direct detector, quantum effects become important when the dc I-V curve has features which are sharp on a voltage scale less than $V_\omega \equiv \hbar\omega/e$. When this is true, (6.75) shows that the quantum-mechanical generalizations of I' and I'' become strongly dependent on frequency, unlike the classical case in which they depend only on the voltage of the operating point. If the differential conductance I' is strongly frequency dependent, then the Kramers-Kronig relations tell us that there must be an important imaginary part of the admittance, which describes quantum-mechanical out-of-phase currents.

In a mixer, several frequencies are present at once, so the simple admittance I' must be replaced by a complex admittance matrix. This matrix yields the linear response of both the in-phase and out-of-phase components of the current at all relevant frequencies (or "ports") in terms of the voltages at these ports, assuming that only the local oscillator must be treated at a nonlinear level. Fortunately, the theory provides a prescription for working this all out, starting from only the dc I-V curve of the unirradiated junction and its Kramers-Kronig transform. In applying this theory in practice, one must also know and be able to control the complex embedding impedance of the junction (i.e., the impedance of the circuit connected across it). Attempts to compare measured mixer performance with predictions of the quantum mixer theory have generally been quite successful.[54] The predicted qualitative effects have been observed, and quantitative comparisons of mixer gain have been quite satisfactory when the underlying tunneling curves are not too sharply featured.

[54]W. R. McGrath, P. L. Richards, D. W. Face, D. E. Prober, and F. L. Lloyd, *J. Appl. Phys.* **63**, 2479 (1988); M. J. Feldman and S. Rudner, "Mixing with SIS arrays," in K. J. Button (ed.), *Infrared and Millimeter Waves*, vol. 1, Plenum, New York (1983), p. 47.

CHAPTER 7

JOSEPHSON EFFECT II: PHENOMENA UNIQUE TO SMALL JUNCTIONS

7.1 INTRODUCTION

In Chapter 6, our discussion of the Josephson effect was carried out in what might be called the *semiclassical limit*, in which the phase of the GL wavefunction was well defined, apart from thermal fluctuation effects. Although we spoke of macroscopic *quantum* interference in the discussion of SQUIDs, for example, the analysis resembled the addition of waves in *classical* optics, as suggested by the reference to Fraunhofer and Airy patterns. This semiclassical behavior was anticipated in Chap. 3, where we noted that the phase-number uncertainty relation put no significant constraint on the precise specification of the phase in macroscopic superconductors in contact with a particle reservoir. With the advent of modern microfabrication techniques, it has now become practical to study samples that are sufficiently small so that the charge of a single electron matters. This has opened up a new field of investigation, in which the quantum conjugate properties of phase and number play an important role, and quantum fluctuations remain important even at $T \approx 0$. This field is the subject of this chapter.

The most definitive studies have involved a nearly isolated, small superconducting grain or island, connected to charge reservoirs only through small tunnel junctions, which have very low capacitance C and high tunneling resistance R. The quantitative constraints on these parameters are easily estimated, as follows.

First, the Coulomb charging energy of a single electron, $E_c = e^2/2C$, must be considerably larger than kT at the experimental temperature, in order that *thermal* fluctuations do not average over the charge numbers. For $T \sim 1\,K$, this requires that $C \lesssim 10^{-15}$ F, which in turn requires a junction area of $\lesssim 10^{-8}\,\text{cm}^2$. This explains why these effects are unique to small junctions.

The second requirement is that the tunneling resistance from the island to the particle reservoirs must exceed the quantum resistance $R_Q = h/4e^2$ to avoid the effects being washed out by *quantum* fluctuations in the particle number. To be observable, the charging energy $e^2/2C$ must exceed the quantum energy uncertainty $\Delta E \gtrsim \hbar/\Delta t \approx \hbar/RC$ associated with the finite lifetime of the charge on the capacitor. Equating the Coulomb energy to the energy uncertainty, we see that the C in both denominators cancels, and the requirement is simply that $R \gtrsim 2\hbar/e^2 \approx R_Q \sim 6\,k\Omega$. This is not a difficult requirement to meet in a small area junction.

There is an important complication, however, when one includes the leads connected to the sample. Since the relevant time scale is set by $RC \sim 10^{-10}$ sec, the effective shunting resistance from the leads is determined by the high-frequency impedance of the leads near the junction (typically 50 to 100 Ω), not simply the dc bias resistance, which can easily be made as large as $10^9\,\Omega$. It is the shunting by the leads which makes charging effects hard to observe in a *single* tunnel *junction*. However, this problem can be circumvented by studying a single *island* isolated by *two* high-impedance tunnel junctions, as mentioned earlier. Nonetheless, we first shall discuss the single, small junction case, where the lead impedance profoundly affects the observed properties. We do this to clarify the physics, e.g., why the supercurrent branch of the *I-V* curve may show a measurable resistance, and also because of the interest in using single junctions in applications. Next we discuss the general implications of the quantum uncertainty between the phase and the number. Finally, we shall discuss a number of concrete examples of tunneling phenomena in which the tunneling of electrons can be controlled one by one, often referred to as involving the "Coulomb blockade."

7.2 DAMPING EFFECT OF LEAD IMPEDANCE

In the discussion of *I-V* curves in Sec. 6.3 based on the RCSJ model, for simplicity we treated the bias circuit as an ideal current source, i.e., a circuit with infinite impedance at all frequencies so that it exerts no damping effect on the junction. This is an excellent approximation at dc, if the current source consists of a voltage source feeding through a sufficiently high fixed resistance such as $10^9\,\Omega$. However, the characteristic frequencies involved in the dynamic processes leading to escape and retrapping are high, typically $\omega_p \sim 10^{11}$ and $\omega_g = 2eV_g/\hbar \sim 10^{12}$. At such frequencies the dc resistance of the remote current source does not play any role. What is important is the shunting impedance of the bias circuit, as seen from the junction, at these high frequencies (at which the wavelength is $\sim$1 mm or less). Unless extremely small high-impedance isolating resistors[1] are inserted in the leads directly at the junction, the leads will typically show a characteristic impedance $Z_1 \sim Z_0/2\pi$, where $Z_0 \approx 377\,\Omega$ is the impedance of free space. For an infinite transmission line, or one terminated in the same Z_1, this impedance is real and independent of frequency. If the line has impedance discontinuities at finite distances from the junction, its impedance shows resonances and becomes complex. For simplicity, we assume that these resonances are largely suppressed by dissipation, so that the line presents a real, nearly constant, impedance $Z_1 \sim 50 - 100\,\Omega$ for high frequencies, while showing nearly infinite impedance for dc. Since Z_1 is orders of magnitude less than the quasi-particle resistance in small, high-resistance junctions, the presence of the transmission line will greatly increase the high-frequency damping. This has a major impact on the retrapping current in such junctions, as discussed below. For typical lower-resistance junctions with $R_n < 100\,\Omega$ and $I_c > 10\,\mu$A, however, the effect of the leads is relatively minor, so that these effects were not given much attention until the advent of submicron junction fabrication techniques.

7.2.1 Effect on Retrapping Current

Recalling our discussion in Sec. 6.3.3, the retrapping current depends on the damping $\sim 1/Q$ and is expected to be given by $I_{r0} = 4I_{c0}/\pi Q$, where $Q = \omega_p RC$ in the RCSJ model. The temperature dependence of the retrapping current $I_r(T)$ measured in small, high-resistance Sn – SnOx – Sn tunnel junctions by Iansiti et al.,[2] and shown by the open circles in Fig. 7.1, is highly anomalous. At $T > 1$ K, $I_r(T)$ follows the expected exponential temperature dependence associated with the freezing out of the quasi-particle damping, roughly as $e^{-\Delta/k_BT}$, with a resulting increase in Q. Below $\sim$1 K, however, $I_r(T)$ *abruptly* stops decreasing and remains constant down to the lowest temperatures ($\sim$20 mK). Because of the direct proportionality of I_r to damping, this observation implies that a new damping mechanism which does *not* freeze out at low temperatures suddenly comes into play at this crossover temperature. This mechanism was identified by Johnson et al.[3] as the onset of pair-breaking tunneling when the voltage from which the junction retraps rises up to reach the gap voltage V_g.

---

[1]See, e.g., L. S. Kuzmin and D. B. Haviland, *Phys. Rev. Lett.* **67**, 2890 (1991).

[2]M. Iansiti, M. Tinkham, A. T. Johnson, W. F. Smith, and C. J. Lobb, *Phys. Rev.* **B39**, 6465 (1989).

[3]A. T. Johnson, C. J. Lobb, and M. Tinkham, *Phys. Rev. Lett.* **65**, 1263 (1990); a more complete exposition of the modified model is given by M. Tinkham in H. Grabert and M. H. Devoret (eds.), *Single Charge Tunneling*, NATO ASI Series, Plenum, New York (1992), p. 144.

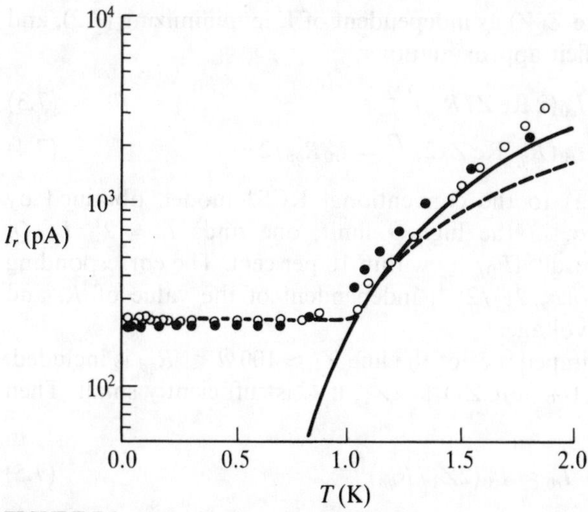

FIGURE 7.1
Temperature dependence of retrapping current in small underdamped Josephson junction. (*After Johnson et al.*) The bottoming out at low temperatures is attributed to the onset of pair-breaking tunneling when the retrapping voltage reaches the gap. The open circles are data; the solid circles result from analog simulations. The dashed line is an analytic approximation to the simulator, whereas the solid line shows the results of a conventional model with frequency-independent damping.

To simplify the analysis, Johnson et al. used an energy-balance argument considering currents at only two frequencies: dc and the Josephson frequency $\omega_J = 2eV/\hbar$, where V is the dc average voltage across the junction. The energy input to the junction from the dc current source is IV. This energy input is accounted for by a sum of dc and ac dissipative terms. The first term, V^2/R_{qp}, is dissipated within the junction by the dc quasi-particle current. The second term is the dissipation stemming from the ac supercurrent $I_{c0} \sin \omega_J t$ generated by the junction and flowing into an impedance $Z(\omega_J)$, i.e., $(I_{c0}^2/2)$ Re $Z(\omega_J)$. This $Z(\omega_J)$ is obtained from the parallel contributions of the junction capacitance C, the junction quasi-particle resistance R_{qp}, and the (real) impedance Z_1 of the line. That is,

$$1/Z(\omega_J) = i\omega_J C + 1/R_{qp} + 1/Z_1 \tag{7.1}$$

Equating the input power IV to the sum of the dc and ac dissipation terms above, recalling that $\omega_J = 2eV/\hbar$, and dividing through by V, we obtain

$$I(V) = V/R_{qp} + (I_{c0}^2/2V)\,\mathrm{Re}Z(V) \tag{7.2}$$

Since the right-hand side has terms which diverge at both $V = 0$ and $V = \infty$, it is clear that there is a smallest value of I for which the equation can be satisfied at some intermediate voltage. We identify this minimum value as the retrapping current I_{r0} and the corresponding voltage as the retrapping voltage V_{r0}. Insofar as $\omega_J CZ_1 < 1$, we can treat Re $Z(V)$ as independent of V in minimizing (7.2), and thus find the convenient explicit approximations

$$I_{r0} \approx I_{c0}(2\,\mathrm{Re}\,Z/R_{qp})^{1/2} \tag{7.3}$$

$$V_{r0} \approx I_{c0}(R_{qp}\,\mathrm{Re}\,Z/2)^{1/2} = I_{r0}R_{qp}/2 \tag{7.4}$$

Applying these formulas to the conventional RCSJ model, obtained by setting $R_{qp} = R$ and $Z_1 = \infty$, in the high-Q limit, one finds $I_{r0} = 2^{1/2}I_{c0}/Q$, which agrees with the exact result $4I_{c0}/\pi Q$ within 10 per cent. The corresponding retrapping voltage is $V_{r0} = (\hbar\omega_p/2e)/2^{1/4}$, independent of the value of R, and typically well below the gap voltage.

But when the shunting impedance of the line $Z_1 \approx 100\,\Omega \ll R_{qp}$ is included, (7.1) reduces to $\mathrm{Re}\,Z = Z_1/[1 + (\omega_J CZ_1)^2] \approx Z_1$, if C is sufficiently small. Then (7.3) becomes

$$I_{r0} \approx I_{c0}(2Z_1/R_{qp})^{1/2} \tag{7.5}$$

which falls as $R_{qp}^{-1/2}$ instead of as $Q^{-1} \sim R^{-1}$ as was the case in the standard RCSJ model. Also, according to (7.4), the retrapping voltage V_{r0} rises as $R_{qp}^{1/2}$, as R_{qp} increases with decreasing T, rather than having a fixed value set by the plasma frequency as we found in the preceding paragraph for the classic RCSJ model. But when V_{r0} exceeds the gap voltage V_g, our solution is no longer self-consistent since *above* V_g the dc dissipation jumps to approximately V^2/R_n, which would be *much* greater than the power input IV; this is only $\sim 2V^2/R_{qp}$ at this point. Thus, when, with rising R_{qp}, V_{r0} reaches V_g, it becomes pinned near that value, the detailed behavior depending on the amplitude of the superposed ac voltage. In this low-temperature limit, we simply equate the dc input power to that dissipated at $\omega_J \approx 2eV_g/\hbar$, and find that I_{r0} is pinned at the value

$$I_{r0} \approx (I_{c0}^2 Z_1)/\{2V_g[1 + (4\Delta CZ_1/\hbar)^2]\} \tag{7.6}$$

which is independent of T at low temperatures, as observed experimentally.

This model accounts for the observed sudden bottoming out of I_r at low temperatures, the temperature at which it occurs, the limiting value of I_r, and the fact that the retrapping in these junctions occurs from a high voltage near V_g, all with a reasonable value for the line impedance Z_1 as the only parameter. This unusual temperature dependence of I_r is not so important in itself. We have gone into some detail to emphasize this experimental confirmation of the reality and importance of the high-frequency damping by the leads, and the fact that the observed behavior *cannot* be explained by the *simple* RCSJ model with *any* frequency-*in*dependent R.

7.2.2 The Phase Diffusion Branch

A more important consequence of the strong high-frequency damping of a high-resistance junction by its leads is the existence of a regime of "phase diffusion," in which the representative point moves steadily down the tilted-washboard potential of Fig. 6.3 in a diffusive motion, without escaping and jumping up to the gap voltage. Such a motion gives rise to a measurable resistive voltage in the nominally zero-resistance state below I_c. The existence of such a regime was pointed out by Martinis and Kautz[4] and elaborated in extensive computer simulations by Kautz and Martinis.[5] A nominally similar regime, having finite thermally activated resistance below I_c, occurs in *non*hysteretic *overdamped* junctions, as treated by Ambegaokar and Halperin within the RCSJ model (see Sec. 6.3.3). The essential point of the new work was the recognition that a similar steady-state phase diffusion without runaway could occur in a small, high-resistance junction even if it is *under*damped at *low* frequencies and shows hysteresis, provided that it is *over*damped at the *high* frequencies relevant for repeated retrapping into the minima of the tilted-washboard potential.

The resistance R_0 due to phase diffusion can be estimated by using the Josephson relation $V = (\hbar/2e)\, d\varphi/dt$ and finding $d\varphi/dt$ by taking the difference in the rate of jumping the barrier in downward and upward directions along the tilted-washboard potential (6.14c). Assuming an attempt frequency ω_A and barrier heights $(E_b \pm hI/4e)$ in the two directions, assuming that high-frequency damping is sufficient so that each jump stops at the nearest adjacent well where $\Delta\varphi = \pm 2\pi$, and assuming that $kT < E_b$, standard transition rate theory yields

$$R_0 = V/I = (h/4e^2)(\hbar\omega_A/kT)e^{-E_b/kT} \tag{7.7}$$

In the underdamped regime, one would take $\omega_A = \omega_p$. However, for the typical values $\omega_p = 10^{11}$, $Z_1 = 100\,\Omega$, and $C = 10^{-15}\,\text{F}$, the quality factor $Q(= \omega_P Z_1 C) \sim 10^{-2} \ll 1$, and one must replace the undamped plasma frequency by the characteristic frequency for an overdamped oscillator, namely, $\omega_A = \omega_p Q = \omega_p^2 Z_1 C$. When this is done, and $2E_J$ is inserted for the barrier height E_b, (7.7) becomes

$$R_0 = 2\pi Z_1 (E_J/kT)e^{-2E_J/kT} \tag{7.8}$$

in agreement with the results of Ingold et al.[6] for the case $E_J \gg kT$.

In the other limit, $E_J \ll kT$, Ingold et al. find

$$R_0 = 2Z_1(kT/E_J)^2 \tag{7.9}$$

In fact, (7.8) and (7.9) are both limits of their more general result for the initial resistive slope,

$$R_0 = \frac{Z_1}{I_0^2(E_J/kT) - 1} \tag{7.10}$$

---

[4] J. M. Martinis and R. L. Kautz, *Phys. Rev. Lett.* **63**, 1507 (1989).

[5] R. L. Kautz and J. M. Martinis, *Phys. Rev.* **B42**, 9903 (1990).

[6] G.-L. Ingold, H. Grabert, and U. Eberhardt, *Phys. Rev.* **B50**, 395 (1994).

where I_0 is a modified Bessel function. If T is high enough so that $E_J \ll kT$ and $eV \ll kT$, and the impedance $Z_1 \ll R_Q$, Ingold et al. obtain an analytic approximation for the I-V curve of the form

$$I(V) = \frac{1}{2} I_{c0}^2 \frac{Z_1 V}{V^2 + [2eZ_1 kT/\hbar]^2} \tag{7.11}$$

This shows that as the bias current is increased, the voltage rises until a maximum current I_m is reached, at which point V must abruptly switch up to the gap voltage on the quasi-particle branch of the I-V curve to carry the increased current. This sort of behavior is illustrated in Fig. 7.2, showing data of Iansiti et al.,[7] on a small junction. In the regime in which (7.11) is valid, I_m is given by

$$I_m = (E_J/4kT)\, I_{c0} \tag{7.12a}$$

which scales with E_J^2 or I_{c0}^2, while the corresponding voltage is

$$V_m = 2eZ_1 kT/\hbar \tag{7.12b}$$

which scales with T and is typically a few microvolts at 1 K.

The switching current condition for the case of a junction with isolating resistors, which reduce the damping to some extent, was analyzed in considerable depth by Kautz and Martinis by using numerical simulations and analytic approximations, and was found to be quite complex. Their detailed simulations show that

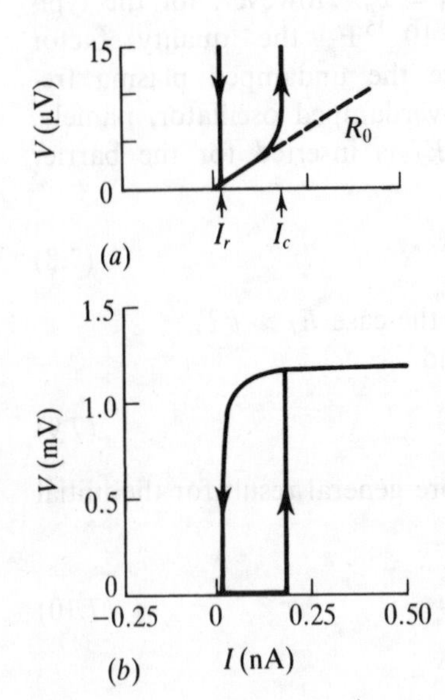

FIGURE 7.2
I-V curve of a Sn-SnOx-Sn tunnel junction with $R_n = 70\,\text{k}\Omega$ and capacitance $\sim 10^{-15}$ F, taken at 1 K and $H = 0$, showing definitions of I_c, I_r, and R_0. R_0 characterizes the slope of the phase diffusion branch. Parts (*a*) and (*b*) have the *same* horizontal scale, but vertical scales differ by a factor of 50. (*After Iansiti et al.*)

[7] M. Iansiti, M. Tinkham, A. T. Johnson, W. F. Smith, and C. J. Lobb, *Phys. Rev.* **B39**, 6465 (1989).

R_0 is significantly underestimated by approximations which neglect phase-slip events in which the phase slips through multiple wells before retrapping.

A less ambitious attack can be made by using the simple energy-balance arguments introduced in the discussion of the model of Johnson et al. One argues that, regardless of details, the junction cannot remain on the phase diffusion branch if the dc input power exceeds the maximum power that can be disposed of by the high-frequency current produced by the Josephson junction. (Here the quasi-particle dissipation V^2/R_{qp} is negligible since $R_0 \ll R_{qp}$.) If more power than this maximum is fed in, it can only be dissipated by having the voltage jump up to the quasi-particle tunneling branch. That is, if the junction is to remain on the phase diffusion branch, IV must be less than $I_{c0}^2 \operatorname{Re} Z/2 \approx I_{c0}^2 Z_1/2$. With the further approximations that R_0 holds all the way to I_m, so that $I_m V_m \approx I_m^2 R_0$, and that the capacitive shunting of the leads is negligible at these low voltages (and hence frequencies), this constraint leads to the general relation $(I_m/I_{c0})^2 R_0 \approx \operatorname{Re} Z/2$, with $\operatorname{Re} Z$ given by (7.1). If instead of this argument one uses (7.11) to make a more quantitative estimate, one finds

$$(I_m/I_{c0})^2 R_0 \approx \operatorname{Re} Z/8 \tag{7.13}$$

which differs only by a factor of $\frac{1}{4}$ from the result of the simple argument. In fact, data from several sources[8] are in semiquantitative agreement with (7.13), with the reasonable value $\operatorname{Re} Z \approx 50\,\Omega$.

Experimentally, it is found that the measured values of I_m, although greatly depressed below I_{c0}, show very little dispersion in the values measured on successive current sweeps. This is completely *inconsistent* with the behavior associated with the depression of I_c by "premature switching" from the zero voltage state, described by (6.18). If this were the mechanism, a large depression of $\langle I_c \rangle$ relative to its unfluctuated value would always be accompanied by a large dispersion of the measured values of I_c. By contrast, in the case of phase diffusion, the phase is constantly slipping and retrapping, and I_m is simply the well-defined current at which the driving force becomes sufficiently great so that this dynamic stability is lost.

We conclude this section by summarizing the conditions under which a hysteretic phase diffusion branch, which is excluded in the classic RCSJ model with constant damping resistance R, is observed. Two conditions must be satisfied: (1) There must be sufficient damping at high frequency to make the phase diffusion branch stable. This means $Q(\omega_p) \approx \omega_p R(\omega_p) C \approx \omega_p Z_1 C < 1$. Since typically $\omega_p \sim 10^{11}$ and $Z_1 \approx 100\,\Omega$, this requires $C < 10^{-13}\,\text{F} = 100\,\text{fF}$, i.e., a junction with area $\lesssim 4\mu\text{m}^2$. (2) The rate of thermally activated phase diffusion must be rapid enough to give a measurable resistance R_0. This requires that the barrier $2E_J$

[8] M. Tinkham "Josephson Effect in Low-Capacitance Tunnel Junctions" in H. Grabert and M. H. Devoret (eds.), *Single Charge Tunneling*, NATO ASI Series, Plenum, New York (1992), p. 152.

should not be too large with respect to thermal energy kT. In Sec. 6.3.3, we noted that $2E_J/kT \approx 1.76(R_Q/R_n)(T_c/T)$ for T well below T_c, so that thermal activation will be strong down to $T/T_c \sim R_Q/R_n$. These two conditions explain why the resistive phase diffusion branch was not observed until the advent of small, high-resistance junctions.

7.3 QUANTUM CONSEQUENCES OF SMALL CAPACITANCE

In the preceding discussion, we have treated the Josephson junction as a purely classical system, in which the phase difference φ across the junction and the charge $Q = CV$ on the junction were treated as classical variables which could be specified to arbitrary precision. (Here we revert to the common usage of φ to represent the gauge-invariant phase difference since $\mathbf{A} = 0$.) In fact, there is a quantum uncertainty relation $\Delta\varphi\,\Delta N \gtrsim 1$, where N is the number of Cooper pairs transferred across the junction, which limits this classical precision and thereby modifies the results of the classical theory of the Josephson effect.

To get some feeling for these effects in the simplest way, we start by considering an *isolated* junction at $T = 0$ with capacitance C, but with no damping resistance and no leads attached, so that $I = 0$. The classical expression for the total energy is then

$$E = -E_J \cos\varphi + \tfrac{1}{2}CV^2 = -E_J \cos\varphi + \tfrac{1}{2}C(\hbar/2e)^2(d\varphi/dt)^2 \tag{7.14}$$

We have used the Josephson frequency relation to get the second form, which clearly has the form of the sum of a potential-energy term and a kinetic-energy term. Using $Q = CV$ and making the operator replacement for the charge in the φ representation, namely, $Q/2e = N \to i\partial/\partial\varphi$, the form of the hamiltonian is

$$H = -E_J \cos\varphi - 4E_c\,\partial^2/\partial\varphi^2 \tag{7.15}$$

where $E_c = e^2/2C$ is the charging energy of the junction for a single electron charge. This H describes only the Cooper pairs, neglecting the quasi-particle degrees of freedom, which should be unimportant at $T = 0$.

Since the isolated junction is unrealistic, it is important to anticipate that in Sec. 7.5.3 we shall show that, with suitable redefinition of parameters, (7.15) also describes quantum phase fluctuations of the island in the *realistic* example of a small superconducting island connected to two macroscopic superconducting leads by small, high-resistance Josephson junctions.

By analogy with the periodic potential problem that leads to the energy bands in crystals, we expect the solution of (7.15) to take the form of Bloch functions $\psi_q = u(\varphi)e^{iq\varphi}$, where q is a "charge" or "pair number" variable and $u(\varphi)$ is periodic with period 2π. If only integral numbers of Cooper pairs were physically relevant, q would take on only integer values and ψ would be periodic over 2π. In fact, q is a continuous variable since it does not represent the total charge on an isolated piece of metal but, rather, the charge on the capacitor formed by the two electrodes. This charge can be controlled continuously, e.g., by a third "gate" electrode which can draw charge away from the junction region, even though the *tunnel* current is restricted to transfer only discrete charges e, or $2e$, if only Cooper pairs are considered.

The parameter $y \equiv 4E_c/E_J$ provides a measure of the relative importance of the charging energy of pairs in forcing a delocalization of the phase, away from the minimum potential-energy point at $\varphi = 0$. To get a feel for the numbers, we recall that $E_c/k \approx 1$ K for $C = 1$ fF, and varies inversely with C, whereas at $T = 0$, $E_J/k = 0.88(R_Q/R_n)T_c$, with $R_Q = h/4e^2$. The ratio $y \equiv 4E_c/E_J = 4.53(R_n/R_Q)(E_c/kT_c)$, so that for small junctions with $C \approx 1$ fF and $T_c \approx 4$ K, $y \approx (R_n/R_Q)$, and $R_n \approx R_Q$ is a reasonable indicator of the crossover point from one regime to the other. For $y \ll 1$, the ground state is a narrowly peaked wavefunction $\psi(\varphi)$ with a width of the order of $y^{1/4}$, meaning that the quantum fluctuations in the phase are small. There are also a number of higher (band) states in the potential minimum, resembling the excited states of a harmonic oscillator. By contrast, when $y \gg 1$, the term in E_c is dominant, and to minimize it in the ground state, $\psi(\varphi)$ approaches a constant, meaning that all values of φ are equally probable.

Since this problem is one dimensional, it is easy to solve exactly by numerical means. However, one gets a bit more insight by a variational approach to find an approximation to the ground state, using appropriate trial functions. For $y \ll 1$, one assumes a gaussian trial function,

$$\psi(\varphi) \sim \exp(-\varphi^2/4\sigma^2) \tag{7.16}$$

where σ, the rms spread in φ, is chosen to minimize the expectation value of (7.15). For $y \ll 1$, the resulting minimum energy is

$$E = -E_J[1 - (y/2)^{1/2}] \tag{7.16a}$$

In the other limiting case of $y \gg 1$, an appropriate trial function which is periodic and satisfies the boundary condition of zero slope at the edge of the cell is

$$\psi(\varphi) \sim (1 + a\cos\varphi) \tag{7.17}$$

which yields the approximate ground-state energy

$$E \approx -E_J/2y = -E_J^2/8E_c \tag{7.17a}$$

The second form shows explicitly that in this limit the binding energy is of *second order* in E_J, in contrast to the *first-order* binding energy in the semiclassical limit (7.16a).

The wavefunctions $\psi(\varphi)$ for values of y ranging from 0.2 to 4 are shown in Fig. 7.3. Qualitatively, it is clear that for $y > 1$ (e.g., for $C \sim 1$ fF and $R_n > R_Q$) the probability density for the phase variable φ is sufficiently delocalized so that it is no longer a good approximation to treat φ as a semiclassical variable as we have been doing in earlier chapters. Accordingly, let us examine these same quantum states of the isolated junction by using the more appropriate, if less familiar, number (charge) representation instead of the φ representation that we have

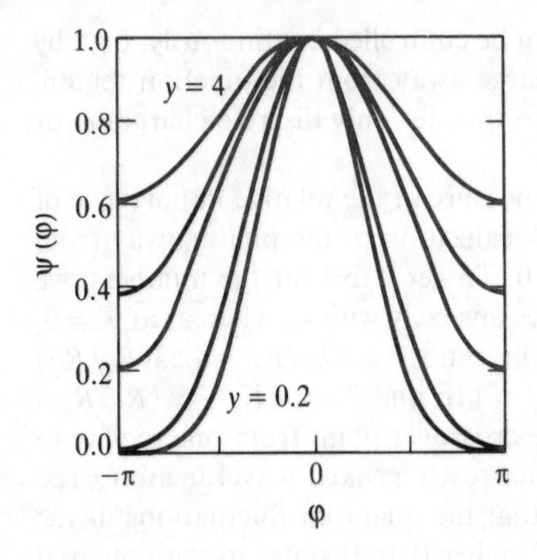

FIGURE 7.3
Wave functions $\psi(\varphi)$ of ground states for low capacitance Josephson junctions with $4E_c/E_J \equiv y = 4$, 2, 1.2, 0.4, and 0.2. (*After Iansiti et al.*)

been using. Again, anticipating Sec. 7.5.3, this charge representation will be used there to find the supercurrent through a realistic double-junction system in which quantum phase fluctuations are large.

7.3.1 Particle Number Eigenstates

Going back to Chap. 3, we pointed out that by introducing in (3.18*a*) a phase factor $e^{i\varphi}$ with each Cooper pair in the BCS ground state, we could set up a state

$$|\psi_\varphi\rangle = \prod_{\mathrm{k}}(|u_k| + |v_k|e^{i\varphi}c^*_{k\uparrow}c^*_{-k\downarrow})|\phi_o\rangle$$

with an indefinite number of particles, but with a definite relative phase between components with a differing number of particles (Cooper pairs). Then in (3.18b) we showed how to use a Fourier transform to project out a state

$$|\psi_N\rangle = \int_0^{2\pi} d\varphi e^{-iN\varphi/2}\prod_{\mathrm{k}}(|u_k| + |v_k|e^{i\varphi}c^*_{k\uparrow}c^*_{-k\downarrow})|\phi_o\rangle = \int_0^{2\pi} d\varphi e^{-iN\varphi/2}|\psi_\varphi\rangle$$

with a definite number of Cooper pairs, but with a completely indefinite phase. These number eigenfunctions are the natural basis choice for discussing problems dominated by charging energy, just as the φ representation is the natural choice for discussing situations dominated by the phase-dependent Josephson coupling energy.

For example, in an isolated superconductor, the minimum energy state would be one with N chosen so that the charge of the electrons exactly cancels the charge of the positive ion cores, in order that the system will be electrically neutral and have no electrostatic self-energy. If a second isolated superconductor is present but far away from the first, it will also be in such a state with M pairs, with M chosen so that it too will be electrically neutral. If we now bring these two superconductors close enough together so that Cooper pairs can tunnel back and forth, the system could lower its energy by the Josephson effect by having the phases of the two superconductors lock together, but this is impossible in a number eigenstate like $|\psi_N\rangle$ in which all phases are equally represented.

To gain the benefit of the Josephson coupling energy, a superposition of $|N\rangle$ states must be set up which has a nonzero expectation value of $\cos\varphi$. (We have simplified the notation from $|\psi_N\rangle$ to $|N\rangle$. To avoid any *net* charge on the combined system, the system state must be made up of a superposition of product states of the form

$$|a\rangle = \sum a_j|N+j\rangle|M-j\rangle \tag{7.18}$$

where in the jth term the charge on the junction capacitance is $2ej$, reflecting the transfer of j Cooper pairs. Since the Coulomb energy in the jth term will be $4j^2E_c$, the energy trade-off is clear—to gain a sufficiently well-defined phase to optimize the Josephson effect requires a number of terms in (7.18), which increases the Coulomb energy. Finding the state of the lowest total energy by choosing the set of a_j is an alternative way of minimizing the expectation value of the hamiltonian (7.15) or solving the equivalent Schrödinger equation.

If the tunneling coupling is weak, only a few terms in (7.18) are important since large charge transfers are energetically prohibitive. For example, the trial function (7.17) can be written as $[1+(a/2)(e^{i\varphi}+e^{-i\varphi})]$, showing its correspondence in the number representation to

$$|a\rangle = |N\rangle|M\rangle + (a/2)[|N+1\rangle|M-1\rangle + |N-1\rangle|M+1\rangle] \tag{7.19}$$

since the factors $e^{\pm i\varphi}$ are associated with a change by ±1 in the number of Cooper pairs. By inspection of (7.19), we see that the probability of a Cooper-pair transfer is $2(a/2)^2/[1+2(a/2)^2]$, so that the uncertainty in the number of pairs on either electrode is

$$\langle(\delta N)^2\rangle = a^2/(2+a^2) \approx a^2/2 \tag{7.20}$$

For $y = 4E_c/E_J \gg 1$, it can be shown that $a \approx 1/y$, so that $\langle(\delta N)^2\rangle \approx 1/2y^2 \ll 1$, and on average only a small fraction of a single Cooper pair is (virtually) exchanged in the entire junction area. In the other, low y, limit ($E_c \ll E_J$), one finds that $\langle(\delta N)^2\rangle \approx (8y)^{-1/2}$, so that to have a single Cooper pair exchanged on average, E_J must be as large as $32E_c$.

7.3.2 Macroscopic Quantum Tunneling

One of the most definitive demonstrations of the quantum aspects of the Josephson effect is probably the observation of *macroscopic quantum tunneling* (MQT) in an underdamped Josephson junction. This refers to the escape of the representative phase point from a minimum of the tilted-washboard potential (see Fig. 6.3*b*) by tunneling *through* the barrier, rather than by thermal activation *over* the barrier as described by (6.18). This process is called *macroscopic* because the tunneling quantity is not a single electron but, rather, the phase variable which describes the collective state of a macroscopic number of electrons. Quantum effects are usually unobservable on the macroscopic scale because they involve microscopic quantities. MQT is a particularly sensitive technique to reveal them because a single tunneling event switches the system from the zero voltage state (with the phase point resting in a local minimum of the washboard potential) to the gap voltage state (with the phase point running with velocity limited by the onset of strong damping when the associated voltage reaches the energy gap). As in a Geiger counter, these states are easily distinguished; the only competing classical process giving "false counts" is thermal activation, which can be "frozen out" in a well-understood way by cooling to a sufficiently low temperature. The experiments have typically been performed on relatively large junctions, where $E_J \gg E_c$, so that the phase φ is a reasonably well-defined semiclassical variable. The presence of damping in the classical picture is another hindrance to the observation of quantum effects. It is reflected in the quantum hamiltonian by coupling the phase degree of freedom to an environmental hamiltonian with infinitely many degrees of freedom. For simplicity, we start by treating the case of zero damping, to avoid this second difficulty, and later summarize the changes induced by finite damping.

In the presence of a dc bias current, the hamiltonian (7.15) is modified by the addition of the term $-\hbar I\varphi/2e$ as in (6.14*c*). As noted before, using the notation $i \equiv I/I_{c0}$, the minimum of the resulting potential occurs at $\varphi = \sin^{-1} i$, where the curvature of the potential is $d^2U/d\varphi^2 = (\hbar I_{c0}/2e)\cos\varphi = (\hbar I_{c0}/2e)(1 - i^2)^{1/2}$ and the classical frequency of small oscillations about that minimum is $\omega_A = \omega_p(1 - i^2)^{1/4}$, as given by (6.16). When solved quantum mechanically, the ground-state wavefunction $\psi(\varphi)$ of a harmonic oscillator in such a potential minimum is approximately gaussian. The lowest state in the tilted washboard is similar, with one critical difference. Because the barrier is of finite width, unlike the harmonic oscillator potential which never returns to zero, there is an exponentially small, but finite, tunneling amplitude through the barrier, which connects to an outgoing wave in the unbounded space beyond the barrier. As a result, strictly speaking, the eigenfunctions form a continuum, of which only those corresponding to the "quasi-bound states" have a high amplitude in the well of the potential. The energy width of these "resonances" gives the inverse lifetime for escape from the well to the continuum.

For making estimates, we shall replace this exact procedure by the much simpler WKB approximation method. Since we seek only semiquantitative results, we shall disregard the algebraic prefactors and concentrate on the exponential factor which dominates the transmission probability. In the usual x-space problem, ψ^2 decreases in the barrier region as

$$\exp -(2/\hbar)\int\{2m[V(x) - E]\}^{1/2}dx \tag{7.21}$$

where the integral extends over the region of width Δx under the barrier. For rough arguments, we can replace this by the schematic expression

$$\exp -(2/\hbar)(2mE_B)^{1/2}\,\Delta x$$

where E_B is a suitable average barrier height. Transcribing to the Josephson junction case, we replace Δx by $\Delta\varphi$, and following Sec. 6.3, we substitute $m = (\hbar/2e)^2 C$. The resulting approximate transmission expression is then

$$\exp -[(E_B/E_c)^{1/2}\,\Delta\varphi] \tag{7.22}$$

Typically, this will be small since $E_B \approx 2E_J \gg E_c$ (except in very small junctions) and $\Delta\varphi \approx \pi$. However, by using a current which approaches the critical current, both E_B and $\Delta\varphi$ can be decreased since $E_B \approx 2E_J(1-i)^{3/2}$ and $\Delta\varphi \sim \pi(1-i)^{1/2}$, and the tunneling probability can be made larger. The question remains, though, how cold the system must be before this probability of MQT exceeds that of thermal activation, which varies as $\exp(-E_B/kT)$, and hence also increases exponentially as I approaches I_{c0}. It turns out that the answer to this question is only weakly dependent on i, as can be seen by the following comparison.

In zero current, we can write $\hbar\omega_p = (8E_JE_c)^{1/2} = 2(E_BE_c)^{1/2}$ and set $\Delta\varphi \approx \pi$, so that (7.22) can be rewritten as $\exp -(2\pi E_B/\hbar\omega_p)$. This clearly will cross over the Boltzmann factor $\exp(-E_B/kT)$ at a crossover temperature (for zero current) given by $kT_{cr} \approx \hbar\omega_p/2\pi$, a result which is confirmed by more exact and detailed analysis. Taking into account the current-dependent factors, we see that the exponent in (7.22) scales with $(1-i)^{5/4}$, whereas the exponent in the Boltzmann factor scales with $(1-i)^{3/2}$. These differ by only $(1-i)^{1/4}$, which corresponds exactly to the current dependence of the frequency of small oscillations ω_A. Thus, our final result for arbitrary current below I_c is

$$kT_{cr} \approx \hbar\omega_A/2\pi = (\hbar\omega_p/2\pi)(1-i^2)^{1/4} \tag{7.23}$$

For a typical value of $\omega_p \approx 10^{11}$, this corresponds to $T \approx 100$ mK, which is easily accessible with dilution refrigeration techniques, although it is below the convenient liquid helium temperature range.

For MQT to be quantitatively confirmed in the laboratory, it must occur with a transition rate which not only exceeds thermal activation, but also exceeds some conveniently measurable rate τ^{-1}, such as ~1 per sec. This imposes a second criterion $\omega_A \exp -[(E_B/E_c)^{1/2}\,\Delta\varphi] \gtrsim \tau^{-1}$. Since the exponent scales with $(1-i)^{5/4}$, this requirement can in principle always be satisfied by sweeping I sufficiently near to I_{c0}, while monitoring the transition probability. If, however, one sets $I = 0.99I_{c0}$, e.g., as a practical upper bound, this puts a limit on the ratio E_J/E_c. In the experiments reported below, the junctions had I_{c0} less than ~10 μA in order to satisfy this requirement.

EFFECT OF DAMPING. The effects of damping on MQT have been treated by a number of authors; a convenient review has been given by Grabert.[9] The key conclusion is that damping strongly suppresses MQT, so that the damped system follows the classical thermal activation theory to lower temperatures. The crossover temperature in the presence of damping is no longer given by (7.23) but, rather, by a similar formula with ω_A replaced by a new, damping-related frequency ω_R, so that

$$kT_{cr} \approx \hbar\omega_R/2\pi \tag{7.24}$$

with

$$\omega_R = \omega_A[(1+\alpha^2)^{1/2} - \alpha] \tag{7.24a}$$

where $\alpha = (2RC\omega_A)^{-1} = 1/2Q$ is a dimensionless damping parameter in the RCSJ model. Clearly, (7.24) reproduces (7.23) for $\alpha = 0$. For $\alpha \ll 1$, $\omega_R \approx \omega_A(1-\alpha)$, whereas for $\alpha \gg 1$, it reduces to $\omega_R \approx \omega_A/2\alpha = \omega_A^2\,\mathrm{RC} \ll \omega_A$. In all cases, increased damping implies a lower crossover temperature to quantum behavior, making quantum effects more difficult to observe. In applying this result to an actual junction with frequency-dependent damping, the appropriate value of R is the one at ω_R, which may be quite different from the shunting resistance for dc. For example, if we take R to be the transmission-line impedance of the leads $Z_1 \approx 100\,\Omega \ll R_n$, and if $C \gg 100$ fF, we typically find the lightly damped case $\alpha \ll 1$ and $T_{cr} \approx \hbar\omega_A/2\pi k$. On the other hand, if $C \ll 100$ fF, we typically find $\alpha \gg 1$, so that we expect $T_{cr} \approx 0.88T_c(Z_1/R_n)$.

The reality of MQT and the predicted crossover temperature has been demonstrated definitively by Clarke and coworkers.[10] The measurements were made by repeatedly sweeping the bias current up toward I_c and recording the current-dependent probability of a transition to the voltage state. In these experiments, C was chosen large enough to put the system in the light-to-moderate damping regime even in the presence of transmission-line damping. The experimental parameters were determined by independent measurements, thus permitting a detailed and unique comparison with theory. Excellent quantitative agreement was found, as shown in Fig. 7.4. In this figure, the data are plotted in the form of an "escape temperature" T_{esc}. This reduces to the actual temperature in the thermal regime, whereas in the low-temperature regime it is a fictitious temperature at which classical thermal activation would yield the same escape rate as the actual MQT.

In such quantitative comparisons of data with theory, one must include refinements to the simple theoretical models that have been described here. For example, even above T_{cr}, classical thermal activation theory is modified by a

---

[9] H. Grabert, "Macroscopic Quantum Tunneling and Quantum Coherence in Josephson Systems" in *Superconducting Quantum Interference Devices and their Applications*, H.-D. Hahlbohm and H. Lübbig (eds.), de Gruyter, Berlin (1985), p. 289.

[10] A comprehensive summary is given by J. M. Martinis, M. H. Devoret, and J. Clarke, *Phys. Rev.* **B35**, 4682 (1987).

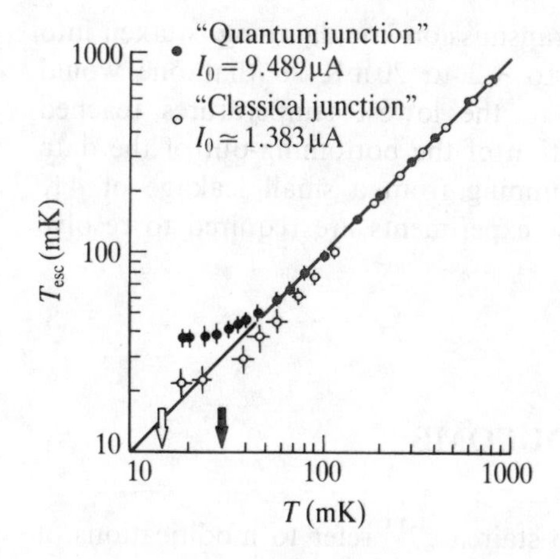

FIGURE 7.4
Temperature dependence of escape rate as described by an effective escape temperature, contrasting "classical" and "quantum" junctions. The arrows mark the predicted crossover temperatures T_{cr} for the two junctions. The straight line is the thermal prediction. (*After Martinis et al.*)

quantum correction prefactor q. This becomes very large as T approaches T_{cr}, and significantly increases the transition rate up to $T \sim 3T_{cr}$. Physically, this factor takes into account the fact that thermal activation to energies which are still below the top of the barrier can give a significant probability of crossing the remaining barrier *by MQT*, a possibility that is not included in the classical theory.

Devoret et al.[11] further confirmed the picture of macroscopic quantum states in these junctions by using microwave radiation to induce transitions between metastable levels in the wells by resonant activation. The resonance condition was identified experimentally by the observation of a peak in the rate of decay to the voltage state, which is induced by microwaves when their frequency corresponds to energy-level differences of the quasi-bound states.

PHASE DIFFUSION BY MQT. An alternative physical situation to that studied in the experiments of Clarke et al. is provided by the experiments of Iansiti et al.,[12] on much smaller junctions that had much smaller capacitance (~1 fF) and higher R_n (15 to 140 Ω). As discussed in Sec. 7.2.2, such junctions are observed to show a finite resistance R_0 at low currents, which is interpreted classically as resulting from phase diffusion. The temperature dependence of R_0 is described at least qualitatively by the classical thermally activated phase diffusion model result (7.8), except that the data appear to bottom out at a finite value at very low temperatures. Qualitatively, this is just what one would expect if MQT took over from thermal activation below a crossover temperature as given by (7.23), and the variation of $R_0(T \approx 0)$ with R_n is much as expected from the MQT probability (7.22), with $E_B = 2E_J$ and $\Delta\varphi = \pi$. The validity of this

[11] M. H. Devoret, J. M. Martinis, D. Esteve, and J. Clarke, *Phys. Rev. Lett.* **53**, 1260 (1984).

[12] M. Iansiti, M. Tinkham, A. T. Johnson, W. F. Smith, and C. J. Lobb, *Phys. Rev.* **B39**, 6465 (1989). See also M. Tinkham, "Josephson Effect in Low-Capacitance Tunnel Junctions" in H. Grabert and M. H. Devoret (eds.), *Single Charge Tunneling*, NATO ASI Series, Plenum, New York (1992), p. 163.

interpretation is unclear, however, when transmission-line damping is taken into account since it would lower T_{cr} down to ~2 to 20 mK so that one would expect the classical theory to be valid to the lowest temperatures reached (~20 mK). A possible alternative explanation of the bottoming-out of the data is that it is an experimental artifact stemming from a small leakage of 4 K blackbody radiation to the junction. New experiments are required to resolve this issue.

7.4 INTRODUCTION TO SINGLE ELECTRON TUNNELING: THE COULOMB BLOCKADE AND STAIRCASE

The "Coulomb blockade" and "Coulomb staircase"[13] refer to modifications of the tunneling *I-V* curves which occur in junctions with capacitance sufficiently low so that the Coulomb charging energy $E_c = e^2/2C$ of a *single* electron is large enough to play a major role. For simplicity, we shall initially restrict our discussion of these effects to small tunnel junctions in the *normal* state, and then discuss the new phenomena which arise when one or more electrodes are in the superconducting state.

The electrostatic energy of an *isolated* capacitor C with charges $Q(\geq 0)$ and $-Q$ on the two electrodes is $Q^2/2C$, or $CV^2/2$, where $V = Q/C$. If an electron tunnels from the negative electrode to the other one, the charge on the capacitor becomes $(Q - |e|)$, so that the energy of the capacitor becomes $(Q - |e|)^2/2C$. This represents an *increase* in system energy *unless* the initial charge $Q \geq |e|/2$, or, equivalently, the initial $V \geq |e|/2C$. Thus, electron transfer is energetically forbidden (at $T = 0$) for voltages such that $|V| < |e|/2C$. This regime of zero tunnel current despite a finite voltage across the junction is a simple example of the Coulomb blockade.

As summarized in Sec. 7.1, two requirements must be met for the blockade to be clearly observed: (1) E_c must exceed kT to avoid having its effects washed out by thermal fluctuations and (2) the resistance seen by the junction capacitance (i.e., the parallel combination of its internal tunneling resistance and any external shunting impedance) must exceed the quantum resistance $R_Q = h/4e^2$ to avoid having its effects washed out by quantum fluctuations. The reason for the second requirement is that, to be observable, the charging energy $e^2/2C$ must exceed the quantum energy uncertainty $\Delta E \gtrsim \hbar/\Delta t \approx \hbar/RC$ associated with the finite lifetime of the charge on the capacitor. Equating the Coulomb energy to the energy uncertainty, we see that the C in both denominators cancels, and the requirement

[13] T. A. Fulton and G. J. Dolan, *Phys. Rev. Lett.* **59**, 109 (1987); D. V. Averin and K. K. Likharev, "Single Electronics: Correlated Transfer of Single Electrons and Cooper Pairs in Systems of Small Tunnel Junctions" in B. L. Altshuler, P. A. Lee, and R. A. Webb (eds.), *Mesoscopic Phenomena in Solids*, Elsevier, Amsterdam (1991), p. 169.

is simply that $R \gtrsim 2\hbar/e^2 \approx R_Q$. Since the relevant time scale is set by $RC \sim 10^{-10}$ sec, the effective shunting resistance is determined by the high-frequency properties of the leads near the junction, not the dc bias resistance, just as in our discussion of the retrapping current in Sec. 7.2.1. Thus, unless ultracompact resistors are inserted right at the junction,[14] the effective impedance seen will again be $Z_1 \approx 100\,\Omega \ll R_Q$, making the blockade effect hard to observe in a single tunnel junction, no matter how low the temperature. Since the ratio $Z_1/R_Q \approx (Z_0/2\pi)/(2\hbar/e^2) = e^2/\hbar c \approx 1/137$, the fine structure constant, this constraint is very fundamental.

To circumvent this problem, experimenters have focused most studies of these phenomena on a configuration involving not a *single* tunnel *junction* but, rather, on a *single* mesoscopic metallic *island* or grain, whose total self-capacitance is sufficiently small, and which is isolated from all low-impedance macroscopic leads by tunnel junctions whose tunneling resistance exceeds the quantum resistance. With this configuration, discrete features appear in the steady-state *I-V* curve at those voltage bias values at which there is a discrete shift by unity in the average number of electrons on the isolated electrode. The resulting regular step structures form the *Coulomb staircase*.

To see how these effects come about, we shall analyze in some detail the typical double-junction circuit shown schematically in Fig. 7.5*a*. Here a mesoscopic metallic island is connected to a bias voltage V by two junctions of

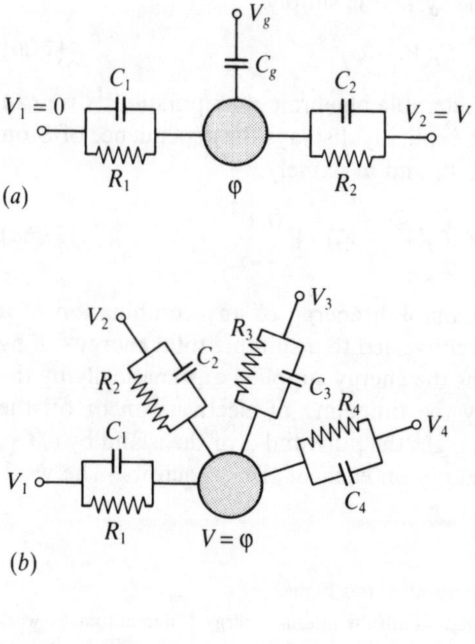

FIGURE 7.5
(*a*) Schematic diagram of the double-junction circuit, with a capacitive gate electrode. (*b*) Generalized single-island circuit, with arbitrarily many linkages to voltage sources. In both figures, R_i represents a *tunneling* resistance; other leakage conductance is assumed to be negligible.

[14] D. B. Haviland, L. S. Kuzmin, P. Delsing, and T. Claeson, *Europhys. Lett.* **16**, 103 (1991).

capacitance C_1, C_2 and tunneling resistance R_1, R_2, and also to a gate voltage V_g by a capacitance C_g, which has zero tunneling conductance. In practice, C_1 and C_2 are usually of similar magnitude $\sim 10^{-16}$ F, whereas C_g is considerably smaller. There may also be significant additional self-capacitance C_0 to ground or "infinity."

7.5 ENERGY AND CHARGING RELATIONS IN QUASI-EQUILIBRIUM

It is convenient to start with a more symmetric generalized circuit, shown in Fig. 7.5*b*, in which the island is connected to any number of voltages V_i by capacitors C_i and tunneling resistances R_i. The reference point for the voltages V_i is arbitrary, and can be chosen later for convenience. Then the potential φ of the island (with respect to the same reference) is given by

$$\varphi = \frac{\sum C_i V_i + ne}{C_\Sigma} \tag{7.25}$$

where n is the number of electrons added to the island from an initially neutral condition, and C_Σ is given by ΣC_i. This plausible result can be verified by noting that the added charge ne must equal the sum of the charges on the inner electrodes of the capacitances, which is $\Sigma C_i(\varphi - V_i)$. (It is important to remember that our sign convention throughout is that e is the charge of the electron including its sign; i.e., e is *not* the charge of a proton!) Knowing the potential of the island, we see that the total electrostatic energy of the system is simply

$$U = \tfrac{1}{2} \sum C_i (V_i - \varphi)^2 \tag{7.26}$$

By substituting (7.25) for φ, after considerable algebraic manipulation,[15] we can transform this into another form which explicitly displays the dependence of U on the appropriate independent variables, V_i and n, namely,

$$U = \frac{1}{2C_\Sigma} \sum_i \sum_{j>i} C_i C_j (V_i - V_j)^2 + \frac{(ne)^2}{2C_\Sigma} \tag{7.26a}$$

Although (7.26*a*) expresses the Coulomb energy of any combination of n and the V_i, this expression needs to be converted to a suitable total energy[16] E by adding a term which takes into account the energy supplied automatically by the voltage sources when n is changed by the tunneling of electrons on or off the island. If an electron tunnels in, this changes the potential φ of the island by e/C_Σ, causing a change by eC_i/C_Σ in the charge on each of the capacitors. The work

[15] I am indebted to J. M. Hergenrother for demonstrating this to me.

[16] More properly, this is an *enthalpy* since it includes both the internal energy U and automatic work terms. Although sometimes referred to as a free energy F or G, this seems inappropriate since no entropy term is included. There is little consensus on notation in the literature. We choose to use E rather than H, a common symbol for enthalpy, to avoid confusion with the hamiltonian.

done by each voltage source is $-eV_iC_i/C_\Sigma$ except at the injecting junction j, where it is $(1-C_j/C_\Sigma)eV_j$. The sum of all these contributions can be written in the symmetrical form

$$W_j = e\sum_i (V_j - V_i)\frac{C_i}{C_\Sigma} \tag{7.27}$$

Since this expression does not involve the parameter n, the work done in inserting n_j electrons through junction j will be just n_jW_j. Although this is a unique expression for any specific tunneling process, the work done in inserting a total of n electrons is not unique but depends, rather, on the junctions through which they enter or leave.

7.5.1 Zero Bias Circuit with Normal Island

We now return to the generic double-junction circuit shown in Fig. 7.5*a*. For the present, we specialize to the limit of zero bias voltage V by setting $V_1 = V_2 = 0$. Since no dissipative currents can flow, the equilibrium energy governs the system. Because $V = 0$, the work term has the unique value $W = -neV_gC_g/C_\Sigma$. After subtracting this from U, we obtain

$$E = \frac{1}{2C_\Sigma}[C_g(C_1+C_2)V_g^2 + 2neC_gV_g + (ne)^2] = \frac{1}{2C_\Sigma}(C_gV_g + ne)^2 + \text{const.} \tag{7.28}$$

where the constant is $\frac{1}{2}C_g(1-2C_g/C_\Sigma)V_g^2$. This constant can be omitted for convenience in applying this formula, even if it is large in magnitude, since it is independent of n, and hence has no effect on the transfer of charges. Minimizing this E with respect to n, we obtain $n = -C_gV_g/e$, corresponding to an induced charge $Q_0 = -C_gV_g$, often referred to as the "gate charge." In practice, naturally occurring random charged impurities near the island shift the polarization charge by an amount Q_{00}, which is independent of V_g, but may drift or change discontinuously in time. We generalize the definition of Q_0 to include this contribution, so that

$$Q_0 \equiv Q_{00} - C_gV_g \tag{7.29}$$

Since n must be an integer, the implication of the above is that the minimum energy for given Q_0 is obtained if n is the integer[17] closest to Q_0/e. That is, the n giving the lowest energy in (7.28) must lie in the range

$$\frac{Q_0}{e} - \frac{1}{2} \le n \le \frac{Q_0}{e} + \frac{1}{2} \tag{7.28a}$$

[17] Recall that our sign convention is that $e = -|e|$ is the charge of the electron including the sign. Thus, $n = Q_0/e = -C_gV_g/e = C_gV_g/|e|$ correctly predicts that a positive V_g will attract a positive number of negative electrons onto the island.

More generally, (7.28) generates a family of parabolic energy curves, one for each integer value of n, as plotted in Fig. 7.6a. Note that the parabolas for n and $n+1$ cross at $Q_0 = (n+\frac{1}{2})e$ at the energy $e^2/8C_\Sigma = E_c/4$, where $E_c = e^2/2C_\Sigma$ is the natural unit of charging energy for this system.[18] Upon sweeping V_g, one electron tunnels into (or out of) the reservoir at each crossing point, changing the number of extra electrons on the island from n to $n-1$ (or $n+1$). At finite temperatures, of course, the changeover from n to $n \pm 1$ in the statistical average number $\langle n \rangle$ of electrons will be broadened by thermal fluctuations in the value of n.

An important consequence of this level-crossing degeneracy is that when the levels are degenerate, an infinitesimal bias voltage V applied between V_1 and V_2

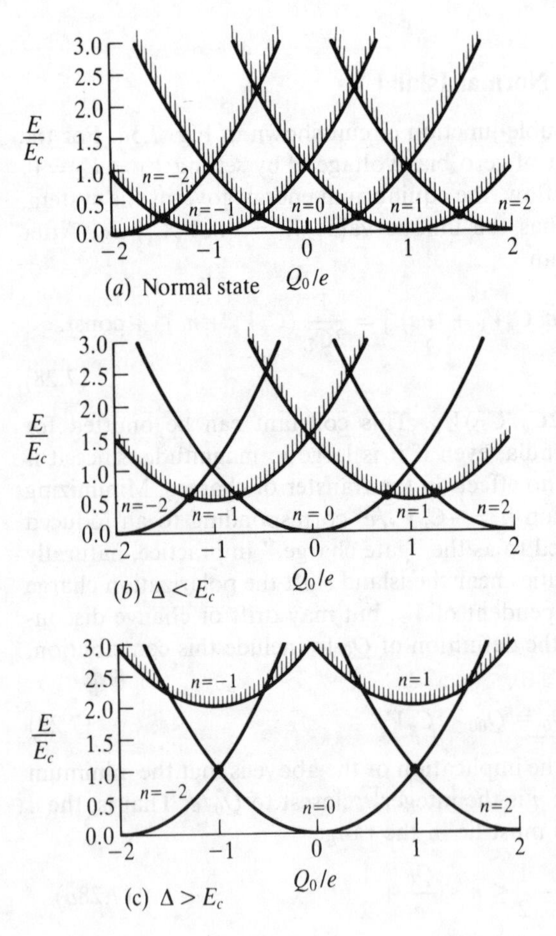

FIGURE 7.6
The n-dependent part of the circuit energy E as a function of the gate charge Q_0 at $T=0$. (a) Normal island, showing e periodicity of energy levels. (b) Superconducting island, with $\Delta < E_c$. (c) Superconducting island, with $\Delta > E_c$. In (b) and (c), the energy-level structure shows $2e$ periodicity. Heavy dots emphasize degeneracy points, where the value of n in the ground state changes, and charge transport can occur at $T=0$ without an energy barrier. Shading indicates schematically the presence of a continuum of low-lying quasi-particle states above the energy gap in the case of $n=$ odd, or in the normal state. (For simplicity, the neutral island is assumed to contain an even number of electrons, so that the parity of N and n is the same.)

[18]Note that this definition differs from that used in discussions of single-junction circuits, where we took $E_c = e^2/2C$.

can transfer electrons one at a time through the island without having to overcome any energy barrier even at $T = 0$, where only the ground state is occupied. Thus, there will be a steady current in the bias circuit whenever V_g is swept through the values where the gate charge Q_0 is a half-integral multiple of e, giving rise to a series of current peaks at V_g intervals corresponding to a change in Q_0 by e.

7.5.2 Even-Odd Number Parity Effect with Superconducting Island

If the island is a superconductor, the above results are modified by the electron pairing. If the total number N of conduction electrons on the island is even, the BCS ground state is fully paired; if N is odd, the ground state must include one quasi-particle above the energy gap Δ. To describe this distinction, Averin and Nazarov[19] introduced an explicit additive energy term, which has the value Δ in odd-N states and zero in even-N states. As can be seen from Figs 7.6*b* and 7.6*c*, this has the effect of introducing a $2e$ periodicity in the energy-level diagram, and hence in the populations of the various possible states. This in turn should be reflected in a $2e$ periodicity in the low-voltage current through the device at low temperatures, but at high temperatures we expect to recover the e periodicity of the normal state. This number parity effect on the tunnel current was first clearly demonstrated by Tuominen et al.[20] They also showed that the $2e$ periodicity changed to an e periodicity upon warming through a temperature T^*, still far below T_c, so that $\Delta(T^*) \approx \Delta(0) \gg kT^*$ and the material is still strongly superconducting. Some more recent data[21] illustrating this effect in an N-S-N device are shown in Fig. 7.7. Empirically, T^* was found to be essentially the temperature at which a *single* thermal quasi-particle is excited in the whole sample. We now examine the physics behind this changeover from a $2e$- to an e-periodic dependence at this particular T^*.

For simplicity, we restrict our attention to the case $\Delta > E_c$, for which Fig. 7.6*c* displays the relevant low-lying energy levels. For all values of Q_0, the ground state has an even number of electrons and is *nondegenerate* (except at the level crossings where Q_0/e is an odd integer). By contrast, the lowest states with an *odd* number of electrons have a high statistical weight N_{eff} because the quasi-particle states form a quasi-continuum above the energy gap. This $N_{\text{eff}} \approx 10^4$ is essentially the total number of quasi-particle states within kT above the gap in the entire island volume (typically, $\sim 10^{-14}\,\text{cm}^3$). Taking into account the multiplicity of

[19] D. V. Averin and Yu. V. Nazarov, *Phys. Rev. Lett.* **69**, 1993 (1992).

[20] M. T. Tuominen, J. M. Hergenrother, T. S. Tighe, and M. Tinkham, *Phys. Rev. Lett.* **69**, 1997 (1992); also *Phys. Rev.* **B47**, 11599 (1993).

[21] M. Tinkham, J. M. Hergenrother, and J. G. Lu, *Phys. Rev.* **B51**, 12649 (1995).

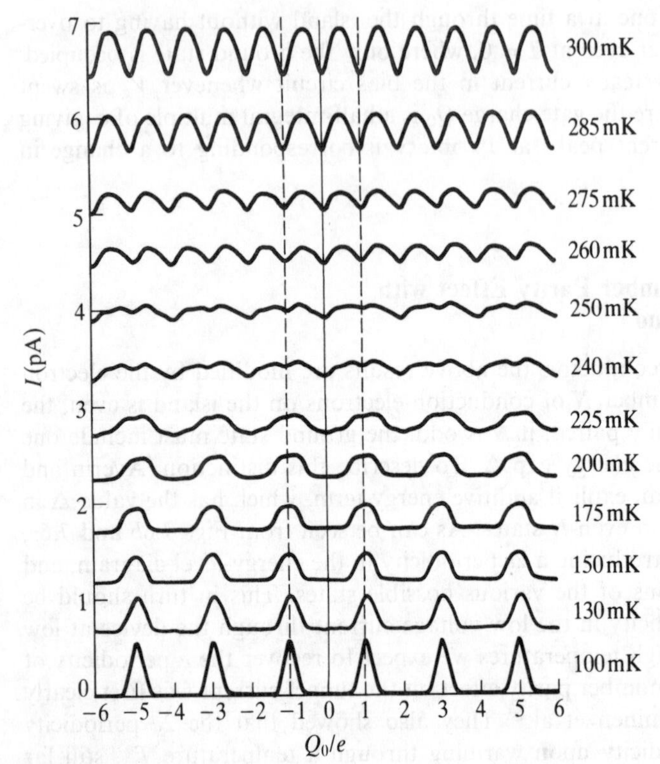

FIGURE 7.7
Current through an N-S(Al)-N double-junction device vs. the gate charge Q_0 at temperatures from 100 to 300 mK with bias voltage $V = 125\ \mu$V. Curves have been displaced upward successively. Note that the transition from dominance of even particle number states (with $2e$ periodicity) to equal likelihood of even and odd numbers (e periodicity) occurs in the rather narrow temperature range 240 to 270 mK near T^*, where the even-odd free-energy difference is going to zero. (*After M. Tinkham, J. M. Hergenrother and J. G. Lu.*)

levels, the probability of finding the system in one of these odd-N levels relative to the probability of being in the even-N ground state is $\sim N_{\text{eff}}\, e^{-\Delta E/kT}$, where $\Delta E(Q_0)$ is the energy difference between the even-N ground state and the lowest odd-N state. Since the average over Q_0 of $\Delta E(Q_0)$ is simply Δ, the heuristic criterion based on approximately one quasi-particle excitation suggests defining a zeroth-order estimate $kT_0^* = \Delta/\ln N_{\text{eff}}$. Upon inserting $N_{\text{eff}} \approx 10^4$ and the BCS value $\Delta = 1.76kT_c$, one finds $T^* = (\Delta/k)/\ln(N_{\text{eff}}) \approx T_c/5$, in good agreement with experimental data.

Although this simple argument gives the correct answer, it is not clear exactly how it relates to the experiment, in which the periodicity of the current with Q_0/e is the issue, while the theoretical estimate is obtained after averaging over Q_0. Physically, two distinct energies are competing with the thermal energy: the energy gap Δ opposes creation of *quasi-particles*, and the Coulomb energy E_c

tries to make the *charge* or *electron* number match the gate charge Q_0. In the normal state, $\Delta = 0$, and Q_0 controls the number of electrons, subject to thermal rounding, as described in the previous section. In the present case, we assume $\Delta > E_c$, so that the ground state must contain the *even* number of electrons closest to that specified by Q_0. At $T > 0$, there is a probability of a number of thermally excited quasi-particles. It is useful to separate the effects of Δ, which is the same for any excitation, and the effect of E_c which enters only when comparing states differing in *charge* by $\pm e$. In such cases, E_c can easily be shown to raise the energy of odd n relative to even n states[22] by $E_0(Q_0) = E_c(1 - 2|Q_0/e|)$ if Q_0/e is restricted to the representative range $-1 < Q_0/e < +1$. This E_0 is $2e$ *periodic* because it has the property that $E_0(Q_0) = -E_0(Q_0 + e)$, and it averages to zero over Q_0, as stated earlier.

The precise criterion for the return to e periodicity is that the probability of an odd number of electrons at Q_0 should be the same as the probability of an even number of electrons at $(Q_0 + e)$, just as would be expected in the normal state. To allow a simple analytic computation of these probabilities, we replace the exact system of energy levels by a model in which, for $-1 \le (Q_0/e) \le 1$, the ground state is the nondegenerate state with $n = 0$, and the first excited state has degeneracy N_{eff}, and $n = 1$ or -1 depending on the sign of Q_0. (See Fig. 7.6*c*.) We ignore all *charge* states with n values other than these, but take into account any number of quasi-particle *excitations* above these even and odd ground states. We also assume that the island is in (weak) tunneling contact with a particle reservoir, so it can contain either an even or odd number of electrons.

By constructing forms in which either the even or odd terms cancel, we can then write down partial partition sums for even and odd n separately as

$$Z_{\text{even}} = [(1 + e^{-\Delta/kT})^{N_{\text{eff}}} + (1 - e^{-\Delta/kT})^{N_{\text{eff}}}]/2 = Z^0_{\text{even}} \tag{7.30a}$$

and

$$Z_{\text{odd}} = e^{-E_0/kT}[(1 + e^{-\Delta/kT})^{N_{\text{eff}}} - (1 - e^{-\Delta/kT})^{N_{\text{eff}}}]/2 = e^{-E_0/kT} Z^0_{\text{odd}} \tag{7.30b}$$

where Z^0 refers to the partition sum with $E_0 = 0$. It follows that the relative probability of odd to even numbers of electrons will be

$$\frac{P_{\text{odd}}(Q_0)}{P_{\text{even}}(Q_0)} = \frac{Z_{\text{odd}}}{Z_{\text{even}}} = e^{-E_0/kT} \frac{Z^0_{\text{odd}}}{Z^0_{\text{even}}} \equiv e^{-E_0/kT} e^{-F_0/kT} \tag{7.31}$$

where F_0 is the odd/even free-energy difference introduced by Tuominen et al., namely,

$$F_0 = kT \ln (Z^0_{\text{even}}/Z^0_{\text{odd}}) \tag{7.32}$$

[22] For simplicity, the neutral island is assumed to contain an even number of electrons, so that the total number of conduction electrons N and the excess number n have the same parity.

At low temperatures the leading terms in the binomial expansions dominate, and we find

$$F_0 \approx \Delta - kT \ln N_{\text{eff}} \tag{7.32a}$$

In this approximation F_0 would drop linearly to zero at the T_0^* defined above, but if more terms in the expansion are included, $F_0(T)$ goes to zero only asymptotically, but very rapidly, as T exceeds T_0^*. A plot of $F_0(T)$ for typical parameters is shown in Fig. 7.8*a*.

As noted earlier, the appropriate test for the presence of e periodicity is the ratio of probabilities $P_{\text{odd}}(Q_0)/P_{\text{even}}(Q_0 + e)$, which reduces to unity if there is e periodicity. Recalling that $E_0(Q_0 + e) = -E_0(Q_0)$, after a little algebra, one obtains

$$\frac{P_{\text{odd}}(Q_0)}{P_{\text{even}}(Q_0 + e)} = \frac{1 + e^{E_0/kT} e^{-F_0/kT}}{1 + e^{E_0/kT} e^{F_0/kT}} \approx 1 - \frac{2F_0}{kT} \frac{1}{(1 + e^{-E_0(Q_0)/kT})} \tag{7.33}$$

where the last form holds only when $(F_0/kT) \ll 1$. Clearly, this ratio approaches unity for any Q_0 as F_0 goes to zero above T_0^*. This analysis also shows that the switchover from $2e$ to e periodicity (for $E_c < \Delta$) largely occurs in the small temperature range from $(1 - 1/\ln N_{\text{eff}})T_0^*$ to T_0^*, in which

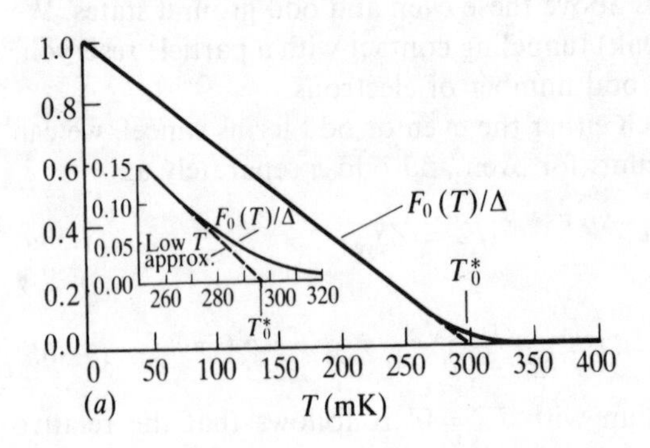

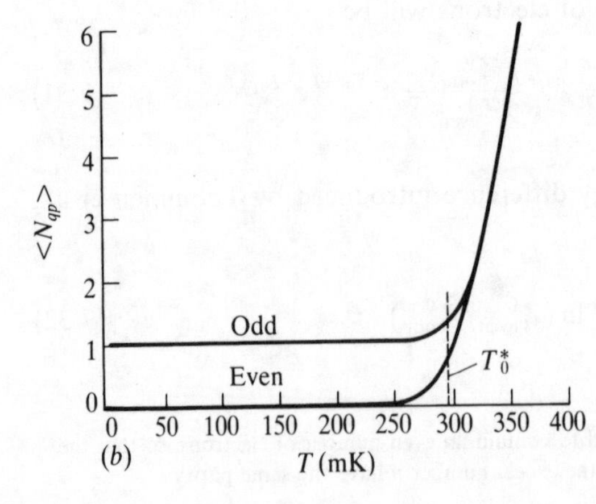

FIGURE 7.8
(*a*) Plot of even-odd free-energy difference $F_0(T)/\Delta$ vs. temperature for typical parameters $N_{\text{eff}} \approx 15{,}000$, $\Delta/k_B = 2.8\,\text{K}$, so that $T_0^* = 296\,\text{mK}$. Inset shows a magnified view of the approach to zero near T_0^*. (*b*) Plot of temperature dependence of the number of quasi-particles for odd and even total numbers of electrons for same parameters as in (*a*). Note the rapid convergence to the macroscopic limit just above T_0^*, where $\langle N_{qp} \rangle \sim 1$. (*After M. Tinkham, J. M. Hergenrother, and J. G. Lu.*)

F_0/kT drops from ~ 1 to ~ 0. For typical numbers, this is only about a 10 per cent range in temperature, which is generally consistent with the data in Fig. 7.7.

By taking into account the Q_0 dependence of $E_0(Q_0)$, one can get more detail. When Q_0/e is an integer, $E_0(Q_0) = \pm E_c$ for the even and odd integers, respectively, showing a strong $2e$ periodicity for the ratio $P_{\text{odd}}(Q_0)/P_{\text{even}}(Q_0 + e)$. When Q_0/e is *any* half-integer, $E_0(Q_0) = 0$ and the ratio is the same from one half-integer to the next. Thus, $2e$ periodicity should be the strongest and last to disappear at integer values of Q_0/e. This prediction is confirmed in the data of Fig. 7.7, in that the current at integer values of (Q_0/e) (i.e., the bottoms of the variations) retains a $2e$-periodic variation at ~ 260 mK, whereas the current at half-integer values of Q_0/e (i.e., the tops of the variations) is already e periodic at that temperature.

The actual calculation of $I(Q_0)$ involves solving a system of rate equations. This approach allows one to study the transition from $2e$ to e periodicity as a function of *bias voltage* as well as of temperature. However, the conclusions about periodicity at infinitesimal bias are the same as those obtained by the more transparent equilibrium analysis given here. In such simulations, tunneling events always take place between states with a definite number of particles, i.e., a *canonical* ensemble. In this fixed n case, the expectation value of the number of thermal quasi-particles for odd and even n is independent of E_c or Q_0 and depends only on T because quasi-particle recombination and generation is assumed fast enough to assure instantaneous equilibrium distributions. Plots of these numbers for typical parameters are shown in Fig. 7.8*b*. This figure illustrates nicely that the crossover from $2e$ to e periodicity occurs in the temperature range in which the number of quasi-particles changes from 0 or 1 for even or odd n, respectively, to a rapidly rising value which is independent of the parity.

Although the exact form of the periodic current dependence $I(Q_0)$ depends on whether one is studying an *S-S-S* or an *N-S-N* system, the e vs. $2e$ period should depend only on T/T^*. In fact, the T^* observed with *S-S-S* devices by Tuominen et al. and by Amar et al.[23] are very similar to that seen for an *N-S-N* device in Fig. 7.7.

If the system follows the minimum energy state while Q_0 is swept, in Fig. 7.6*c* the system always stays in states of even N, with two electrons entering or leaving at each crossing point of the parabolas. On the other hand, in Fig. 7.6*b*, electrons enter one at a time, so that the ground state visits both even and odd N states, but the system is in even N states for a greater fraction of the sweep. Lafarge et al.[24] used this effect to extract from their data the temperature dependence of $F_0(T)$, confirming that the dependence (7.32) expected from this analysis, and plotted in Fig. 7.8*a*, is a good description.

---

[23] A. Amar, D. Song, C. J. Lobb, and F. C. Wellstood, *Phys. Rev. Lett.* **72**, 3234 (1994).

[24] P. Lafarge, P. Joyez, D. Esteve, C. Urbina, and M. H. Devoret, *Phys. Rev. Lett.* **70**, 994 (1993).

7.5.3 Zero Bias Supercurrents with Superconducting Island and Leads

In the *S-S-S* configuration, in which the leads as well as the island are superconducting, a supercurrent at zero bias voltage is possible. For simplicity, we restrict our treatment to $T = 0$. One can distinguish three regimes: If $E_c \ll E_J$, the charging energy effects are small, and one has essentially the classical Josephson effect. If $E_c \gg \Delta$, single electron charging energies dominate, observed phenomena show e periodicity, and Cooper-pair phenomena are unimportant. The most interesting regime is $E_J < E_c < \Delta$, in which we shall now show that there is a supercurrent which is modulated by the imposed gate charge Q_0 with period $2e$.

If we consider junctions in which $R_n \gg R_Q$, then $E_J \approx (R_Q/R_n)\,\Delta \ll \Delta$, and there is a considerable range of E_c for which the two inequalities $E_J \ll E_c < \Delta$ hold. The fact that Δ is larger than all other energies allows us to restrict our attention (at $T \approx 0$) to states of the island containing only an even number of electrons, which form Cooper pairs. The net charge Q then can be written as $2eN$, where N is an integer. Since we assume zero bias voltage, we can build on (7.28) and describe the total energy of the system by the hamiltonian

$$H = -E_1 \cos \varphi_1 - E_2 \cos \varphi_2 + \frac{(Q - Q_0)^2}{2C_\Sigma} \tag{7.34}$$

where E_1 and E_2 are the Josephson coupling energies of the two junctions which couple the island to the superconducting leads, and φ_1 and φ_2 are the phase differences across them.

As shown by Averin and Likharev,[25] it is convenient to make the change of variables

$$\theta = \varphi_1 + \varphi_2$$

and

$$\varphi = (\varphi_2 - \varphi_1)/2 \tag{7.35}$$

where θ is the total phase difference across the *S-S-S* device and φ is the departure of the phase on the island from the midpoint between the phases of the two leads. Since the leads are macroscopic superconductors, we can assume that they have a well-defined phase difference, so that θ is a semiclassical variable. On the other hand, φ is free to fluctuate and show quantum-mechanical properties. If one uses (7.35) to eliminate φ_1 and φ_2 in (7.34) in favor of θ and φ, and uses trigonometric identities to simplify, one can reduce this equation to

$$H = -E_J(\theta)\cos(\varphi - \chi) + \frac{(Q - Q_0)^2}{2C_\Sigma} \tag{7.36}$$

[25] D. V. Averin and K. K. Likharev, in B. L. Altshuler, P. A. Lee, and R. A. Webb (eds.), *Mesoscopic Phenomena in Solids*, Elsevier, New York (1991), p. 213.

where the parameters

$$E_J(\theta) = (E_1^2 + E_2^2 + 2E_1E_2 \cos\theta)^{1/2} \tag{7.36a}$$

and

$$\chi = \tan^{-1}\left[\frac{(E_1 - E_2)}{(E_1 + E_2)} \tan\frac{\theta}{2}\right] \tag{7.36b}$$

are determined by the fixed values of E_1, E_2, and θ. Since the eigenvalues of (7.36) do not depend on the reference phase χ, we shall suppress it in the following equation and rewrite (7.36) as

$$H = -E_J(\theta) \cos\varphi + \frac{(Q - Q_0)^2}{2C_\Sigma} \tag{7.37}$$

In this equation, it is important to note that $Q/2e$ is the number variable that is conjugate to the phase φ, whereas Q_0 is a semiclassical charge controlled by the gate voltage V_g. One should also note that, apart from inclusion of a gate charge Q_0, this expression is equivalent to (7.15), which we developed to describe the quantum fluctuations in a single *isolated* junction. Thus, the quantum phase fluctuations graphed in Fig. 7.3 directly describe the fluctuations of the quantum variable φ, which here describes the phase of the "internal" island in our double-junction model. The great advantage of the double-junction model over the single junction is that it describes a realistic physical structure with macroscopic leads, which allows physical measurements to be made. Physically, it resembles the isolated junction because each junction is effectively isolated from the low-impedance environment by the high tunnel resistance and low capacitance of the "other" junction.

To illustrate our approach, we first treat the case in which C_Σ is so large that the charging energy term in (7.37) can be neglected. Then the ground state has $\varphi = 0$ [or $\varphi = \chi$ in (7.36)], so that $E = -E_J(\theta)$. The supercurrent is then calculated[26] to be

$$I_s = \frac{2e}{\hbar}\frac{\partial E}{\partial\theta} = -\frac{2e}{\hbar}\frac{\partial E_J}{\partial\theta} = \frac{2e}{\hbar}\frac{E_1E_2 \sin\theta}{E_J} \tag{7.38}$$

In the special case in which $E_1 = E_2 = E_i$, this reduces to

$$I_s = \frac{2e}{\hbar} E_i \sin\frac{\theta}{2} \tag{7.38a}$$

which agrees with the classical result for two identical Josephson junctions in series, splitting equally the total phase difference θ. In the opposite limiting case of very asymmetric junctions, with $E_1 \gg E_2$, (7.36a) shows that $E_J \approx E_1$, and

[26]The relation $I_s = (2e/\hbar)\,\partial E/\partial\theta$ follows from a generalization of the energetic argument leading to (6.3).

(7.38) reduces to $I_s = (2e/\hbar)E_2 \sin\theta$. Physically, in this limit the entire phase difference is dropped across the weaker junction, which determines the current.

We now treat (7.37) in the other limit, $E_c > E_J$, in which the charging energy term is dominant. In this case, as anticipated in Sec. 7.3.1, it is useful to reexpress (7.37) in a representation based on eigenstates of N, the excess Cooper pair number, in which $Q = 2eN$ is sharp. In this representation, the Josephson coupling term $\sim \cos\varphi = (e^{i\varphi} + e^{-i\varphi})/2$ is the sum of two terms which, respectively, raise and lower the number of Cooper pairs by one, as can be seen to be plausible from the discussion in Sec. 7.3.1. Hence, we can rewrite (7.37) as

$$H = -E_J \sum_N \frac{|N+1\rangle\langle N| + |N-1\rangle\langle N|}{2} + E_c\left(2N - \frac{Q_0}{e}\right)^2 \tag{7.39}$$

with $E_c = e^2/2C_\Sigma$. This defines a hamiltonian matrix whose diagonal elements are

$$H_{NN} = E_c\left(2N - \frac{Q_0}{e}\right)^2 \tag{7.39a}$$

which describe the even-n parabolas ($n = 2N$) of Fig. 7.6. The off-diagonal elements of the matrix are

$$H_{N,N\pm1} = -E_J/2 \tag{7.39b}$$

Since one is assuming that $E_J \ll E_c$, second-order perturbation theory might be used to find approximate eigenvalues. However, although the second-order term is in general smaller than the leading term by a factor of order $(E_J/E_c)^2 \ll 1$, it diverges when the gate charge Q_0 differs from $2Ne$ by $\pm e$ because one approaches a degeneracy between states with N and $N \pm 1$ Cooper pairs. (See Fig. 7.6.) To avoid this problem near the degeneracy point, one can solve *exactly* the 2×2 matrix connecting the two nearly degenerate states. In this way, one obtains

$$E_N = E_c\left\{\frac{(2N - Q_0/e)^2 + (2N+2-Q_0/e)^2}{2} \pm 2\left[(2N+1-Q_0/e)^2 + \left(\frac{E_J}{4E_c}\right)^2\right]^{1/2}\right\} \tag{7.40}$$

For example, at the former degeneracy point $Q_0/e = 2N+1$, (7.40) simplifies to

$$E = E_c(1 \pm E_J/2E_c) \tag{7.40a}$$

From this we see that the degeneracy is lifted, with an energy gap of E_J between the two eigenenergies, as pointed out in a slightly different context by Likharev and Zorin.[27]

[27]K. K. Likharev and A. B. Zorin, *J. Low Temp. Phys.* **59**, 347 (1985).

Having worked out the eigenenergies, we can again compute the supercurrent I_s by differentiation with respect to the overall phase difference θ. Using (7.40) and (7.36*a*), we find

$$I_s = \frac{2e}{\hbar}\frac{E_1 E_2}{8E_c}\frac{\sin\theta}{[(2N+1-Q_0/e)^2+(E_J/4E_c)^2]^{1/2}} \tag{7.41}$$

From this we see that the limiting (maximum) value occurs at the degeneracy points, where

$$I_s = \frac{2e}{\hbar}\frac{E_1 E_2}{2E_J}\sin\theta \tag{7.41a}$$

This is exactly half the value found in (7.38) for the classical case in which E_c is negligible. For example, in the symmetric special case in which $E_1 = E_2 = E_i$, (7.41*a*) reduces to

$$I_s = \frac{2e}{\hbar}\frac{E_i}{2}\sin\frac{\theta}{2} \tag{7.41b}$$

Midway between the degeneracy points the supercurrent decreases to a minimum value[28] of

$$I_s = \frac{2e}{\hbar}\frac{E_1 E_2}{4E_c}\sin\theta \tag{7.41c}$$

if we neglect $(E_J^2/16E_c^2)$ compared to unity.

Summing up, in this section, we have shown that the *S-S-S* double-junction superconducting tunneling transistor acts as a novel sort of Josephson junction in which the critical current I_c varies periodically with the gate charge $Q_0/2e$. At the resonant or degenerate values of Q_0 [i.e., $Q_0 = (2N \pm 1)e$] it reaches a maximum value given by (7.41*a*), comparable to the classical values for the individual junctions. In between these resonant values of Q_0, however, (7.41*c*) shows that the critical current is depressed by a factor of the order of $E_J/E_c \ll 1$. It should be noted, of course, that at finite temperatures these very small predicted supercurrents may be washed out by *thermal* fluctuations, which have been ignored for simplicity in our discussion of quantum effects. Moreover, as discussed in Sec. 7.2.2, the supercurrent branches of small Josephson junctions are marked by finite resistance due to phase diffusion, and by switching currents I_m lower than the nominal critical current I_{c0}. Despite these complications, the predicted behavior has been semiquantitatively confirmed experimentally by Eiles and Martinis.[29]

---

[28] Equation (7.41) yields only half this value because the two-level approximation used to derive it includes only one of two adjacent N states which contribute equally when Q_0/e is an even integer.

[29] T. M. Eiles and J. M. Martinis, *Phys. Rev.* **B50**, 627 (1994).

A more general treatment of this problem has been given by Matveev et al.[30] So long as $\Delta > E_c$, their work yields only quantitative refinements to the results described earlier in this section. However, when Δ drops below E_c, a case not treated above, each of the resonances, which occur at the values of Q_0 at which the parabolas in Fig. 7.6 cross, splits into two. Finally, when $\Delta \ll E_c$, the resonances become uniformly spaced with period e, instead of $2e$. Their work shows that the occupation of states with odd as well as even numbers of electrons on the island (which is the origin of the doubling of the number of resonances) also reduces the peak Josephson critical current of the device to much lower values since the presence of a quasi-particle "poisons" the Josephson tunneling. After a steep initial drop when Δ first becomes less than E_c, I_c eventually goes to zero linearly with Δ/E_c. The splitting of the resonance peaks and other predictions of Matveev et al. have been confirmed in recent experiments by Joyez et al.[31]

7.6 DOUBLE-JUNCTION CIRCUIT WITH FINITE BIAS VOLTAGE

In this section, we relax the restriction to infinitesimal bias voltages made in the previous section and allow the bias voltage V in Fig. 7.5*a* to be arbitrarily large. This enables us to treat the Coulomb blockade and Coulomb staircase. Again, we shall start with the normal state, and then consider the new features which arise when the island is superconducting.

With V finite, the work term in E depends separately on the numbers of electrons n_1 and n_2 entering the island via the two inequivalent channels. Using (7.27), we find $W_1 = -(e/C_\Sigma)(C_2V - Q_0)$, and $W_2 = (e/C_\Sigma)(C_1V + Q_0)$. In both cases, the change in electrostatic energy U on going from n to $n \pm 1$ electrons is $(\frac{1}{2} \pm n)e^2/C_\Sigma$. Combining the two contributions, we find

$$\Delta E_1^{\pm} = \frac{e^2}{C_\Sigma}\left\{\left[\frac{1}{2} \pm \left(n - \frac{Q_0}{e}\right)\right] \pm \frac{C_2V}{e}\right\} \tag{7.42a}$$

$$\Delta E_2^{\pm} = \frac{e^2}{C_\Sigma}\left\{\left[\frac{1}{2} \pm \left(n - \frac{Q_0}{e}\right)\right] \mp \frac{C_1V}{e}\right\} \tag{7.42b}$$

$\Delta E_1^{\pm}$ and $\Delta E_2^{\pm}$ are the energy changes for $n \rightarrow (n \pm 1)$ by tunneling of an electron through junctions 1 and 2, respectively. For such a transition to occur at $T \approx 0$, it is necessary that the relevant ΔE be negative. The sum[32] of the two ΔE values for a *through* passage of charge is *always* exactly eV, so that the energy would always be lowered by $|eV|$ *overall.* The essence of the Coulomb blockade is that the

---

[30] K. A. Matveev, M. Gisselfält, L. I. Glazman, M. Jonson, and R. I. Shekhter, *Phys. Rev. Lett.* **70**, 2940 (1993).

[31] P. Joyez, P. Lafarge, A. Fiolipe, D. Esteve, and M. H. Devoret, *Phys. Rev. Lett.* **72**, 2458 (1994).

[32] In making the sum, one must be careful to evaluate the expressions for the instantaneously correct value of n; e.g., $\Delta E^+(n) + \Delta E^-(n+1)$.

energy must be lowered on *each* of the two successive transitions of charging and discharging the island. Actually, the only requirement is that the *first* step should have no barrier because the second step can be easily shown to be always energetically favorable at bias voltages which permit the first step to occur.

For example, if the voltage V is positive and we start with n electrons on the island, an electron current will flow through both junctions 1 and 2 (see Fig. 7.5*a*) if $\Delta E_1^+(n)$ is negative *or* if $\Delta E_2^-(n)$ is negative. In the first case, an electron can charge the island through junction 1, bringing n to $n+1$, and then discharge it readily through junction 2, returning n to its original value. In the second case, an electron first charges the island (in the opposite sense) by tunneling off the island through junction 2, bringing n to $n-1$, and then the island is discharged by an electron entering through junction 1. Either cycle returns the system to its original state with n electrons, so that it can be repeated over and over, transferring one electron per cycle. These two alternative cycles have different threshold voltages, and the Coulomb blockade extends to a voltage limit determined by whichever of these is lower.

Looking in detail at the two alternatives mentioned above, we find from (7.42) that $\Delta E_1^+(n)$ for the charging step will become negative for positive bias voltages above the threshold given below in (7.43*a*), whereas $\Delta E_2^-(n)$ will become negative for positive bias voltages above the threshold given in (7.43*b*). The extent of the Coulomb blockade is determined by whichever is the lesser of the two:

$$V_{th,1}^{+} = \frac{-e}{C_2}\left[\frac{1}{2} + \left(n - \frac{Q_0}{e}\right)\right] \tag{7.43a}$$

and

$$V_{th,2}^{-} = \frac{-e}{C_1}\left[\frac{1}{2} - \left(n - \frac{Q_0}{e}\right)\right] \tag{7.43b}$$

The physical reason that C_2 appears in the first of these and C_1 in the second, is that a fraction C_2/C_Σ of the bias voltage appears across C_1 to assist the transfer across junction 1, which governs the threshold in (7.43*a*), and vice versa for the other case. The *magnitudes* of the corresponding expressions for a negative bias voltage can be obtained by reversing the sign between the two terms in the square bracket.

To evaluate (7.43), we need to know the appropriate value of $(n - Q_0/e)$. Recalling our generalization (7.29) of $-C_g V_g$ to Q_0, the fact that n is an integer, and (7.28*a*), we see that in *equilibrium* at $T = 0$, the quantity $(n - Q_0/e)$ has a unique value (for given Q_0) lying in the range $-\frac{1}{2} \leq (n - Q_0/e) \leq \frac{1}{2}$ because n adjusts its value so that ne is as near as possible to the induced charge Q_0. If Q_0 is zero, this implies that $n = 0$ and V_{th} is $|e|/2C_1$ or $|e|/2C_2$, whichever is less. In the other limit, when Q_0/e is a half-integer, $V_{\text{th}} = 0$, and *there is no Coulomb blockade*. To obtain reliable results, we must find the proper value for n under general conditions. In the presence of dissipation, the minimization of energy by itself is no longer a reliable guide because kinetics also enters. It is necessary to take into account the transition probabilities for the charging and discharging steps to work out the probability of various values of $n(V)$, and the corresponding I-V curve, in the biased, dissipative circuit.

7.6.1 Orthodox Theory and Determination of the I-V Curve

When energetically allowed, the tunnel probability is proportional to $1/R_j$, where R_j is the resistance of the junction at temperatures or voltages high enough to make Coulomb energies unimportant. At nonzero temperatures, one must take into account the finite probability of reverse tunneling processes which *increase* the system energy. In this way, one finds that the directed tunneling rate for the jth junction can be written as

$$\Gamma_j^{\pm}(n) = \frac{1}{R_j e^2}\left(\frac{-\Delta E_j^{\pm}}{1-\exp(\Delta E_j^{\pm}/k_B T)}\right) \tag{7.44}$$

At $T = 0$ this reduces correctly to zero for $\Delta E > 0$ and to $|\Delta E|/R_j e^2$ for $\Delta E < 0$. The significance of Γ in the latter case is that $\Gamma = 1/\tau$, where τ is the time per electron in the current flow described by Ohm's law. If the conduction is not ohmic, as in the superconducting case, this prescription is generalized by replacing the factor $-\Delta E/R_j e^2$ by $I(-\Delta E/|e|)/|e|$.

The steady-state current (which must be the same in both junctions) can be written as

$$I(V) = e\sum_{n=-\infty}^{\infty}\sigma(n)\left[\Gamma_2^{+}(n)-\Gamma_2^{-}(n)\right] = e\sum_{n=-\infty}^{\infty}\sigma(n)\left[\Gamma_1^{-}(n)-\Gamma_1^{+}(n)\right] \tag{7.45}$$

where $\sigma(n)$ is the ensemble distribution of the number of electrons on the center electrode. The distribution $\sigma(n)$ is obtained by noting that the net probability for making a transition between any two adjacent values of n in steady state is zero; thus

$$\sigma(n)\left[\Gamma_1^{+}(n)+\Gamma_2^{+}(n)\right] = \sigma(n+1)\left[\Gamma_1^{-}(n+1)+\Gamma_2^{-}(n+1)\right] \tag{7.46}$$

Since the $\Gamma_j^{\pm}$ are known from (7.42) and (7.44), this allows us to solve for the distribution $\sigma(n)$, subject to the normalization condition, $\sum_{n=-\infty}^{\infty}\sigma(n) = 1$. We can thus numerically solve for $I(V)$ from (7.45). The approach outlined here is often referred to as the "orthodox theory."[33]

One can gain some physical insight into this formidable procedure by examining simple cases. At low temperatures and in the voltage range just above the blockade voltage, only a single charging cycle is energetically allowed. For example, if (7.43*a*) gives the threshold at some Q_0, then a charging tunneling event

[33] D. V. Averin and K. K. Likharev, "Single Electronics: Correlated Transfer of Single Electrons and Cooper Pairs in Systems of Small Tunnel Junctions" in B. L. Altshuler, P. A. Lee, and R. A. Webb (eds.), *Mesoscopic Phenomena in Solids*, Elsevier, New York (1991), p. 169.

occurs across junction 1 with tunneling rate $\Gamma_1^+(n)$, and the discharging event occurs across junction 2 with a rate $\Gamma_2^-(n+1)$. The time required for the complete cycle is the sum of the times $\tau_i = 1/\Gamma_i$ for each step. Thus, the average current will be

$$\bar{I} = \frac{e}{\tau_1 + \tau_2} = \frac{e}{\Gamma_1^{-1} + \Gamma_2^{-1}} = e\frac{\Gamma_1\Gamma_2}{\Gamma_1 + \Gamma_2} \tag{7.47}$$

where the Γ_i are determined by inserting into (7.44) the appropriate values of tunneling resistances R_i and the voltage-dependent energy differences ΔE_i given by (7.42). At higher voltages or temperatures, it gets more complicated since several tunneling cycles may operate and the island may change charge by more than one electron. In such cases, we cannot avoid numerical simulations to get quantitative results.

7.6.2 The Special Case $R_2 \gg R_1$

Fortunately, one can obtain useful semiquantitative analytic results over a wide voltage range in the special case in which the two junctions have very different tunneling resistances, as in an STM configuration.[34] To be concrete, we assume that $R_2 \gg R_1$, so that in general $\Gamma_1 \gg \Gamma_2$ for allowed transitions. Accordingly, (7.47) implies that the current will be controlled by Γ_2.

The key point is this: In the limit in which $R_2/R_1 \to \infty$, the circuit involving C_1 and C_2 can be mapped into the one treated in Sec. 7.5.1, with $C_1 \to (C_1 + C_2)$ and $C_2 \to C_g$. In that case, minimizing the energy *does* determine n, as given by (7.28*a*). Transcribing this result to the present context, we replace $C_g V_g$ by $C_2 V$. If we also include Q_0 to take into account an actual V_g or other source of charge offset, we obtain the result

$$-\frac{C_2 V}{e} - \frac{1}{2} \le \left(n - \frac{Q_0}{e}\right) \le -\frac{C_2 V}{e} + \frac{1}{2} \tag{7.48}$$

This equation determines the "resting" value of n, which can be seen to be the integer nearest to $(C_g V_g + C_2 V - Q_{00})/|e|$. This is the value of n between charging/discharging cycles, which occur at the infrequent rate $\Gamma_2(n, V)$ set by the slower transition. Using the $T = 0$ limit of (7.44) for ΔE negative, we have $\Gamma_2(n, V) = |\Delta E_2^-|/R_2 e^2$, with ΔE_2^- given by (7.42*b*). Since the transitions are all in one direction, and since $\Gamma_1 \gg \Gamma_2$, the magnitude of the current will be simply $I = |e|\Gamma_2$. In this way, we obtain

$$I = \frac{|e|}{R_2 C_\Sigma}\left[\frac{C_1 V}{|e|} + \left(n - \frac{Q_0}{e}\right) - \frac{1}{2}\right] \qquad (V > 0) \tag{7.49}$$

[34] A. E. Hanna and M. Tinkham, *Phys. Rev.* **B44**, 5919 (1991).

where the value of $(n - Q_0/e)$ is constrained by (7.48). From the form of (7.49) we see that the slope of the I-V curve outside the blockade region is always given by

$$\frac{dI}{dV} = \frac{1}{R_2}\frac{C_1}{C_\Sigma} \tag{7.49a}$$

reflecting the voltage divider action of the capacitors and the fact that the current is limited by the larger resistance R_2. More dramatic is the fact that whenever V reaches a value such that (7.48) requires a change from n to $n+1$, there will be an abrupt step in current by $\Delta I = |e|/R_2C_\Sigma$, leading to the appellation "Coulomb staircase." Focusing on the top and bottom endpoints of these discontinuities, we find that the current levels are

$$I = \frac{1}{R_2}\left[V - \frac{e}{2C_\Sigma}(1 \pm 1)\right] \tag{7.49b}$$

so that the overall slope of the envelope is just $1/R_2$, as would be the case if charging effects were negligible. However, extrapolated back to $I = 0$, this equation shows that there is a voltage offset which averages to $e/2C_\Sigma$.

One should note that the slope (7.49*a*) does not apply within the Coulomb blockade at low voltages, where $dI/dV = 0$, but only after the blockade is overcome. The onset of current above the blockade voltage will be either (*a*) by a ramp of slope (7.49*a*) if the threshold (7.43) is crossed "continuously" by increasing V or (*b*) by a current step of amount $|e|/R_2C_\Sigma$ if the threshold is crossed "discontinuously" by a discrete change in n driven by (7.48). Examples of both behaviors are shown in Fig. 7.9. This figure shows experimental data compared with three levels of approximation for the theory. The poorest is the analytic model discussed here, with $T = 0$ and $R_2/R_1 = \infty$; next poorest are the curves calculated with the full theory for $T = 4$ K, the experimental temperature, and $R_2/R_1 = \infty$, showing the modest thermal smearing. The best curves also include the effect of a finite resistance ratio. These best curves can barely be distinguished from the experimental data, demonstrating the success of the orthodox theory in fitting even quite complex I-V curves. Clearly, the largest error in the simple approximation presented here stems from the assumption of a discontinuous jump in n at each step, where in reality there is a statistical mixture near the changeovers.

By controlling Q_0 with a gate electrode, these I-V characteristics can be manipulated to create a sort of single electron transistor. By combining several such circuits on a chip, one opens a rich potential for applications, such as forming an "electron turnstile"[35] or an "electron pump,"[36] which can be used to make a current standard based on the transfer of a single electron per cycle of a precision frequency standard. Another possibility is the creation of a digital logic system based on single electron transfers.

---

[35] L. J. Geerligs, V. F. Anderegg, P. Holweg, J. E. Mooij, H. Pothier, D. Esteve, C. Urbina, and M. H. Devoret, *Phys. Rev. Lett.* **64**, 2691 (1990).

[36] H. Pothier, P. Lafarge, P. F. Orfila, C. Urbina, D. Esteve, and M. H. Devoret, *Physica* **B169**, 573 (1991); also H. Pothier et al., *Europhys. Lett.* **17**, 249 (1992).

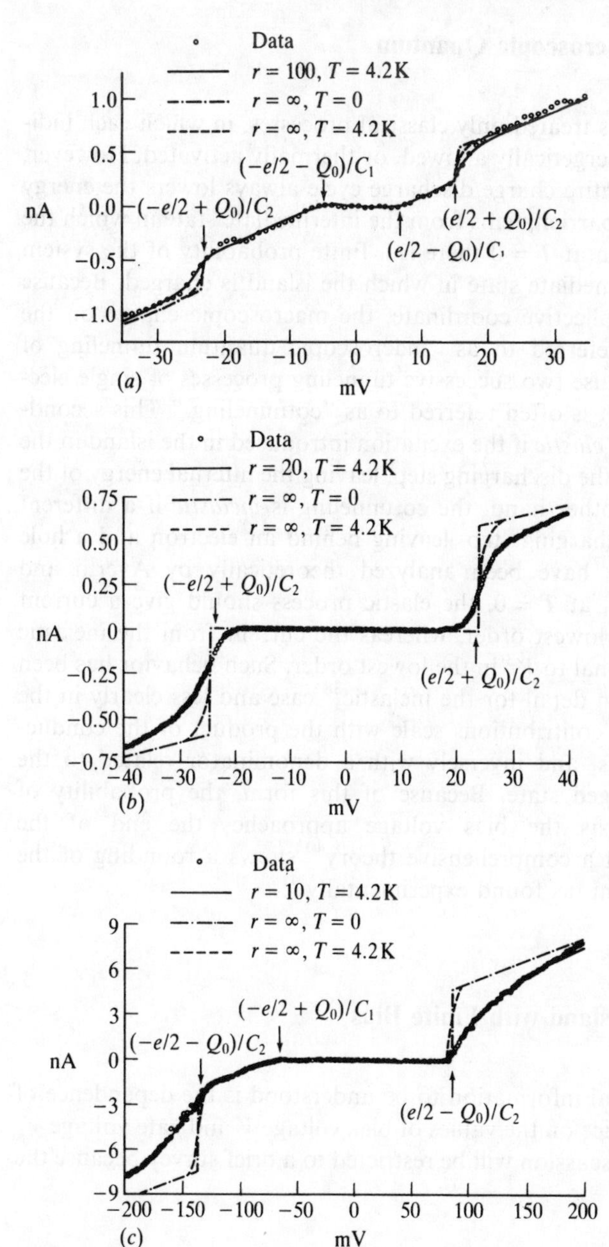

FIGURE 7.9
Representative experimental data showing continuous and discontinuous current onsets at edge of the Coulomb blockade and first step in the Coulomb staircase, compared with the simple analytic approximation for $T = 0$ and $r \equiv R_2/R_1 = \infty$ discussed in the text. Also shown are model calculations using full orthodox theory, first with $R_2/R_1 = \infty$ but the actual experimental temperature, and finally with actual temperature and a fitted value for R_2/R_1. The latter fit is almost indistinguishable from the data points. (*After Hanna and Tinkham.*)

7.6.3 Cotunneling or Macroscopic Quantum Tunneling of Charge

The preceding discussion has treated only classical processes, in which each individual electron transfer is energetically allowed, or thermally activated. However, as emphasized earlier, the entire charge/discharge cycle always lowers the energy by $|eV|$ overall. The energy barrier stems from the intermediate state in which the island is charged. Thus, even at $T = 0$ there is a finite probability of the system *tunneling* through this intermediate state in which the island is charged. Because the barrier stems from a collective coordinate, the macroscopic charge on the island, this is sometimes referred to as "macroscopic quantum tunneling of charge." Alternatively, because two successive tunneling processes of single electrons are involved overall, it is often referred to as "cotunneling." This second-order cotunneling process is *elastic* if the excitation introduced in the island in the charging step is removed in the discharging step, leaving the internal energy of the island unchanged. On the other hand, the cotunneling is *inelastic* if a different state is emptied by the discharging step, leaving behind an electron and a hole excitation. These processes have been analyzed theoretically by Averin and coworkers,[37] who find that, at $T = 0$, the elastic process should give a current which is linear in V to the lowest order, whereas the current from the inelastic process should be proportional to V^3 in the lowest order. Such behavior has been confirmed experimentally, in detail for the inelastic[38] case and less clearly in the elastic[39] case. Both current contributions scale with the product of the conductances of the two junctions, and inversely with a denominator related to the energy barrier in the charged state. Because of this form, the probability of cotunneling rises rapidly as the bias voltage approaches the end of the Coulomb blockade, so that a comprehensive theory[40] shows a rounding of the onset of current at this point, as found experimentally.[41]

7.6.4 Superconducting Island with Finite Bias Voltage

Here again, the experimental information to be understood is the dependence of the current through the device on the values of bias voltage V and gate voltage V_g (or gate charge Q_0). This discussion will be restricted to a brief survey because the

[37] D. V. Averin and A. A. Odintsov, *Phys. Lett.* **A140**, 251 (1989); D. V. Averin and Yu. V. Nazarov, *Phys. Rev. Lett.* **65**, 2446 (1990).

[38] L. J. Geerligs, D. V. Averin, and J. E. Mooij, *Phys. Rev. Lett.* **65**, 3037 (1990).

[39] A. E. Hanna, M. T. Tuominen, and M. Tinkham, *Phys. Rev. Lett.* **68**, 3228 (1992).

[40] D. V. Averin, *Physica* **B194–196**, 979 (1994).

[41] L. J. Geerligs, M. Matters, and J. E. Mooij, *Physica* **B194–196**, 1267 (1994).

presence of superconductivity adds complications and new features which are still being worked out at the time of this writing.

In the *N-S-N* case, the most obvious effect of having a superconducting island is that the threshold voltage (7.43) of the *I-V* curve, which is periodically modulated by V_g, must be increased by $2\Delta/e$ to supply the energy to create excitations at both junctions. Since such high voltages drive the system far from equilibrium, it is difficult to analyze all the processes which become possible. For that reason, considerable attention has been paid instead to higher-order mechanisms which allow (much smaller) currents to flow at much lower voltages than implied by the combined Coulomb blockade and energy gap. These processes are generalizations of the cotunneling processes described earlier, in which the elementary processes can now involve Cooper pairs in the superconducting island as well as individual quasi-particles.

For example, consider the case in which $\Delta > E_c$, for which energy levels are shown in Fig. 7.6*c*. If a small bias voltage is applied, this diagram remains essentially correct. [The term (7.27) representing the work done by the applied voltage shifts the vertical positions of the parabolas by amounts depending on which transfers have taken place.] For V_g (or Q_0) values giving the crossing points of the parabolas, there is degeneracy between the states of the island in which the electron number n differs by two. This means that there is no *energetic* barrier to transferring a steady current of *pairs* of electrons on and off the island through the bias circuit at such values of V_g, which show a period corresponding to $2e$.

A pair cannot simply be transferred one electron at a time, however, since there *is* a barrier of $(\Delta - E_c)$ to transferring a *single* electron on or off the island. The required charge-transfer mechanism is supplied by an analog of the Andreev reflection (discussed in Sec. 11.5) which occurs at an interface between the normal and superconducting phases. In this process, an electron incident from the normal metal at an energy below the superconducting gap is reflected as a hole, with the creation of a Cooper pair in the superconductor, all in a single step. The analog for the tunneling case was worked out by Hekking et al.[42] and gave a good account of the *shape* of $I(V, V_g)$. For example, the sharp current peak exactly at the degeneracy point $Q_0 = (2n+1)e$, expected at infinitesimal bias V, broadens into a peak whose width at the base, $\Delta Q_0 = C_\Sigma |V|$, is proportional to the bias voltage. However, the experimental *magnitude* of the current was much greater than expected for ballistic electron motion. In further work, Hekking and Nazarov[43] showed that this enhancement stems from the *coherent* addition of probability amplitudes for Andreev reflection in repeated attempts, as the electrons execute diffusive trajectories in the vicinity of the interfacial contact.

[42]F. W. J. Hekking, L. I. Glazman, K. A. Matveev, and R. I. Shekhter, *Phys. Rev. Lett.* **70**, 4138 (1993).

[43]F. W. J. Hekking and Yu. V. Nazarov, *Phys. Rev. Lett.* **71**, 1625 (1993).

Excellent agreement between this theory and experimental data was demonstrated by Hergenrother et al.[44]

In the S-S-S case, where there are superconductors on both sides of the tunnel junctions, transfer of Cooper pairs is the dominant charge transfer mechanism at sufficiently low temperatures. In Sec. 7.5.3, we discussed the supercurrent behavior (which can occur either at zero voltage or show a finite phase diffusion resistance) which dominates at zero or very low voltages. At finite voltages we must take into account the fact that the Cooper pairs lack the continuous density of states which facilitates energy conservation in the transfer of normal electrons or quasi-particles across finite potential differences. Since all Cooper pairs are effectively at the chemical potential, the transfer of pairs at finite voltage is dependent on the energy exchange with the electromagnetic environment of the leads, or on finding special bias voltage values at which "resonant" transfer of energy can occur to other electronic degrees of freedom, or combinations of the two. Such processes have been analyzed extensively, e.g., by Maassen van den Brink et al.[45]

[44] J. M. Hergenrother, M. T. Tuominen, and M. Tinkham, *Phys. Rev. Lett.* **72**, 1742 (1994); J. M. Hergenrother, M. T. Tuominen, J. G. Lu, D. C. Ralph, and M. Tinkham, *Phys, Rev.* **B51**, 9407 (1995).

[45] A. Maassen van den Brink, G. Schön, and L. J. Geerligs, *Phys. Rev. Lett.* **67**, 3030 (1991).

CHAPTER 8

FLUCTUATION EFFECTS IN CLASSIC SUPERCONDUCTORS

In our development and use of the Ginzburg-Landau theory, we have concentrated on finding the properties of that $\psi_0(\mathbf{r})$ which has the minimum free energy. However, thermodynamic fluctuations allow the system to sample other functions $\psi(\mathbf{r})$, and there will be significant statistical weight for any ψ which raises the free energy by only $\sim kT$. We have made use of this concept in our discussion of thermally activated flux creep, e.g., where it gave rise to a finite resistance below T_c. In the present chapter, we shall examine more closely the simple case of a thin wire or film of type I material to see how perfect the expected absence of resistance is as a function of temperature. This is equivalent to the question of how long-lived is the metastable persistent current in a ring against quantum jumps in which the fluxoid quantum number decreases by one or more units. We shall then examine the region *above* T_c, where thermodynamic fluctuations give rise to superconducting effects because $\langle\psi^2\rangle \neq 0$ although $\langle\psi\rangle = 0$. A more comprehensive review of these fluctuation effects can be found in the literature.[1]

[1]W. J. Skocpol and M. Tinkham, *Repts. Prog. Phys.* **38**, 1049 (1975).

8.1 APPEARANCE OF RESISTANCE IN A THIN SUPERCONDUCTING WIRE

In terms of the Ginzburg-Landau (GL) theory, the requirement for a persistent current in a ring is that the line integral $\oint \nabla\varphi \cdot d\mathbf{s}$ around the ring remain an invariant integral multiple of 2π. This corresponds to retaining the same fluxoid quantum number. In a singly connected superconducting wire, fed with current from normal leads, perfect conductivity requires that the potential difference V between the ends be zero. Applying the ideas of Josephson, this implies that the relative phase φ_{12} of the two ends retains a constant value, which will depend on the strength of the supercurrent. More precisely, φ_{12} will fluctuate about a constant mean value, as the supercurrent fluctuates to keep the total current constant by compensating for Johnson noise normal currents. Thus, at any nonzero frequency there will be an ac noise voltage reflecting the real part of the ac impedance of the superconductor, which increases as ω^2, as discussed in Sec. 2.5. At best a superconductor is really a perfect conductor only for direct current. Consequently, we shall confine our attention to direct current in investigating the appearance of resistance below T_c in superconductors. Even so, there is a subtlety involved since any real measurement is limited to a finite time span; hence, these ac fluctuations will not average out perfectly, and $\bar{V}$ will not be *exactly* zero, even though it may be unobservably small. An appropriate operational definition of zero dc resistance is that V have no measurable average value proportional to a dc applied current. In terms of the phase difference φ_{12}, this means that there should be no measurable secular progression of a short-term average of φ_{12}.

On the other hand, if there *is* resistance, this averaged φ_{12} will increase with time; this would appear to be inconsistent with a steady state. The resolution of this apparent inconsistency is that phase-slip events occur, in which the phase coherence is momentarily broken at some point in the superconductor, allowing a phase slip to occur before phase coherence is reestablished. These events can be spatially localized so long as the phase slip is through an integral multiple of 2π since a uniform phase change by $2n\pi$ outside the phase-slip region has no physical significance. In fact, we may concentrate on simple phase slips by 2π since phase slips by multiples of 2π turn out to be most easily accomplished as multiple slips, each of 2π. To maintain a steady state, such events must occur with an average frequency $2e\bar{V}/h$. If V is constant, φ_{12} increases steadily at the rate $2eV/h$ but instantaneously snaps back by 2π when each phase slip occurs. Thus, φ_{12} executes an irregular sawtooth which is equivalent, modulo 2π, to a uniform ramp.

To go beyond this qualitative picture to a quantitative calculation of the frequency of these resistive phase slips, it is convenient to restrict our attention to a one-dimensional superconductor. By this, we mean a wire with transverse dimension $d \ll \xi$, so that variations of ψ over the cross section of the wire are energetically prohibited. Then ψ is a function of a single coordinate x, running along the wire. We also assume $d \ll \lambda$, in which case magnetic energies can be neglected compared to kinetic energies.

If we neglect fluctuation effects, $|\psi(x)|$ is constant; this is the problem treated in Sec. 4.4, where we found a nonlinear relation (4.35) between supercurrent density and velocity, and also a critical-current density (4.36). Above the critical current density, dissipative processes set in strongly, and the resistance rapidly approaches the normal value, as discussed in Sec. 11.6. Here we seek the description of the resistive processes when $J < J_c$, so that in the absence of fluctuations there would be perfect conductivity. This regime was first treated in detail by Langer and Ambegaokar.[2]

To visualize the evolution of the complex function $\psi(x)$ during the phase-slip process, we need to depict $|\psi(x)|e^{i\varphi(x)}$ in polar form in a plane perpendicular to the x axis. Then the usual solutions discussed in Sec. 4.4, which have the form $\psi_0 e^{iqx}$, are represented by helices of pitch $2\pi/q$ and radius ψ_0. (See Fig. 8.1*a*.) These solutions are stationary, equilibrium solutions, representing the flow of supercurrent at zero voltage.

What happens to this picture when a voltage exists between the ends of the wire? The relative phase at the ends of the wire increases at the Josephson rate

$$\frac{d\varphi_{12}}{dt} = \frac{2eV}{\hbar} \tag{8.1}$$

We can visualize this as occurring by the phase at one end being steadily "cranked" around the Argand diagram, while the other end is held fixed, thus tightening the helix. So far this simply describes the accelerative supercurrent of the London equation $\mathbf{E} = \partial(\Lambda \mathbf{J}_s)/\partial t$. That is, the presence of the voltage increases q at such a rate that the total phase difference $\varphi_{12} = qL$ along the wire obeys (8.1). More locally, this is equivalent to

$$\frac{dv_s}{dt} = \frac{eE}{m} \tag{8.2}$$

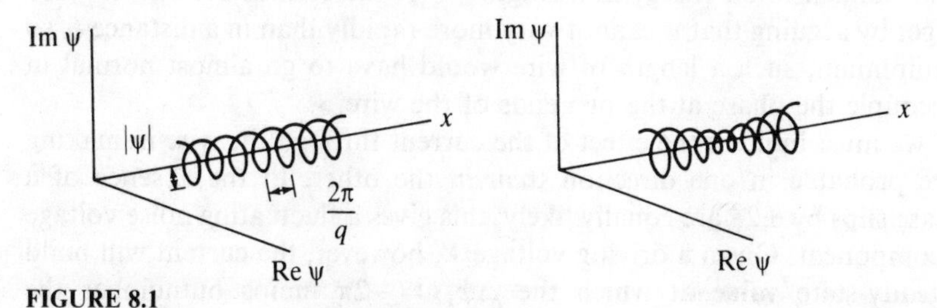

FIGURE 8.1
Graphical representation of complex current-carrying GL wavefunction in one-dimensional superconductors. (*a*) Uniform solution. (*b*) Nonuniform solution just before phase-slip event.

---

[2] J. S. Langer and V. Ambegaokar, *Phys. Rev.* **164**, 498 (1967).

But we know that there is a critical velocity v_c beyond which the simple uniform solution is impossible. Thus, this picture must break down when v_s reaches v_c, if not before. The phase-slip process of Langer and Ambegaokar maintains a steady state with $v_s < v_c$, in the presence of a nonzero voltage V, by annihilating turns of the helix in the interior of the wire at the same average rate as new ones are being cranked in at the end. By this means the energy that is supplied at a rate IV is dissipated as heat rather than converted into kinetic energy of supercurrent, which would otherwise soon exceed the condensation energy.

Although we shall not work through the details of their calculation, we shall review some of the salient points. First, so long as we neglect the normal current (which is nonzero when $E \neq 0$), conservation of current requires that $J_s(x) \propto |\psi(x)|^2 v_s(x)$ be constant. In other words, if $\psi(x) = |\psi(x)|e^{i\varphi(x)}$, then

$$|\psi(x)|^2 \frac{d\varphi}{dx} = \text{const} \propto I \tag{8.3}$$

serves as a constraint on possible variations of the complex function $\psi(x)$. In particular, if $|\psi|$ becomes very small in some region, $d\varphi/dx$ must become large there. (See Fig. 8.1*b*.) As first emphasized by Little,[3] as we approach the limit $|\psi| \to 0$, it is relatively easy to add or subtract a turn.

What Langer and Ambegaokar did was to find that path through function space, between two uniform solutions with different numbers of turns, which had the lowest intermediate free-energy barrier to overcome. By definition, at this saddle point in the barrier the GL free energy is again stationary with respect to small changes in ψ, so that ψ should satisfy the usual GL equations which were derived variationally by setting $\delta F = 0$. Using the constraint (8.3), they were able to calculate the saddle-point free-energy increment, namely,

$$\Delta F_0 = \frac{8\sqrt{2}}{3}\frac{H_c^2}{8\pi} A\xi \tag{8.4}$$

where A is the cross-sectional area of the conductor. This result is very plausible since it is the condensation energy in a length $\sim\xi$ of the conductor. This is what one would get by arguing that ψ cannot vary more rapidly than in a distance ξ, so that, as a minimum, such a length of wire would have to go almost normal in order to decouple the phase at the two ends of the wire.

Next, we must build in the effect of the current through the wire in making jumps more probable in one direction than in the other. In the absence of a current, phase slips by $\pm 2\pi$ are equally likely; this gives a fluctuating noise voltage but no dc component. Given a driving voltage V, however, the current will build up to a steady-state value at which the $\Delta\varphi_{12} = -2\pi$ jumps outnumber the $\Delta\varphi_{12} = +2\pi$ jumps by an amount $2eV/h$ per second. The different jump rates arise from a difference δF in the energy barrier for jumps in the two directions,

[3] W. A. Little, *Phys. Rev.* **156**, 398 (1967).

and this difference stems from the electrical work $\int IV\,dt$ done in the process, as in the tilted-washboard model discussed in Sec. 6.3. In view of (8.1), for a phase slip of 2π, the energy difference is

$$\delta F = \Delta F_+ - \Delta F_- = \frac{h}{2e} I \tag{8.5}$$

As shown by McCumber,[4] all these arguments carry over exactly to the usual experimental situation, where a constant current rather than a constant voltage source is used.

To complete the theory, we need to introduce an attempt frequency or "prefactor" Ω, so that the mean net phase-slip rate is

$$\frac{d\varphi_{12}}{dt} = \Omega\left[\exp\left(-\frac{\Delta F_0 - \delta F/2}{kT}\right) - \exp\left(-\frac{\Delta F_0 + \delta F/2}{kT}\right)\right]$$

$$= 2\Omega e^{-\Delta F_0/kT} \sinh \frac{\delta F}{2kT} \tag{8.6}$$

Substituting (8.5) into (8.6) for δF and equating (8.6) to the Josephson frequency, we find

$$V = \frac{\hbar\Omega}{e} e^{-\Delta F_0/kT} \sinh \frac{hI}{4ekT} \tag{8.7}$$

In the limit of very small currents, the hyperbolic sine can be replaced by its argument, and one obtains Ohm's law, with

$$R = \frac{V}{I} = \frac{\pi\hbar^2\Omega}{2e^2kT} e^{-\Delta F_0/kT} \tag{8.8}$$

However, this approximation is valid only for $I \lesssim I_0$, where

$$I_0 = \frac{4ekT}{h} = 0.013\,\mu\text{A/K} \tag{8.9}$$

In this regime, the numbers of jumps with $\Delta\varphi_{12} = \pm 2\pi$ are approximately equal, with the current being a small perturbation. At higher currents a preponderance of jumps occur in the direction which removes turns from the helix. It is then useful to approximate (8.7) by

$$V = \frac{\hbar\Omega}{2e} e^{-\Delta F_0/kT} e^{I/I_0} \tag{8.10}$$

where, in the full theory, ΔF_0 is found to decrease from (8.4) as I^2. In this regime, the superconductor acts like a nonlinear resistor.

These results leave open the value of the prefactor Ω. Evidently, it should be proportional to the length of the wire since we would expect the jump to be able to occur independently at sites all along the wire. This causes the voltage

[4]D. E. McCumber, *Phys. Rev.* **172**, 427 (1968).

drop to be proportional to the length of the wire for a given current, so that the resistance is an extensive variable. In the original work of Langer and Ambegaokar, the attempt frequency was taken rather arbitrarily as nAL/τ, where τ is the electronic relaxation time in the normal state and n is the electron density. Subsequently, McCumber and Halperin[5] reexamined the problem by using the time-dependent GL theory and found a temperature-dependent prefactor of the form

$$\Omega = \frac{L}{\xi}\left(\frac{\Delta F_0}{kT}\right)^{1/2}\frac{1}{\tau_s} \tag{8.11}$$

where $1/\tau_s = 8k(T_c - T)/\pi\hbar$ is the characteristic relaxation rate of the superconductor in the time-dependent GL theory, discussed further in Secs. 8.6 and 10.3. This form is plausible since L/ξ is the number of nonoverlapping locations in which the fluctuations might occur. The factor $(\Delta F_0/kT)^{1/2}$ has little numerical importance. This McCumber-Halperin prefactor is typically smaller than the Langer-Ambegaokar one by a factor of 10^{10}, and it goes to zero as one approaches T_c. Despite the enormous size of this correction factor, its absence was not noticed at first since it corresponds to a change of only a few millidegrees in the temperature scale because of the exponential dependence of the voltage (8.7) on $\Delta F_0/kT$.

These ideas were directly tested in the experiments of Lukens, Warburton, and Webb[6] and those of Newbower, Beasley, and Tinkham,[7] both of which were done on tin "whiskers." These are single-crystal, cylindrical specimens, typically $\sim$0.5 μm in diameter and a fraction of a millimeter long, grown by applying pressure to a sandwich of tin-plated steel such as that used in tin cans. Even for samples of such small diameter, $\Delta F_0/kT \approx 6 \times 10^6(1-t)^{3/2}$, so that the probability of a phase slip becomes astronomically small unless one is within about 1 mK of T_c, where $(1-t) \sim 0.0003$. The very satisfactory fit between this LAMH (Langer-Ambegaokar-McCumber-Halperin) theory and the experimental data of Newbower et al. is displayed in Fig. 8.2. The LAMH theory is expected to fail, as it does, very near T_c, where its model of isolated phase slips in a superconducting medium is inappropriate since both the attempt frequency and the free-energy barrier go to zero at T_c. In that case, it may be more appropriate to start with the normal state and to consider superconducting fluctuations from it rather than the reverse.[8] But once the resistance has fallen significantly, the LAMH fit appears to be quantitatively accurate over six orders of magnitude.

---

[5] D. E. McCumber and B. I. Halperin, *Phys. Rev.* **B1**, 1054 (1970).

[6] J. E. Lukens, R. J. Warburton, and W. W. Webb, *Phys. Rev. Lett.* **25**, 1180 (1970)

[7] R. S. Newbower, M. R. Beasley, and M. Tinkham, *Phys. Rev.* **B5**, 864 (1972).

[8] For attempts to work down through T_c from above, see, e.g., W. E. Masker, S. Marcelja, and R. D. Parks, *Phys. Rev.* **188**, 745 (1970); J. Tucker and B. I. Halperin, *Phys. Rev.* **B3**, 378 (1971); R. J. Londergan and J. S. Langer, *Phys. Rev.* **B5**, 4376 (1972).

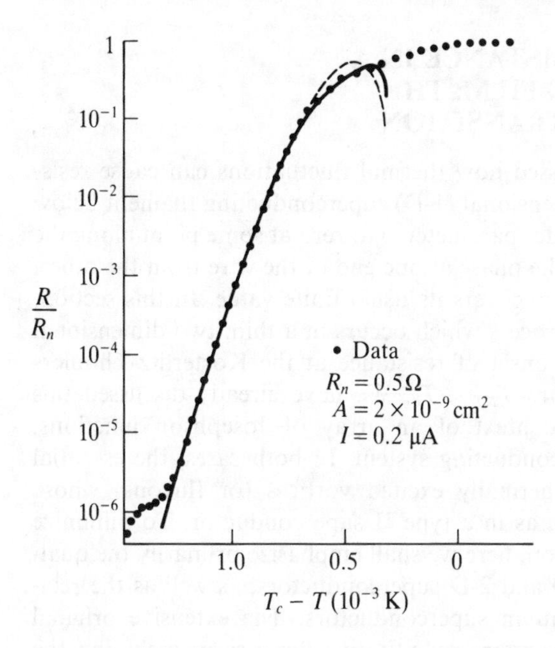

FIGURE 8.2
Decrease of resistance below T_c in a superconducting tin "whisker," as measured by Newbower et al. The solid curve is LAMH theory, with only T_c as adjustable parameter. The dashed curve is LAMH theory if parallel normal conduction channel is omitted. "Foot" at $R/R_n \approx 10^{-6}$ is believed to be caused by contact effects.

(The "foot" in Fig. 8.2 near $R/R_n = 10^{-6}$, where the resistance falls more slowly before resuming its rapid fall, is sample-dependent and believed to be caused by contact effects.)

Given this excellent fit between theory and experiment, it is interesting to use the theory to extrapolate beyond the observable range of resistance. At the lowest temperature shown in Fig. 8.2, with the measuring current of 0.2×10^{-6} A, the resistive voltage is about 10^{-13} V; this corresponds to about 100 phase slippages (by 2π) per second. Extrapolating down another millidegree, the rate is 10^{-11} per second, or 1 in 1,000 years; in another millidegree, it is 1 in 10^9 years. Thus, in three millidegrees, we have gone from the normal resistance to a regime in which no resistive event would be expected to occur in the age of the universe! Of course, the disappearance of resistance should be even faster in a thicker wire.

In view of these time scales, it is clear that time-average results must be used with care. According to (8.8), there is a finite resistance at all nonzero temperatures, although it becomes astronomically small well below T_c. But this refers to a statistical average over a period that is long enough for many phase slips to occur. Given an *infinitely* long wire, this would be no problem, and a small resistance should be measurable. With any finite-length wire, however, one rapidly reaches the situation in which *no* phase slip would be expected to occur in any feasible experimental time scale. In that case, the dc resistance would appear to be *strictly* zero, not just small. Thus, the quantized nature of the phase slips provides the key needed to get from a very small to a zero resistance.

8.2 APPEARANCE OF RESISTANCE IN A THIN SUPERCONDUCTING FILM: THE KOSTERLITZ-THOULESS TRANSITION

In the previous section, we discussed how thermal fluctuations can cause resistance to appear in a thin, one-dimensional (1-D) superconducting filament below T_c. The fluctuation reduces the order parameter ψ to zero at some point along the wire, momentarily disconnecting the phase at one end of the wire from the other, allowing it to slip by 2π before ψ recovers its usual finite value. In this section, we briefly review the analogous process which occurs in a thin, two-dimensional (2-D) film, where it leads to the onset of resistance at the Kosterlitz-Thouless transition at a specific temperature $T_{KT} < T_c$. We have already discussed this transition in Sec. 6.6.1, in the context of an array of Josephson junctions, which form a discrete 2-D superconducting system. In both cases, the essential physics is the same, involving thermally excited vortices (or fluxons) whose motion gives the resistive voltage, as in a type II superconductor. To minimize duplication of that earlier discussion, here we shall emphasize primarily the qualitative difference between the 1-D and 2-D superconductors, as well as the relation between discrete and continuum superconductors. The extensive original literature[9] can be consulted for a more quantitative discussion, including the statistical mechanics of the problem when many vortices are present.

Our discussion of the 1-D case in Sec. 8.1 centered on the minimum energy cost of a fluctuation in $\psi(x)$ which would take $\psi(x)$ to zero at some point, to allow phase slippage by 2π. Detailed calculations show that this ΔF_0 is $\sim(H_c^2/8\pi)A\xi$, which can be interpreted physically as the condensation energy per unit volume times a minimum volume, based on the cross-sectional area of the filament times the coherence length ξ. A straightforward extension of this argument to the 2-D case of a film of thickness $d \ll \xi$ would replace the volume $A\xi$ by $Wd\xi$, the volume of a ξ-wide strip running across the width W of the film perpendicular to the direction of current flow. W can be arbitrarily large, so that the energy $(H_c^2/8\pi)Wd\xi$ can quickly become prohibitive for thermal fluctuations, and such a fluctuation is not very effective in broadening the transition of a macroscopic film.

To reduce this activation energy, the quasi-normal volume must be made smaller. This can be done by considering a 2-D *vortex* excitation, within which ψ

---

[9] See, e.g., J. M. Kosterlitz and D. J. Thouless, *J. Phys.* **C6**, 1181 (1973) for the original discussion in the context of superfluid helium films; M. R. Beasley, J. E. Mooij, and T. P. Orlando, *Phys. Rev. Lett.* **42**, 1165 (1979) and S. Doniach and B. A. Huberman, *Phys. Rev. Lett.* **42**, 1169 (1979) for the translation of the effect to thin superconducting films with $\lambda_\perp$ larger than the film dimension; B. I. Halperin and D. R. Nelson, *J. Low Temp. Phys.* **36**, 599 (1979) for a more detailed discussion of the temperature and frequency dependence of the resistive transition in near zero external field; J. E. Mooij, "Two-Dimensional Transition in Superconducting Films and Junction Arrays" in A. M. Goldman and S. A. Wolf (eds.), *Percolation, Localization, and Superconductivity*, NATO ASI series, Plenum, New York (1984), p. 325, for a review of experiments and theory; P. Minnhagen, *Revs. Mod. Phys.* **59**, 1001 (1987) for a review of the 2-D Coulomb gas version of the theory.

goes to zero at the center of a core of volume $\sim \pi\xi^2 d$, which is small and independent of the macroscopic system size. Such a vortex would move under the influence of a transport current and cause flux-flow resistance, as discussed in Sec. 5.5. As we showed in Sec. 5.1, inclusion of the kinetic energy of the circulating current multiplies the core energy by a numerical factor, so that the vortex energy is $(H_c^2/8\pi)4\pi\xi^2 d \ln \kappa$. Using the GL relation (4.20), we can rewrite this energy as $(\Phi_0^2/16\pi^2)(d/\lambda^2) \ln \kappa$. Recalling our discussion of the penetration depths in thin films in Sec. 3.11.4, we recognize that $d/\lambda^2 = 1/\lambda_\perp$, so that the activation energy to create such a pancake vortex is proportional to $\Phi_0^2(\ln \kappa)/\lambda_\perp$. Finally, we recall that the $\ln \kappa$ factor simply reflects the result of integrating the kinetic-energy density of the vortex, which scales as $1/r$, from ξ to λ. In the 2-D case, the cutoff of the circulating vortex current is not at λ but, rather, at $\lambda_\perp = \lambda^2/d$ (which diverges as the film thickness d goes to zero). In this case, the $\ln \kappa$ factor is replaced by $\ln(\lambda_\perp/\xi)$.

We now restrict our attention, as we did in Sec. 6.6.1, to the limiting case in which $\lambda_\perp$ is larger than the width of the film. In that case, the cutoff of the integration of the kinetic-energy density is simply set by the sample size, which we characterize by a radius R. In this way, we find that the energy of a single vortex in a finite size sample is of the order of $(\Phi_0^2/\lambda_\perp) \ln(R/\xi)$, which is obviously parallel to the relation (6.65) found for the case of the Josephson junction array.

Similarly, by the reasoning used there, two vortices of opposite circulation sense will have twice the core energy as one vortex. However, since the integration of kinetic-energy density will be cut off beyond the intervortex separation $\sim R_{12}$, instead of the sample size R, the total energy $\sim 2(\Phi_0^2/\lambda_\perp) \ln(R_{12}/\xi)$ will be finite, even in an infinitely extended film. Since this energy increases as R_{12} increases, there is an attractive force between the two opposing vortices that tends to pull them closer together until a distance of closest approach $\sim\xi$, when the vortex cores are touching, whereupon they can annihilate each other. The process of generation of vortex pairs is the reverse of this, and in thermal equilibrium both processes occur with the same frequency. Below T_{KT}, thermal energy is insufficient to allow the vortices to unbind. Above T_{KT} they can unbind because the free energy $U - TS$ is lowered sufficiently by the gain in entropy that is associated with two independently moving entities relative to a single bound pair. As described in Sec. 6.6.1, this balance between energy and entropy is what determines T_{KT}, by the relation $kT_{KT} \approx (\Phi_0^2/32\pi^2\lambda_\perp)$, when the numerical factors are reinserted. This expression is one of the equivalent expressions quoted in (6.67b).

As a final remark, we note that in this 2-D system the phase slip by 2π can occur continuously by the motion of a vortex from one edge of the film to the other. However, a more typical process would be the creation of a vortex-antivortex pair in the interior of the film, followed by their motion in opposite directions driven by the current, until both disappear at opposite edges. Either process leads to a 2π phase slip between the electrodes at the ends of the film. Of course, when many vortices are present, these motions can be described only statistically.

8.3 SUPERCONDUCTIVITY ABOVE T_c IN ZERO-DIMENSIONAL SYSTEMS

In the GL theory, T_c is defined as the temperature at which the coefficient $\alpha(T)$ (in the leading term $\alpha|\psi|^2$ in the free-energy expansion) changes sign. Thus, above T_c, F is a minimum when $|\psi| = 0$. However, thermal fluctuations from ψ to $\psi + \delta\psi$ raising the free energy by an amount $\sim kT$ are common since the probability falls only as $e^{-F/kT}$. This leads to the existence of fluctuation-induced superconducting effects above T_c. These fluctuations are largest in amplitude if confined to small volumes since the total energy increase must be only $\sim kT$.

We can get a useful orientation on this problem by considering first a particle which is small compared to ξ, so that we can treat ψ as constant over its volume V. This might be called the *zero-dimensional limit*. Then the GL free energy relative to the normal state (in the absence of any fields) is

$$F = V(\alpha|\psi|^2 + \tfrac{1}{2}\beta|\psi|^4) \tag{8.12}$$

where $\alpha \equiv \alpha_0(t-1)$.

Below T_c this leads to the usual result that the minimum free energy is

$$F_0 = -\frac{\alpha^2}{2\beta}V = -\frac{\alpha_0^2}{2\beta}(1-t)^2 V = -\frac{H_c^2}{8\pi}V \tag{8.13}$$

and this occurs for

$$|\psi_0|^2 = -\frac{\alpha}{\beta} = \frac{\alpha_0(1-t)}{\beta} \tag{8.14}$$

The fluctuations about this ψ_0 can be estimated by computing

$$\left.\frac{\partial^2 F}{\partial \psi^2}\right|_{\psi_0} = -4\alpha V = 4\alpha_0(1-t)V \tag{8.15}$$

and setting

$$\langle F - F_0 \rangle = \frac{1}{2}\left.\frac{\partial^2 F}{\partial \psi^2}\right|_{\psi_0} (\delta\psi)^2 \approx kT \tag{8.16}$$

This leads to

$$\frac{(\delta\psi)^2}{\psi_0^2} \approx \frac{kT}{4|F_0|} = \frac{2\pi kT}{H_c^2 V} \approx \frac{10^{-20}}{(1-t)^2 V} \tag{8.17}$$

using numerical values for tin. From this we see that the fluctuations cause a very small fractional change in ψ except *very* near T_c or in a very small sample. Therefore, we have generally been well justified in using the "mean field" ψ_0 in our previous work. However, by the use of very small particles ($d < 1{,}000$ Å), it has been possible to probe the so-called "critical region," where $(\delta\psi/\psi_0)^2$ is *not* necessarily small, and the mean-field results become inaccurate. In this connection, it is important to note that the apparent divergence of (8.17) at T_c is actually cut off by the anharmonic terms in the free-energy expansion, so that even at T_c, $(\delta\psi)^2$ has the finite value

$$(\delta\psi)^2_{T_c} \approx \left(\frac{2kT_c}{V\beta}\right)^{1/2} \tag{8.18}$$

Now let us examine the situation *above* T_c. Here $\alpha > 0$, so that by inspection of (8.12) we see that the minimum free energy is $F_0 = 0$ (relative to the normal state) which occurs for $\psi_0 = 0$. The fluctuations are limited by

$$\left.\frac{\partial^2 F}{\partial\psi^2}\right|_{\psi=0} = 2\alpha V = 2\alpha_0(t-1)V \tag{8.19}$$

which we see differs only by a factor of 2 from the value given by (8.15) for a temperature that is an equal distance *below* T_c. The corresponding fluctuation level here is

$$(\delta\psi)^2 \approx \frac{kT}{\alpha V} = \frac{kT}{\alpha_0(t-1)V} \tag{8.20}$$

Again this is of the same order as the fluctuations below T_c, but since ψ_0 is now zero, *all* the superconducting effects arise from the fluctuations. Note that $(\delta\psi)^2$ tends to diverge as $(T - T_c)^{-1}$, as in the familiar Curie-Weiss law in the statistical mechanics of paramagnetism, but this divergence is cut off very near T_c by the quartic term which leads to (8.18). If we replace ψ by $\delta\psi$ in (8.12), equate the free energy to kT, and solve, we obtain

$$(\delta\psi)^2 = \frac{\alpha}{\beta}\left[\left(1 + \frac{2\beta kT}{\alpha^2 V}\right)^{1/2} - 1\right] \qquad T > T_c \tag{8.21}$$

which reduces to (8.20) and (8.18) in the appropriate limits.

This brings us to the question of how these superconducting fluctuations above T_c can be observed. The most direct way is a susceptibility measurement on tiny particles since χ depends on λ, which is a measure of $\langle\psi^2\rangle$. More explicitly, for spherical particles of radius $R \ll \lambda$, the London equations lead to a susceptibility

$$\chi = -\frac{1}{40\pi}\frac{R^2}{\lambda^2} = -\frac{1}{40\pi}\frac{4\pi e^{*2}}{m^* c^2}\langle\psi^2\rangle R^2 \tag{8.22}$$

(If $R < \xi_0$, an additional factor of $\sim R/\xi_0$ enters to account explicitly for the nonlocal electrodynamics.) Thus, if $\langle\psi^2\rangle$ is given by (8.21), χ should rise as $(t-1)^{-1}$ as the temperature is reduced, but then rise more slowly once the critical region is entered. Finally, well below T_c, χ and $\langle\psi^2\rangle$ should rise as $(1-t)$, after fluctuation effects are swamped by the mean-field superconductivity. These dependences are shown in Fig. 8.3. Exactly this behavior was observed by Buhrman and Halperin[10] in measurements on fine aluminum powders. In their

[10] R. A. Buhrman and W. P. Halperin, *Phys. Rev. Lett.* **30**, 692 (1973).

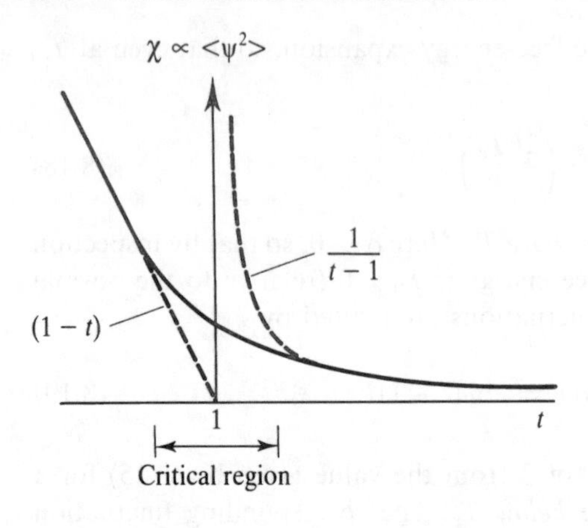

FIGURE 8.3
Temperature dependence of pair density and susceptibility of zero-dimensional superconductor near T_c.

sample containing the finest particles ($R \approx 250$ Å), the critical region was found to cover the range $0.95 < t < 1.05$, in agreement with theory. Altogether, they obtained such a quantitative fit between their data and exact calculations[11] in the GL framework that we can conclude that the GL free-energy expression (8.12) is satisfactory both inside and outside the critical region, so far as zero-dimensional systems are concerned. However, χ fell below the predicted value for $T \gtrsim 1.5T_c$. This is not surprising since the GL theory is expected to be reliable only near T_c.

8.4 SPATIAL VARIATION OF FLUCTUATIONS

Although the zero-dimensional case just discussed is simple and permits a rather exact theoretical analysis, it cannot be applied directly to the usual experimental situation in which one or more sample dimensions exceed ξ since ψ cannot be treated as constant over the sample. Nonetheless, the qualitative ideas can be carried over to some extent by treating a macroscopic sample as if it were composed of tiny, independent particles, whose size is limited by the correlation length of the fluctuations, typically $\sim\xi$. For example, since $|\alpha| = \hbar^2/2m^*\xi^2$, (8.20) and (8.22) lead in this way to a diamagnetic susceptibility of a bulk superconductor above T_c which is proportional to $kT\xi(T)/\Phi_0^2$. This simple result is confirmed by exact calculations, discussed below. Let us now examine the spatial variations of the fluctuations more closely.

We first consider the case of a bulk sample, far enough above T_c so that the effects of the quartic term in the free energy can be neglected. The GL free-energy density relative to the normal state can then be written as

[11] See, e.g., B. Mühlschlegel, D. J. Scalapino, and R. Denton, *Phys. Rev.* **B6**, 1767 (1972). In these calculations a proper thermally weighted average is taken over ψ^2, rather than simply equating the free-energy increase to kT as we have done here.

$$f = \alpha|\psi|^2 + \frac{\hbar^2}{2m^*}\left|\left(\frac{\nabla}{i} - \frac{2\pi\mathbf{A}}{\Phi_0}\right)\psi\right|^2 \tag{8.23}$$

Since $\alpha > 0$, both terms are positive, so that the free energy must exceed that of the normal state for any nonzero ψ. The corresponding linearized GL equation is

$$\left(\frac{\nabla}{i} - \frac{2\pi\mathbf{A}}{\Phi_0}\right)^2\psi = -\frac{2m^*\alpha}{\hbar^2}\psi = -\frac{1}{\xi^2}\psi \tag{8.24}$$

Since $1/\xi^2 \equiv 2m^*|\alpha|/\hbar^2$ its sign is changed relative to (4.56) reflecting the sign change of α.

Let us first consider the case of $\mathbf{A} = 0$, and expand $\psi(\mathbf{r})$ in Fourier series, so that

$$\psi(\mathbf{r}) = \sum_{\mathbf{k}} \psi_{\mathbf{k}} e^{i\mathbf{k}\cdot\mathbf{r}} \tag{8.25}$$

Inserting this into (8.23), integrating over unit volume, and using the orthogonality of the terms in a Fourier series, we find

$$f = \sum_{\mathbf{k}}\left(\alpha + \frac{\hbar^2k^2}{2m^*}\right)|\psi_{\mathbf{k}}|^2 \tag{8.26}$$

If we assign an energy k_BT to each orthogonal mode,[12] i.e., to each $\mathbf{k}$ value, then (in unit volume)

$$|\psi_{\mathbf{k}}|^2 = \frac{k_BT}{\alpha + \hbar^2k^2/2m^*} = \frac{2m^*}{\hbar^2}\frac{k_BT}{k^2 + 1/\xi^2} \tag{8.27}$$

From this we see that Fourier components describing variations in distances less than ξ come in with reduced weight, as expected. However, the density of modes goes as $k^2\,dk$, so that a stronger cutoff is required to give a finite value for the sum over $\mathbf{k}$ in $\langle\psi^2\rangle = \Sigma_{\mathbf{k}}|\psi_{\mathbf{k}}|^2$. Presumably such a cutoff must occur when $k \approx 1/\xi(0)$ since the GL theory is not valid for more rapid variations than that.

At this point we must recognize that the presence of a finite value for $|\psi|^2$ alone does imply superconducting properties. Only insofar as the *phase* of $\psi(\mathbf{r})$ retains a coherent spatial variation will its response to an imposed vector potential or phase gradient give a supercurrent. The spatial extent of phase coherence is unlimited in the solution of the GL equation below T_c. Its extent above T_c is limited by the fact that ψ is a sum of independent contributions $\psi_{\mathbf{k}}$ with different wavelengths. Thus, even if they add constructively at point $\mathbf{r}$, they will tend to

[12] In this section, we use k_B for the Boltzmann constant to avoid confusion with the wave vector $\mathbf{k}$.

cancel at $\mathbf{r}' \neq \mathbf{r}$. The extent of the coherence is described by the two-point correlation function

$$g(\mathbf{r},\mathbf{r}') \equiv \langle \psi^*(\mathbf{r})\psi(\mathbf{r}')\rangle = \left\langle \sum_{\mathbf{k}} \psi_{\mathbf{k}}^* e^{-i\mathbf{k}\cdot\mathbf{r}} \sum_{\mathbf{k}'} \psi_{\mathbf{k}'} e^{i\mathbf{k}'\cdot\mathbf{r}'} \right\rangle \tag{8.28}$$

where the angular brackets indicate an average. Changing variables to mean and relative coordinates $\bar{\mathbf{r}} = (\mathbf{r}+\mathbf{r}')/2$ and $\mathbf{R} = \mathbf{r}' - \mathbf{r}$, we can write (8.28) as

$$g(\mathbf{r},\mathbf{r}') = \left\langle \sum_{\mathbf{k}\mathbf{k}'} \psi_{\mathbf{k}}^*\psi_{\mathbf{k}'} \exp\left[\frac{i(\mathbf{k}+\mathbf{k}')}{2}\cdot\mathbf{R}\right] \exp[-i(\mathbf{k}-\mathbf{k}')\cdot\bar{\mathbf{r}}] \right\rangle$$

Carrying out the average over the mean coordinate $\bar{\mathbf{r}}$ gives zero unless $\mathbf{k} = \mathbf{k}'$, when it gives unity. Thus,

$$g(\mathbf{r},\mathbf{r}') = g(\mathbf{R}) = \sum_{\mathbf{k}} |\psi_{\mathbf{k}}|^2 e^{i\mathbf{k}\cdot\mathbf{R}} \tag{8.29}$$

where $|\psi_{\mathbf{k}}|^2$ is given by (8.27). By the symmetry of this formula (and the underlying physics) it is clear that this depends only on the magnitude of $\mathbf{R}$. Replacing the sum by an integral, we have

$$g(R) = \frac{2m^* k_B T}{\hbar^2} \int\!\!\int \frac{e^{ikR\cos\theta}}{k^2 + 1/\xi^2} \sin\theta \, d\theta \, k^2 \, dk$$

The integral on θ is elementary, and that on k can be easily done by contour integration, leading to the result

$$g(R) = \frac{m^* k_B T}{2\pi\hbar^2} \frac{e^{-R/\xi(T)}}{R} \tag{8.30}$$

Thus, in the fluctuation regime, at $H = 0$, the local values of ψ are correlated over a distance $\sim\xi(T)$, as anticipated above. [The divergence of (8.30) as $R \to 0$ is nonphysical, arising from carrying the integration on k to infinity rather than imposing a cutoff at $\sim 1/\xi(0)$.]

Now let us see what effect the presence of a magnetic field has on these results. Before choosing a specific gauge, we note in general that it will be convenient to work with orthonormal eigenfunctions ψ_ν of the pseudohamiltonian operator $\mathscr{H}$ defined by

$$\mathscr{H}\psi_\nu = \frac{\hbar^2}{2m^*}\left[\left(\frac{\nabla}{i} - \frac{2\pi\mathbf{A}}{\Phi_0}\right)^2 + \frac{1}{\xi^2}\right]\psi_\nu = \epsilon_\nu\psi_\nu \tag{8.31}$$

This is simply the hamiltonian of a free charged particle in a magnetic field, apart from the additive constant $1/\xi^2$, which reflects the $\alpha(T)$ in the GL equation. Comparing this with (8.24), we see that eigenfunctions with $\epsilon_\nu = 0$ satisfy the linearized GL equation. However, above T_c, all the ϵ_ν are positive, and ψ_ν are simply used as basis functions. Returning to the argument used in Sec. 4.8 in the derivation of H_{c2} [where the same operator appears as in (8.31), apart from the sign of $1/\xi^2$], we see that

$$\epsilon_\nu = \epsilon_{n,k_z} = \frac{\hbar^2}{2m^*}\left(\frac{1}{\xi^2} + k_z^2\right) + (n + \tfrac{1}{2})\hbar\omega_c \tag{8.32}$$

where $\omega_c = 2eH/m^*c$ is the cyclotron frequency of pairs in the applied field.

If we expand a general $\psi(\mathbf{r})$ in this set as

$$\psi(\mathbf{r}) = \sum_\nu c_\nu \psi_\nu(\mathbf{r}) \tag{8.33}$$

and calculate the free energy using (8.23), we find by a partial integration using orthogonality that $F = \Sigma_\nu |c_\nu|^2 \epsilon_\nu$. Assigning an energy k_BT to each normal mode (as we did earlier in the absence of a field), we have $|c_\nu|^2 = k_BT/\epsilon_\nu$.

We can now compute the correlation function

$$g(\mathbf{r}, \mathbf{r}') \equiv \langle \psi^*(\mathbf{r})\psi(\mathbf{r}')\rangle = \sum_{\nu\nu'} c_\nu^* c_{\nu'} \langle \psi_\nu^*(\mathbf{r})\psi_{\nu'}(\mathbf{r}')\rangle \tag{8.34}$$

to see how (8.30) is modified by the field. At this point we must make a specific gauge choice. Since the physical problem has axial symmetry about the field, it is convenient to choose

$$\mathbf{A} = \tfrac{1}{2}\mathbf{H} \times \mathbf{r} = \tfrac{1}{2}Hr\hat{\varphi}$$

where $\hat{\varphi}$ is a unit vector. Since there is also translational invariance along the field in the z direction, the eigenfunctions must have the form

$$\psi_\nu = f_{mn}(\rho)e^{im\varphi}e^{ik_z z} \tag{8.35}$$

where $\rho = (x^2 + y^2)^{1/2}$. Putting this in the differential equation (8.31), we find that the asymptotic form of f_{mn} for all m, n is

$$f(\rho) \to f_1(\rho)e^{-a\rho^2} \qquad \rho \to \infty \tag{8.36}$$

where $f_1(\rho)$ is a polynomial and $a = \pi H/2\Phi_0$. This exponential cutoff shows that the wavefunctions are confined to cylindrical regions whose area is such that the flux threading them is of the order of Φ_0. A particularly simple solution is the lowest one, for which $m = 0$ and $f_1(\rho)$ is constant. The eigenvalue is

$$\epsilon_0 = \frac{\hbar^2}{2m^*}\left(\frac{2\pi H}{\Phi_0} + k_z^2 + \frac{1}{\xi^2}\right) \tag{8.37}$$

in agreement with the $n = 0$ case of (8.32).

Now let us return to the evaluation of the correlation function (8.34). Transforming to relative and center-of-mass coordinates as above, we find that the average over the center-of-mass coordinate vanishes unless $m = m'$ and $k_z = k_z'$ in the general indices ν and ν'. A further simplification results because $f_1(0) = 0$ unless $m = 0$, so that

$$g(\rho, Z) = \sum_{n,n',k_z} c_{nk}^* c_{n'k} f_{0n}(0) f_{0n'}(\rho) e^{ik_z Z} \tag{8.38}$$

The one-dimensional integration over k_z can be carried out by noting that $|c_{nk_z}|^2 = k_B T/\epsilon_{nk_z}$, with ϵ_{nk_z} given by (8.32). Thus, for given n, $|c_{nk_z}|^2 \propto (k_{0n}^2 + k_z^2)^{-1}$, where

$$k_{0n}^2 = \xi^{-2} + \frac{(2n+1)2\pi H}{\Phi_0} \tag{8.39}$$

With this dependence on k_z, the various terms in $g(\rho, Z)$ will fall off roughly as $e^{-k_{0n}|Z|}$. The asymptotic behavior is governed by the smallest k_{0n}, namely, $k_{00} = (\xi^{-2} + 2\pi H/\Phi_0)^{1/2}$. Combining this dependence with (8.36), we expect that the correlation function will fall off at large distances roughly as

$$g(\rho, Z) \propto e^{-k_{00}|Z|} e^{-\pi H\rho^2/2\Phi_0} \tag{8.40}$$

Thus, the "radius" of the correlated fluctuations shrinks below ξ as H gets larger than $\sim\Phi_0/\pi\xi^2$. Note that this characteristic field value is essentially $H_{c2}(\tilde{T})$, where $T_c - \tilde{T} = T - T_c$.

In view of (8.22), we might expect that the shrinkage of the size of the fluctuations with increasing H would cause the susceptibility to be less in a finite field than in the limit of zero field. As discussed in more detail below, this is, in fact, the case.

8.5 FLUCTUATION DIAMAGNETISM ABOVE T_c

Before discussing the more detailed theory, let us start by reviewing the physical essence of the phenomenon along the lines suggested by A. Schmid.[13] We model the superconductor crudely as a collection of independent fluctuating droplets of superconductivity, with χ given by (8.22). We then estimate $|\psi|^2$ for a typical fluctuation of volume V as in (8.20). This leads to

$$\chi \approx -\frac{\pi^2 k_B T \xi^2 \langle r^2 \rangle}{\Phi_0^2 V} \tag{8.41}$$

where $\langle r^2 \rangle$ is a mean-square radius, and the numerical coefficient is approximate. For a three-dimensional bulk sample in weak fields, we have seen that the correlation function for fluctuations dies out in a length $\sim\xi$. Thus, it is reasonable to take $V = 4\pi\xi^3/3$ and $\langle r^2 \rangle = (\xi/2)^2$. With a minor adjustment of the numerical coefficient, this leads to Schmid's exact result based on the GL theory:

$$\chi = -\frac{\pi}{6} \frac{k_B T}{\Phi_0^2} \xi(T) \approx -10^{-7} \left(\frac{T_c}{T - T_c} \right)^{1/2} \tag{8.42}$$

---

[13] A. Schmid, *Phys. Rev.* **180**, 527 (1969).

Note that this susceptibility is of the same order of magnitude as the Landau diamagnetism of normal metals, apart from the temperature-dependent enhancement factor.

Although the susceptibility is formally divergent at T_c, in practice the enhancement factor never gets very large before being limited either by the first-order transition in a magnetic field or by the width of the transition in a real sample. Thus, the susceptibility is always extremely small compared to the Meissner regime, where $\chi = -1/4\pi$. Moreover, only small fields can be used without destroying the effect by shrinking and weakening the fluctuations. Nonetheless, the susceptibility is substantial compared with the background, and it can be isolated by measuring the temperature-dependent part of the magnetization in a magnetic field held absolutely constant by a superconducting coil in the persistent current mode. Such experiments were first carried out by Gollub et al.,[14] using a SQUID magnetometer.

Since the magnetization must be measured in a finite field, the temperature at which the fluctuation diamagnetism would diverge is decreased from T_c to the nucleation temperature $T_{c2}(H)$, which is defined by the fact that a solution of the linearized GL equation exists or, equivalently, the temperature at which $H = H_{c2}(T)$. As discussed in Sec. 4.8, in a type I superconductor, this $T_{c2}(H)$ is the supercooling temperature for an ideal sample, and is less than the thermodynamic $T_c(H)$, at which a first-order transition would be expected in the absence of supercooling. In typical bulk samples, however, little supercooling is observed because nucleation at defects takes place at a temperature higher than the ideal $T_{c2}(H)$, and closer to $T_c(H)$. Thus, the divergence point at T_{c2} is experimentally inaccessible, as can be seen in Fig. 8.4.

On the other hand, in type II superconductors, the second-order phase transition at $T_{c2}(H)$ can be approached without discontinuity since $T_{c2}(H) > T_c(H)$. Unfortunately, the breadth of the transitions ($\sim 5 \times 10^{-3}$ K) in most type II materials obscures the detailed behavior at T_{c2} since the linear increase in magnetization below T_{c2} in a few millidegrees becomes orders of magnitude larger than the fluctuation diamagnetism a few millidegrees above T_{c2}.

Some typical data on indium are shown in Fig. 8.4. The upper part shows results in relatively low fields; M' increases with H but less than linearly. The lower part shows higher-field data; here M' *decreases* as H increases because the higher fields are rapidly extinguishing the fluctuations. Note the discontinuous jump indicated at the left end of the curve for $H = 34.9$ Oe. At this point M jumps by five orders of magnitude to the Meissner-effect value. But since it is a first-order transition, there is no divergence anticipating the jump. As suggested by this figure, a temperature-dependent M' can be observed out to about $2T_c$.

[14]J. P. Gollub, M. R. Beasley, R. S. Newbower, and M. Tinkham, *Phys. Rev. Lett.* **22**, 1288 (1969); J. P. Gollub, M. R. Beasley, and M. Tinkham, *Phys. Rev. Lett.* **25**, 1646 (1970); J. P. Gollub, M. R. Beasley, R. Callarotti, and M. Tinkham, *Phys. Rev.* **B7**, 3039 (1973).

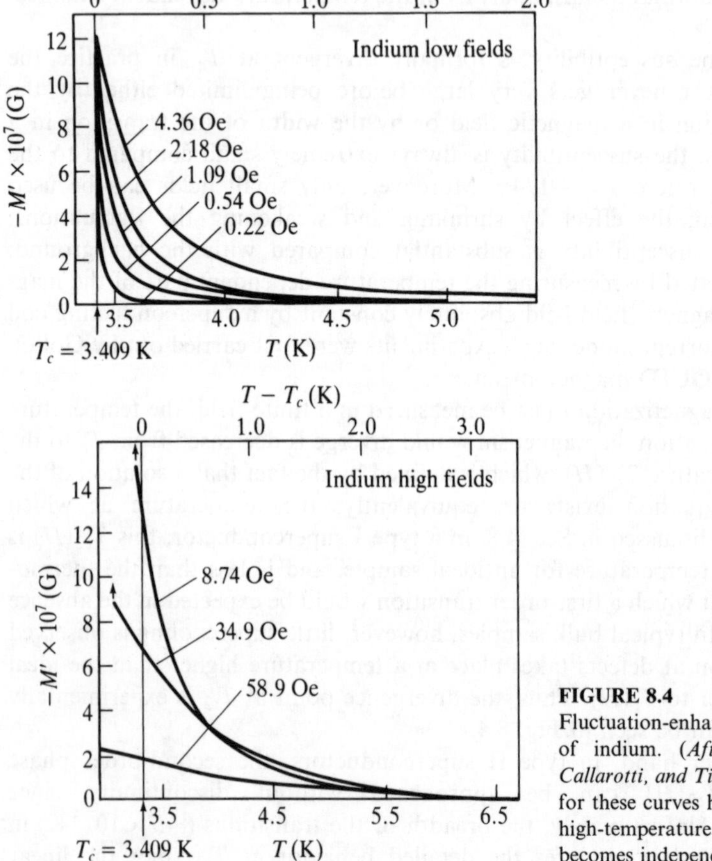

FIGURE 8.4
Fluctuation-enhanced diamagnetism of indium. (*After Gollub, Beasley, Callarotti, and Tinkham.*) The baseline for these curves has been taken as the high-temperature limit, where M becomes independent of T.

To compare these results with theory, one obviously needs to generalize the Schmid result (8.42), which actually had been obtained first by Schmidt,[15] to the case of finite fields. This was done exactly in the framework of the GL theory by Prange.[16] He found that M' should indeed diverge as $(T - T_{c2})^{-1/2}$, and that it should be a universal result if scaled variables were used. That is, he found that

$$\frac{M'}{H^{1/2}T} = f_{\mathrm{P}}(x) \tag{8.43}$$

[15]H. Schmidt, *Z. Phys.* **216**, 336 (1968). (A numerical error of a factor of 4 occurs in this calculation.)

[16]R. E. Prange, *Phys. Rev.* **B1**, 2349 (1970).

where f_P is a function of the single variable $x = (T - T_c)/H \times (dH_{c2}/dT)_{T_c}$. Unfortunately, when the data for several materials were plotted in terms of these variables, they did not fall on the theoretical universal curve but, instead, fell systematically well below it, especially for the higher-field values. This was initially disturbing since it represented serious disagreement between an exact consequence of GL theory and experimental fact.

The explanation of this disagreement was first suggested by Patton, Ambegaokar, and Wilkins.[17] They pointed out that since the GL theory is based on an expansion of the free energy in derivatives of ψ, it is restricted to treating slow variations in space. Since the vector potential as well as the gradient operator enters in the canonical momentum, the GL theory is also limited to reasonably weak fields. Thus, one might expect it to give a poor account of the short-wavelength [$\lesssim \xi(0)$] fluctuations which dominate far above T_c and in strong magnetic fields. For example, at $2T_c$, $\xi(T) \approx \xi(0)$, while even at T_c the fluctuation size in a field $\sim H_{c2}(0)$, as governed by (8.40), is also of order $\xi(0)$. Considerations of this general sort led them to attempt to correct the Prange calculation by cutting off the short-wavelength fluctuations. In this way, they were led to generalize Prange's result (8.43) to

$$\frac{M'}{H^{1/2}T} = f_{\mathrm{PAW}}(x, H/H_s) \tag{8.44}$$

where x is the same reduced temperature variable as before, and H_s is a material-dependent scaling field to be determined by the model.

Gollub et al. tested this idea by plotting data on many materials taken at T_c (where $x = 0$) as functions of H, as shown in Fig. 8.5. According to the Prange form (8.43), $M'(T_c)/H^{1/2}T_c$ would be a universal numerical constant for all materials. The data fall progressively below this value with increasing H in what appears to be a universal dependence on the scaled-field variable H/H_s introduced in (8.44). [H_s is defined as the field for each material for which the observed $M'(T_c)$ has fallen to half the Prange value.] The specific form of falloff predicted by Patton et al. turned out to be qualitatively, but not quantitatively, correct.

Shortly after this universal behavior had been demonstrated experimentally, Lee and Payne[18] (LP) and independently Kurkijärvi, Ambegaokar, and Eilenberger[19] (KAE) produced a theoretical curve (shown dashed in Fig. 8.5) giving good agreement. This theoretical result was based on the microscopic Gor'kov theory in the clean limit. In working out this theory, the theorists found that nonlocal electrodynamic effects played an unexpectedly important role. In effect, they reduce H_s by about an order of magnitude below the value

[17]B. R. Patton, V. Ambegaokar, and J. W. Wilkins, *Solid State Commun.* **7**, 1287 (1969).

[18]P. A. Lee and M. G. Payne, *Phys. Rev. Lett.* **26**, 1537 (1971); *Phys. Rev.* **B5**, 923 (1972).

[19]J. Kurkijärvi, V. Ambegaokar, and G. Eilenberger, *Phys. Rev.* **B5**, 868 (1972).

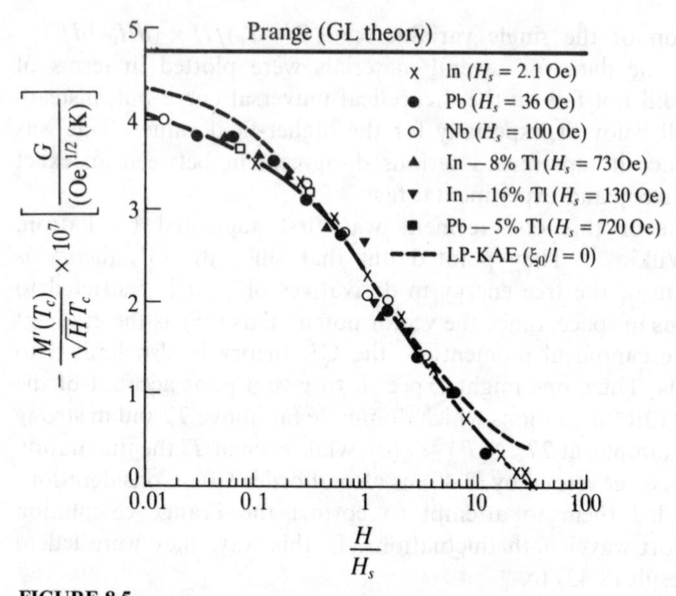

FIGURE 8.5
Universal dependence of $M'(T_c)/H^{1/2}T_c$ on H/H_s. The solid curve is an empirical curve drawn through the data of Gollub, Beasley, Callarotti, and Tinkham; the dashed curve is that of LP-KAE in clean limit (see text).

expected from our qualitative arguments, which had suggested that H_s should be of the order of $H_{c2}(0)$. Note that this nonlocality occurs although the field **B** is everywhere uniform (unlike the usual incidence of nonlocal effects only when fields are confined to a thin penetration layer). Although **B** is uniform, **A** is not, and **A** is what matters in the superconducting electrodynamics.

For alloys, it appears experimentally that $H_s/H_{c2}(0)$ approaches a limiting value $\sim\frac{1}{2}$. This seems intuitively reasonable since nonlocal effects should drop out with a short mean free path. But the calculations of LP-KAE, based on a straightforward application of the Gor'kov theory, appeared to give $H_s/H_{c2}(0)$ increasing without limit as the mean free path was reduced. On the other hand, an alternative calculation by Maki and Takayama[20] give a finite limit for $\ell \to 0$ which seems to be in satisfactory agreement with the experiment. As Maki[21] has shown, the results of LP-KAE were distorted by the inclusion of a zero-point term which reflects only the normal properties of the metal.

We may summarize this discussion by contrasting two regimes: In a type II superconductor, fluctuation diamagnetism can be observed as we approach very close to T_{c2}. In this case, it is dominated by the very lowest energy, long-

---

[20]K. Maki and H. Takayama, *J. Low. Temp. Phys.* **5**, 313 (1971).

[21]K. Maki, *Phys. Rev. Lett.* **30**, 648 (1973).

wavelength modes, which diverge at T_{c2} and are well described by the GL theory. As a result, we expect the Prange (or GL) results to work well near T_{c2}. This is in fact the case, as shown by measurements of Gollub et al. on such type II samples as Pb-5%Tl. But as one goes up in temperature, all fluctuation modes are excited to a comparable extent; statistical weight then favors the short-wavelength modes which are poorly described by the GL theory, and large discrepancies should, and do, appear. It is satisfying that developments in the microscopic theory, spurred by these discrepancies, have largely succeeded in explaining them.

8.5.1 Diamagnetism in Two-Dimensional Systems

A superconducting film of thickness $d \ll \xi$ is considered to be two dimensional since ψ cannot vary in the direction perpendicular to the plane without incurring undue energy cost. Similarly, fluctuations $\delta\psi$ also vary only in the two dimensions of the film plane. The fluctuation-induced diamagnetism of such a film above T_c can be easily estimated by using (8.41). Assuming a correlation range of $\sim\xi$ in the plane and d in the normal direction, one finds that the volume V of such a fluctuation is $\sim\pi\xi^2 d$, so that $\chi \approx -\pi k_B T \langle r^2 \rangle / \Phi_0^2 d$. Because the fluctuating volume is not spherical, the definition of $\langle r^2 \rangle$ requires care. The following physical argument may be used: The energy density $\chi H^2/8\pi$ can also be written as $\mathbf{J}\cdot\mathbf{A}/2c \propto A^2$ in the London gauge. But $\oint \mathbf{A}\cdot d\mathbf{s} \approx B\mathscr{A}$, where $\mathscr{A}$ is the area of the fluctuating region as viewed along the field. Thus, $A \approx B\mathscr{A}/s$, where s is the perimeter of the area. Since $B \approx H$ in the present case of weak-fluctuation diamagnetism, it follows that $\chi \propto (\mathscr{A}/s)^2$. For a sphere, this is simply $\langle r^2 \rangle$. For the disk shape of the fluctuating region, $\langle r^2 \rangle_{\text{eff}} \approx \xi^2$ for $H_\perp$, and $\langle r^2 \rangle_{\text{eff}} \approx (d/2)^2$ for $H_\parallel$. [This general qualitative argument can be confirmed for the parallel-field case by reference to (4.54*a*), which describes the screening of a parallel field in a thin film. For $d \ll \lambda$, one finds $h/H = B/H = 1 + 4\pi\chi$, with $4\pi\chi = -d^2/12\lambda^2 \propto d^2$, as found here.] Thus, in order of magnitude

$$\chi'_\perp \approx \frac{-k_B T \xi^2}{\Phi_0^2 d} \approx \frac{\xi}{d}\chi'_{3D} \propto \frac{T_c}{T - T_c} \tag{8.45a}$$

whereas

$$\chi'_\parallel \approx \frac{-k_B T d}{\Phi_0^2} \approx \frac{d}{\xi}\chi'_{3D} \propto \text{const.} \tag{8.45b}$$

In these expressions, χ_{3D} refers to (8.42).

From these results we conclude that $\chi'_\parallel$ will be essentially unobservable since it is extremely small and temperature-independent. On the other hand, $\chi'_\perp$ will be ξ/d times larger than χ'_{3D} per unit volume; but since the volume falls as d, this gives a thickness-independent total susceptibility equal only to that of a bulk superconductor of thickness $\sim\xi$. For a single film, this small moment would again be hard to observe. However, by constructing a multilayer stack of such films, one could get a larger volume. In this way, one could test the predicted different temperature dependence $\chi'_\perp \propto (T - T_c)^{-1}$ of (8.45*a*) compared to the dependence of $\chi'_{3D} \propto (T - T_c)^{-1/2}$ of (8.42).

In fact, there exist superconducting layered compounds, such as TaS_2, into which organic compounds such as pyridine can be intercalated to separate each metallic conducting layer ($d \approx 6$ Å) from its neighbors. In an ideal sample, the layers are coupled together only by Josephson-like tunneling through the pyridine intercalate, and one might imagine that this would give a two dimensional behavior. Actually, as shown by Lawrence and Doniach,[22] even weak Josephson coupling leads to behavior which is better described near T_c as three dimensional with anisotropic effective mass than as two dimensional. (In fact, in Chap. 9 we shall base our treatment of the anisotropy of the layered high-temperature superconductors on this Lawrence–Doniach model.) In particular, the temperature dependence of the fluctuation susceptibility does not appear[23] to differ much from that of a three-dimensional system, at least until well above T_c.

8.6 TIME DEPENDENCE OF FLUCTUATIONS

Since diamagnetism is an equilibrium property, we were able to compute it above using only the time-average quantities $|\psi_{\mathbf{k}}|^2$. When we discuss a nonequilibrium property such as electrical conductivity, however, we need a model of the time dependence since the contribution of a given fluctuation to the conductivity above T_c is proportional to its lifetime, as that time limits the period available for acceleration in an applied field. Such a model is provided by the time-dependent Ginzburg-Landau (TDGL) equations developed by various workers.[24] According to this model, which will be discussed in more detail in Chap. 10, the superconducting ψ function relaxes exponentially toward its instantaneous equilibrium value; above T_c this value is zero. The linearized TDGL equation is then a simple generalization of (8.24), namely,

$$\frac{\partial\psi}{\partial t} = -\frac{1}{\tau_0}(1 - \xi^2\nabla^2)\psi \qquad T > T_c \tag{8.46}$$

if electromagnetic potentials are neglected. The parameter

$$\tau_0 = \frac{\pi\hbar}{8k_B(T - T_c)} \tag{8.47}$$

[22] W. E. Lawrence and S. Doniach, in E. Kanda (ed.), *Proc. 12th Int. Conf. Low Temp. Phys.* (Kyoto, Japan, 1970; Keigaku Publ. Co., 1971, p. 361); see also T. Tsuzuki, *J. Low. Temp. Phys.* **9**, 525 (1972).

[23] D. E. Prober, M. R. Beasley, and R. E. Schwall, *Phys. Rev.* **B15**, 5245 (1977).

[24] See, e.g., A. Schmid, *Phys. Kondens. Mat.* **5**, 302 (1966); C. Caroli and K. Maki, *Phys. Rev.* **159**, 306, 316 (1967); E. Abrahams and T. Tsuneto, *Phys. Rev.* **152**, 416 (1966); and J. W. F. Woo and E. Abrahams, *Phys. Rev.* **169**, 407 (1968). A critical review has been given by M. Cyrot, *Repts. Prog. Phys.* **36**, 103 (1973).

is the temperature-dependent relaxation time of the uniform ($k = 0$) mode. According to (8.46), the higher-energy modes with $k > 0$ decay more rapidly, with relaxation rate

$$\frac{1}{\tau_k} = \frac{1 + k^2\xi^2}{\tau_0} \tag{8.48}$$

By themselves, these equations imply that (above T_c) any nonzero value of $\psi_{\mathbf{k}}$ will die out exponentially in a time τ_k. To maintain the nonzero thermal average of $|\psi_{\mathbf{k}}|^2$ found in (8.27), one invokes a so-called Langevin force, a completely random (i.e., "white spectrum") driving force which represents the interaction between the superconducting electrons and the rest of the thermodynamic system with which they are in equilibrium. The magnitude of this force is fixed by the requirement that it maintain the appropriate value of $\langle|\psi_{\mathbf{k}}|^2\rangle$, as calculated in (8.27) using only *equilibrium* statistical mechanics. Adding a Langevin force $F_{\mathbf{k}}$ on the right-hand side of (8.46), and choosing the value of $F_{\mathbf{k}}$ to give the correct time average,

$$\langle|\psi_{\mathbf{k}}|^2\rangle = \frac{1}{2\pi}\int_{-\infty}^{\infty} \langle|\psi_{\mathbf{k},\omega}|^2\rangle \, d\omega$$

we have

$$\langle|\psi_{\mathbf{k},\omega}|^2\rangle = \langle|\psi_{\mathbf{k}}|^2\rangle \frac{2\tau_k}{1 + \omega^2\tau_k^2} \tag{8.49}$$

It is easily verified (Wiener-Khintchine theorem)[25] that this power spectrum of $\psi_{\mathbf{k}}$ in frequency space corresponds to exponential decay in time of the correlation function

$$\langle\psi_{\mathbf{k}}^*(0)\psi_{\mathbf{k}}(t)\rangle = \langle|\psi_{\mathbf{k}}|^2\rangle e^{-t/\tau_k} \tag{8.50}$$

Finally, substituting for $\langle|\psi_{\mathbf{k}}|^2\rangle$ from (8.27) and for τ_k from (8.48), we obtain, after simplifying,

$$\langle|\psi_{-\mathbf{k},\omega}|^2\rangle = \frac{16}{\pi}\frac{k_B(T - T_c)}{\hbar\alpha(T)}\frac{k_B T\tau_k^2}{1 + \omega^2\tau_k^2} \tag{8.51}$$

Since $\alpha(T) \propto (T - T_c)$, the entire dependence on T, as well as on k and ω, is in the last factor.

8.7 FLUCTUATION-ENHANCED CONDUCTIVITY ABOVE T_c

In the absence of superconducting fluctuations, the normal dc conductivity is given by

[25] See, e.g., C. Kittel, *Elementary Statistical Physics*, Wiley, New York (1958), p. 136.

$$\sigma_n = \frac{ne^2\tau}{m} \tag{8.52}$$

where τ is the mean scattering time of the normal electrons and n is the number of them per unit volume. By analogy, we might expect the fluctuations to contribute an extra term

$$\sigma' \approx \frac{(2e)^2}{m^*} \sum_{\mathbf{k}} \frac{\langle |\psi_{\mathbf{k}}|^2 \rangle \tau_k}{2} \tag{8.53}$$

since ordinary scattering processes are ineffective until a given fluctuation relaxes, and $|\psi_{\mathbf{k}}|^2$ will relax twice as fast as $\psi_{\mathbf{k}}$. Using the values of $\langle |\psi_{\mathbf{k}}|^2 \rangle$ and τ_k from (8.27) and (8.48), and integrating over k space, we can show that this prescription gives results which differ from the exact calculations only by small numerical factors. In particular, the temperature dependence of σ' is correctly found to be $(T - T_c)^{-(4-d)/2}$, where d $(= 1, 2, 3)$ is the dimensionality of the system.

Since the calculation is quite tractable and gives some additional insight, let us now compute σ' exactly, within the framework of the GL linearized fluctuation theory. This is done most easily by using the Kubo formalism, which relates linear response coefficients to the fluctuations in the unperturbed system, as required by the fluctuation-dissipation theorem. We confine our attention to uniform fields and currents. Then our general starting point is the Kubo result

$$\sigma_{xx}(\omega) = \frac{1}{k_B T} \int_0^\infty \langle J_x(0) J_x(t) \rangle \cos \omega t \, dt \tag{8.54}$$

[If this approach is unfamiliar, it may be helpful to note that the integral gives the power spectrum of $J_x(t)$; (8.54) is then equivalent to the Nyquist noise current formula; $J_x^2(\omega) = 4k_B T \sigma_{xx}(\omega)$ per unit bandwidth. Considering a unit cube, this formula is equivalent to the even more familiar expression for the thermal noise voltage in bandwidth B of a resistance R, namely, $V^2 = 4k_B TRB$.] We assume that the normal quasi-particle current fluctuations are unchanged by the superconducting fluctuations. (This is not strictly correct very near T_c, where the fluctuations are strong.) Thus, to compute σ'_{xx}, we include only the fluctuating supercurrent in (8.54). For $\psi = \sum_{\mathbf{k}} e^{i\mathbf{k}\cdot\mathbf{r}}$, the current is given by

$$\begin{aligned} \mathbf{J} &= \frac{e\hbar}{m^* i}(\psi^* \nabla \psi - \psi \nabla \psi^*) \\ &= \frac{e\hbar}{m^*} \sum_{\mathbf{k},\mathbf{q}} (2\mathbf{k} + \mathbf{q}) \psi_{\mathbf{k}}^* \psi_{\mathbf{k}+\mathbf{q}} e^{i\mathbf{q}\cdot\mathbf{r}} \end{aligned} \tag{8.55}$$

Restricting attention to uniform ($\mathbf{q} = 0$) currents in the x direction, we find that this reduces to

$$J_x = \frac{2e\hbar}{m^*} \sum_{\mathbf{k}} k_x |\psi_{\mathbf{k}}|^2 \tag{8.56}$$

We now want to compute the current-current correlation function

$$\langle J_x(0)J_x(t)\rangle = \left(\frac{2e\hbar}{m^*}\right)^2 \left\langle \sum_{\mathbf{k},\mathbf{k}} k_x k_x' |\psi_{\mathbf{k}}(0)|^2 |\psi_{\mathbf{k}'}(t)|^2 \right\rangle \tag{8.57}$$

Since $\psi_{\mathbf{k}}$ and $\psi_{\mathbf{k}'}$ are statistically independent, the cross terms average out, and this can be written as

$$\langle J_x(0)J_x(t)\rangle = \left(\frac{2e\hbar}{m^*}\right)^2 \sum_{\mathbf{k}} k_x^2 \langle \psi_{\mathbf{k}}^*(0)\psi_{\mathbf{k}}(t)\rangle^2 \tag{8.58}$$

Inserting the exponential time decay (8.50) of the correlation function, and carrying out the cosine Fourier transform (8.54), we obtain

$$\sigma_{xx}'(\omega) = \left(\frac{2e\hbar}{m^*}\right)^2 \frac{1}{k_B T} \sum_{\mathbf{k}} k_x^2 \langle |\psi_{\mathbf{k}}|^2 \rangle^2 \frac{\tau_k/2}{1+(\omega\tau_k/2)^2} \tag{8.59}$$

Specializing to the dc case ($\omega = 0$), and inserting $\langle |\psi_{\mathbf{k}}|^2 \rangle$ and τ_k from (8.27) and (8.48), we obtain

$$\sigma_{xx}'(0) = \frac{\pi e^2}{\hbar} \left(\frac{T}{T-T_c}\right) \sum_{\mathbf{k}} \frac{k_x^2 \xi^4}{(1+k^2\xi^2)^3} \tag{8.60}$$

8.7.1 Three Dimensions

In a three-dimensional bulk sample, k can be taken as a continuous variable, allowing the sum to be performed by integration. Averaged over a sphere, $k_x^2 = k^2/3$; the density of states in unit volume is given by $4\pi k^2\, dk/(2\pi)^3$. After an elementary integration we have

$$\sigma'(0)\bigg|_{3D} = \frac{1}{32} \frac{e^2}{\hbar\xi(0)} \left(\frac{T}{T-T_c}\right)^{1/2} \tag{8.61}$$

where as usual $\xi(T) \equiv \xi(0)[T/(T-T_c)]^{1/2}$. Since the conductivity is isotropic, we have dropped the subscripts. Although this result is formally divergent as we approach T_c, the coefficient is less than the normal conductivity σ_n by a factor of the order of $1/(k_F^2 \ell\xi(0)) \sim 10^{-5}$. Thus, the fractional change in conductivity at any meaningful temperature interval above T_c will be very small. Note the contrast with the diamagnetic susceptibility (8.42), where the coefficient is comparable to the background normal-state diamagnetism, so that large fractional changes in χ are observable.

8.7.2 Two Dimensions

Now let us consider the case of a film that is thin enough to justify a two-dimensional approximation, i.e., one in which the thickness $d \ll \xi$. This is the case on which the greatest amount of experimental work has been done. In this case, the variation of ψ perpendicular to the film is limited to a discrete set of standing waves with $k_\perp = \nu\pi/d$, with $\nu = 0, 1, 2$, etc. If the film is thin enough to drop all except the $\nu = 0$ term, the summation in (8.60) becomes a two-dimensional integration in the plane of the film. The average of k_x^2 is then $k^2/2$, and the density-of-states factor becomes $2\pi k\,dk/(2\pi)^2 d$. Carrying out the integration, we obtain the remarkably simple result

$$\sigma'(0)\bigg|_{2D} = \frac{e^2}{16\hbar d}\frac{T}{T - T_c} \tag{8.62}$$

Note that this result contains no adjustable parameters (apart from T_c). It is also important to realize that the film thickness d is not critical since the quantity actually measured experimentally is the conduct*ance* $\sigma' d$ per square, not the conductivity. This result was first derived by Aslamasov and Larkin[26] and is in excellent agreement with the measured results of Glover[27] on thin amorphous films. Such films are particularly favorable since they have a high normal resistance (low normal conductance) in parallel with the fluctuation term (8.62), which has a universal value, independent of the normal resistance. Thus, the fluctuation conductivity is a larger fractional effect when the background normal conductance is lower.

Before making a detailed comparison between (8.62) and experimental data, one must assess the size of the error made by dropping all except the $k_\perp = 0$ term in the sum (8.60). The integration over k in the plane can be carried out for arbitrary $k_\perp$, with the value of the integral reduced by a factor $(1 + k_\perp^2 \xi^2)^{-1}$. Thus, for finite d/ξ, (8.62) should be multipled by a factor

$$\frac{\sigma'}{\sigma'_{2D}} = \sum_{\nu=0}^{\infty} \frac{1}{1 + (\nu^2 \pi^2 \xi^2 / d^2)} \tag{8.63}$$

Thus, so long as $d \leq \xi$, the simple two-dimensional result is accurate to ~10 percent. On the other hand, if $d \gg \xi$, this sum can be evaluated by integration. The result is $d/2\xi$, just the factor required to recover the three-dimensional result (8.61) from the two-dimensional result (8.62). Because of the temperature dependence of ξ, films may change from two-dimensional to three-dimensional behavior as one goes farther above T_c. Experimental data appear to follow the transitional behavior predicted by (8.63).

8.7.3 One Dimension

To complete our survey of the important special cases, we now consider the case of a thin wire of cross-sectional area $A \ll \xi^2$, so that a one-dimensional

---

[26] L. G. Aslamasov and A. I. Larkin, *Phys. Lett.* **26A**, 238 (1968).

[27] R. E. Glover, III, *Phys. Lett.* **25A**, 542 (1967).

approximation may be made. In this case, only the component of k along the wire (k_x) has a continuous distribution. Retaining only the terms in (8.60) with zero transverse momentum, we find that the density-of-states factor is $dk_x/2\pi A$. Integrating on k_x from $-\infty$ to $+\infty$, we obtain

$$\sigma'(0)\bigg|_{1D} = \pi \frac{e^2\xi(0)}{16\hbar A}\left(\frac{T}{T-T_c}\right)^{3/2} \tag{8.64}$$

with corrections for finite A which are down by a factor of the order of $(A/\pi^2\xi^2)^{3/2}$. As with the two-dimensional case, the measured quantity is the conductance $\sigma' A$, so that the area does not need to be known accurately to test this theoretical prediction. Nonetheless, the agreement between theory and experiment in the tin whisker crystals which were used in the experiments discussed in Sec. 8.1 is not very good because of the importance of the anomalous contributions (Maki terms) discussed next.

8.7.4 Anomalous Contributions to Fluctuation Conductivity

While the initial comparison between the Aslamasov-Larkin (AL) result (8.62) and the experimental data of Glover and coworkers[28] on thin films showed excellent agreement, later measurements[29] showed σ' values as much as 10 times larger than the AL prediction. These large values were typically found in clean (i.e., low-resistance) aluminum films; lead and bismuth films were close to AL, as found earlier, whereas tin films showed σ' up to about 4 times the AL value.

At about the same time these anomalously large conductivity enhancements were found experimentally, Maki[30] noted the existence of another term (or Feynman diagram) in the theoretical calculation which had been omitted in the work of AL. Physically, this "Maki term" appears to reflect the increase in the normal-electron conductivity induced by the superconducting fluctuations. A similar increase in σ' above σ_n for $\hbar\omega < \Delta < kT$ is familiar just *below* T_c in the presence of weak but stable superconductivity. In fact, the discussion of (3.93) indicated that, for $\omega = 0, \sigma'$ has a logarithmic infinity unless the peaked BCS-state density is limited by lifetime effects or gap anisotropy. Although the analogy between this effect and the Maki term is rather superficial, it *is* the case that in one-dimensional or two-dimensional systems the Maki term appears to give an infinite conductivity at all temperatures *above* T_c.

---

[28]See, e.g., D. C. Naugle and R. E. Glover, III, *Phys. Lett.* **28A**, 110 (1968).

[29]See, e.g., J. E. Crow, R. S. Thompson, M. A. Klenin, and A. K. Bhatnagar, *Phys. Rev. Lett.* **24**, 371 (1970).

[30]K. Maki, *Prog. Theor. Phys. (Kyoto)* **39**, 897 (1968); **40**, 193 (1968).

As shown by Thompson,[31] this nonphysical divergence is prevented by the presence of any pair-breaking effect, such as a magnetic field or magnetic impurities, which effectively limits the lifetimes of the evanescent Cooper pairs and cuts off the divergence. A considerable amount of data, especially on films in magnetic fields, can be accounted for by adding this so-called Maki-Thompson term to the AL term treated above.

Patton[32] and also Keller and Korenman[33] subsequently reexamined this problem and showed that only a finite conductivity is obtained for the Maki term even for the simple BCS model without any extraneous pair breakers, if the so-called vertex corrections are calculated sufficiently carefully. These theories account for an effective Maki-Thompson pair breaker of strength proportional to the resistance per square ($R_\square$) of the film. However, even these improved theories require an intrinsic pair breaker to account for the difference in data obtained on different materials with the same $R_\square$, especially in the low-resistance films. The strength of this pair breaker is usually specified by a parameter τ_{c0}, which is the fractional depression of T_c due to its action. For aluminum, the data fit with $\tau_{c0} = 2 \times 10^{-4}$, while for tin, $\tau_{c0} \approx 0.02$. For the strong coupling materials lead and bismuth, it appears that τ_{c0} may be as large as 0.1; this, together with the high $R_\square$, probably accounts for the agreement of the data of Glover with the simple AL theory without Maki-Thompson corrections. The trend of the values of τ_{c0} with different materials is consistent with the suggestion of Appel[34] that pair breaking by thermal phonons should give $\tau_{c0} \propto \lambda(T/\theta_D)^2$, where λ is the electron-phonon coupling constant and θ_D is the Debye temperature. For further details on the fluctuation-enhanced conductivity problem, including many different regimes, the reader is referred to the comprehensive review of Craven, Thomas, and Parks.[35]

8.7.5 High-Frequency Conductivity

Measurements of the frequency dependence of σ' allow a more specific test of the TDGL theory than the simple dc conductivity measurements described earlier since from (8.59) it is clear that σ' will fall off for $\omega > \tau_k^{-1}$. More quantitatively, each term in the sum in (8.60) is multiplied by $[1 + (\omega\tau_k/2)^2]^{-1}$ before the integration over k. Specializing to the two-dimensional case, one finds that the integration can still be carried out exactly, with the result

$$\sigma'(\omega)\bigg|_{2D} = \frac{e^2}{16\hbar d}\left(\frac{T}{T - T_c}\right)\left[\frac{4}{\omega\tau_0}\tan^{-1}\frac{\omega\tau_0}{2} - \frac{4}{\omega^2\tau_0^2}\ln\left(1 + \frac{\omega^2\tau_0^2}{4}\right)\right] \tag{8.65}$$

---

[31] R. S. Thompson, *Phys. Rev.* **B1**, 327 (1970).

[32] B. R. Patton, *Phys. Rev. Lett.* **27**, 1273 (1971).

[33] J. Keller and V. Korenman. *Phys. Rev. Lett.* **27**, 1270 (1971); *Phys. Rev.* **B5**, 4367 (1972).

[34] J. Appel, *Phys. Rev. Lett.* **21**, 1164 (1968).

[35] R. A. Craven, G. A. Thomas, and R. D. Parks, *Phys. Rev.* **B7**, 157 (1973).

The prefactor is recognized as the dc result. If the expression in the square bracket is expanded at low frequencies, one finds

$$\sigma'(\omega) = \sigma'(0)\left(1 - \frac{\omega^2\tau_0^2}{24} + \dots\right) \tag{8.66}$$

Thus, as expected, the conductivity rolls off when ω exceeds some average τ_k^{-1}, which in turn is somewhat greater than τ_0^{-1} because of the k dependence of τ_k. Another interesting limit is right at T_c, where $\tau_0 = \infty$. Then for any finite frequency (8.65) reduces to

$$\sigma'(\omega)\bigg|_{T_c} = \frac{e^2}{\hbar d}\frac{k_B T_c}{\hbar\omega} \qquad \omega > 0 \tag{8.67}$$

Note that this is the same as the dc value $\sigma'(0)$ for $(T - T_c) = \hbar\omega/16k_B$. Clearly, this fluctuation conductivity is finite even at T_c for all except zero frequency.

These predictions of the theory were tested by microwave transmission measurements on thin lead films by Lehoczky and Briscoe.[36] They were able to make measurements at frequencies of 24, 37, and 69 GHz, as well as dc measurements. Below T_c the microwave transmission data agreed well with that computed by using the ordinary Mattis-Bardeen complex conductivity function $[\sigma_1(\omega) - i\sigma_2(\omega)]$ of the BCS theory, discussed in Sec. 3.10. Above T_c the data fit well with the predicted transmission based on (8.65) for σ_1, with $\sigma_2 = 0$. The conductivity increase at T_c was found to be inversely proportional to ω, as expected from (8.67), reaching ~11 percent at the lowest frequency used. The good agreement between theory and experimental data as functions of both frequency and temperature indicates that the TDGL model is accurate, at least in these high-resistance lead films. The frequency dependence of the conductivity when the Maki terms are important was not accessible with the films studied.

[36] S. L. (A.) Lehoczky and C. V. Briscoe, *Phys. Rev. Lett.* **23**, 695 (1969).

CHAPTER 9

THE HIGH-TEMPERATURE SUPERCONDUCTORS

9.1 INTRODUCTION

After the discovery of superconductivity in mercury at 4 K by Kamerlingh Onnes in 1911, the search for new superconducting materials led to a slow increase in the highest known transition temperature T_c over the decades, reaching a plateau at 23 K with the discovery of the superconductivity of Nb_3Ge by Gavaler.[1] After 13 more years, the path to radically higher transition temperatures was opened by the discovery in 1986 of superconductivity at ~35 K in "LBCO" (a mixed oxide of lanthanum, barium, and copper) by Bednorz and Müller,[2] for which they were awarded the Nobel prize in 1987.

The discovery was surprising and exciting, not simply because of the large increase in T_c, but also because it revealed that the oxides formed an unsuspected new class of superconducting materials with great potential. Another big jump to $T_c \sim 90$ K followed quickly, with the discovery of the "123" class of materials, exemplified by $Y_1Ba_2Cu_3O_{7-\delta}$ ("YBCO"), which was made nearly simultaneously by groups in the United States, Japan, and China.[3] [In this structure, the Y

---

[1]J. R. Gavaler, *Appl. Phys. Lett.* **23**, 480 (1973).

[2]J. G. Bednorz and K. A. Müller, *Z. Phys.* **B64**, 189 (1986).

[3]M. K. Wu et al. *Phys. Rev. Lett.* **58**, 908 (1987); S. Hikami et al., *Jpn. J. Appl. Phys.* **26**, L314 (1987); Z. X. Zhao et al., Kexue Tongbao **33**, 661 (1987).

(yttrium) can be replaced by many other rare earth elements, e.g., La, Nd, Sm, Eu, Gd, Ho, Er, and Lu, with similarly high T_c.] Shortly thereafter, still higher T_c values were found[4] in the "BSCCO" system (mixed oxides of bismuth, strontium, calcium, and copper) and the "TBCCO"[5] system (mixed oxides of thallium, barium, calcium, and copper).

In all of these systems, copper oxide planes form a common structural element, which is thought to dominate the superconducting properties. Depending on the choice of stoichiometry, the crystallographic unit cell contains varying numbers of CuO_2 planes. In addition, the 123 compounds contain CuO "chains," which are thought to serve largely as reservoirs to control the electron density in the planes. The exact T_c depends on these particulars but, roughly speaking, the highest T_c achieved in the YBCO, BSCCO, and TBCCO systems are 93, 110, and 130 K, respectively.

These very high transition temperatures were of obvious technical interest because they opened the way to applications which required only liquid N_2 cooling (77 K), rather than liquid helium. They also posed intriguing fundamental questions: What is the mechanism responsible for the high T_c? Whatever the mechanism, is the nature of the superconducting state basically the same Cooper-paired state as in BCS, or is it fundamentally different? If it is basically the same sort of state, how profoundly are the phenomenological properties of the materials, described by the Ginzburg-Landau theory, modified by the radically modified parameter values?

At the time of this writing there is still no real consensus as to the mechanism causing the high T_c in these materials. However, it does appear that, whatever the mechanism, the magnetic properties of the materials can be well described by the familiar BCS/GL concepts. Nonetheless, as we shall see, important changes arise from the extreme anisotropy which is caused by the layered structure, and from the extremely short coherence length and prominent fluctuation effects, both associated with the high T_c itself. In keeping with the approach followed in the rest of this text, we shall set aside material-specific questions and focus on the relatively universal physical properties which stem from the high transition temperatures and the existence of a strongly layered crystal structure in all of the materials.

As will be discussed in Sec. 9.2, this structure causes the materials to be highly anisotropic magnetically, and in extreme cases to approach the two-dimensional behavior expected from a stack of decoupled superconducting film planes. From this starting point we shall proceed to discuss the anisotropic magnetization properties and the nature of the resistive transition of these materials, which is the single most important property for practical applications. We shall find that the

---

[4]H. Maeda et al., *Jpn. J. Appl. Phys.* **27**, L209 (1988).

[5]Z. Z. Sheng and A. M. Hermann, Nature **232**, 55 (1988).

resistive transition can only be understood by a discussion of the elastic properties of the flux-line lattice, which will determine its melting temperature and also the effectiveness of material defects in pinning the flux-line lattice.

Next we shall discuss the effect of granular structure in polycrystalline samples and the effect of penetrating fluxons on high-frequency losses. These topics have relevance to the classic superconductors as well but have received heightened attention in connection with the high-temperature superconductors.

We conclude the chapter with a survey of certain anomalous properties of the high-temperature superconductors, which suggest unconventional *d*-wave pairing. This work gives clues about the fundamental difference between the superconductivity in the high-temperature and the classic superconductors.

9.2 THE LAWRENCE-DONIACH MODEL

A convenient model for the analysis of the consequences of a layered structure in a superconducting material was proposed by Lawrence and Doniach[6] (LD) and extensively applied[7] in the context of layered transition-metal dichalcogenides such as TaS_2 with organic molecules intercalated between the metallic layers. Although these materials had transition temperatures of only a few Kelvins (K), the formalism is equally useful for the high-temperature superconductors. In their model, layered superconductors are viewed as a stacked array of two-dimensional superconductors [within each of which the GL order parameter $\psi_n(x, y)$ is a 2-D function], coupled together by Josephson tunneling between adjacent layers. As we shall show below, this model reduces to the anisotropic 3-D GL theory for long-wavelength phenomena which dominate near T_c, but it can yield new results for short-wavelength phenomena at lower temperatures after effectively crossing over to 2-D behavior.

As in the conventional GL theory discussed in Chap. 4, we define the model in terms of a free-energy expression for the stack of layers. Following conventional crystallographic nomenclature, we see that the layers define the *ab* planes and the *c* axis is normal to them. We take z as the coordinate along **c**, with s as the interplanar distance, and x, y as coordinates in the planes. Then omitting the vector potential for simplicity, we can generalize (4.1) and write the free energy as

$$F = \sum_n \int \alpha|\psi_n|^2 + \frac{1}{2}\beta|\psi_n|^4 + \frac{\hbar^2}{2m_{ab}}\left(\left|\frac{\partial\psi_n}{\partial x}\right|^2 + \left|\frac{\partial\psi_n}{\partial y}\right|^2\right) + \frac{\hbar^2}{2m_c s^2}|\psi_n - \psi_{n-1}|^2 \tag{9.1}$$

[6]W. E. Lawrence and S. Doniach, in E. Kanda (ed.) *Proc. 12th Int. Conf. Low Temp. Phys.* (Kyoto, 1970; Keigaku, Tokyo, 1971), p. 361.

[7]R. A. Klemm, M. R. Beasley, and A. Luther, *J. Low Temp. Phys.* **16**, 607 (1974); D. E. Prober, M. R. Beasley, and R. E. Schwall, *Phys. Rev.* **B15**, 5245 (1977).

where the sum runs over the layers and the integral is over the area of each layer. We have introduced different masses m_{ab} and m_c to reflect the different modes of charge transport in the two principal directions but, for simplicity, we ignore the relatively small anisotropy in the *ab* plane. Also, we have replaced the *z* derivative by the finite difference form that is appropriate to a discrete system. Note that if we write $\psi_n = |\psi_n|e^{i\varphi_n}$, and assume that all $|\psi_n|$ are equal, the last term of (9.1) becomes

$$(\hbar^2/m_c s^2)|\psi_n|^2[1 - \cos(\varphi_n - \varphi_{n-1})] \tag{9.1a}$$

which makes clear the equivalence of this term to a Josephson coupling energy ($\sim 1/m_c$) between adjacent planes.

Minimizing (9.1) with respect to a variation in ψ_n^* in the usual way, we obtain the LD equation for ψ_n:

$$\alpha\psi_n + \beta|\psi_n|^2\psi_n - \frac{\hbar^2}{2m_{ab}}\left(\frac{\partial^2}{\partial x^2} + \frac{\partial^2}{\partial y^2}\right)\psi_n - \frac{\hbar^2}{2m_c s^2}(\psi_{n+1} - 2\psi_n + \psi_{n-1}) = 0 \tag{9.2}$$

Note that the last term is the discrete form of the second derivative. Finally, inserting the vector potential term, we can generalize this to the full form of the LD equation:

$$\alpha\psi_n + \beta|\psi_n|^2\psi_n - \frac{\hbar^2}{2m_{ab}}\left(\nabla - i\frac{2e}{\hbar c}\mathbf{A}\right)^2\psi_n$$
$$- \frac{\hbar^2}{2m_c s^2}(\psi_{n+1}e^{-2ieA_z s/\hbar c} - 2\psi_n - \psi_{n-1}e^{2ieA_z s/\hbar c}) = 0 \tag{9.3}$$

where in the m_{ab} term ∇ and $\mathbf{A}$ are two-dimensional vectors in the *xy* plane.

9.2.1 The Anisotropic Ginzburg-Landau Limit

From the structure of (9.1) it is clear that the free energy reduces directly to a GL form with ellipsoidal anisotropy in the long-wavelength limit in which the variation along *z* is smooth enough so that $(\psi_n - \psi_{n+1})/s$ can be replaced by $\partial\psi/\partial z$. In this same limit, (9.3) reduces to

$$\alpha\psi + \beta|\psi|^2\psi - \frac{\hbar^2}{2}\left(\nabla - i\frac{2e}{\hbar c}\mathbf{A}\right)\cdot\left(\frac{1}{m}\right)\cdot\left(\nabla - i\frac{2e}{\hbar c}\mathbf{A}\right)\psi = 0 \tag{9.4}$$

where ∇ and $\mathbf{A}$ are now 3-D quantities, and $(1/m)$ is the reciprocal mass tensor with principal values $1/m_{ab}$, $1/m_{ab}$, and $1/m_c$. If the interlayer coupling is weak, then $m_c \gg m_{ab}$.

The mass anisotropy causes the coherence length ξ to be anisotropic. Generalizing (4.17) to the anisotropic case, we have

$$\xi_i^2(T) = \frac{\hbar^2}{2m_i|\alpha(T)|} \tag{9.5}$$

where the subscript i identifies a particular principal axis. Since $\alpha(T)$ is isotropic and proportional to $(T - T_c)$, ξ_i scales with $1/\sqrt{m_i}$ and diverges as $|T - T_c|^{-1/2}$. Hence, sufficiently close to T_c, ψ will indeed vary sufficiently smoothly to justify this continuum GL approximation.

In this GL regime, we can also apply the relation (4.20), which can be written as

and
$$2\sqrt{2}\pi H_c(T)\xi_i(T)\lambda_i(T) = \Phi_0 \tag{9.6}$$

to infer that the anisotropy of the penetration depth λ_i will be inverse to that of ξ_i since H_c is isotropic. In applying this rule, we must remember that λ_i describes the screening by *supercurrents* flowing along the ith axis, *not* the screening of a *magnetic field* along the ith axis. For example, consider an Abrikosov vortex in a sample with the magnetic field along the a axis. In an isotropic superconductor, this vortex would have circular symmetry. In an anisotropic superconductor, however, the core radius along the plane direction will be ξ_{ab}, whereas the core radius in the c direction will be $\xi_c \ll \xi_{ab}$. On the other hand, the flux-penetration radius will be λ_c along the plane direction, whereas it will be the smaller value λ_{ab} in the c direction. Thus, both the core and the current streamlines confining the flux are flattened into ellipses with long axes parallel to the planes (b axis), with aspect ratio $(m_c/m_{ab})^{1/2}$, as depicted in Fig. 9.1.

Going a step further, we can now work out the anisotropy of the upper critical field in this GL regime. Since $H_{c2} = \Phi_0/2\pi\xi^2$ is determined by vortices of current flowing in a plane perpendicular to the field, the relevant values of ξ_i are those for the two axes perpendicular to the field direction. For fields along the two distinct principal axes, we have

$$H_{c2\|c} = \Phi_0/2\pi\xi_{ab}^2 \tag{9.7a}$$

$$H_{c2\|ab} = \Phi_0/2\pi\xi_{ab}\xi_c \tag{9.7b}$$

so that $H_{c2\|ab} \gg H_{c2\|c}$ since $\xi_{ab} \gg \xi_c$. Because $H_{c1} \sim 1/\lambda^2$, which is inversely related to $1/\xi^2$, the anisotropy in H_{c1} will be inverse to that for H_{c2}; i.e., $H_{c1\|ab} \ll H_{c1\|c}$.

For convenient future reference, we collect these relations into the following equalities:

$$\gamma \equiv \left(\frac{m_c}{m_{ab}}\right)^{1/2} = \frac{\lambda_c}{\lambda_{ab}} = \frac{\xi_{ab}}{\xi_c} = \left(\frac{H_{c2\|ab}}{H_{c2\|c}}\right) = \left(\frac{H_{c1\|c}}{H_{c1\|ab}}\right) \tag{9.8}$$

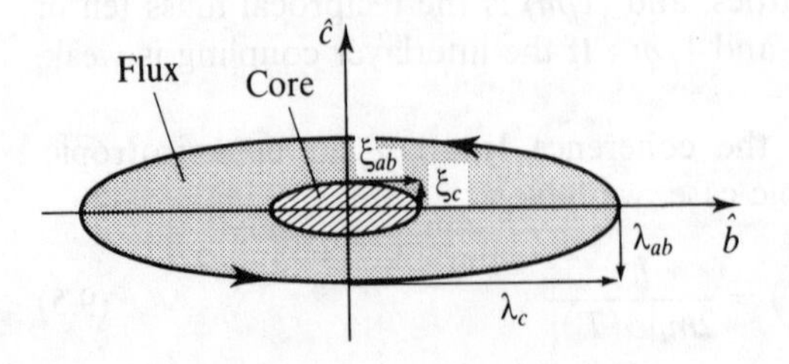

FIGURE 9.1
Schematic cross section of a vortex along the **a** axis in an anisotropic superconductor. The various dimensions are related by $\gamma = \lambda_c/\lambda_{ab} = \xi_{ab}/\xi_c = (m_c/m_{ab})^{1/2}$.

Here we have introduced the conventional dimensionless anisotropy parameter γ. To give a feel for the magnitude of the anisotropy of these materials, we note that the value of the mass ratio m_c/m_{ab} is found to be approximately 50 for YBCO and $\gtrsim 20,000$ for BSCCO,[8] corresponding to γ values of ~ 7 and $\gtrsim 150$. This great anisotropy is one of the decisive factors in making the high-temperature superconductors act so differently from the classic superconductors.

Within the anisotropic GL approximation, the angular dependence interpolating H_{c2} between the limiting values given in (9.7) can be worked out by an anisotropic generalization of the harmonic oscillator calculation used in Sec. 4.8. The result is the simple ellipsoidal form:

$$\left(\frac{H_{c2}(\theta)\sin\theta}{H_{c2\|c}}\right)^2 + \left(\frac{H_{c2}(\theta)\cos\theta}{H_{c2\|ab}}\right)^2 = 1 \tag{9.9}$$

or, equivalently,

$$H_{c2}(\theta) = \frac{H_{c2\|ab}}{(\cos^2\theta + \gamma^2\sin^2\theta)^{1/2}} \tag{9.9a}$$

where θ is the angle between the magnetic field *and the ab plane*.

Recent work of Blatter, Geshkenbein, and Larkin[9] has made the important point that, instead of solving each similar problem a second time for the more difficult anisotropic case, a very general scaling approach can be used to map almost all of the results obtained for isotropic superconductors to anisotropic ones, *so long as the continuum GL approximation is applicable*. One begins by introducing rescaled coordinates, vector potential, and magnetic field by

$$\mathbf{r} = (\tilde{x}, \tilde{y}, \tilde{z}/\gamma) \qquad \mathbf{A} = (\tilde{A}_x, \tilde{A}_y, \gamma\tilde{A}_z) \qquad \mathbf{B} = (\gamma\tilde{B}_x, \gamma\tilde{B}_y, \tilde{B}_z) \tag{9.10a}$$

which makes isotropic the gauge-invariant derivative term, at the expense of introducing anisotropy in the magnetic energy terms. But if we restrict ourselves to the regime in which the flux density is great enough so that the vortices overlap sufficiently to make the field quite uniform in space (i.e., $B \gg H_{c1}$), the macroscopic magnetic energy can be readily computed by using this average field, which basically just fixes the average density of vortices. Proceeding along this line, Blatter et al. obtain their scaling rule, which can be written as

$$Q(\theta, H, T, \xi, \lambda, \gamma, \delta) = s_Q\tilde{Q}(\epsilon_\theta H, \gamma T, \xi, \lambda, \gamma\delta) \tag{9.10b}$$

where θ is the angle between the field and the *ab* plane and δ is the scalar disorder strength. In (9.10*b*), Q is the desired quantity for which the isotropic result $\tilde{Q}$ is known. The scaling factors are defined by $\gamma^2 = m_c/m_{ab} > 1$ and

[8]This value of γ for BSCCO is a bit schematic because the effect of the layered structure is so strong that the anisotropic GL model does not give a quantitative fit to data.

[9]G. Blatter, V. B. Geshkenbein, and A. I. Larkin, *Phys. Rev. Lett.* **68**, 875 (1992).

$\epsilon_\theta^2 = \gamma^{-2}\cos^2\theta + \sin^2\theta$, while $s_Q = 1/\gamma$ for the volume, energy, temperature, and action, but $s_Q = 1/\epsilon_\theta$ for the magnetic field. In (9.10*b*), ξ and λ refer to the values in the *ab* plane. It is easily verified that (9.10*b*) reproduces the result (9.9). Later in this chapter, we shall use (9.10) to analyze the effect of anisotropy on other physical parameters.

9.2.2 Crossover to Two-Dimensional Behavior

Sufficiently near T_c, $\xi_c(T) \approx \xi_c(0)(1-t)^{-1/2}$ will always be large enough to justify the GL approximation discussed in the preceding section. But as one lowers the temperature, $\xi_c(T)$ shrinks toward a limiting value. If that limit is smaller than the interplanar distance, it is clear that the smooth variation assumption used to obtain the GL limit (9.4) will break down at some intermediate temperature T^*. At lower temperatures new features can be expected to appear as the 3-D continuum approximation is replaced by the 2-D behavior of the individual layers. Aspects of this crossover behavior were studied by Klemm et al. and Prober et al. in connection with experiments on the layered dichalcoginides. According to the simplest Lawrence-Doniach model, in a strictly parallel field, $H_{c2\|ab}$ is predicted to diverge at that temperature T^* for which $\xi_c(T^*) = s/\sqrt{2}$, if such a temperature exists. At lower temperatures the vortex cores lie between the layers, and each layer, treated as a 2-D superconductor of zero thickness, has an infinite critical field. Of course, this unphysical infinity is eliminated by taking into account the finite layer thickness, pair breaking due to Pauli paramagnetism, and spin-orbit coupling effects.

To see how the simple model works out, let us apply the general LD equation (9.3) to find the condition for the second-order phase transition $H_{c2}(T)$. As in the solution for H_{c2} in the isotropic case treated in Sec. 4.8, we seek the solution of the linearized version of (9.3) with the lowest eigenvalue, which will determine the highest field value for which superconductivity can exist for a given $\alpha(T)$. We take **H** to lie along **y** (i.e., parallel to the *ab* planes) and for convenience represent it by the vector potential $A_z = Hx$. By analogy with Sec. 4.8, we expect that, with this gauge choice for **A**, the lowest-energy solution will depend on x only and be independent of y and z (i.e., the layer index n). With these assumptions, and using (9.5), we can reduce (9.3) to the eigenvalue equation

$$-\frac{d^2\psi}{dx^2} + \frac{2m_{ab}}{m_c s^2}\left[1 - \cos\frac{2\pi Hsx}{\Phi_0}\right]\psi = \frac{1}{\xi_{ab}^2(T)}\psi \tag{9.11}$$

First, we check that this general equation reduces properly in the GL continuum limit. We can do that by letting the interplanar spacing $s \to 0$. Then we can expand the cosine about $x = 0$ and obtain

$$\frac{d^2\psi}{dx^2} + \frac{m_{ab}}{m_c}\left(\frac{2\pi Hx}{\Phi_0}\right)^2 \psi = \frac{1}{\xi_{ab}^2(T)}\psi \tag{9.12}$$

By comparison with Sec. 4.8, we see that this implies a critical field

$$H_{c2} = \frac{\Phi_0}{2\pi\xi_{ab}^2}\left(\frac{m_c}{m_{ab}}\right)^{1/2} = \frac{\Phi_0}{2\pi\xi_{ab}\xi_c} \tag{9.13}$$

which agrees with (9.7*b*) as it should.

In the other limit, when $m_c s^2 \to \infty$, so that the layers are completely decoupled, the second term on the left-hand side of (9.11) drops out, and an acceptable uniform solution to the linearized equation can only be found at T_c, where $\xi_{ab}(T) \to \infty$. If, instead, we let $H \to \infty$, then the cosine term in (9.11) oscillates so rapidly as to average effectively to zero on the relevant (infinite) length scale. In that case, (9.11) becomes

$$-\frac{d^2\psi}{dx^2} = \left[\frac{1}{\xi_{ab}^2(T)} - \frac{2m_{ab}}{m_c s^2}\right]\psi \tag{9.14}$$

showing that the uniform infinitesimal solution found at T_c for separated layers can be found at $H = \infty$ if $\xi_{ab}^2 = m_c s^2/2m_{ab}$, or, using (9.8), $\xi_c(T) = s/\sqrt{2}$. This result fixes the criterion for going from 3-D to 2-D solutions on individual planes of zero thickness and an infinite critical field in this approximation. We can work out an improved approximation for the case where H is large but not infinite, so that the cosine does not average out completely. The result is

$$H_{c2,ab}(T) \approx \frac{(\Phi_0/2\pi s^2)(m_{ab}/m_c)^{1/2}}{[1 - s^2/2\xi_c^2(T)]^{1/2}} \tag{9.15}$$

This expression diverges as $(T - T^*)^{-1/2}$ as T is reduced toward T^*, the crossover temperature where $\xi_c(T^*) = s/\sqrt{2}$. Of course, the divergence to *infinity* is an artifact of this simple model, which ignores Pauli paramagnetism, finite layer thickness, etc. Nonetheless, it gives a useful qualitative guide to the difference in behavior which stems from the existence of discrete layers if ξ_c approaches the interlayer spacing s. It should be noted, however, that we have been restricting our attention to finding critical fields by using the linearized equations. In other contexts, the relevant length scale may exceed ξ_c, and the anisotropic GL approximation may be more widely usable.

9.2.3 Discussion

Although the Lawrence-Doniach theory was created to describe the properties of naturally occurring layered superconductors, the model has been tested more completely by fabricating artificial layered materials, in which the separation and thickness of the layers can be controlled over a wider range. A classic example is the work of Ruggiero et al.[10] on Nb-Ge multilayers, illustrated in Fig. 9.2. By

[10] S. T. Ruggiero, T. W. Barbee, and M. R. Beasley, *Phys. Rev. Lett.* **45**, 1299 (1980).

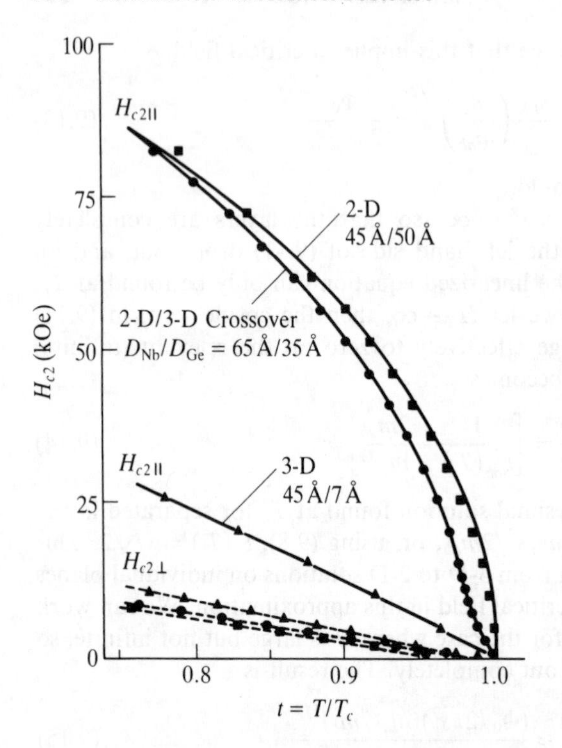

FIGURE 9.2
Upper critical fields of layered Nb/Ge composites with layer thickness D_{Nb} and D_{Ge} as indicated. Decreasing the Ge thickness effects the progression from anisotropic 3-D behavior $[H_{c2} \sim (T_c - T)]$ to "crossover" to "decoupled," or 2-D behavior with $H_{c2} \sim (T_c - T)^{1/2}$. The solid lines are from the Lawrence-Doniach model. ($H_{c2\perp}$ is essentially independent of the thickness parameters since it is determined solely by the coherence length in the plane.) [*After Ruggiero et al.*, Phys. Rev. Lett. *45, 1299 (1980).*]

systematically increasing (from 7 to 50 A) the thickness D_{Ge} of the Ge layers separating the 2-D ($D_{Nb} \approx 50\,A \ll \xi_{Nb}$) Nb superconducting layers, they were able to demonstrate: (1) anisotropic 3-D behavior $[H_{c2} \propto (T_c - T)]$, (2) 2-D behavior $[H_{c2} \propto (T_c - T)^{1/2}]$, and (3) crossover behavior, in which H_{c2} starts as $(T_c - T)$ very near T_c, where ξ_c is large and crosses over to the $(T_c - T)^{1/2}$ dependence at lower temperatures where ξ_c is smaller than s. In an important recent extension of this work to simulate the high-temperature superconductors intentionally, White et al.[11] prepared multilayer samples of superconducting $Mo_{77}Ge_{23}$ separated by insulating layers of Ge. By varying D_{Ge}, they were able to obtain mass ratios similar to those found in YBCO and in BSCCO, and they found resistive transitions very similar to those found in these materials.

Also relevant to the high-temperature superconductors is work[12] on samples grown in the form of superlattices with alternating layers of superconducting YBCO and insulating PrBCO. This work showed, e.g., that compared to a T_c of 90 K for bulk YBCO, single-unit cell layers of YBCO have $T_c \approx 20$ K when separated by many layers of PrBCO, but T_c rises to ~56 K when the intervening

[11] W. R. White, A. Kapitulnik, and M. R. Beasley, *Phys. Rev. Lett.* **66**, 2826 (1991).

[12] See, e.g., D. H. Lowndes, D. P. Norton, and J. D. Budai, *Phys. Rev. Lett.* **65**, 1160 (1990).

layers of PrBCO are only a single-unit cell thick. These depressions of T_c can be accounted for qualitatively by application of the linearized version of (9.2). The same argument indicates that much less reduction would be expected for BSCCO because of its smaller value of ξ_c.

To conclude this section, let us review some numerical implications of our results when applied to YBCO and BSCCO. Quantitative comparisons between the LD theory and the real materials are problematic because the crystallographic unit cell typically contains two or more inequivalent CuO_2 planes which, in principle, would require a generalization of the LD model; thus, even the appropriate choice of interplane spacing s in the model is not well defined. Moreover, it has proved difficult to obtain a consensus on accurate experimental values of ξ_i because they are inferred from $H_{c2}(T)$ data, and H_{c2} is poorly defined because of fluctuation rounding of the transition in high-temperature superconductors. Roughly speaking, we can take[13] $\xi_{ab}(0) \approx 20\,\text{A}$ as representative for all the layered compounds. Using (9.8) and mass ratios of 50 and 20,000 for YBCO and BSCCO, respectively, this implies $\xi_c(0)$ values of roughly 2.8 and 0.1 A, respectively. Taking an interlayer distance of ~10 A, and estimating the crossover temperature T^* by setting $\xi_c(T^*) = s/\sqrt{2}$, we find $T^*/T_c = 0.84$ and 0.999, respectively. This implies that YBCO acts like a 3-D GL superconductor down to about 78 K, whereas BSCCO should cross over to two-dimensional behavior in less than 0.1 K below T_c in fields exactly parallel to the *ab* plane.

Although not rigorous or precise, these simple numbers account for much of the difference between the magnetic properties of YBCO and BSCCO. For example, for YBCO, the resistive transition field typically displays the ellipsoidal angular dependence (9.9) expected for H_{c2} from the 3-D anisotropic GL model, whereas, except *very* near T_c, for BSCCO, the resistive transition field shows[14] a cusp at $\theta = 0$, which is better described by the formula (4.70), which was derived for an isolated 2-D thin film. Further evidence comes from torque measurements of the equilibrium magnetization: For YBCO, the 3-D angular dependence holds down to 80 K, below which anomalous behavior sets in at angles within a few degrees of the *ab* plane[15]; for BSCCO, results are sample dependent, but it appears[16] that the discrete layers play an important role even very near T_c.

---

[13] A critical discussion of the difficulties in determining H_{c2} and ξ for YBCO is given by D. K. Finnemore, "Critical Fields in High-Temperature Superconductors," and the appended discussion, in K. S. Bedell, M. Inui, D. Meltzer, J. R. Schrieffer, and S. Doniach (eds.), *Phenomenology and Applications of High Temperature Superconductors*, Addison Wesley, Reading, MA (1992), p. 164. This work suggests taking $\xi_{ab}(0) \approx 15\,\text{A}$, and $\xi_c(0) \approx 2\,\text{A}$ for YBCO. For BSCCO, less definitive results are available because effects of fluctuations and lower dimensionality are even more severe than for YBCO.

[14] R. Marcon et al. *Phys. Rev.* **B46**, 3612 (1992); M. J. Naughton et al., *Phys. Rev.* **B38**, 9280 (1988).

[15] D. E. Farrell et al., *Phys. Rev. Lett.* **64**, 1573 (1990).

[16] J. C. Martinez et al., *Phys. Rev. Lett.* **69**, 2276 (1992).

It is interesting to follow the notion of an isolated thin film a bit further by estimating the parallel critical field of a thin film of atomic thickness. The conventional result (4.51), namely, $H_{c\parallel} = 2\sqrt{6}H_c\lambda/d$, where d is the film thickness, can be rewritten by using (9.6) in the form $H_{c\parallel} = [\sqrt{3}\Phi_0/\pi d\xi_{ab}(0)](1-t)^{1/2}$. ($\xi_{ab}$ is appropriate since the screening currents flow in the plane.) Comparing this expression with (9.7*b*) for $H_{c2\parallel ab}$, we find that they differ by only a factor of $2\sqrt{3}\xi_c/d$, which is of the order of unity well below T_c if $\xi_c \sim 0.4$ A and $d \sim 1$ A. Thus, the low-temperature limit of $H_{c2\parallel ab}$ estimated from experimental data on the BSCCO-type materials is close to the value that we might estimate for a superconducting film of atomic layer thickness.

9.3 MAGNETIZATION OF LAYERED SUPERCONDUCTORS

We now summarize the implications of the anisotropy of these layered superconductors for the observable magnetization. In doing so, we shall first treat the uniaxial 3-D anisotropic GL regime because of its mathematical simplicity, and because it is quantitatively applicable to YBCO down to ~80 K and at least qualitatively applicable even to BSCCO. We follow this with a brief discussion of the lock-in transition and other predicted consequences of the discrete layered structure.

9.3.1 The Anisotropic Ginzburg-Landau Regime

Because of the anisotropy, the vector quantities **H**, **B**, and **M** are no longer collinear unless they lie along a principal axis. This requires a generalization of the approach used in Chap. 5 in treating isotropic type II superconductors. Fortunately, in the broad field range where $H_{c1} \ll H \ll H_{c2}$, one can assume both nearly a constant order parameter (because the high values of κ imply that the *cores* are small and widely separated) and nearly uniform field penetration (because of *vortex* overlap). These assumptions allow the use of the London approximation, as in Sec. 5.2, and also assure that $B \approx H \gg |M| \sim H_{c1}$. As shown by Kogan and collaborators,[17] the latter inequality allows the problem to be simplified by neglecting demagnetizing fields (which are of order M and depend on the shape of the sample), and also, to a good approximation, allows neglect of all terms that are higher than linear in M. By starting with the London equations with anisotropic mass tensor instead of the isotropic (5.32), these authors traced through the analysis leading to an anisotropic generalization of (5.36). In this way, they showed that the Helmholtz free energy per unit volume can be written (after a change in notation) as

[17] V. G. Kogan, *Phys. Rev.* **B24**, 1572 (1981); L. J. Campbell, M. M. Doria, and V. G. Kogan, *Phys. Rev.* **B38**, 2439 (1988); V. G. Kogan, *Phys. Rev.* **B38**, 7049 (1988).

$$F = \frac{B^2}{8\pi} + \frac{H^*}{4\pi}(B_{ab}^2 + \gamma^2 B_c^2)^{1/2} \tag{9.16}$$

where B_{ab} and B_c, respectively, are the components of **B** in the plane and perpendicular to it, and where

$$H^*(\theta) \approx \frac{\Phi_0}{8\pi\gamma\lambda_{ab}^2} \ln \frac{\eta H_{c2}(\theta)}{B} \tag{9.16a}$$

is a quantity of order H_{c1} with only the weak logarithmic dependence on the angle. ($\eta \sim 1$ is a fitting parameter.) Given this expression for $F(\mathbf{B})$, we can find $\mathbf{H} = 4\pi(\partial F/\partial \mathbf{B}) = \mathbf{B} - 4\pi\mathbf{M}$, so that

$$H_{ab} = B_{ab} + H^* \frac{B_{ab}}{(B_{ab}^2 + \gamma^2 B_c^2)^{1/2}} \tag{9.17a}$$

$$H_c = B_c + H^* \frac{\gamma^2 B_c}{(B_{ab}^2 + \gamma^2 B_c^2)^{1/2}} \tag{9.17b}$$

[In (9.17), we have omitted a factor $[1 - \ln^{-1}(H_{c2}/B)]$ which would appear in the second terms from the formal differentiation of the logarithmic term in H^* since that term itself is only an approximation.]

From these expressions we confirm that indeed $M = (B - H)/4\pi \sim H^* \ll H$. More striking is the fact that the magnetization is not parallel to the field but, rather, is strongly tilted away from the plane. Quantitatively, we can write $M_{ab}/M_c = \gamma^{-2}(B_{ab}/B_c)$, so that the angle θ_M made by M relative to the planes is given by

$$\tan\theta_M = \gamma^2 \tan\theta \tag{9.18}$$

where θ is the angle between the field and the planes. Since $\gamma^2 \equiv m_c/m_{ab} \sim 50 - 20{,}000$, this result (9.18) shows that the magnetic moment is nearly perpendicular to the planes even for fields applied within a few degrees of the planes. The physical reason for this result is that the mass anisotropy reflects the fact that the currents prefer to circulate in the planes, forming a stack of "pancake vortices" along each flux line, which produces a moment perpendicular to the planes, as contrasted with a moment opposite the field direction. This geometry is sketched in Fig. 9.3 for a field inclined at an arbitrary angle to the planes. A consequence of this anisotropy is that the response is *not* purely diamagnetic, i.e., longitudinally opposed to the applied field, as in an isotropic superconductor. There is also a transverse component of the moment M_T, perpendicular to the applied field, which is of the same order as the diamagnetic response M_L and may even exceed it, especially when the field direction is near the *ab* plane. Tuominen et al.[18] have shown that the ratio M_T/M_L is a function *only* of the mass ratio γ and the field angle θ. They used this observation as the basis for determining the

[18] M. Tuominen, A. M. Goldman, Y. Z. Chang, and P. Z. Jiang, *Phys. Rev.* **B42**, 412 (1990).

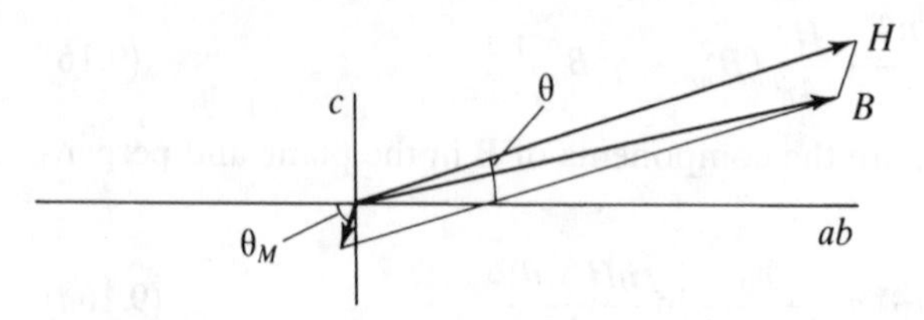

FIGURE 9.3
Schematic diagram illustrating the fact that **M** is not collinear with **B** unless **B** lies along a principal axis, and that $|\mathbf{M}| \ll |\mathbf{B}|$, so that $\mathbf{B} \approx \mathbf{H}$.

mass ratio from measurements of M_T/M_L.

These predictions have been tested particularly carefully by measuring the torque per unit volume, $\tau = \mathbf{M} \times \mathbf{H}$, which results from the noncollinearity of the moment and field. Evaluating the cross product, we see that the magnitude of the torque is

$$\tau(\theta) = V(M_c H_{ab} - M_{ab} H_c) = V(\gamma^2 - 1)\frac{HH^*(\theta)}{4\pi}\frac{\sin\theta\cos\theta}{(\cos^2\theta + \gamma^2\sin^2\theta)^{1/2}} \tag{9.19}$$

where V is the sample volume. The excellent agreement of this predicted angular dependence with experimental data of Farrell et al.[19] on YBCO is shown in Fig. 9.4*a*. Note that the torque vanishes at the principal directions $\theta = 0, \pi/2$. Since the expression (9.16) shows that the lowest free energy occurs when the field lies in the *ab* plane ($\theta = 0$, so that $B_{ab} = B, B_c = 0$), it follows that the sense of the torque must be such as to turn the sample toward this orientation, in which the **c** axis is perpendicular to the field. The orientation with the **c** axis parallel to the field is a point of unstable equilibrium.

Insofar as one can neglect the logarithmic angular dependence of $H^*(\theta)$, one can make a few simple observations about the angular dependence of the torque in (9.19). For small anisotropy ($\gamma \approx 1$), the torque is small [$\sim(\gamma^2 - 1)$] and reaches its maximum near $\pi/4$, where $\sin\theta\cos\theta$ is maximum. For large anisotropy ($\gamma \gg 1$), the scale of the torque becomes independent of γ (recall that $H^* \sim 1/\gamma$) and increases approximately as $\cos\theta$ as $\theta \to 0$. Differentiation of (9.19) shows that the torque reaches a maximum at $\theta_m \sim \gamma^{-1/2}$ and then drops to half-maximum at $\theta_{1/2} \sim (1/\sqrt{3})\gamma^{-1} \ll 1$ on the way to zero torque at $\theta = 0$. Note that the scaling of θ_m and $\theta_{1/2}$ with different powers of γ implies that the torque curve changes shape as γ gets larger, with the final plunge to zero becoming steeper compared to the position of the maximum torque.

Although these estimates for θ_m and $\theta_{1/2}$ capture the qualitative change in the shape of $\tau(\theta)$ with increasing anisotropy, the angular dependence of H^* must

[19] D. E. Farrell et al., *Phys. Rev. Lett.* **64**, 1573 (1990); D. E. Farrell et al. *Phys. Rev. Lett.* **63**, 782 (1989).

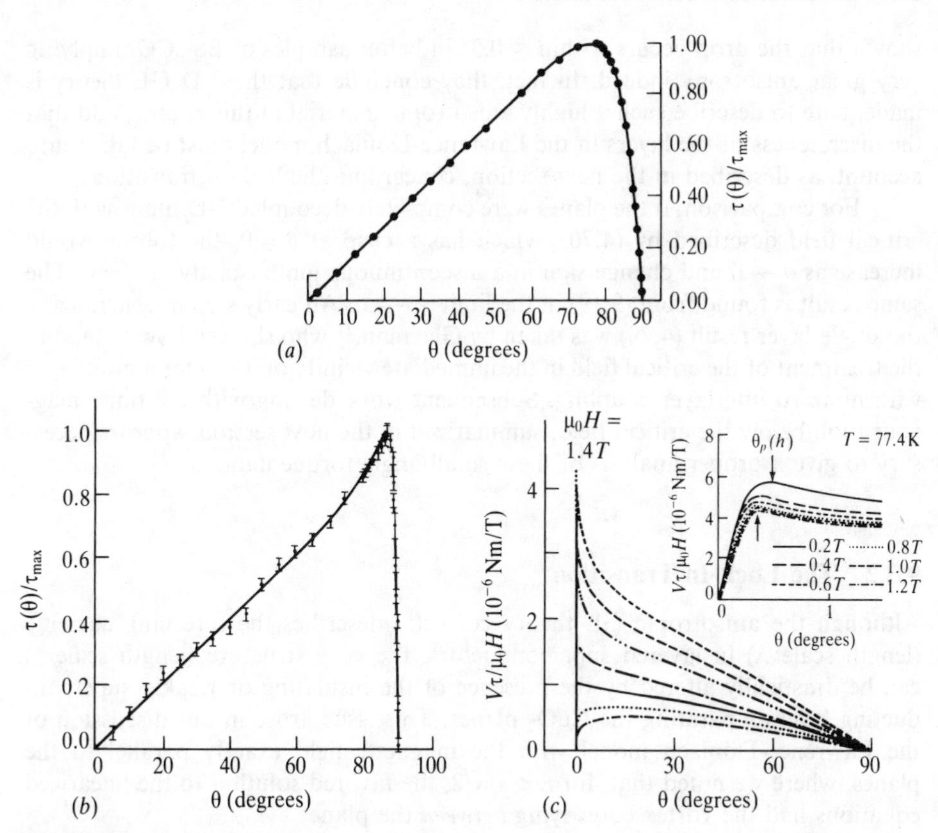

FIGURE 9.4
Comparison of measured normalized torque data (at $B = 1$T) with prediction of (9.19) for (*a*) YBCO and (*b*) BSCCO. [The data in (a) are taken from Farrell et al., *Phys. Rev. Lett.* **64**, *1573 (1990)*; they were taken at 80 K and fitting parameters are $\gamma = 7.9$ and $H_{c2\|} = 23\,T$. The data in (*b*) are taken from Farrell et al., *Phys. Rev. Lett.* **63**, *782 (1989)*; they were taken at 78.5 K and fitting parameters are $\gamma = 55$ and $H_{c2\|} = 35$T]. (*c*) Angular dependence of scaled torque of a BSCCO crystal at 1.4T and various temperatures. [*After Martinez et al.*, Phys. Rev. Lett. **69**, *2276 (1992)*.] From top to bottom, the plots correspond to $T = 77.4$, 78.6, 79.7, 81.5, 82.6, 83.1, and 84.4 K. The inset shows the behavior with **H** near the *ab* plane for various field values at 77.4 K, showing that the torque drops from maximum to zero in ~0.3° of rotation. [*Note*: In parts (*a*) and (*b*), θ is measured from the *c* direction and is equivalent to $(90° - \theta)$ in the text, whereas in part (*c*), θ is measured from the *ab* plane, as in the text.]

be included in any quantitative fit since it is the cause of the continued rise in τ even at small angles where $\cos\theta$ is essentially constant at unity. In fact, in early work with BSCCO, the drop in torque was found to take place within 1 to 2° of $\theta = 0$. [See Fig. 9.4*b*, but note that θ in the figure is the complement $(90° - \theta)$ of the angle used in the text.] Later work by Martinez et al.[20] (see Fig. 9.4*c*) has

[20] J. C. Martinez, S. H. Brongersma, A. Koshelev, B. Ivlev, P. H. Kes, R. P. Griessen, D. G. de Groot, Z. Tarnavski, and A. A. Menovsky, *Phys. Rev. Lett.* **69**, 2276 (1992); also, L. Mihaly, private communication.

shown that the drop occurs within $\leq 0.3^\circ$ in better samples of BSCCO, implying very great anisotropy indeed. In fact, they conclude that the 3-D GL theory is inadequate to describe such a highly anisotropic material in this regime, and that the discreteness of the layers in the Lawrence-Doniach model must be taken into account, as described in the next section, concerning the lock-in transition.

For comparison, if the planes were completely decoupled 2-D films, with the critical field described by (4.70), which has a cusp at $\theta = 0$, the torque would increase as $\theta \to 0$ and change sign in a discontinuous jump exactly at $\theta = 0$. The same result is found from (9.19) in the limit $\gamma \to \infty$. An early step in generalizing the single-layer result (4.70) was taken by Glazman,[21] who showed how to modify the treatment of the critical field in the immediate vicinity of $\theta = 0$ for a multilayer with nonzero interlayer coupling. Subsequent work dealing with the finite magnetization below the critical field, summarized in the next section, appears necessary to give a proper analysis of these small-angle torque data.

9.3.2 The Lock-In Transition

Although the anisotropic GL theory correctly describes the screening currents (length scale λ) in layered superconductors, the core structure (length scale ξ) can be drastically altered by the presence of the insulating or weakly superconducting layers separating the CuO_2 planes. This issue arose in our discussion of the Lawrence-Doniach model with the magnetic field exactly parallel to the planes, where we noted that, for $\xi < s/\sqrt{2}$, the favored solution to the linearized equations had the vortex cores lying *between* the planes.

Feinberg and Villard[22] have given a phenomenological treatment of the more general case in which H is applied at a finite angle θ_H from the planes. They predict that there is a finite lock-in angle θ_c, such that when $\theta_H < \theta_c$, the flux lines (and **B**) run *strictly parallel* to the planes, remaining "locked in" between the layers. This implies a "transverse Meissner effect," with the component of **B** perpendicular to the planes being *zero*, not just reduced in magnitude by the magnetization, as in the continuum GL theory. Their calculation takes into account the lower line tension and the absence of pancake vortices when the core runs between the planes, as well as the dependence of magnetic Gibbs energies on the angle between **B** and **H**.

This problem has been addressed more recently by Bulaevskii, Ledvij, and Kogan,[23] who obtained qualitatively similar results. They find the plausible result than an applied field first penetrates the planes when its perpendicular component $H \sin\theta$ exceeds a threshold value H_J, which is of the order of $H_{c1\|c}$. In other

[21] L. I. Glazman, *Sov. Phys. JETP* **66**, 780 (1987).

[22] D. Feinberg and C. Villard, *Phys. Rev. Lett.* **65**, 919 (1990).

[23] L. N. Bulaevskii, M. Ledvij, and V. G. Kogan, *Phys. Rev.* **B46**, 366 (1992).

words, the lock-in angle θ_c should fall roughly as $1/H$ at small angles. Quantitative tests of these predictions are made difficult by the complications of demagnetization effects, field penetration at sample corners, and extrinsic pinning effects. Nonetheless, Martinez et al., were able to show that the torque in BSCCO at 77 K ($T/T_c \approx 0.92$) increased linearly with the *internal perpendicular* field H_z (after correction for the demagnetizing factor of the sample), independent of the absolute field strength or the field orientation, up to $H_z \sim 100$ Oe. This field value agrees well with the value of $H_{c1\|c}$ estimated for this temperature by using nominal values of $\lambda_{ab}(0) \approx 1500$ A and $\kappa \approx 70$ in (5.18). The observed linear increase of torque with H_z up to $\sim H_{c1\|c}$ strongly supports the notion of a transverse Meissner effect for lower fields. Further support for the model comes from ac susceptibility experiments on the organic layered superconductor $(BEDT\text{-}TTF)_2Cu(SCN)_2$ by Mansky, Chaikin, and Haddon,[24] who detected a threshold field which follows the expected dependence $H_{th} \sin\theta \approx H_J$, where H_J has the same temperature dependence as $H_{c1\|c}$ but is about half as large.

On the other hand, in the torque measurement on YBCO shown in Fig. 9.4, no discontinuity of the angular dependence, as would be predicted by this model, is visible at any angle. Presumably, this sharp lock-in phenomenon is washed out by thermal fluctuations in the region of reversible magnetization near T_c where those data were taken, whereas the data of Mansky et al. were mostly taken at lower temperatures, and the fitted H_J was found to go strongly to zero as T approached T_c. This view is borne out by further theoretical work of Bulaevskii, Ledvij, and Kogan,[25] taking into account the fluctuation effects. Their results are found to give a good fit to the torque data of Martinez et al.[26] near the zero angle. Martinez et al. conclude that the extreme anisotropy of BSCCO cannot be adequately described by the anisotropic GL model; only a sample-dependent lower bound $\gamma > 150$ can be given.

9.4 FLUX MOTION AND THE RESISTIVE TRANSITION: AN INITIAL OVERVIEW

Our discussion so far has been greatly simplified by focusing on equilibrium properties treated in the GL mean-field approximation, including anisotropy and planar structure, but ignoring fluctuations and extrinsic flux pinning (due to material inhomogeneties and sample boundaries), and in the absence of transport current. At this level of approximation the thermodynamic phase transition to superconductivity occurs at a sharply defined curve $H_{c2}(T)$ in the (H, T)

[24] P. A. Mansky, P. M. Chaikin, and R. C. Haddon, *Phys. Rev. Lett.* **70**, 1323 (1993).

[25] L. N. Bulaevskii, M. Ledvij, and V. G. Kogan, *Phys. Rev. Lett.* **68**, 3773 (1992).

[26] J. C. Martinez, S. H. Brongersma, A. Koshelev, B. Ivlev, P. H. Kes, R. P. Griessen, D. G. de Groot, Z. Tarnavski, and A. A. Menovsky, *Phys. Rev. Lett.* **69**, 2276 (1992).

plane. However, in this model there would be no sharp change in electrical resistance at this H_{c2} because, in the absence of pinning, the flux-flow resistance of the superconducting phase at H_{c2} joins smoothly with the normal-state resistance. In the classic low-temperature superconductors, pinning is reasonably effective while fluctuation effects are relatively unimportant, so that there is a rather sharply defined resistive transition, in which the resistance drops to unobservably small values in a narrow temperature or field range at $H_{c2}(T)$. In fact, that resistive transition is the traditional experimental method for measuring $H_{c2}(T)$.

The high-temperature superconductors (HTSC) present a remarkably different behavior. The resistive transition broadens markedly in a magnetic field, making ambiguous the identification of $H_{c2}(T)$. Qualitatively, this can be understood as follows. Because T_c is large and v_F is small, the coherence lengths are smaller (even in the clean limit) than in the classic superconductors, resulting in a small coherence volume $\sim\xi^3$. Thus, the characteristic energy F_0 of point pinning centers tends to be somewhat smaller than in classic superconductors, even though the condensation energy density $H_c^2/8\pi$ is greater. Given comparable values of F_0 and temperatures an order of magnitude higher, it is clear that thermally activated processes with rates proportional to $e^{-F_0/kT}$ will be *exponentially* more important in the HTSC. To illustrate the power of this argument, consider a typical pinning energy $F_0 = kT_p$, where $T_p = 800$ K, and a typical attempt frequency $\Omega = 10^8$ per second. Then $\Omega e^{-F_0/kT}$ is $\sim 10^3$ per second at 77 K, but is $\sim 10^{-79}$ per second at 4 K. Thus, relaxation processes which would not occur in the age of the universe at 4 K, will occur in a millisecond at 77 K.

As a consequence of this rapid relaxation, the magnetization of HTSC remains reversible (as described by equilibrium GL theory) over a much wider temperature range below T_c than in the classic superconductors, where hysteretic magnetization associated with screening currents (as in the critical state described in Sec. 5.6) sets in just below T_c. This difference explains why the torque magnetometry experiments described in the previous section work so well in the HTSC compared to the classic superconductors. This wide reversible region below $H_{c2}(T)$ in the H-T plane in LaBCO was pointed out by Müller et al.,[27] who identified the boundary line below which the magnetization showed hysteresis on laboratory time scales as the "irreversibility line," and proposed interpreting it in terms of a superconductive glass state by analogy with similar phenomena observed in spin glasses. This irreversibility line, which they found could be fitted to the approximate form $H_{\mathrm{irr}} \propto (T_c - T)^{1.5}$, was the first of a number of proposals to describe and interpret intermediate stages in the transition to zero resistance and persistent currents.

Since magnetic irreversibility implies persistent currents and hence zero resistance, this line must also mark the locus in the H-T plane at which the resistance has been reduced by *at least* some large factor ($\sim 10^8$) below the normal

[27] K. A. Müller, M. Takashige, and J. G. Bednorz, *Phys. Rev. Lett.* **58**, 1143 (1987).

value, so that typical L/R decay times of currents have reached ~ 1 sec. Yeshurun and Malozemoff[28] showed that the $(T_c - T)^{1.5}$ temperature dependence of the irreversibility line could be explained in terms of an activation energy $U \propto (T - T_c)^{3/2}/B$ for flux motion. Tinkham[29] then showed that the insertion of such an activation energy into the Ambegaokar-Halperin formula (see Sec. 6.3.3) for the linear resistance due to thermally activated phase slippage in an overdamped Josephson junction (analogous to fluxons slipping past each other) yielded a temperature- and field-dependent resistance which gave an impressive phenomenological fit (shown in Fig. 9.5) to the measurements of Iye et al.[30] on

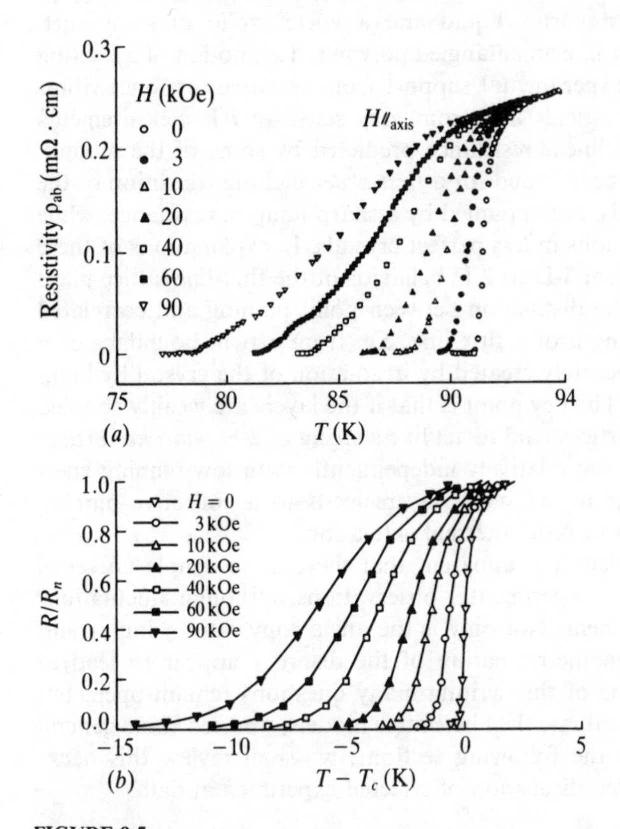

FIGURE 9.5
(*a*) $\rho_{ab}(T)$ of YBCO crystal for various values of H (*reported by Iye et al.*) showing the broadening of the resistive transition by magnetic field. (*b*) Comparison curves computed by a model of Tinkham. [Phys. Rev. Lett. *61*, *1658 (1988)*.] Although agreement is good over the top 90 percent of the resistive transition, the experimental resistance cuts off more sharply as $R \to 0$.

---

[28]Y. Yeshurun and A. P. Malozemoff, *Phys. Rev. Lett.* **60**, 2202 (1988).

[29]M. Tinkham, *Phys. Rev. Lett.* **61**, 1658 (1988).

[30]Y. Iye, T. Tamegai, H. Takeya, and H. Takei, *Jpn. J. Appl. Phys.* **26**, L1057 (1988).

YBCO single crystals. In particular, its uncertain theoretical justification notwithstanding, this version of the thermally activated flux flow (TAFF) model accounted nicely for the hitherto surprising field-dependent broadening of the resistive transition.

Subsequent measurements with higher sensitivity showed, however, that the fit was quantitative only down to $R \sim 0.1R_n$. Below that point the measured value dropped more rapidly than the exponential predicted by the TAFF model, suggesting the likelihood of a crossover to another regime, or perhaps an actual phase transition to a regime with ideally *zero* resistance, rather than an exponentially *small* resistance. A number of authors suggested that this might be a freezing/melting transition between a vortex liquid and a vortex solid of some sort, whether crystalline, glassy, or like an entangled polymer. This notion of a melting or glass transition received experimental support from sensitive torsional vibration experiments in magnetic fields and from very sensitive I-V measurements probing the onset of the nonlinear resistance predicted by some of the models. More recently, evidence has been found for a *first-order* melting transition of the flux-line solid in good crystals, accompanied by a sharp jump in resistance, while the transition may be continuous in less perfect crystals. In explanations of these phenomena, the crossover from 3-D to 2-D behavior of the flux-line lattice plays an important role, as does the distinction between point pinning and correlated pinning over an extended length of a flux line, e.g. from a twin boundary or a columnar defect that is deliberately created by irradiation of the crystal by high-energy beams of heavy ions. The key point is that if the layers are weakly coupled (large value of γ), the 3-D vortices tend to act like a string of 2-D *pancake vortices* in the individual layers, moving relatively independently with low-pinning energies, whereas correlated pinning on many layers leads to a collective pinning energy that is strong enough to resist thermal activation.

From this quick overview it is apparent that there is no simple universal explanation of the numerous experimental observations, although fluctuations play a central role in all of them. Not only is the anisotropy ratio γ important, but also the strength and geometric nature of the disorder appear to lead to different regimes. At the time of this writing many questions remain open, but in order to understand and evaluate the alternative models, we need some general background in any case. In the following sections, we shall review this background and give more detailed discussion of selected experimental data.

9.5 THE MELTING TRANSITION

We start our more complete description of the properties of the HTSC by investigating the implications of thermal vibrations of the flux lines away from their ideal Abrikosov lattice configuration of rigidly straight lines parallel to the field. At this point we still consider an ideal homogeneous sample and defer a discussion of pinning effects until the next section. By analogy with other systems, one might expect that the long-range crystalline order would eventually be destroyed discontinuously by a first-order melting transition as the temperature is raised. In the mean-field treatment of the classic superconductors, this melting transition is obscured because it occurs so near $T_{c2}(H)$, where the magnitude of $|\psi|^2$ goes to zero. It was shown by Brezin et al.,[31] however, that the first-order nature of the transition is made manifest if fluctuation effects are taken into account.

In the complete absence of pinning, a superconductor is expected to be highly resistive whether the flux lines are in a solid or liquid state. However, pinning is expected to be more effective in reducing resistance if the pinning force can be transmitted from one flux line to another by the rigidity of a flux-line solid. Accordingly, such phenomena as the more rapid drop in resistance deeper into the resistive transition have been attributed to a freezing of the flux-line liquid, as modified by the presence of inhomogeneity and pinning. The great practical importance of such a drop in resistance has motivated much work devoted to clarifying the reality of such a melting/freezing transition.

On general grounds, the total entropy reduction on freezing is expected[32] to be of the order of Boltzmann's constant k_B per *vortex* per layer. (If the transition is "strongly" first order, most of the entropy change occurs right at the transition temperature; if "weakly" first-order, much of it is taken up on the approach to the transition.) Because of the *mesoscopic* size of the vortices which are ordering, their number of degrees of freedom, and hence the associated entropy reduction, is very small compared to the entropy drop associated with the ordering of the truly *microscopic* electronic degrees of freedom into the superconducting state, which begins at $T_{c2}(H)$. Consequently, the melting transition will be difficult to observe thermodynamically, by specific heat measurements, although evidence for a thermodynamic transition has been detected in magnetic measurements.[32a] Nonetheless, it should have relatively dramatic consequences for *transport* properties.

In outline, in this section, we shall do three things: (1) Work out a simple model which predicts the melting line $B_M(T)$ in an anisotropic 3-D superconductor, (2) summarize the experimental evidence showing that such a transition line is observed in YBCO, and (3) work out the conditions under which the melting becomes a 2-D process in BSCCO.

9.5.1 A Simple Model Calculation

Before considering experimental data in any detail, let us briefly review the theoretical basis for the prediction of flux-line melting in ideal crystals. Although a careful treatment of the melting transition requires integrating over

---

[31] E. Brezin, D. R. Nelson, and A. Thiaville, *Phys. Rev.* **B31**, 7124 (1985).

[32] For a more detailed argument, see R. E. Hetzel, A. Sudbø, and D. A. Huse, *Phys. Rev. Lett.* **69**, 518 (1992).

[32a] H. Pastoriza et al., *Phys. Rev. Lett.* **72**, 2951 (1994); E. Zeldov et al., *Nature* **375**, 373 (1995).

an entire Fourier spectrum of fluctuations, using the elastic constants of the flux-line lattice (FLL) to take into account the long-range interaction and continuity of the flux lines,[33] it is perhaps more illuminating to start with an elementary argument based on nearest neighbor interactions. In this model calculation, we shall use the Lindemann criterion to estimate a melting temperature by finding the temperature at which the amplitude of the thermal vibrations of a typical flux line equals some fraction c_L of the interline spacing. Such an approach gives a way of estimating the melting temperature T_m, but, of course, it does not answer the question of whether the transition is first order or continuous. For simplicity, we develop our argument by considering an isotropic superconductor, and then insert the important effects of anisotropy by using the prescription (9.10*b*) of Blatter et al.

As shown in Sec. 5.2, the force between two parallel flux lines oriented along the z axis is repulsive, and its x component can be written as $f_{2x} = (\Phi_0/4\pi)[\partial h_1(\mathbf{r}_2)/\partial x_2]$, where $h_1(\mathbf{r}_2)$ is the field of vortex 1 at the position of vortex 2. In a regular array of vortices in which each vortex is at a center of inversion symmetry, these forces exactly cancel, and there is no net force. If the vortex line of interest is displaced by δx with respect to the rest of the array, the cancellation is no longer exact, and there is a net force proportional to δx. This must be a *restoring* force if the FLL is stable, as is the triangular Abrikosov lattice. The magnitude of the force (per unit length) on the line will be $(\Phi_0/4\pi)\Sigma[\partial^2 h_i(\mathbf{r})/\partial x^2]\,\delta x$. From the expressions for $h(r)$ in (5.14), we see that at very low flux densities these forces are exponentially small ($\sim e^{-r/\lambda}$), and the FLL is very soft. However, in the usual regime ($H_{c1} \ll H \ll H_{c2}$), where the vortices are overlapping but the cores are not, $h(r)$ falls as $(\Phi_0/2\pi\lambda^2)\ln(\lambda/r)$, and the second derivative is $(\Phi_0/2\pi\lambda^2 r^2)$. Summing vectorially over the nearest neighbors, inserting the factor of $(\Phi_0/4\pi)$, and using the fact that $r^2 = a_\Delta^2 = 2\Phi_0/\sqrt{3}B$ in the triangular lattice, we estimate the restoring force constant per unit length to be

$$K \approx (\sqrt{3}\Phi_0/4\pi^2\lambda^2)B \approx H_{c1}B \tag{9.20}$$

To go further, we must identify the length of the displaced segment. If the sample is sufficiently thin, the line might displace rigidly, and the appropriate length would then be the thickness of the sample. Of, if we were applying this analysis to a pancake vortex in a single atomic layer in a totally decoupled HTSC, we would take the interlayer spacing s. In a bulk sample, however, a length shorter than the sample dimension must give the dominant effect. This dominant length scale can be found by assuming a continuous displacement of amplitude A over a length L_z and by minimizing the sum of the elastic displacement energy $\sim KA^2L_z$ and the energy $\sim\epsilon_1 A^2/L_z$ required to stretch the line against the line tension $\epsilon_1 \sim (\Phi_0/4\pi\lambda)^2$. [See (5.17).] It is easily seen from this argument that the

[33] A. Houghton, R. A. Pelcovits, and A. Sudbø, *Phys. Rev.* **B40**, 6763 (1989).

optimal length L_z is $\sim(\epsilon_1/K)^{1/2}$. Substituting the quoted value of ϵ_1 and that of K from (9.20), we find that the optimal length L_z is of the order of $(\Phi_0/B)^{1/2}$, i.e., the FLL spacing. Inserting this optimal length, and equating the energy increase to $\sim kT$, we find a mean-square thermal vibration amplitude $A^2 \sim kT(K\epsilon_1)^{-1/2}$. Finally, if we define the melting temperature T_m to be that at which $A^2 = c_L^2 a_\Delta^2$, where $c_L \approx 0.15$ is the empirical Lindemann parameter, we obtain

$$kT_m = Cc_L^2 \Phi_0^{5/2} \lambda^{-2} B^{-1/2} \tag{9.21}$$

where C is a numerical constant; a more careful analysis yields $C \approx 1/4\pi^2$.[34]

Despite the numerous approximations involved in obtaining such a simple result as (9.21), the physical basis underlying the functional dependence on λ and B should be reliable, even if the numerical coefficient is less certain. Solving (9.21) for B_m, the melting field at temperature T, we obtain

$$B_m = C^2 c_L^4 \Phi_0^5 (kT)^{-2} \lambda^{-4} \tag{9.22}$$

We now insert the effects of anisotropy, using (9.10*b*), the general prescription of Blatter et al., and obtain

$$B_m = \frac{C^2 c_L^4 \Phi_0^5}{(kT)^2 \lambda_{ab}^4 \gamma (\cos^2\theta + \gamma^2 \sin^2\theta)^{1/2}} \tag{9.22a}$$

Recalling that $\lambda^{-2} \sim (T_c - T)$ in the mean-field approximation near T_c, we find that this implies that near T_c

$$B_m \propto (T_c - T)^2 \tag{9.22b}$$

The fit to data can be extended farther below T_c by using the "two-fluid" temperature dependence $\lambda^{-2} \sim (T_c^4 - T^4)$ or some alternative form, instead of using only its leading term $\sim(T_c - T)$. Equation (9.22*a*) accounts explicitly for the dependences on T and θ of the sharp feature found in the torsional oscillator experiments of Beck et al., discussed in the next section.

Comparing (9.22*a*) with (9.9*a*), we see that the angular dependences of B_m and H_{c2} are the same, so that the ratio of the two is independent of angle, namely,

$$\frac{B_m}{H_{c2}} = \frac{2\pi C^2 c_L^4 \Phi_0^4 \xi_{ab}^2}{(kT)^2 \lambda_{ab}^4 \gamma^2} \tag{9.23}$$

This form shows clearly that the large anisotropy factors γ and the high temperatures are the primary factors which depress B_m below H_{c2} in the HTSC, relative to the classic superconductors. The values of λ_{ab} and ξ_{ab} are not radically different from those found in dirty conventional superconductors. But the fact that the

[34]For completeness, it should be noted that (9.21) was derived specifically for the intermediate field range $H_{c1} \ll H \ll H_{c2}$. For H just above H_{c1}, the restoring force constant K is exponentially rather than algebraically small in $1/B$, leading to a narrow $[\Delta H \sim H_{c2}(T)/\kappa^2 \sim 10^{-4} H_{c2}(T)]$ slice of liquid phase just above $H_{c1}(T)$.

temperatures are typically 10 times higher gives a factor of $\sim 10^{-2}$, and anisotropy contributes factors $1/\gamma^2 \sim 2 \times 10^{-2}$ for YBCO and $\sim 10^{-4}$ or lower for BSCCO. Taken together, these two factors readily account for the striking difference in importance of FLL melting phenomena in HTSC and classic superconductors. In the latter, $B_m(T)$ and $H_{c2}(T)$ are usually indistinguishable within the width of the resistive transition, and (9.23) is not applicable since it was derived on the assumption that $H \ll H_{c2}$.

9.5.2 Experimental Evidence

The first experiments interpreted as giving direct evidence of a possible melting transition of the flux-line solid were the vibrating reed ($f \sim 1$ kHz) experiments of Gammel et al.[35] on HTSC samples in a magnetic field. These showed a rather sharp peak in the damping of the oscillatory motion at a field-dependent temperature $T_m(H)$, qualitatively similar to the irreversibility line in the H-T plane, which these authors identified as a melting transition. Their data are shown in Fig. 9.6, in comparison with the results of the quantitative calculations of Houghton et al.[36] However, it was quickly pointed out that such a peak could also be explained rather naturally by a frequency-dependent thermal depinning crossover. The essential point is that dissipation in general has its peak value when $\omega\tau = 1$, where ω is the oscillation frequency and τ is the relaxation time. Since, as noted earlier, we expect τ to vary exponentially rapidly with T (and even discontinuously near T_m), the condition $\tau(T, H) \approx 1/\omega$ will yield a well-defined locus $T(H)$, only logarithmically dependent on ω, and related to $T_m(H)$. The details would, however, be expected to depend on pinning strengths, and so the loss peak would not necessarily identify the position of a phase transition in ideal material or in the static sense.

One piece of experimental evidence for a sharp melting transition is found in very low-frequency ($\sim$0.1 Hz) torsional oscillator data on untwinned single-crystal YBCO by Beck et al.[37] They found a very sharp dissipation peak (width $\leq$0.3 K) at a temperature which varied systematically with field orientation θ following the dependence $(T_c - T_m) = AB^{1/2}\epsilon^{1/2}$, where A is a constant and $\gamma^2\epsilon^2 = \cos^2\theta + \gamma^2 \sin^2\theta$ is the now familiar factor describing the effect of anisotropy within the GL theory. The angular dependence is exactly what is predicted by (9.22*a*), and the fit does not involve any new parameters that are not already known from the static torque measurements described in Sec. 9.3. This suggests that the position of this damping peak is determined by fundamental

[35] P. L. Gammel, L. F. Schneemeyer, J. V. Waszczak, and D. J. Bishop, *Phys. Rev. Lett.* **61**, 1666 (1988).

[36] A. Houghton, R. A. Pelcovits, and A. Sudbø, *Phys. Rev.* **B40**, 6763 (1989).

[37] R. G. Beck, D. E. Farrell, J. P. Rice, D. M. Ginsberg, and V. G. Kogan, *Phys. Rev. Lett.* **68**, 1594 (1992); also D. E. Farrell, J. P. Rice, and D. M. Ginsberg, *Phys. Rev. Lett.* **67**, 1165 (1991).

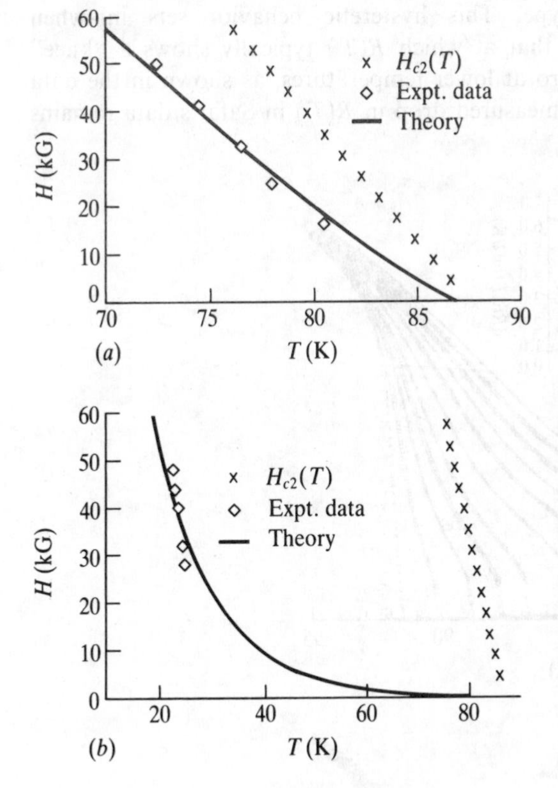

FIGURE 9.6
Phase diagrams for flux-line lattice melting in (*a*) YBCO and (*b*) BSCCO. The squares denote experimental estimates of the melting transition [From the vibrator data of Gammel et al., *Phys. Rev. Lett.* ***61****, 1666 (1988).*] The crosses indicate an estimate for $H_{c2}(T)$, namely, 440 kG × $(1 - T/T_c)$. The solid curves [Calculated by Houghton et al., *Phys. Rev.* ***B40****, 6763 (1989)*] are in the framework of a 3-D anisotropic GL model, with c_L, γ, and κ as fitting parameters. As noted in Sec. 9.5.3, the nearly vertical rise of the melting field in BSCCO is probably better explained in terms of a 2-D melting temperature, as in (9.26).

thermodynamic quantities, not by pinning or imperfections. It is perhaps surprising that the temperature dependence of H_m implied by the experimental result is $(T_c - T_m)^\beta$, with $\beta = 1.99 \pm 0.16$, rather than the lower powers $\frac{3}{2}$ or $\frac{4}{3}$ found for other "transition lines." However, $\beta = 2$ is, in fact, the dependence predicted in (9.22*b*) by the melting theory, provided one inserts the mean-field temperature dependence (near T_c) for the penetration depth, i.e., $\lambda \sim (T_c - T)^{-1/2}$.

Further important experimental support for a first-order melting transition has been provided by resistivity measurements on high-quality crystals by several groups.[38] For example, the measurements on untwinned, defect-free single-crystal YBCO by Safar et al., made with picovolt sensitivity and millikelvin temperature resolution in fields up to 7 T, revealed reproducible hysteresis in the *I-V* curves, whether sweeping the field or the temperature. Since continuous phase transitions are usually reversible, this hysteresis is evidence for a first-order transition,

[38] M. Charalambous, J. Chaussy, and P. Lejay, *Phys. Rev.* **B45**, 5091 (1992); H. Safar, P. L. Gammel, D. A. Huse, D. J. Bishop, J. P. Rice, and D. M. Ginsberg, *Phys. Rev. Lett.* **69**, 824 (1992); W. K. Kwok, S. Fleshler, U. Welp, V. M. Vinokur, J. Downey, G. W. Crabtree, and M. M. Miller, *Phys. Rev. Lett.* **69**, 3370 (1992).

presumably of the melting type. This hysteretic behavior sets in when $R/R_n \lesssim 0.12$, a level similar to that at which $R(T)$ typically shows a "knee" before falling rapidly toward zero at lower temperatures, as shown in the data of Kwok et al. in Fig. 9.7. The measured drop in $R(T)$ in Safar's data remains

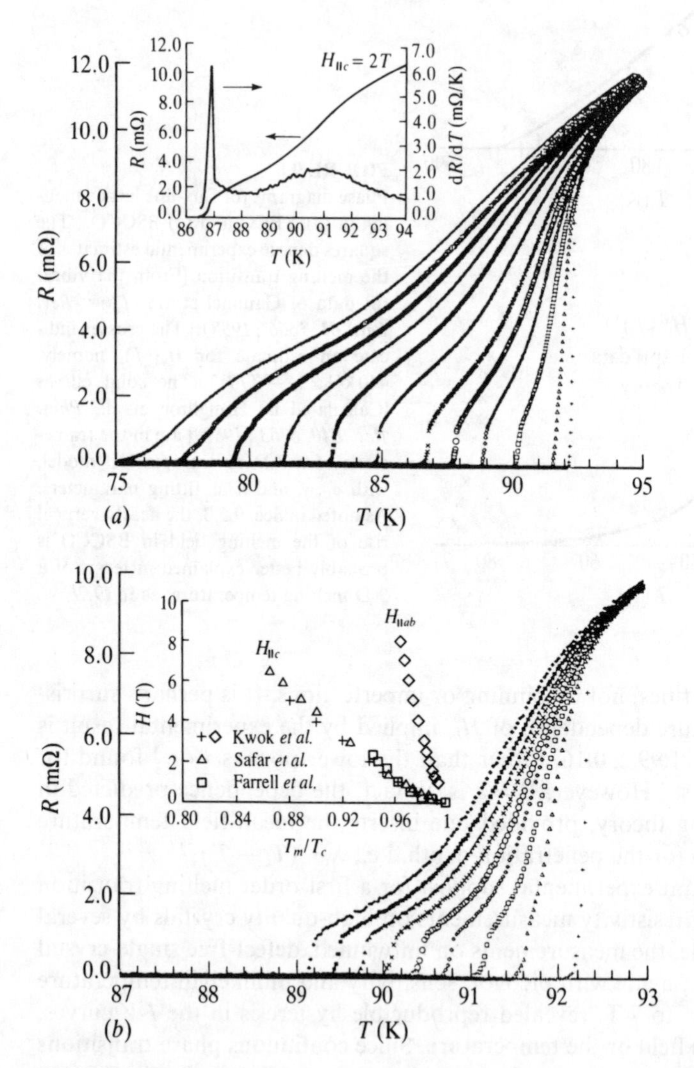

FIGURE 9.7
(a) Resistive transition in magnetic fields of 0, 0.1, 0.5, 1, 1.5, 2, 3, 4, 5, 6, 7, and 8 Tesla for $H_{||c}$ in an untwinned YBCO crystal, showing a drop at the melting transition. (Inset) Determination of T_m from peak of dR/dT for $H = 2T$. (b) Similar data for $H_{||ab}$, at $H = 0$, 1, 2, 3, 4, 5, 6, 7, and 8T. (Inset) Phase diagram of the melting transition for both field orientations, showing comparison with torsional oscillator data of Farrell et al., and hysteretic I-V criterion of Safar et al. The "zero resistance" data points of Iye et al. in Fig. 9.5a also coincide closely with these results. [*All other data plotted here are from W. K. Kwok et al.*, Phys. Rev. Lett. *69*, *3370 (1992)*.]

apparently continuous, presumably because the resistive transition is the superposition of a large number of small jumps occurring in subunits of the sample which have slightly different critical temperatures. As shown in the inset in Fig. 9.7*b*, the temperature dependence of the transition lines found by Kwok et al. and by Safar et al. appear to coincide closely with each other and with that found in the torsional oscillator experiments of Farrell et al. over the temperature range of overlap $\sim T \approx 87$–$92\,\mathrm{K}$. In the range from 80 to 87 K, however, H_m rises more slowly than implied by a quadratic dependence on $(T_c - T)$, approaching a linear temperature dependence.

Let us now focus on the question of the temperature dependence of H_m. Although the predicted dependence $\sim(T_c - T_m)^2$, found in (9.22) is in good agreement with the work of Farrell et al. and Beck et al., other experimenters have reported obtaining better fits to other phenomena, such as the irreversibility line, by using lower powers, e.g., $H_{\mathrm{irr}} \propto (T_c - T)^{3/2}$ or $(T_c - T)^{4/3}$. As mentioned in the preceding paragraph, the H_m data of Safar et al. and Kwok et al. even suggest an asymptotic linear dependence extrapolating back to a depressed T_c. If one assumes that these diverse phenomena are manifestations of the same basic physics, this apparent discrepancy appears puzzling. We make two remarks concerning this issue.

First, we note that (9.22*a*) is only expected to be valid when $B_m \ll H_{c2}(T_m)$ and T is near T_c. Insofar as B_m rises as $(T_c - T)^2$ while H_{c2} rises only as $(T_c - T)$, it is clear that as we go farther below T_c, B_m tends to rise more rapidly and to approach $H_{c2}(T)$. But when B approaches the second-order transition to the normal state at H_{c2}, we know that $\lambda^{-2} \sim n_s \sim |\psi|^2 \propto (1 - b)$, where $b \equiv B/H_{c2}$. Given the form (9.22), we see that this brings in a factor $(1 - b_m)^2$ on the right-hand side of the equation. This acts to reduce the computed value of B_m so as to prevent it from approaching too closely to H_{c2}, which is rising only linearly with $(T_c - T)$. This factor will cause B_m to rise in a way that is intermediate between the first and second powers of $(T_c - T)$ at temperatures where B_m is becoming comparable to H_{c2}. This consideration is built into the quantitative evaluation of the detailed theory by Houghton et al.[39] and they find remarkably good agreement with the very different measured temperature dependences in YBCO and BSCCO, using reasonable material parameters. (See Fig. 9.6.) That is, for YBCO they find that $B_m(T)$ becomes almost linear in $(T_c - T)$ well below T_c, whereas with BSCCO they find a $B_m(T)$ which shows much curvature, remains very small near T_c, and rises steeply at $T \sim T_c/4$. In their calculation, this latter rise stems primarily from the factor $1/T^2$ in (9.22). In the next section, we shall see that 2-D effects probably offer a better interpretation of the steep rise of $B_m(T)$ in BSCCO at low temperatures.

The second consideration is expected to modify the power law in the other extreme of the temperature range, very near T_c, as one approaches the critical

[39] A. Houghton, R. A. Pelcovits, and A. Sudbø, *Phys. Rev.* **B40**, 6763 (1989).

region, where the mean-field temperature dependences break down. Although of an entirely negligible width ($\sim 10^{-6}$ K) in classic superconductors, the width of the critical region in the HTSC has been estimated[40] to be of the order of 1 K. In the critical region, λ^{-2} is expected to vary as $(T_c - T)^{2/3}$ instead of as $(T_c - T)$, so that, according to (9.22), we should expect B_m (and H_{c2}) $\propto (T_c - T)^{4/3}$.

When we combine these two considerations, it does not seem unreasonable that a quantity which should vary as $(T_c - T)^{4/3}$, very near T_c, then changes over to $(T_c - T)^2$ in the mean-field region, yet always stays below a curve rising as $(T_c - T)$, will not have a unique power-law fit, but will give reasonable fits to various powers that are intermediate between 1 and 2, depending on the temperature range and the parameters of the material.

9.5.3 Two-Dimensional vs. Three-Dimensional Melting

In the foregoing discussion, we have taken for granted that the material acts as an anisotropic but three-dimensional superconductor, so that we could assume continuity of the flux lines. For YBCO, this gave a satisfactory description, but in the more anisotropic BSCCO, the interplane coupling is so weak that it is necessary to return conceptually to the general Lawrence-Doniach model, recognizing the importance of discrete superconducting planes.[41] We consider the configuration in which **H** is along the **c** (or **z**) axis, *perpendicular* to the planes, so that each vortex line can be thought of as a string of pancake vortices in the planes through which it passes. In this case, the vortex system is considered to be three dimensional if the (x, y) positions of the pancake vortices in successive planes are strongly correlated so as to define a continuous vortex line, or it can be two dimensional if the pancake vortices in successive planes move essentially independently. These issues have been discussed extensively by several authors;[42] here we shall only summarize the most salient features. We first examine the physical reason for the existence of a field-dependent crossover from vortex lines in 3-D to vortex points in 2-D, and then we consider its consequences for the melting transition.

To find the crossover criterion, we estimate the relative strength of the restoring forces exerted on a given pancake vortex by adjacent pancake vortices in the same plane and by pancake vortices in the planes above and below the given

[40] See, e.g., C. J. Lobb, *Phys. Rev.* **B36**, 3930 (1987).

[41] In Sec. 9.2.2, we considered the effect of this discreteness on the mean-field phase transition at H_{c2} when the magnetic field was applied exactly *parallel* to the *ab* plane and found a regime in which the vortices ran between the discrete planes, which acted as 2-D superconductors. Here we consider a field *perpendicular* to the planes.

[42] D. S. Fisher, M. P. A. Fisher, and D. Huse, *Phys. Rev.* **B43**, 130 (1991); L. I. Glazman and A. E. Koshelev, *Phys. Rev.* **B43**, 2835 (1991).

one. There are two sources of the force between vortices in adjacent planes: the Josephson coupling and the magnetic coupling of the current loops. Clearly, the latter must dominate in the limit $\gamma \to \infty$, reached by having $E_J \to 0$; the crossover to Josephson-dominated coupling is estimated[43] to occur when $\gamma \lesssim \lambda_{ab}/s \sim 100$ for typical parameters. For the moment, however, we consider only the Josephson coupling term, which will dominate in all but the most weakly coupled cases.

We can estimate the force constant for relative motion in the plane by taking the force constant K per unit length (9.20) between lines, multiplied by the interplanar distance s for the length L_z. Apart from numerical factors, this gives an intraplane force constant of $Ks \sim B\Phi_0 s/\lambda_{ab}^2$. To find the Josephson interplane term, we note that (9.1a) implies that E_J per unit area is given by

$$E_J = \Phi_0^2/(16\pi^3\lambda_c^2 s) \tag{9.24}$$

Letting a be the intervortex separation, the coupling energy per vortex will be $\sim a^2 E_J$. Since this energy will be lost for a relative displacement by $\sim a/2$ because the phases in the two layers will no longer match, this implies a force constant of order E_J, regardless of a (and hence B).

Since the interplane force constant is independent of B while the intraplane force constant is proportional to B, these force constants will be equal at some characteristic crossover field B_{cr}. Inserting the above expressions for the two force constants, we find

$$B_{cr} \sim \Phi_0(\lambda_{ab}/s\lambda_c)^2 \sim \Phi_0/(s^2\gamma^2) \tag{9.25}$$

in agreement with the result of Fisher et al.[43] This formula explicitly shows the critical importance of the anisotropy factor γ. After a more detailed analysis, integrating over a Fourier spectrum of fluctuations using the FLL elastic constants, Glazman and Koshelev[44] find the same functional dependence on parameters with an estimated numerical coefficient of order 10. From these results we estimate $B_{cr} \approx (10^3 - 10^4 \text{ Tesla})/\gamma^2$. For YBCO, this is an unobtainably high field, but for BSCCO, it is probably $\lesssim$1 Tesla. Whatever the numerical value of B_{cr}, its physical significance is that when $B \gg B_{cr}$, the interaction between adjacent pancake vortices in the same layer is stronger than the interaction between those in adjacent layers but on the same vortex line. This causes the thermal fluctuations to have a quasi-two-dimensional character in the high-field regime $B \gg B_{cr}$.

We can then estimate the 2-D melting temperature in this regime by following the same method used to obtain the 3-D result (9.21). That is, we estimate the mean-square displacement of the pancake vortices in a plane by equating the in-plane elastic energy $\frac{1}{2}Ks(\delta x)^2$ to $\frac{1}{2}kT$ and use the Lindemann melting criterion to

---

[43] D. S. Fisher, M. P. A. Fisher, and D. Huse, *Phys. Rev.* **B43**, 130 (1991).

[44] L. I. Glazman and A. E. Koshelev, *Phys. Rev.* **B43**, 2835 (1991).

identify T_m^{2D} as the temperature at which this $(\delta x)^2$ equals some fraction c_L^2 of the square of the vortex-lattice separation $\sim\Phi_0/B$. In this way, we obtain

$$kT_m^{2D} = Cc_L^2\Phi_0^2 s/\lambda_{ab}^2 \tag{9.26}$$

This expression agrees with the familiar Kosterlitz-Thouless result for the vortex unbinding transition in a 2-D system if the numerical scale factor $Cc_L^2 = 1/(128\pi^3\sqrt{3})$. In that case, the numerical value of $T_m^{2D} \sim 40$ K. Note that in this approximation, the melting temperature is independent of B because the increasing stiffness with increasing B just compensates for the smaller lattice spacing. Also, note that when B is less than B_{cr}, so that the interplane restoring forces are more important, this melting temperature will increase and approach the 3-D dependence. Combining the 3-D and 2-D results for a highly anisotropic material like BSCCO with $H\|\mathbf{c}$, we expect that the melting line will start near T_c with B_m following (9.22*a*) with $\theta = \pi/2$, rising initially as $(T_c - T)^2$ until B_m reaches a value $\sim B_{cr}$ at $T \sim T_m^{2D}$, after which B_m rises almost vertically. This scenario is the basis for the schematic phase diagram for a very anisotropic material in Fig. 9.8*b*, which contrasts this behavior with that of a less anisotropic material shown in Fig. 9.8*a*, for which B_{cr} is above the accessible range. This conceptual perspective accounts qualitatively, at least, for the difference in the experimental observations in YBCO and BSCCO, which were plotted in Fig. 9.6, in comparison with detailed numerical calculations of Houghton et al.

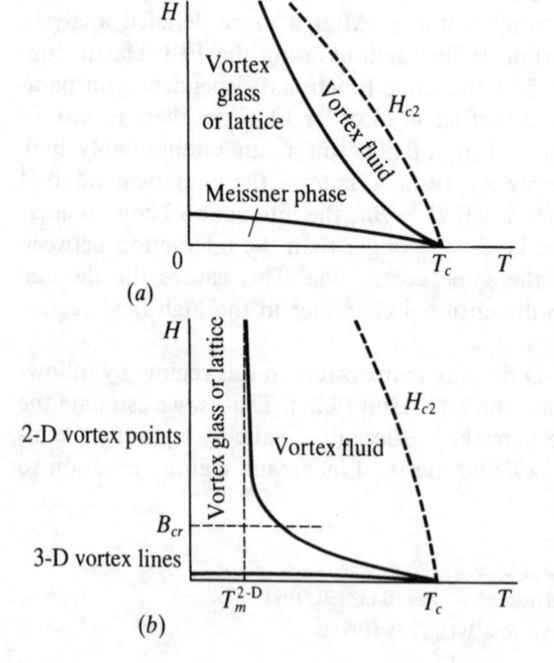

FIGURE 9.8
Schematic phase diagrams for melting of the flux-line solid for (*a*) 3-D material (e.g., YBCO) and (*b*) highly layered material (e.g., BSCCO). The latter shows the crossover from 3-D to 2-D melting at the crossover flux density B_{cr}, which is at inaccessibly high fields for the 3-D material in (*a*).

9.6 THE EFFECT OF PINNING

Having developed some understanding of the properties of flux lines in an ideal homogeneous material, including the effects of anisotropy and thermal vibration, we now add the important factor of spatial inhomogeneity. Such inhomogeneities can be localized defects, such as departures from stoichiometry at the atomic scale, or more extended defects, such as dislocations, grain boundaries, inclusions of second phases, twin planes, etc. The energy of a flux line will be different if it passes through such an inhomogeneity, instead of passing to the side. The resulting force tends to "pin" flux lines in particularly favorable positions relative to the underlying material. It is this pinning which allows the system to sustain the Lorentz force between the flux and the current without flux motion and dissipation, thus giving the material a nonzero critical current, an essential property of practical superconductors. None of this is specific to high-temperature superconductivity, but because the higher temperatures make thermally activated depinning and flux motion more prominent, and because the melting line in ideal crystals of HTSC lies much farther below $H_{c2}(T)$ than in classic superconductors due to the anisotropy, the discovery of HTSC triggered a resurgence of the study of these properties.

From a practical standpoint the goal is to find a means to introduce sufficient pinning to raise significantly the flux-line lattice melting temperature, and therefore the effective resistive transition temperature, from that of the ideal crystal, which was discussed in the previous section. From a conceptual standpoint there is much interest in the question of whether the *linear* resistance becomes truly zero below some sort of glass transition or only exponentially small at low temperatures. In this section, we shall extend the brief discussion of pinning and flux-creep effects given in Sec. 5.7 and give a more detailed discussion of these topics, with special reference to the high-temperature superconductors.

9.6.1 Pinning Mechanisms in HTSC

The collective pinning model of Larkin and Ovchinnikov (discussed in Sec. 9.6.2) is the appropriate treatment of the effect of pinning by randomly distributed weak point defects, but other methods are appropriate for treating strong or correlated defects such as twin boundaries. Accordingly, it is useful to begin by describing some of the sources of pinning which are thought to be important in the high-temperature superconductors. This "catalog" will provide a basis for judging the relevance of various theories of the effect of pinning.

POINT DEFECTS. Because of the short coherence length in the HTSC, departure from stoichiometry at even a single atomic site is sufficient to depress locally the superconducting order parameter. An obvious candidate is oxygen in the CuO_2 planes which dominate the superconductivity in these materials. For example, in YBCO, it is known that an overall 10 percent oxygen deficiency is sufficient to destroy superconductivity entirely, and even high-quality material is usually a few percent off ideal stoichiometry. This implies a considerable density of oxygen vacancies, which are probably close to randomly distributed, and form an example of the sort of point defect weak pinning centers envisioned in the collective pinning model. The work of Thuneberg et al.[45] provides a useful basis for making a quantitative evaluation of the pinning effect of a specific defect of this sort. It shows that the pinning force on a vortex (i.e., the restoring force trying to pull the vortex core back onto the defect) initially rises linearly with the distance from a vacancy. This rise is followed by a gaussian cutoff at a distance of the order of $\xi(T)$, where the order parameter $\psi(\mathbf{r})$ has reached its constant value between vortices. With reasonable assumptions, Kes[46] estimates $\sim$8 vacancies per core of each pancake vortex, whose pinning forces add incoherently (as in the Larkin-Ovchinnikov theory). In this way, he estimates a depinning current density for BSCCO of 5×10^6 A/cm^2 and a pinning energy of $\sim$34 K in temperature units. The latter value is so small, that, if correct, it would imply rapid flux creep except at very low temperatures.

TWIN PLANES AND OTHER EXTENDED DEFECTS. Because YBCO has an orthorhomic crystal structure, in which the *a* and *b* directions in the plane are not exactly equivalent, twin planes are a prominent form of defect in this class of material. These planes are parallel to the (110) and $(1\bar{1}0)$ lattice planes and serve to separate domains in which the *a* and *b* directions interchange roles. Because point defects and impurities inevitably accumulate along such a twin boundary and weaken the superconductivity there, it will attract vortices. Moreover, because of its extended planar structure, the pinning acts in a coherent rather than a random way on a flux line lying in the plane, making the pinning unusually effective. In this case, the Larkin-Ovchinnikov collective pinning model is *not* appropriate. The importance of pinning by twin planes in YBCO was demonstrated experimentally by Kwok et al.[47] They measured the critical-current density as a function of the orientation of the FLL relative to crystal axes and found a sharp peak when the field was within $\sim$1° of the twin plane. In further work, they also observed *intrinsic* pinning by the discreteness of the planes in the layered superconductor, when the field was aligned within $\sim$0.5° of the *ab* planes. Unlike all the other mechanisms enumerated here, this pinning mechanism does not involve any crystallographic defect; on the other hand, it is only effective in a narrow range of field orientations in crystals of high quality.

---

[45] E. V. Thuneberg, J. Kurkijärvi, and D. Rainer, *Phys. Rev.* **B29**, 3913 (1984).

[46] P. H. Kes, "Flux Pinning and the Summation of Pinning Forces," in J. Evetts (ed.), *Concise Encyclopedia of Magnetic and Superconducting Materials*, Pergamon, Oxford (1992), p. 170.

[47] W. K. Kwok, U. Welp, G. W. Crabtree, K. G. Vandervoort, R. Hulscher, and J. Z. Liu, *Phys. Rev. Lett.* **64**, 966 (1990); W. K. Kwok, U. Welp, V. M. Vinokur, S. Fleshler, J. Downey, and G. W. Crabtree, *Phys. Rev. Lett.* **67**, 390 (1991).

Stacking faults are among the other extended defects occurring naturally in HTSC crystals. These consist of intergrowths of other phases with additional or fewer CuO planes per unit cell. Such defects are commonly observed in high-resolution electron microscope images of these crystals, and clearly will contribute to the flux pinning.

Another type of defect, first observed by scanning tunneling microscopy on oriented YBCO films,[48] is the presence of a large number of screw dislocations aligned in the direction of growth and surrounded by a characteristic spiral staircase growth pattern. Such defects pin vortices not only by the internal structure of the dislocation itself, but also by the surface roughness which is induced by the growth pattern.

ARTIFICIAL DEFECTS. Because effective pinning is crucial for obtaining materials with high critical currents, the possibility of increasing pinning by artificial means has been studied by many workers. For example, Murakami et al.[49] have developed a *melt-process melt-quench* heat-treatment procedure for YBCO which is designed to introduce a high density of inclusions of off-stoichiometric second phases. In this way, substantial increases in critical current density have been obtained, even though the typical dimension of the inclusions is larger than would be optimal.

More recently there have been a number of studies of the effectiveness of radiation-induced pinning defects. In general, it is found that electron irradiation is relatively ineffective, but neutron or proton irradiation provides more substantial deformations and more pinning. Most impressive have been the results obtained with bombardment by a high-energy beam of heavy ions. Because of the high momentum associated with their mass, these ions can penetrate through a substantial thickness of material along a nearly ballistic trajectory. In doing so, they create an extended set of correlated defects lying along a straight line. Such defects have been shown experimentally[50] to be extremely effective in pinning flux lines which are parallel to the damage tracks but less so for arbitrary field directions. A theoretical treatment of this pinning process by analogy with boson localization has been given by Nelson and Vinokur,[51] and is discussed further in Sec. 9.6.5. Subsequent work by Krusin-Elbaum et al.[52] has shown that the randomly directed tracks of fission fragments also greatly increase critical currents and significantly raise the irreversibility temperature, as shown in Fig. 9.9.

---

[48] C. Gerber, D. Anselmetti, J. G. Bednorz, J. Mannhart, and D. G. Schlom, *Nature* **350**, 279 (1991); M. Hawley, J. D. Raistrick, J. G. Berry, and R. J. Houlton, *Science* **251**, 1587 (1991).

[49] M. Murakami, M. Morita, K. Doe, and K. Miyamoto, *Jpn. J. Appl. Phys.* **28**, 1189 (1989).

[50] L. Civale et al., *Phys. Rev. Lett.* **67**, 648 (1991).

[51] D. R. Nelson and V. M. Vinokur, *Phys. Rev. Lett.* **68**, 2398 (1992).

[52] L. Krusin-Elbaum et al., *Appl. Phys. Lett.* **64**, 3331 (1994).

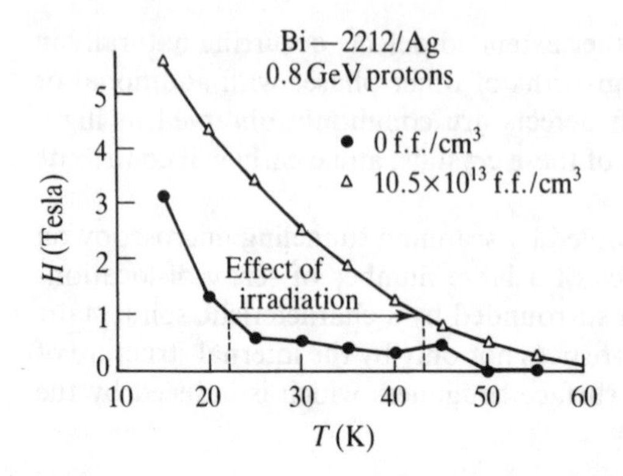

FIGURE 9.9
The irreversibility line in a BSCCO tape before and after introduction of defects by fission fragments induced by proton bombardment. The data at zero flux reflect the usual melting transition in a good sample. The data from the irradiated sample reflect a glass transition. The practical significance of the data is that the resistive transition in a field of $1T$ has been increased from ~22 to ~43 K, thus expanding the parameter range in which the material is effectively superconductive. (*After Krusin-Elbaum et al.*)

We now begin our review of pinning theories by examining the classic theory of Larkin and Ovchinnikov, which demonstrates the crucial role played by the *elasticity* of the flux-line lattice in determining the effectiveness of a random collection of weak point pinning sites.

9.6.2 Larkin-Ovchinnikov Theory of Collective Pinning

We begin with the observation that if the flux-line lattice (FLL) were perfectly *periodic* and *rigid*, it would not be effectively pinned by any *random* collection of pinning sites. The reason is that, for any position of the FLL relative to the material, an equal number of random pinning sites would exert forces f adding to the Lorentz force as opposing it, summing to zero net force. More precisely, the random pinning forces would add as in a random walk, giving a net total force which increases only as $\sqrt{N}$ or $\sqrt{V}$, if $N = nV$ is the total number of pinning centers in volume V, and n is the number of pinning centers per unit volume. On the other hand, for any given current density J, the Lorentz force scales with JV. Thus, the critical depinning current density J_c, at which the Lorentz force equals the maximum pinning force, will scale with $V^{-1/2}$ and vanish in the limit of large volumes.

Once we allow for the elasticity of the FLL, however, the paths of individual flux lines can deviate from the ideal periodic arrangement of the Abrikosov lattice to take advantage of the opportunity to lower their energy by passing through favorable pinning sites, but at the expense of increasing the elastic energy of the FLL by deforming it. The equilibrium flux-line configuration will be that distorted arrangement which minimizes the sum of these two energies. This logic was developed as the theory of *collective pinning* by Larkin and Ovchinnikov.[53] Our

[53]A. I. Larkin, *Zh. Eksp. Teor. Fiz.* **58**, 1466 (1970); A. I. Larkin and Yu. V. Ovchinnikov, *J. Low Temp. Phys.* **34**, 409 (1979).

discussion here is adapted from their treatment, with the aim of bringing out the main physical ideas as simply as possible.

The key idea of this approach is to describe the distortion of the FLL in terms of correlation volumes, within which the FLL is reasonably undistorted, but between which there are pinning-motivated shear and tilt distortions on a relevant scale. Thus, as illustrated in Fig. 9.10, the macroscopic sample volume is considered to be subdivided into correlation volumes V_c with length $\sim L_c$ along the field direction and transverse dimensions $\sim R_c$, with R_c and L_c chosen to optimize the trade-off between pinning and elastic energies described above. For example, if V_c is made smaller, one sees that the FLL can adjust more frequently from region to region over the sample to optimize pinning energy, but at the expense of increased elastic energy. Because the FLL is periodic with period $a \sim (\Phi_0/B)^{1/2}$, *lattice correlation* is lost as soon as the distortion distance is $\sim a$. However, if one is interested in finding the effect of pinning on the *critical current density*, the appropriate distortion distance is only the range ξ of the pinning force, which is typically less than a. When this distortion is accommodated over distances R_c or L_c for shear and tilt, respectively, it gives strains (fractional distortions) s_s and s_t of the order of ξ/R_c and ξ/L_c, respectively.

It is customary to describe the resulting increase in elastic free energy per unit volume as

$$\tfrac{1}{2}[C_{66}s_s^2 + C_{44}s_t^2],$$

where C_{66} and C_{44} are the elastic moduli of the FLL for shear and tilt, respectively. (We do not discuss the modulus C_{11} for uniaxial compression.) An early discussion of these moduli was given by de Gennes and Matricon[54] in 1964. A

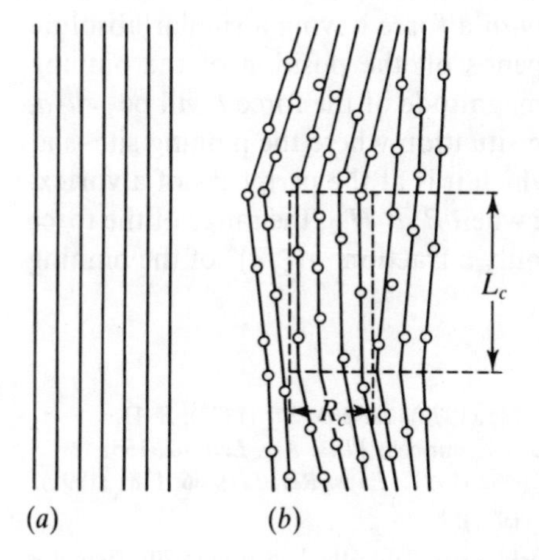

FIGURE 9.10
Schematic diagram illustrating the coherence volume concept of the Larkin-Ovchinnikov theory. (*a*) With no pinning, the flux-line lattice (FLL) is periodic and exactly parallel to the magnetic field. (*b*) With random attractive pinning sites, the local direction of the FLL is modulated slightly within each coherence volume (defined by R_c and L_c) so as to combine optimally the energy reduction from the pins with energy increase from distortion of the FLL.

[54] P. G. de Gennes and J. Matricon, *Revs. Mod. Phys.* **36**, 45 (1964).

number of authors[55] have treated them more recently, taking into account the anisotropy of the HTSC and also the nonlocal nature of the intervortex interactions when the vortex-line directions are distorted from strictly straight and parallel. For our purposes, however, the moduli can be thought of approximately as manifestations of the two energies that we combined in our model calculation of the melting criterion in Sec. 9.5.1.

According to GL theory, for uniform shear of a triangular FLL in an isotropic superconductor with $\kappa \gg 1$ and $H_{c2} \gg B \gg H_{c1}$, one finds[56] $C_{66} \approx (H_c^2/16\pi)b(1-b)^2 \approx (BH_{c1}/16\pi)(1-b)^2$. (Here $b \equiv B/H_{c2}$.) Since this modulus reflects the energy increase when flux lines are shifted in rigid translational motion relative to their equilibrium triangular configuration while remaining parallel to the original orientation, it involves the same energy scale as that which determined the restoring force constant per unit length K found in (9.20). In fact, by working through the geometry, we find that the shear modulus C_{66} should be of the same order as K, namely, $\sim\Phi_0 B/\lambda^2$, so long as $H_{c1} \ll B \ll H_{c2}$, in agreement with the result quoted earlier.

We can also estimate the tilt modulus C_{44} by adapting the line tension argument used to obtain (9.20). The increase in the line length for a tilt strain (or tilt angle) $s_t = \delta x/L$ is $(L^2 + \delta x^2)^{1/2} - L \approx (\delta x)^2/2L$, the number of flux lines per unit area is B/Φ_0, and the line energy is $\Phi_0 H/4\pi$ using (5.20) for the case of highly overlapping vortices and nearly uniform flux density, where $B \approx H$.[57] Combining these factors, and equating the energy increase per unit volume to $\frac{1}{2}C_{44}s_t^2$, we obtain $C_{44} = BH/4\pi$, in agreement with more general thermodynamic arguments.

We now must examine the pinning energy more closely. If the magnetic field B is of the order of H_{c2}, the intervortex distance $a \approx (\Phi_0/B)^{1/2}$ is of the order of the core radius ξ, so that the cores fill most of the volume. In this case, most local pinning centers will interact with the FLL with a force having a similar absolute magnitude f, but with a direction that depends on the position of the pinning center relative to the nearest flux line. The magnitude of the force f will be $\sim u/a$, where u is the energy difference between the situation where the pinning site is at the center of a vortex and the situation in which it is at the periphery of a vortex. On the other hand, in the more usual regime when $B \ll H_{c2}$, the range of the force derived from each flux line is $\sim\xi \ll a$, and only a fraction $\sim(\xi/a)^2$ of the pinning

---

[55] See, e.g., E. H. Brandt, *J. Low Temp. Phys.* **26**, 709, 735 (1977); **28**, 263, 291 (1977); E. H. Brandt, *Phys. Rev.* **B34**, 6514 (1986); V. Kogan and L. J. Campbell, *Phys. Rev. Lett.* **62**, 1552 (1989); A. Sudbø and E. H. Brandt, *Phys. Rev.* **B43**, 10482 (1991); *Phys. Rev. Lett.* **66**, 1781 (1991).

[56] See, e.g., E. H. Brandt, *J. Low Temp. Phys.* **26**, 709 (1977).

[57] The line energy is always proportional to the local field h parallel to the line as in (5.20). Here it is $B \approx H$ because the lines are all parallel, but it was $\sim H_{c1}$ in connection with (9.21) because the added line length there was perpendicular to the average flux. The fact that more careful analysis for the present argument gives H rather than B is not surprising since the vortex core position is a point of maximum h within the unit cell of the Abrikosov vortex array.

centers will be in a position to interact with a flux line, but with a stronger force $\sim u/\xi$, instead of u/a. Taking into account the fact that a stronger force is felt over a fraction $\sim(\xi/a)^2$ of the volume, the mean square of the force is $\sim(u/a)^2$ in both cases, and we interpret f to be the root-mean-square (rms) force $\sim u/a$ for a single pin on a given flux line.

To find the corresponding pinning *energy*, we recall the argument given earlier that the sum of N such forces in the correlation volume V_c, adding randomly, will be of the order of $N^{1/2}f$, where $N = nV_c$. Because this force acts only through a distance $\sim\xi$ before changing randomly, the amplitude of the associated potential energy will be $\sim\xi f N^{1/2}$, or $\xi f n^{1/2}/V_c^{1/2}$ per unit volume. If we subtract this energy gained by accommodating the pinning sites from the (positive) elastic distortion energy, we obtain the net free-energy change per unit volume from the presence of the pinning sites:

$$\delta F = \frac{1}{2}C_{66}\left(\frac{\xi}{R_c}\right)^2 + \frac{1}{2}C_{44}\left(\frac{\xi}{L_c}\right)^2 - f\xi\frac{n^{1/2}}{V_c^{1/2}} \tag{9.27}$$

Although this expression is rather schematic, let us formally minimize it with respect to R_c and L_c to learn its qualitative implications. Recalling that $V_c = R_c^2 L_c$ in our approximation, we obtain

$$L_c = \frac{2C_{44}C_{66}\xi^2}{nf^2} \qquad R_c = \frac{2^{1/2}C_{44}^{1/2}C_{66}^{3/2}\xi^2}{nf^2} \qquad V_c = \frac{4C_{44}^2 C_{66}^4 \xi^6}{n^3 f^6} \tag{9.28}$$

These results confirm the physical expectation that the FLL will be more distorted (i.e., smaller correlation volume) if there are more and/or stronger pinning sites, and if the elastic moduli are smaller. Also, we note that the correlation volume is elongated along the field direction since $L_c/R_c = \sqrt{2}(C_{44}/C_{66})^{1/2} \gg 1$ in the case at hand. The longitudinal correlation length L_c has recently been measured[58] by neutron diffraction and found to agree well with predictions of this theory. Substituting (9.28) into (9.27), we obtain the magnitude of the net pinning energy of the correlation volume, namely,

$$\delta F_{\min} = -n^2 f^4/(8C_{44}C_{66}^2\xi^2) \tag{9.28a}$$

Finally, the pinning force per unit volume $f(n/V_c)^{1/2}$ determines the maximum sustainable Lorentz force density, so that the critical-current density J_c is given by

$$J_c B/c = \frac{n^2 f^4}{2C_{44}C_{66}^2\xi^3} \tag{9.29}$$

Insofar as these results hold, we see that a softer FLL (i.e., smaller elastic moduli) implies a larger pinning energy and larger critical current, all else being equal, because the FLL can better conform to the pinning sites. Because this argument is

[58] U. Yaron, et al., *Phys. Rev. Lett.* **73**, 2748 (1994).

based on linear elasticity theory, it breaks down, of course, when the effective shear modulus C_{66} goes all the way to zero at the melting temperature.

It is worth noting that because these results are based on the incoherent addition of a large number of pinning forces, one can characterize the "pinning strength" by a *single* parameter $W = nf^2$, which takes into account both the density and strength of the pins. Then both L_c and R_c are proportional to W^{-1}, while J_c scales with W^2.

The above discussion is appropriate if there is a large density of weak pinning centers acting in a random statistical manner. However, the formulation can also be used to identify the crossover condition at which individual pinning sites are sufficiently strong so as to define individual coherence volumes. This will occur if $nV_c \approx 1$. From (9.28) we see that this can happen if the pinning forces f are sufficiently large or the elastic moduli sufficiently small, as near H_{c2}. In this case, each pinning center plastically deforms the FLL in order to obtain the full value f of the pinning force. Accordingly, the total pinning force per unit volume is simply proportional to the defect concentration, and $J_cB/c = nf$. This will greatly exceed the random-walk sum of pinning forces, and will give a larger J_c. This phenomenon is thought to be one explanation of the "peak effect," in which J_c rises to a sharp peak just before falling to zero at H_{c2}, or, more precisely, at the irreversibility or melting line.

COLLECTIVE PINNING IN TWO DIMENSIONS. In the case of a thin film or decoupled layers in a three-dimensional crystal, with the magnetic field normal to the plane, the collective pinning argument is modified because L_c is effectively set by d, the thickness of the film or layer. Accordingly, the C_{44} term in (9.27) is omitted, and $V_c = R_c^2 d$. With these changes and replacement of ξ by a, minimization of (9.27) yields

$$R_c = \frac{C_{66} a d^{1/2}}{n^{1/2} f} \tag{9.30}$$

as the radius of correlation of the pancake vortices in a given plane. Equating the pinning force per unit volume to the Lorentz force, we find that the critical current density is given by

$$J_c B/c = \frac{nf^2}{C_{66}\xi d} \tag{9.31}$$

In this 2-D case, we see that R_c scales with the pinning strength $W(= nf^2)$ as $W^{-1/2}$, whereas J_c scales with W. In both cases, this dependence is weaker than in the 3-D case (9.28) and (9.29), where they scale as W^{-1} and W^2, respectively. For notational consistency, (9.30) and (9.31) have been written in terms of the 3-D quantities used earlier, with the thickness d appearing explicitly. If desired, the formulas can be reexpressed in terms of 2-D versions of C_{66}, J_c, and n, in which case the factors involving d are absorbed in the definitions of the 2-D quantities.

Good agreement with the 2-D Larkin-Ovchinnikov model was found by Kes and Tsuei[59] in experiments on amorphous Nb_3Ge films.

While discussing the effect of dimensionality in the Larkin-Ovchinnikov model, it is interesting to note that if the model is generalized to D dimensions with correlation volume L^D and generalized moduli C, one finds $D = 4$ is a critical dimensionality. That is, for $D > 4$, the physically meaningful minimum free energy occurs for $L = \infty$, which implies that long-range order in the FLL is retained, whereas for $D < 4$, long-range order is destroyed by any density of small random defects, as we have seen in two and three dimensions.

An important *caveat* in comparing the Larkin-Ovchinnikov collective pinning theory with a broad range of data on real systems is that it presupposes that the correlation lengths are determined solely by *elastic* deformations of the FLL. This ignores the importance of dislocations in the FLL, which can yield rather similar results, and which dominate the distortion at length scales larger than R_c and L_c. In two dimensions, vacancies and interstitial defects can be even more important additions to the model. These considerations do not disturb the qualitative usefulness of the correlated volume concept but, rather, lessen the quantitative reliability of specific predictions.

9.6.3 Giant Flux Creep in the Collective Pinning Model

In the discussion above, we have found a prescription for estimating the critical current density J_c resulting from a random distribution of weak point pinning sites, taking into account the elastic properties of the FLL but ignoring the effect of thermal activation in assisting flux motion. Since these thermally activated processes give rise to "giant flux creep" in the high-temperature superconductors, it is necessary to generalize the classic Anderson-Kim flux-creep theory discussed in Chap. 5 to take into account more carefully the current dependence of the activation energy U. For example, if we infer the "persistent current" in a superconducting sample by measuring its magnetic moment after a sudden change of an external field, the initial surface current density will be set by J_c, but the logarithmic time-dependence characteristic of creep leads to a fast initial decay, which may quickly (e.g., in less than a second) relax the current to, say, half J_c, followed by continuing slower decay. This makes it essential to determine how the activation energy for flux creep depends on J/J_c over a wide current range, if such experiments are to be interpreted properly.

By definition, the activation energy U for flux motion goes to zero at $J = J_c$. In the Anderson-Kim model, it was assumed for simplicity that $U = U_0[1 - (J/J_c)]$, although this assumption has no rigorous justification. For example, in Chap. 6, we noted that the tilted-washboard cosine potential of the

[59] P. H. Kes and C. C. Tsuei, *Phys. Rev. Lett.* **47**, 1930 (1981).

Josephson junction yields $U = U_0[1 - (J/J_c)]^{3/2}$ to a good approximation. These two forms can be generalized to $U = U_0[1 - (J/J_c)]^{\alpha}$. Such forms focus on the detailed behavior near J_c, which is appropriate for the classic superconductors where fluctuation effects cause only slight degradation of J_c, but they all simply assume that U is a constant U_0 well below J_c. Although plausible for individual strong pinning sites, this turns out to be fundamentally incorrect for the case of collective pinning by many weak pinning sites. As we shall see, for $J \ll J_c$, the collective pinning regime leads to $U(J) \approx U_0(J_c/J)^{\mu}$, with $\mu \leq 1$. Such a form has been found experimentally in flux-creep measurements of Maley et al.[60] It has the dramatic and important consequence that as $J \to 0$, the activation energy $U \to \infty$, and flux motion and *linear* resistance vanish, even if the voltage remains finite at *finite* current density. In contrast to this essentially nonlinear response even to small forces, which we consider the hallmark of "glassy" behavior, the conventional Anderson-Kim flux-creep picture with $U \approx U_0$ for $J \ll J_c$ always gives a *nonzero linear* resistance, even if it is exponentially small. Let us now explore the origin of this crucial difference.

An inverse dependence of U upon J is expected to arise in a variety of models and regimes. For conceptual simplicity, we consider the regime in which the flux density is low enough and the current density high enough (though still much below J_c) so that one can ignore the forces between vortices and consider the thermally activated response of a single flux line to the combination of only its elastic line tension ϵ_1, the pinning forces, and the Lorentz force from the current. We start with the results of the fluctuation-free analysis, which yields a characteristic length scale L_c for collective pinning and a characteristic transverse displacement scale set by the range $\sim\xi$ of the pinning force. Balancing the pinning force $\sim U_0/\xi$ on the segment against the Lorentz force $J\Phi_0 L_c/c$ on it, we find a critical current density $J_c \approx cU_0/\Phi_0 L_c\xi$. For any lower J, there will be an energy barrier to flux motion, requiring thermal activation. It turns out that this barrier energy can be minimized by considering alternative metastable configurations extending over length scales longer than L_c and with displacements δx longer than ξ. The physical reason for this is that such larger displacement configurations gain more from the Lorentz force and are restrained primarily by increasing elastic energy since the net pinning force increases sublinearly with L and tends to average out for displacements beyond ξ.

We can formalize this argument by postulating that the optimum displacement δx scales up from ξ by a factor of $(L/L_c)^{\zeta}$ with increasing L. Then the elastic energy $\sim\epsilon_1(\delta x)^2 L$, which is of the order of U_0 by the sort of argument leading to (9.28), scales as $(L/L_c)^{2\zeta-1}$. This allows us to write the total free-energy increment of the displaced vortex loop for $L > L_c$ as

$$\begin{aligned}\delta F(L) &\approx U_0(L/L_c)^{2\zeta-1} - J(\Phi_0/c)L_c\xi(L/L_c)^{\zeta+1} \\ &= U_0[(L/L_c)^{2\zeta-1} - (J/J_c)(L/L_c)^{\zeta+1}]\end{aligned} \tag{9.32}$$

[60] M. P. Maley, J. O. Willis, H. Lessure, and M. E. McHenry, *Phys. Rev.* **B42**, 2639 (1990).

This describes a nucleation problem, in which the energy of the displaced flux-line segment first increases and then decreases with increasing L (since $J < J_c$ and $\zeta < 2$). The barrier energy is then the maximum value of δF, which occurs for the L value $[\sim L_c(J_c/J)^{1/(2-\zeta)}]$ separating these two regimes. This activation energy then has the form

$$U(J) \sim U_0(J_c/J)^{\mu} \tag{9.33}$$

where $\mu = (2\zeta - 1)/(2 - \zeta)$. This μ has the value $\frac{1}{7}$ if $\zeta = \frac{3}{5}$ as estimated from numerical simulations, and for more general models, it is thought that $\mu \lesssim 1$.

This simple line of argument establishes the physical basis for the inverse current dependence of the activation energy, but it is not quantitative. A more complete argument must also take into account the interaction between vortices, which becomes more important at higher flux densities and at lower transport current densities. It can be shown[61] that the current density below which we must consider the correlated motion of flux *bundles* containing more than one fluxon should scale as $B^{(2-\zeta)/2}$, but we refer the reader to more specialized discussions for details.

The current dependence (9.33) of the barrier U to current-driven flux motion implies a nonohmic current-voltage characteristic of the form

$$V \propto \exp\left[-\frac{U_0}{kT}\left(\frac{J_c}{J}\right)^{\mu}\right] \tag{9.34}$$

Note that in this model there is no correction term for "backward" flux jumps because for them both terms in (9.32) are positive for any L, and hence all such configurations are *unstable* and must return to the initial configuration. There is no *metastable* configuration with a simple energy barrier, as for the forward jumps treated above.

Recognizing that in an inductive circuit, $V \propto dI/dt$, one sees that this exponential dependence of dJ/dt on J implies an approximately logarithmic dependence of J upon t, as in the Anderson-Kim theory, where

$$J(t) \approx J_c[1 - (kT/U_0)\ln(1 + t/t_0)] \tag{9.35}$$

Here $t_0 \ll 1$ sec is a microscopic time, so that in the argument of the logarithm, 1 can normally be dropped compared to t/t_0. However, the power-law dependence of U upon J/J_c in (9.34) modifies the Anderson-Kim result to the approximate dependence

$$J(t) \approx J_c\left(\frac{kT}{U_0}\ln\frac{t}{t_0}\right)^{-1/\mu} \tag{9.36}$$

[61] See extensive review article by G. Blatter, M. V. Feigel'man, V. B. Geshkenbein, A. I. Larkin, and V. M. Vinokur, *Revs. Mod. Phys.* **66**, 1125 (1994).

for $J \ll J_c$, and neglecting 1 compared to t/t_0. An interesting consequence of this formula is that when substituted into (9.33), it implies that, within its regime of validity, the relevant barrier energy U can be approximated by

$$U \approx kT \ln(t/t_0) \tag{9.37}$$

Since t_0 is a microscopic time, this in turn implies that observed activation energies U will typically be of the order of 10 to 20 times kT at the temperature of observation. This may account for the fact that the observed values of U/k are typically $\sim$1,000 K in samples of high-temperature superconductors of quite a variety of qualities. Although it seems surprising that the intrinsic U_0 has dropped out in (9.37), this result is only expected to hold for collective pinning and when U_0/kT is small enough that creep is so fast that J falls well below J_c during the experimental time scale.

Although useful for $J \ll J_c$, (9.36) does not describe the behavior for $J \approx J_c$, where the barrier, in fact, goes to zero rather than to U_0 and the Anderson-Kim form (9.35) is expected to hold. A useful interpolation formula between the two limits is

$$J(t) \approx J_c \left[1 + \frac{\mu kT}{U_0} \ln\left(1 + \frac{t}{t_0}\right)\right]^{-1/\mu} \tag{9.38}$$

This reduces properly to the Anderson-Kim form when $J \approx J_c$, in which case one can simplify by using the binomial theorem, causing μ to drop out of the formula. It also reduces to (9.36) for the case when $J \ll J_c$, apart from a numerical factor of μ. It is interesting that over a certain range of parameters this formula implies that J falls approximately exponentially with T (for given t) and approximately as an inverse power of t (for given T).

Although we have written these expressions in terms of a logarithmically decaying current density J, essentially the same dependence should describe the directly observable nonequilibrium magnetization $[M(t) - M_{\text{eq}}]$, which results from the nonequilibrium circulating currents in a bulk superconductor.

9.6.4 The Vortex-Glass Model

The work of Larkin and Ovchinnikov, described above, shows that in the presence of a density of pinning centers, no matter how weak, the crystalline long-range order of the FLL is destroyed beyond a correlation volume V_c, within which there is short-range order. In the previous section, we have shown in addition that the collective pinning model predicts a highly nonlinear glassy response to currents far below J_c, which differs from the classic Anderson-Kim flux-creep model (discussed in Sec. 5.7) in which the thermally activated *linear* mobility of the flux-line fluid remains *finite* at all temperatures but becomes exponentially small at low temperatures. In the absence of long-range spatial order in the FLL the operational question is: Is there still a phase transition at a well-defined glass-melting temperature T_g which separates a vortex-fluid phase with *linear* resistance from a vortex-solid (glass) phase, with *zero* linear resistance? The possible existence of a "vortex-glass" phase transition was proposed by M. P. A. Fisher,[62] and it has been discussed extensively by Fisher et al.[63] This model is summarized below.[64]

The postulated glass temperature T_g is defined as the dividing point between temperatures for which the linear resistance (i.e., R in the limit of $I \to 0$) is zero, and that in which it is not zero. Obviously, this is essentially the same as the resistive transition temperature determined at low current, so that $T_g(H)$ is operationally very much the same as the melting temperature $T_m(H)$ measured in (nearly) ideal crystals. Since experimental sensitivity excludes measurements at $I \equiv 0$, the data to be fitted are measurements of $V(I)$ made over a range of currents down to the lowest current for which satisfactory accuracy can be obtained. The test of the model is how well all these data can be accounted for in terms of a fitted $T_g(H)$ and a limited number of other parameters.

Fisher et al. approach the glass transition by means of scaling arguments. They define an exponent ν to describe the divergence of the vortex-glass phase-correlation length as

$$\xi_G \sim |T - T_g|^{-\nu} \tag{9.39a}$$

and another exponent z to describe the critical slowing down of the relaxation time

$$\tau_G \sim \xi_G^z \tag{9.39b}$$

Arguing that the electric field E should scale as 1/(length × time) and that J should scale as $1/(\text{length})^{D-1}$, where D is the spatial dimension, they reach the scaling hypothesis that

$$\xi_G^{z+1} E \approx \mathscr{E}_{\pm}(\xi_G^{D-1} J) \tag{9.40}$$

where $\mathscr{E}_{\pm}$ are (different) scaling functions for temperatures above (+) and below (−) the glass temperature T_g.

In the vortex-glass state below T_g, they expect that $\mathscr{E}_{-}(x)$ should be proportional to $\exp(-\text{const.}/x^{\mu})$, as noted above, so that the nonlinear electric-field response to a current density J is of the form

$$E(J) \sim \exp[-(J_T/J)^{\mu}] \tag{9.41}$$

where J_T is a temperature-dependent characteristic current density and $\mu \leq 1$. This form is equivalent to (9.34) and clearly goes strongly to zero as $J \to 0$, with no linear resistance term, but it cannot be expressed in terms of a power law.

---

[62] M. P. A. Fisher, *Phys. Rev. Lett.* **62**, 1415 (1989).

[63] D. S. Fisher, M. P. A. Fisher, and D. A. Huse, *Phys. Rev.* **B43**, 130 (1991).

[64] It should be noted, however, that rather similar predictions have been obtained by Nelson and coworkers based on a "Bose glass" model, discussed in the next section. That model should be more appropriate for the case of correlated disorder, as opposed to the random point disorder which underlies the collective pinning and vortex-glass models.

Exactly at T_g, Fisher et al. predict a power-law *I-V* characteristic, in which

$$E \sim J^{(z+1)/(D-1)} = J^{(z+1)/2} \tag{9.42}$$

where the final form is specialized for the 3-D case.

Finally, for $T > T_g$, one expects ohmic behavior $E \sim J$ for sufficiently low current levels, but from the form of the scaling hypothesis (9.40) we see that the appropriate comparison current level shrinks as $\xi_G^{-2} \sim |T - T_g|^{2\nu}$ near T_g, so experiments at very low current levels are required to bring out the linear behavior near T_g. This is a critical issue in testing the model.

These predictions have been carefully tested experimentally by Koch et al.[65] on thin film samples of YBCO, and subsequently by Gammel et al.[66] on single crystals, using a SQUID voltmeter to obtain higher sensitivity so as to be able to work at lower current levels. Both groups made measurements of the *I-V* curves at a series of closely spaced temperatures (~0.1 K interval) and found *I-V* curves which scaled as predicted by (9.40) above and below T_g. T_g itself was identified as the temperature at which a plot of log V vs. log I was a straight line, corresponding to the power-law form predicted by (9.42) with $z \sim 4 \pm 1$. As anticipated, this fitted $T_g(H)$ was similar in form to other measures of the resistive transition temperature. At lower temperatures the logarithmic plots curve downward with the voltage dropping rapidly as some temperature-dependent critical current is approached. On the other hand, at $T > T_g$ the curvature is positive, as the slope changes from unity (ohmic) at sufficiently low currents to a power approaching that in (9.42) for currents exceeding the linear range.

By rescaling the data according to (9.40), the experimenters were able to cause voltage data taken at many different temperatures to "collapse" onto a single universal functional dependence on the scaled current density, as shown in Fig. 9.11. In making these comparisons, it is convenient to rewrite the scaling equation (9.40) by transferring a factor of $(\xi_G^{D-1}J)$ from the function $\mathscr{E}_\pm(\xi_G^{D-1}J)$ to the left-hand side of the equation, to obtain a factor E/J, which becomes independent of J in the fully ohmic regime. Specializing to $D = 3$, (9.40) becomes

$$\xi_G^{z-1}E/J \approx \mathscr{E}_\pm(\xi_G^2 J)/(\xi_G^2 J) \tag{9.43}$$

Together with the temperature dependence of $\xi_G J$ given by (9.39*a*), this implies that $E/J|T - T_g|^{\nu(z-1)}$ should be a universal function of $J/|T - T_g|^{2\nu}$. Noting that the measured voltage V and current I are proportional to E and J, this accounts for the choice of variables used in Fig. 9.11. The data collapse shown there, and found by other groups on other samples, strongly supports the concept of a glass

[65] R. H. Koch, V. Foglietti, W. J. Gallagher, G. Koren, A. Gupta, and M. P. A. Fisher, *Phys. Rev. Lett.* **63**, 1511 (1989); R. H. Koch, V. Foglietti, and M. P. A. Fisher, *Phys. Rev. Lett.* **64**, 2586 (1990).

[66] P. L. Gammel, L. F. Schneemeyer, and D. J. Bishop, *Phys. Rev. Lett.* **66**, 953 (1991).

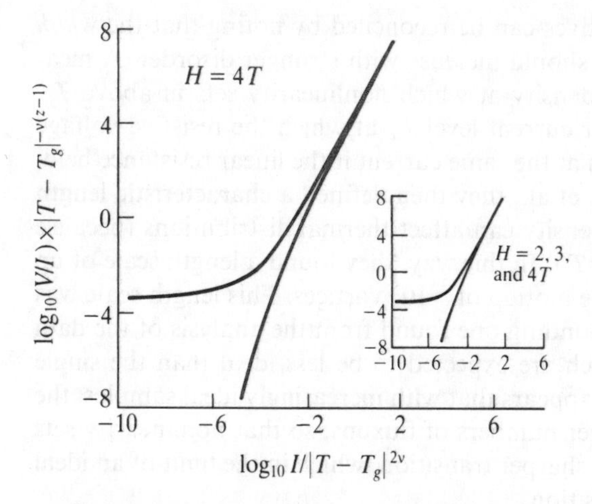

FIGURE 9.11
Empirical scaling functions for scaled nonlinear resistance vs. scaled current density for temperatures above and below the vortex-glass transition temperature T_g. This is a collapsed data plot of 119 *I-V* curves on a YBCO sample [*After Koch et al.*, Phys. Rev. Lett. *63, 1511 (1989); ibid. 64, 2586 (1990)*], with temperatures ranging from 84.5 to 72.7 K in 0.1 K intervals at $H = 4T$. The inset shows these data superimposed with similarly collapsed data at $H = 2$ and $3T$. In each case, $\nu = 1.7$ and $z = 4.8$ were used. The upper curve is for $T > T_g$; the flat part at the left corresponds to linear resistance at low current. The lower curve is for $T < T_g$ and shows no sign of approaching a nonzero linear resistance. The data points from all I–V curves superpose within the width of the plotted curves.

transition of the sort predicted by this model, or an alternative model (such as the Bose glass) with similar predictions.

Another implication of the scaling model is the temperature dependence of the linear resistance above T_g. This can be seen by noting that, in a range where $\mathscr{E}_+$ is linear in its argument as it must be to give an ohmic form, (9.40) implies that (in 3-D) E must be proportional to ξ_G^{1-z}. Upon inserting (9.39*a*), we obtain

$$R \sim (T - T_g)^{\nu(z-1)} \tag{9.44}$$

This implies that a plot of $(\partial \ln R/\partial T)^{-1}$ vs. T should be a straight line which extrapolates to zero at T_g with a slope $1/\nu(z-1)$. In this way, Gammel et al. found that $\nu(z-1) \approx 6.5$. Combining this result with the estimate for z from the power-law *I-V* relation at T_g, they obtained $\nu \approx 2 \pm 1$. These values are in agreement with the conclusions of Koch et al. based on YBCO films ($\nu \approx 1.7$ and $z \approx 4.8$), indicating that at least a modest degree of consistency exists over a considerable range of samples of this material.

Let us conclude this section by returning to the general question of the effect of pinning disorder on the melting transition of the vortex solid. As we noted in Sec. 9.5.2, experimental evidence supports the notion that the melting of the FLL in the best crystals is a *first-order* transition. On the other hand, we have been discussing the data of Koch et al. and of Gammel et al. on the basis of a continuous transition. These perspectives can be reconciled by noting that the *width* of the vortex-glass critical regime should increase with stronger disorder. A measure of this width is the current density at which nonlinearity sets in above T_g. Gammel et al. defined a crossover current level J_{sc} at which the resistive voltage was twice what it would have been at the same current if the linear resistance held. Following the approach of Fisher et al., they then defined a characteristic length scale ξ_d over which this current density can affect thermal distributions (because the coupling energy $J_{sc}\xi_d^2\Phi_0/c \sim kT$). In this way, they found a length scale of up to $\sim$15 μm, implying a cooperative motion of $\sim 10^5$ vortices. This length scale was $\sim$50 times larger than the corresponding one found from the analysis of the data of Koch et al. on thin films, which are expected to be less ideal than the single crystals of Gammel et al. Thus, it appears that with increasingly ideal samples, the current acts cooperatively on larger numbers of fluxons, so that nonlinearity sets in at lower current levels, giving a sharper transition, which in the limit of an ideal sample becomes the melting transition.

VORTEX GLASS IN TWO DIMENSIONS. According to the work of Fisher and of Fisher et al., cited above, the lower critical dimension for the vortex-glass state lies between 2 and 3. In other words, in a two-dimensional sample, there should be no phase transition into a zero-linear-resistance vortex-glass state except at $T = 0$. The results of experiments on a 16 A thick film of YBCO by Dekker et al.[67] are consistent with these predictions, and we summarize them here. Rather than diverging at a *finite* glass temperature T_g as in 3-D, in 2-D the vortex-glass correlation length is predicted to diverge as

$$\xi_{2D} \propto T^{-2\nu_{2D}} \tag{9.45}$$

as T approaches *zero*. Combining this dependence with the current scaling expected in 2-D from (9.40), we find that the characteristic 2-D current density at which nonlinearity sets in should scale as

$$J_{nl} \approx k_B T/\Phi_0\xi_{2D} \propto T^{1+\nu_{2D}} \tag{9.46}$$

Arbitrarily defining the measured J_{nl} to be the current density at which the logarithmic slope of E vs. J has increased to 1.2 from the linear resistance value of 1, Dekker et al. found that the J_{nl} measured in several fields from $0.5T$ to $5T$ gave an excellent fit to a power law $T^{3.0\pm0.3}$. This implies that $\nu_{2D} = 2.0 \pm 0.3$, in agreement with expectations from theoretical models. At all $T > 0$, there is a linear resistance T_{lin} below this J_{nl}, in contrast to the 3-D case where the linear resistance vanishes below a *nonzero* T_g. However, because of barriers to fluxon motion, R_{lin} is expected theoretically to freeze out exponentially as

$$R_{\text{lin}} \propto \exp[-(T_0/T)^p] \tag{9.47}$$

[67] C. Dekker, P. J. M. Wöltgens, R. H. Koch, B. W. Hussey, and A. Gupta, *Phys. Rev. Lett.* **69**, 2717 (1992).

where T_0 is a characteristic temperature and p is of the order of unity. The experiments of Dekker et al. found $p \approx 0.6$, whereas T_0 was found to depend on magnetic field, ranging from 80 to 230 K. Finally, while Dekker et al. found that their data could not be fitted to the 3-D scaling form (9.43), they found that the data collapsed nicely when $E/(JR_{\text{lin}})$ was plotted against J/J_{nl}, with J_{nl} and R_{lin} varying with T as given by (9.46) and (9.47), respectively.

9.6.5 Correlated Disorder and the Boson Glass Model

In the preceding three sections, we have outlined how the Larkin-Ovchinnikov theory of collective pinning by a *random* collection of weak point pinning sites can be used to predict critical currents, flux creep, and a possible vortex-glass phase. Although all real materials contain such pinning sites, their effects can be overshadowed in some circumstances by the effects of *correlated* disorder such as twin planes and columnar damage tracks left after bombardment by heavy ions. The basic reason is that the forces from random point pins add incoherently, whereas correlated disorder can give forces which add coherently over an extended pinning structure. This distinction is especially important in the high-temperature materials because operation at high temperatures requires a corresponding increase in pinning energies for pinning to be effective. For example, the major impact on the irreversibility line of defects induced by fission fragments in BSCCO tape was shown in Fig. 9.9. We have already mentioned the experiments of Kwok et al., which clearly demonstrated that twinned single crystals of YBCO have a peak in critical current when the magnetic field is aligned with the twin planes, and even with the *ab* planes of the ideal crystal. Here we shall sketch a model proposed by Fisher et al.[68] and developed further by Nelson and Vinokur,[69] which leads to predictions which resemble those of the vortex-glass model, but are based on *correlated* disorder, specifically columnar defects aligned in the direction of the applied magnetic field.

For a magnetic field aligned sufficiently parallel with these defects, the vortices are strongly attracted to lie in them. Nonetheless, given sufficient thermal energy, vortices wander from one columnar pin to another in an unconfined diffusive path, allowing them to respond to the Lorentz force of currents, giving linear resistance. At sufficiently low temperatures, however, the vortices are *all* localized, each to the vicinity of at most a few columnar pins, and the linear resistance is zero. This latter regime is called the *Bose glass phase* because the analysis is based on a mapping of the localization of flux lines along columnar

[68] M. P. A. Fisher, P. B. Weichman, G. Grinstein, and D. S. Fisher, *Phys. Rev.* **B40**, 546 (1989).

[69] D. R. Nelson and V. M. Vinokur, *Phys. Rev. Lett.* **68**, 2398 (1992).

pins in three dimensions onto the localization of bosonic particles in potential minima in two dimensions.

The model starts with the classical free energy F_N for N fluxons, defined by their trajectories $\mathbf{r}_j(z)$ as they traverse a sample of thickness L with columnar pins and magnetic field both aligned along the z axis, which is perpendicular to the CuO_2 planes. This free energy contains three terms, each integrated over the length of the fluxon: (1) a term proportional to $|d\mathbf{r}_j(z)/dz|^2$ representing the increase in line energy due to meander away from the z direction, (2) a term describing the defect pinning potential $U_D[\mathbf{r}_j(z)]$, and (3) an interaction potential depending on the separation of pairs of fluxons at the same height z. By replacing z by t, one finds that the first term becomes a kinetic-energy term, the second remains an attractive potential energy centered on the position of each defect in the xy plane, and the interaction term represents an equal time interaction between particles in the xy plane. The classical statistical mechanics of the fluxons in three dimensions is then equivalent to the quantum mechanics of interacting bosons in two dimensions with a random static potential $U_D(\mathbf{r})$.

This problem can be solved in a tight-binding approximation, involving three parameters: a chemical potential to control the number of particles (fluxons), a repulsive energy for two fluxons on the same site, and hopping matrix elements connecting nearby sites. It can be shown by standard theoretical methods that the partition function is determined by the ground-state energy of this fictitious quantum problem. Localization of a fluxon on a single columnar pin is equivalent to the binding of a particle to a single potential minimum. It is the presence of many fluxons with interaction terms among them which makes the transition between the glassy and resistive states occur as a sharp phase transition at a Bose glass temperature T_{BG}, instead of as a continuous crossover.

As $T \to T_{BG}$ from below, the transverse localization length $l_\perp$ is taken to diverge as $(T_{BG} - T)^{-\nu_\perp}$ with $\nu_\perp \gtrsim 1$. There is also a parallel correlation length along the z axis, $l_\| \sim l_\perp^2/D_0$, which is the distance along z that it takes a fluxon to "diffuse" across a tube of diameter $l_\perp$. If we assume that the short-distance "diffusion" constant D_0 remains finite at T_{BG}, then $l_\| \sim (T_{BG} - T)^{-\nu_\|}$ with $\nu_\| = 2\nu_\perp$.

Without going into the detailed arguments, we now summarize some of the observable predictions of this model. As in the previously discussed models, the Bose glass phase has the highly nonlinear *I-V* characteristic $V \sim \exp[-(J_0/J)^\mu]$, but with a definite prediction of $\mu = \frac{1}{3}$ at low current values and $\mu = 1$ at higher current values. A scaling theory of the dynamics near T_{BG} can be constructed paralleling that for the vortex-glass model described in the previous section. Here, however, there are *two* lengths $l_\perp$ and $l_\|$, divering with exponents $\nu_\perp \equiv \nu'$, and $\nu_\| = 2\nu_\perp \equiv 2\nu'$, instead of just one length, ξ_G, diverging with exponent ν according to (9.39*a*). The characteristic relaxation time τ is assumed to diverge with critical exponent z' as $\tau \sim l_\perp^{z'}$, instead of as ξ_G^z, as in (9.39*b*). The assumed scaling relation between the electric field and the current (9.40) is changed by replacing ξ_G^{z+1} in the left member by $l_\perp^{z'+1}$, and ξ_G^2 (for $D = 3$) by $l_\perp l_\|$ in the right member. This implies that in the ohmic regime above T_{BG}, the resistivity

$\rho \sim (T - T_{BG})^{\nu'(z'-2)}$, as compared to $(T - T_{VG})^{\nu(z-1)}$. Similarly, exactly at the glass temperature, the power-law *I-V* relation is expected to be $E \sim J^{(1+z')/3}$ as compared to $\sim J^{(1+z)/2}$ for the vortex-glass model. Given the limited consistency of empirical determinations of the critical exponents, comparable fits to the *I-V* data can be made with either Bose or vortex-glass models. Nelson and Vinokur argue, however, that the directional character of the correlated pinning, whether from columnar defects or twin boundaries, should lead to a characteristic *sharp* angular dependence in the position of the irreversibility line, apparently visible in data of Worthington et al.,[70] which would not be expected in the essentially isotropic vortex-glass model.

9.7 GRANULAR HIGH-TEMPERATURE SUPERCONDUCTORS

The discussion of pinning effects in the previous section treated the inhomogeneities as isolated weak spots in otherwise ideal crystalline material. Difficulties in preparation, however, limit the size of actual single-crystal samples to dimensions of millimeters or less. Even these "single-crystal" samples are typically broken up by twin planes with a separation of only ~1,000 A. Truly macroscopic samples of the sort required for large-scale applications for superconducting transmission lines and magnets are inevitably polycrystalline, with grain sizes typically on the scale of microns. Such structure is particularly conspicuous in granular material prepared by reacting together a compressed pellet of mixed oxide powders. In such samples, grains of relatively good stoichiometric crystalline material are separated by off-stoichiometry material which may be superconducting at a lower temperature, nonsuperconducting but metallic, or even semiconducting. In discussing the properties of materials of these sorts, it is useful to consider them as an array of superconducting grains weakly coupled by Josephson junctions, a view which is complementary to the usual perspective of local pinning sites in an ideal matrix. The Josephson junctions model the weak linkage between grains, whether they have the character of *S-N-S* or *S-I-S* junctions or of metallic weak links. So long as we focus on static properties, any of these can be modeled adequately by simply specifying the critical current of the link.

It actually seems likely that intergranular coupling is better modeled as an *S-N-S*-type weak link than as a tunnel junction that is directly connecting two bits of *ideal* material because the short coherence length allows a continuous depression of the superconducting order parameter near the interface. Thus, the Josephson coupling energy E_J and the critical current I_c are expected in general to be *less* than would be implied for the same normal resistance by the ideal Ambegaokar-Baratoff (AB) relation (6.10). To get a feel for the numbers, if the intergranular

[70]T. K. Worthington et al., *Physica* **153C**, 32 (1988), Fig. 4.

resistance adds $\sim 10^{-4}\ \Omega$-cm to the normal resistivity, and if the grain size a is $\sim 1\ \mu$m, it follows that the resistance of a typical intergranular link is $\sim 1\ \Omega$. For ideal AB junctions of YBCO, this would imply a link critical current (at $T = 0$) of $I_c(0) \sim 10^{-2}$ A, or $J_c(0) \sim 10^6$ A/cm^2. This value is comparable with J_c in crystals, but ceramic samples usually have J_c of only 10^3 to 10^4 A/cm^2, indicating a sizable reduction below ideal values. Another unfortunate experimental manifestation of a weak-linked material is that the critical current is further depressed strongly by the application of a magnetic field, as discussed for individual junctions in Sec. 6.4. Yet another complication is that if there is considerable variation in the strength of the Josephson coupling from grain to grain, the properties of the sample may be dominated by relatively few percolating paths of particularly strong junctions.

9.7.1 Effective Medium Parameters

In analyzing the response of the granular composite medium, we must separate two regimes depending on the strength of the field H relative to Φ_0/a^2, the field which puts a flux quantum into the characteristic area of the structure. For fields higher than this, the intrinsic parameters of individual grains play an explicit role. For fields lower than this, the electromagnetic response of the material can be treated as that of a homogeneous effective medium with suitably chosen parameters, just as the Josephson coupling between planes in YBCO is described by an effective mass parameter in the anisotropic Ginzburg-Landau model. The following analysis is devoted to finding these macroscopic parameters in terms of the parameters of the underlying granular model.[71]

We adopt a highly simplified model consisting of grains of an ideal crystalline superconductor connected in a three-dimension cubic array (with lattice spacing a) by weak links, each having critical current I_c. In real samples, these parameters will presumably have a distribution of values, but we ignore this complication as well as the large anisotropy of the grains, using angular averages of parameters for making estimates. When illustrative numerical values are quoted, they will refer to T well below T_c.

Two limiting regimes occur in granular superconductors, depending on how tightly coupled the grains are. A useful measure of this coupling strength is the ratio of the macroscopic critical current density set by the Josephson effect coupling, namely, $J_{cJ} = I_c/a^2$, to that set by the critical current density J_{cg} inside the grains. Since J_c values in excess of 10^6 A/cm^2 are measured in crystalline material, we may presume that granular material with typical measured values of $J_c \sim 10^3$ to 10^4 A/cm^2 will correspond to the loosely coupled limit, in which the measured J_c is limited by J_{cJ}. For simplicity, we consider only such loosely

[71]The treatment given here is adapted from M. Tinkham and C. J. Lobb, "Physical Properties of the New Superconductors" in H. Ehrenreich and D. Turnbull (eds.), *Solid State Phys.* **42**, 91 (1989).

coupled grains. We emphasize that the regime treated here is totally different from that found in, e.g., granular aluminum, where the grain size is *small* compared to the coherence length. In that case, Clem[72] has shown that the effect of the weak links is simply to modify the parameter values in the standard dirty-limit Ginzburg-Landau theory.

We have already noted that the (zero-field) macroscopic critical current density will be set by the intergranular I_c through the relation

$$J_{cJ} = I_c/a^2 \tag{9.48}$$

In the following analysis, we shall take J_{cJ} as a given, empirical parameter which, together with the grain size a, characterizes the material.

In response to a magnetic field, a three-dimensional array of junctions sets up screening currents, analogous to those in a bulk sample, which prevent the field from penetrating deeply into the sample. To determine the effective penetration depth λ_J, we start with the expression for the current in a single Josephson junction. According to (6.11) and (6.12), a junction which connects grains i and j in the presence of a vector potential $\mathbf{A}$ carries a supercurrent given by

$$I = I_c \sin\left(\varphi_i - \varphi_j - \frac{2\pi}{\Phi_0}\int_i^j \mathbf{A} \cdot d\mathbf{s}\right) \tag{9.49}$$

where the expression in parentheses is the gauge-invariant phase difference between the grains i and j. Following the discussion of the London equations in Chap. 1, we can choose a gauge such that all of the φ_i's are zero, the so-called London gauge. Furthermore, if fields and currents are assumed to be small, we can approximate the sine by its argument. We also replace the integral by $\mathbf{A}a$, where a is the spacing between nearest neighbors, neglecting any phase gradient within the grain since it will be small if the grains are weakly coupled. Defining the macroscopic current density by $J = I/a^2$, we obtain

$$\mathbf{J} = -2\pi \frac{J_{cJ}a}{\Phi_0}\mathbf{A} \tag{9.50}$$

Taking the curl of this equation, and combining it with Ampere's law, gives

$$\nabla^2 \mathbf{h} = \frac{8\pi^2 J_{cJ}a}{c\Phi_0}\mathbf{h} \tag{9.51}$$

where $\mathbf{h}$ is the local value of the magnetic flux density. From this one sees that the magnetic field is screened exponentially in a characteristic screening length given by

$$\lambda_J = (c\Phi_0/8\pi^2 a J_{CJ})^{1/2} \tag{9.52}$$

[72] J. R. Clem, *Physica* **C153–155**, 50 (1988).

Comparing this with the London penetration depth $\lambda_L = (mc^2/4\pi n_s e^2)^{1/2}$, one can define an effective density of superconducting electrons n_s that is proportional to aJ_{cJ}.

Note that (9.52) reduces to the usual expression (6.38) for the Josephson penetration depth in a classic tunnel junction if a is replaced by $(d+2\lambda)$, the thickness of the region penetrated by flux. It also reduces essentially to the usual penetration depth if one replaces J_{cJ} by the GL critical current and replaces a by 2ξ; this allows easy conversion of our granular results to continuum results.

Taking the representative values $J_{cJ} = 10^3$ A/cm^2 and $a = 1$ μm, we find $\lambda_J = 5$ μm. This λ_J is the screening length for a magnetic field which is sufficiently weak that the induced intergranular currents remain well below I_{cJ}, so that our linear approximation holds. Since λ_J is 5 times the assumed grain size, this result is consistent with our assumption of slow spatial variations of the field on the scale of a.

The other characteristic length of the GL theory is the GL coherence length $\xi(T)$. In conventional superconductors, $\xi(T)$ reduces, for $T \ll T_c$, essentially to the BCS ξ_0 if the metal is "clean" and to $(\xi_0\ell)^{1/2}$ if it is "dirty," i.e., if the mean free path $\ell < \xi_0$; in both cases, it diverges as $(1 - T/T_c)^{-1/2}$ near T_c. The question is: What is the appropriate value of $\xi(T)$ to use in the effective medium treatment of the granular material? We recall that in the standard GL theory the maximum phase gradient $\nabla\varphi$ (i.e., at J_c) is $1/\sqrt{3}\xi$. For the granular model, the maximum phase gradient at J_{cJ} is $\nabla\varphi = \pi/2a$. Equating these two expressions, we obtain

$$\xi_J = 2a/\sqrt{3}\pi \sim 0.4a \tag{9.53}$$

The thermodynamic critical field H_c is defined by equating $H_c^2/8\pi$ with the condensation energy per unit volume. In conventional superconductors, H_c varies as $(1 - T/T_c)$ near T_c, and it reflects the free-energy gain of the phase-coherent BCS ground state relative to the normal state. In the weak-field loosely coupled granular case, the most nearly parallel energy is that resulting from the phase-locking of the Josephson coupling between grains. In our cubic array model, each grain has six neighbors. Summing over these six links, dividing by 2 to cancel double counting, and normalizing to a unit cell volume of a^3, we have $H_{cJ}^2/8\pi = 3E_J/a^3$, or

$$H_{cJ} = (12\Phi_0 J_{cJ}/ca)^{1/2} \tag{9.54}$$

Inserting our standard set of representative parameters, we obtain a value of ~1.6 Oe.

In GL theory, H_c, ξ, and λ are related by (4.20), namely, $\Phi_0 = 2\sqrt{2}\pi H_c\lambda\xi$. The physics underlying this relation is that the linear current response characterized by λ must break down at a maximum phase gradient defined by ξ^{-1}, at which the energy increase in the linear approximation reaches the condensation energy defined by H_c. If we use this formula to check the consistency of our estimates, we find that the product relation holds apart from a factor of $4/\pi$, which presumably stems from our rather crude definitions of ξ and H_c.

Using the values of λ_J and ξ_J found above, one can write the GL parameter κ_J as

$$\kappa_J = \lambda_J/\xi_J = (3\Phi_0 c/32 J_{cJ} a^3)^{1/2} \tag{9.55}$$

For our standard numerical values, this yields $\kappa_J \sim 10$.

According to the GL theory, the field for first fluxon penetration in a high-κ superconductor is $H_{c1} = (\Phi_0/4\pi\lambda^2)\ln\kappa$. Inserting the values of λ_J and κ_J from above, we find

$$H_{c1J} = (2\pi a J_{cJ}/c)\ln(\lambda_J/\xi_J) \tag{9.56}$$

which has the value ~0.5 Oe for our representative parameters. H_{c1J} sets the limit for the strength of external fields which are screened in λ_J by reversible surface screening currents. To screen fields $H > H_{c1J}$ in λ_J, one notes that the surface current density would need to exceed J_{cJ}, causing a breakdown of the Josephson weak links, allowing flux to penetrate farther. In this way, a sort of "Bean model" critical-state penetration[73] occurs for $H > H_{c1J}$, in which the flux penetrates *between* the grains,[74] with field gradient $4\pi J_{cJ}/c \sim 1{,}000$ Oe/cm for our representative parameters.

In conventional type II superconductors, the upper critical field $H_{c2} = \Phi_0/2\pi\xi^2$ is the field at which the material makes a second-order phase transition into the normal state; it goes linearly to zero as $T \to T_c$. If we apply the same formula to the granular system, using ξ_J, we obtain

$$H_{c2J} = 3\pi\Phi_0/8a^2 \tag{9.57}$$

dependent only on grain size and having a value ~24 Oe for $a = 1\,\mu$m. This temperature-independent field is essentially the field at which one flux quantum Φ_0 fits in each cell of area a^2. Since the superconductivity of the grains is essentially unaffected by such small fields, this field does *not* have the conventional significance of marking the extinction of all superconductivity. Rather, it marks the point at which the flux enclosed in each unit cell is sufficient to change by 2π the sum of the four gauge-invariant phase differences γ_i around the perimeter of a square plaquette. In the absence of any flux, all γ_i are zero in the ground state of the system to obtain the full binding energy E_J from each junction; accordingly, each can carry its full critical current I_c by increasing γ_i to $\pi/2$, so long as self-field effects can be neglected. The effect of an external magnetic field in introducing essentially random initial values of γ_i is to reduce the capacity of the ("frustrated") network to carry a net macroscopic supercurrent by suitable subsequent readjustment of the γ_i. Thus, H_{c2J} gives a measure of the field at which the

[73] See Sec. 5.6.

[74] Throughout this discussion, the exclusion of flux by the diamagnetism of the grains should be taken into account by the introduction of an effective permeability of the medium, as done by Clem. We continue to omit this factor for maximum conceptual simplicity, but it *will* significantly affect quantitative results.

macroscopic critical current will be substantially reduced. Allowing for randomness in the effective areas of the various junctions, one expects a nonoscillatory falloff of the macroscopic critical current, with H_{c2J} setting the scale for the roll-off to begin.

We now recapitulate our description of the behavior of a virgin granular sample as an external field is applied and increased from zero. In doing so, we emphasize that the stated numerical results are based on assumed parameter values of $J_{cJ} = 10^3$ A/cm^2 and $a = 1$ μm; these parameters are *illustrative only*. (1) For $H < H_{c1J}(\sim 0.5$ Oe), the field is screened exponentially over a distance $\lambda_J \sim 5\,\mu$m. (2) For $H_{c1J} < H < H_{c2J}(\sim 25$ Oe), the field penetrates (between the grains) a distance $\sim cH/4\pi J_{cJ}$ in a Bean-type critical state, leaving the grains as partially diamagnetic inclusions. (3) As H increases toward and above H_{c2J}, the effective J_{cJ} is reduced by a factor $\sim H_{c2}/H$ by the phase randomization, allowing even deeper field penetration. (4) So long as $H < H_{c1g}(\sim 500$ Oe), the field penetrates into each grain only to a depth $\lambda_g \sim 1,500$ A. (5) When $H > H_{c1g}$, fluxons enter the grains, setting up another "Bean model" screening *within* each grain, but with the penetration depth determined by J_{cg} instead of by J_{cJ}. (6) Finally, at H_{c2g}, all superconductivity is extinguished.

9.7.2 Relationship between Granular and Continuum Models

Although the model of Josephson coupled grains has a clear structural basis in the ceramic materials, in crystalline material it seems more natural to consider a continuum model, with inhomogeneities which pin vortices in place. In fact, one can consider these two models as limiting cases of the same picture, as discussed below.

We start by noting that the continuum model can be described within the granular framework by effectively discretizing the continuum using a length $\sim\xi$ as the mesh interval. The physical basis for this choice is simply that ξ sets the distance scale between points in the continuum at which ψ can take on independent values. We then identify the two parameters J_{cJ} and a of the granular model with continuum variables as follows: We set $J_{cJ} = J_c = cH_c/3\sqrt{6}\pi\lambda$, the usual GL intrinsic maximum current density, and invert (9.53) and set $a = \sqrt{3}\pi\xi/2$. Within this framework, the Josephson coupling energy associated with the link between two adjacent blocks of volume $\sim\xi^3$ is $E_J = \hbar I_c/2e = (\hbar/2e)J_c\xi^2 \sim (H_c^2/8\pi)\xi^3$, using the standard GL identity $\Phi_0 = 2\sqrt{2}\pi\lambda\xi H_c$. Evidently, this energy is essentially the same as the characteristic energy associated with a length ξ of a flux-line vortex, which is the characteristic energy from which the usual pinning energies are derived.

In the conventional continuum pinning view, the vortex pinning energy results from the fractional modulation of this characteristic energy by the modulation of physical parameters of the material over a length scale ξ. In the granular view, the pinning arises because of variations of E_J from link to link, which modulates the increase in Josephson energy associated with the presence of a vortex center within a plaquette of junctions as discussed in Sec. 6.6.2. Since we have just seen that, for the continuum, the energy scale is the same in both points of view, and, in both, the depth of energy modulation giving pinning comes from the degree of disorder in the material, it is clear that the two descriptions are closely entwined. It may sometimes be convenient to use concepts from the discrete granular model for nominally continuum cases, as well. For example, we noted in connection with fabricated arrays of Josephon junctions in Sec. 6.6.2 that the critical current against flux motion scales with the amount of inhomogeneity in the array. It can be raised to over half the theoretical maximum that would exist in the absence of fluxons by the insertion of sufficiently strong disorder, such as missing junctions or islands. In the context of the continuum model, the corresponding statement is that the ideal GL depairing critical current sets the limit for J_c, which can be approached in the presence of fluxons in proportion to the fractional modulation of the superconducting order parameter on a length scale of ξ.

9.7.3 The "Brick-Wall" Model

Tape conductors made of polycrystalline BSCCO drawn and rolled in a silver matrix have been shown[75] to have remarkably high critical current densities of up to $\sim 3 \times 10^5$ A/cm^2 in zero field and as high as $\sim 10^5$ A/cm^2 even in a field of 23T. This behavior is very promising for applications. It is also surprising because the weak links between crystallites in granular materials usually limit the critical current density, especially in the presence of a magnetic field, which gives the Fraunhofer-pattern reduction in critical current discussed in Sec. 6.4. Bulaevskii, Clem, Glazman, and Malozemoff[76] have proposed a grain-structure model to account for these properties. It is based on the strongly laminar character of these materials, in which the *ab* planes predominantly lie in the tape plane. Specifically, it is postulated that grains of thickness D and length $2L \gg D$ are laid on top of one another in a staggered manner reminiscent of a brick wall. Such an arrangement resembles the experimentally observed microstructure, and its implications for macroscopic critical current density had been noted earlier by Mannhart and Tsuei.[77]

The current travels along the length of a grain, gradually transferring to the grains above and below, from which the current again gradually transfers to other grains, etc. Because of the highly elongated grain shape, the current density in the c direction across the intergranular junctions is only a fraction $\sim D/L \sim 0.1$–0.01 times the current density along the *ab* planes of the grains. Thus, even if the Josephson J_c in the c direction is much less than the in-grain J_c in the *ab* plane,

[75] J. Tenbrink, L. K. Heine, and H. Krauth, *Cryogenics* **30**, 422 (1990).

[76] L. N. Bulaevskii, J. R. Clem, L. I. Glazman, and A. P. Malozemoff, *Phys. Rev.* **B45**, 2545 (1992).

[77] J. Mannhart and C. C. Tsuei, *Z. Physik B-Condensed Matter* **77**, 53 (1989).

a high *macroscopic* in-plane J_c is possible because this geometrical factor assists the intergrain transfer of current.

The more surprising result in the data is the insensitivity to magnetic fields. This is explained in terms of an assumption that the Josephson J_c between platelets is strongly modulated on a short length scale (<1,000 A) determined by irregularities in the interfacial layer. As pointed out by Yanson[78] and discussed in Sec. 6.4.2, this can account for the observed type of $J_c(H)$. After an initial drop at low fields J_c remains on a plateau which extends to a field value that is inversely proportional to the product of the modulation length scale and the effective flux-bearing thickness. This plateau can be thought of as resulting from a sort of pinning of the Josephson vortices by the inhomogeneity of J_c. The key contribution of Bulaevskii et al. was to work out the effective thickness of the flux-bearing region. In the classic case of a Josephson junction between thick superconducting films, this effective thickness is $(d + 2\lambda)$, where d is the thickness of the barrier and 2λ accounts for the penetration into the films. In the present case, vortices are present in the grains because $H \gg H_{c1}$ of the grain, and further analysis is needed. Bulaevskii et al. conclude that the experimental data are consistent with the model using plausible parameter values.

9.8 FLUXONS AND HIGH-FREQUENCY LOSSES

In the early part of this book, we discussed the high-frequency electromagnetic properties of superconductors, first from the perspective of the simple London two-fluid model, and then using the BCS theory to provide a microscopic rationale for it as well as bringing in the new features related explicitly to the onset of dissipation at frequencies above the energy gap. This analysis must be generalized to describe the high-frequency losses in a type II superconductor in a static field $H > H_{c1}$, so that it is permeated by B/Φ_0 vortices per unit area. Various aspects of this problem have been treated by many authors, but a particularly comprehensive and unified approach has been presented in a series of papers by Coffey and Clem.[79] The general case is complicated by the many possible geometrical relationships among the direction of the ac field, the dc field, the surface normal, and the anisotropy axis of the superconductor. Moreover, for any given geometry, many règimes exist, depending on frequency, temperature, pinning strength, etc.

In order to introduce the main ideas as simply as possible, we shall restrict ourselves to the simplest geometry: ac and dc magnetic field parallel to each other and to the surface of a half-space of *isotropic* superconductor. We shall also

[78] I. K. Yanson, *Sov. Phys. JETP* **31**, 800 (1970).

[79] See, e.g., M. W. Coffey and J. R. Clem, *Phys. Rev. Lett.* **67**, 386 (1991); *Phys. Rev.* **B45**, 9872 (1992); *Phys. Rev.* **B45**, 10527 (1992); J. R. Clem and M. W. Coffey, *J. Supercond.* **5**, 313 (1992); M. W. Coffey, *Phys. Rev.* **B46**, 567 (1992).

restrict ourselves to frequencies below the energy gap to eliminate the possibility of pair-breaking processes, so we can focus on the dissipation associated with the normal fluid and with the driven motion of fluxons. In addition, we shall restrict ourselves to amplitudes low enough so that all responses can be assumed to be linear. [This excludes the glassy response regime but includes the thermally assisted flux-flow (TAFF) regime.] We can then completely describe the response of the medium to the incident wave by its complex surface impedance $Z_s = R_s - iX_s$. Here the surface resistance R_s is proportional to the absorptivity of the surface, as discussed in Sec. 2.4.2, and the surface reactance X_s describes the out-of-phase, or nondissipative, response to the drive field. The surface impedance in turn is uniquely determined by the *complex* penetration depth $\tilde{\lambda}$ by the relation

$$Z_s = R_s - iX_s = -i(4\pi\omega/c)\tilde{\lambda}(\omega, B, T) \tag{9.58}$$

This $\tilde{\lambda}(\omega, B, T)$ describes the exponential decay and phase evolution of the rf magnetic field $\mathbf{b}$ by the defining relation

$$\mathbf{b}(x, t) = \hat{\mathbf{z}} b_0 e^{-x/\tilde{\lambda}} e^{-i\omega t} \tag{9.59}$$

in the superconductor which occupies the half-space $x > 0$.

Coffey and Clem determine $\tilde{\lambda}(\omega, B, T)$ by finding a self-consistent solution for $\mathbf{B}(x, t)$ and the vortex density $n(x, t)$, which does not need to be exactly B/Φ_0. Coarse-grained over several lattice spacings, the London equation in the presence of vortices [see (5.10)] is

$$\operatorname{curl} \mathbf{J}_s = -\frac{c}{4\pi\lambda^2}(\mathbf{B} - n\Phi_0\hat{\mathbf{z}}) \tag{9.60}$$

Note that this equation allows B to deviate from $n\Phi_0$, in which case screening currents flow over a distance $\sim\lambda$. Given a time dependence, $\mathbf{B}$ is also screened by the current $\sigma_{nf}\mathbf{E}$ of normal fluid driven by the induced electric field. Combining these two responses with Maxwell's equations, we have the generalized diffusion-London equation for $\mathbf{B}$:

$$\nabla^2\mathbf{B} = \frac{4\pi\sigma_{nf}}{c^2}\dot{\mathbf{B}} + \frac{1}{\lambda^2}(\mathbf{B} - n\Phi_0\hat{\mathbf{z}}) \tag{9.61}$$

Turning now to the vortices, the equation of motion for their displacement u (along x) from equilibrium at pinning sites (assumed periodic for simplicity) is

$$\eta\dot{\mathbf{u}}(x, t) + k\mathbf{u}(x, t) = \mathbf{J}(x, t) \times \Phi_0\hat{\mathbf{z}}/c \tag{9.62}$$

where η is a viscous drag coefficient and k is the restoring force constant, both per unit length. By integrating this with the vortex continuity equation, one can obtain the driven change in density $\delta n(x)e^{-i\omega t}$ of vortices, given $\mathbf{J}(x, t)$. $\mathbf{J}$, on the other hand, is given by curl $\mathbf{B}$, which is influenced by the vortex density $n(x, t)$, as described by (9.61).

Solving in a self-consistent manner, Coffey and Clem found that, with certain restrictions, $\tilde{\lambda}$ can be written in terms of three other lengths as

$$\tilde{\lambda}(\omega, B, T) = \lambda\left(\frac{1 + i\tilde{\delta}_{\nu}^{2}/2\lambda^{2}}{1 - 2i\lambda^{2}/\delta_{nf}^{2}}\right)^{1/2} \tag{9.63}$$

We now review the significance of these three lengths, which characterize the ingredients in the response, and then mention some limiting cases of (9.63).

The first length is the London penetration depth $\lambda(B, T)$, which is real and taken to diverge both as $T \to T_c$ and as $B \to B_{c2}(T)$ to reflect the fact that $n_s \to 0$ at the second-order phase transition. For the purpose of making explicit calculations, Clem and Coffey assume the simple analytic dependence

$$\frac{1}{\lambda^{2}(B, T)} = \frac{(1 - t^{4})[1 - B/B_{c2}(T)]}{\lambda^{2}(0, 0)} \tag{9.64}$$

which builds in the conventional two-fluid temperature dependence, and the linear vanishing of n_s or $|\psi|^2$ at B_{c2}. This $\lambda(B, T)$ describes the superfluid response.

The next length is $\delta_{nf}(\omega, B, T)$, which is the generalization of the normal skin depth $\delta_n^{-2} = (2\pi\omega\sigma_n)/c^2$. Building in the (nonrigorous) two-fluid assumption that normal conductivities can be scaled by $n_{nf}/n = 1 - n_s/n$, and the requirement that σ_{nf} reduce to σ_n in the normal state above T_c or above B_{c2}, Coffey and Clem adopt the model dependence

$$\frac{1}{\delta_{nf}^{2}(\omega, B, T)} = \frac{1 - (1 - t^{4})[1 - B/B_{c2}(T)]}{\delta_{n}^{2}(\omega)} \tag{9.65}$$

to describe the strength of the normal fluid response.

The explicit effect of the *motion* of vortices, as distinct from the effect of their presence on the background conduction, is described by the (complex) third length $\tilde{\delta}_{\nu}(\omega, B, T)$. For simplicity, we consider only the case in which the vortices are pinned sufficiently strongly that activated motion from one pin to another can be neglected. (Clem and Coffey propose a generalization which attempts to take into account the thermally activated flux motion.[80]) Even a pinned vortex can still respond to the rf drive as described by (9.62). As pointed out by Gittleman and Rosenblum (see Sec. 5.5.4), the pinning and viscous terms define a crossover frequency $\omega_c = k/\eta = 1/\tau$, above which the pinning force becomes ineffective and the motion becomes pure flux flow,

---

[80] They generalize these results by introducing a flux-creep factor $\epsilon = 1/I_0^2(U_0/2kT)$, where U_0 is the height of the energy barrier which gives the pinning, and I_0 is the modified Bessel function found in the work of Ambegaokar and Halperin, and mentioned in Sec. 9.4. This factor ϵ reduces to unity when $U_0/2kT$ is small and goes exponentially to zero when $U_0/2kT$ is large, allowing it to be used to interpolate the values of ρ_f and τ between values that are appropriate to these limits.

whereas below ω_c pinning becomes dominant. More precisely, the fluxons add an effective complex resistivity

$$\tilde{\rho}(\omega, B, T) = \frac{\rho_f(B, T)}{1 + i/\omega\tau} \tag{9.66}$$

[This is expressed as a resistivity rather than as a conductivity, since it describes the voltage (from flux motion) caused by a drive current.] Inserting (9.66) into the ordinary skin depth analysis, we obtain the complex penetration depth given by

$$\frac{1}{\tilde{\delta}_\nu^2} = \frac{1}{\delta_f^2(\omega, B, T)} + \frac{i}{2\lambda_C^2(B, T)} \tag{9.67}$$

Here δ_f is the flux-flow skin depth (which would result at $\omega \gg \omega_c$, where the resistance is that of ideal flux flow ρ_f) and $\lambda_C = (B\Phi_0/4\pi k)^{1/2}$ is the Campbell penetration depth (which would result at $\omega \ll \omega_c$, when the flux motion is controlled by the restoring force due to pinning).

Inserting all these partial response depths into (9.63) yields the final result for the complex penetration depth (and hence the surface impedance) for the case of no activated motion of flux from pinning site to pinning site. Note that at T_c or $T_{c2}(H)$, where $\lambda \to \infty$, $\tilde{\lambda}$ reduces continuously to $\delta_n\sqrt{(i/2)}$ in the normal state, as it should. In the other, more typical, limit, when λ is much smaller than the other lengths, then $\tilde{\lambda}^2$ reduces to $i\delta_\nu^2/2$, which is wholly determined by the fluxon motion because the screening is so complete. At low frequencies ($\omega\tau \ll 1$), the Campbell depth dominates, and one has a purely reactive response, with $\tilde{\lambda}$ real; at high frequencies δ_f dominates, and, as in the normal state, $\tilde{\lambda}$ has equal real and imaginary parts.

9.9 ANOMALOUS PROPERTIES OF HIGH-TEMPERATURE AND EXOTIC SUPERCONDUCTORS

Although our choice of topics for coverage in this chapter has been guided by the desire to avoid topics on which consensus has not yet developed, rigorous application of this principle would force omission of some of the most exciting and challenging issues in high-temperature superconductivity today, namely, those having to do with the microscopic nature of the superconducting state and the mechanism causing it. The status of these issues will be briefly reviewed in this section, with the understanding that the present perspective is definitely open to evolution and change.

The open questions are often condensed into a single one: Is the superconducting pairing in these materials of the familiar *s*-wave type on which conventional BCS theory is based, or is it some other form of pairing, of which the most likely appears to be *d*-wave pairing. This question is central because microscopic mechanisms based on the exchange of antiferromagnetic spin fluctuations have been proposed[81] which would lead to $d_{x^2-y^2}$ pairing. If *d*-wave pairing were to be proven experimentally, it would lend support to such theories and exclude theories which were incompatible with *d*-wave pairing.

On the other hand, the bulk of this chapter has been devoted to demonstrating that most of the properties of the HTSC can be accounted for perfectly well in terms of the ordinary BCS and GL theories, if one simply inserts the observed high T_c value, the natural anisotropy due to the planar crystal structure, and estimates of the Fermi velocity and density of states. In particular, there is no doubt that the superconductivity is based on *Cooper pairs* of electrons with *zero net momentum* because the usual ac Josephson effect frequency $2eV/h$ is observed,[82] the observed[83] flux quantum is of the usual size $hc/2e$, and Andreev reflection along time-reversed trajectories is seen[84] as with conventional superconductors. Moreover, the fact that the Knight shift is observed[85] to go smoothly to zero at $T = 0$ implies that the pairing is again of the *spin singlet* form. This leads to the question: What experimental properties are there that are *not* compatible with conventional *s*-wave superconductivity with material anisotropy, such as was studied years ago in TaS_2 and other layered low-temperature superconductors.

This question can be addressed along two directions: (1) superconducting phenomena directly reflecting the symmetry of the paired state and (2) phenomena reflecting the density of states for quasi-particle excitations, i.e., whether there is a clean energy gap, or whether there are states in the "gap" as would be predicted for *d*-wave pairing. Historically, most evidence has been of the second, or indirect, type. Unfortunately, it is less convincing because imperfect samples can have "states in the gap" which are not intrinsic. Nonetheless, an enormous number of samples have been examined in various ways, and there is sufficient consistency in the deviations from "clean gap BCS" that these results cannot be ignored. Recently, the situation was transformed by measurements in several laboratories of flux in superconducting loops with Josephson junctions involving HTSC. These experiments have given evidence of the first kind, specifically supporting the *d*-wave symmetry. Although there is as yet no unanimity on this point, the evidence is quite strong. Accordingly, after a brief review of the nature of unconventional pairings, we shall reverse the historical order and first discuss these recent experiments.

---

[81]See, e.g., N. E. Bickers, D. J. Scalapino, and S. R. White, *Phys. Rev. Lett.* **62**, 961 (1989); P. Monthoux, A. V. Balatsky, and D. Pines, *Phys. Rev. Lett.* **67**, 3448 (1991) and *Phys. Rev.* **B46**, 14803 (1992); P. Monthoux and D. Pines, *Phys. Rev.* **B49**, 4261 (1994); P. Monthoux and D. J. Scalapino, *Phys. Rev. Lett.* **72**, 1874 (1994).

[82]D. Esteve et al., *Europhys. Lett.* **3**, 1237 (1987).

[83]C. E. Gough et al., *Nature* **326**, 855 (1987); R. H. Koch et al., *Appl. Phys. Lett.* **51**, 200 (1987); P. Gammel et al., *Phys. Rev. Lett.* **59**, 2592 (1987).

[84]H. F. C. Hoevers et al., *Physica* **C152**, 105 (1988).

[85]S. E. Barrett et al., *Phys. Rev. Lett.* **66**, 108 (1991) and *Phys. Rev.* **B41**, 6283 (1990); M. Takigawa, P. C. Hammel, R. H. Heffner, and Z. Fisk, *Phys. Rev.* **B39**, 7371 (1989).

9.9.1 Unconventional Pairing

From our discussion of the BCS theory in Chap. 3 we recall that the gap parameter in state $\mathbf{k}$ was defined in a self-consistent manner in terms of the paired occupancy of all states $\mathbf{k}'$ as

$$\Delta_{\mathbf{k}} \propto \langle c_{-\mathbf{k}\downarrow} c_{\mathbf{k}\uparrow} \rangle \propto -\sum_{\mathbf{k}'} V_{\mathbf{k}\mathbf{k}'} \langle c_{-\mathbf{k}'\downarrow} c_{\mathbf{k}'\uparrow} \rangle \tag{9.68}$$

In our discussion there, we assumed an isotropic system, so that $\Delta_{\mathbf{k}}$ was independent of $\mathbf{k}$, and because of the spherical symmetry, this is often referred to as *s*-wave pairing. However, if the material is anisotropic (weakly, like Sn, or strongly, like YBCO), then one expects that $\Delta_{\mathbf{k}}$ will no longer be isotropic, but that its dependence on $\mathbf{k}$ will have the same symmetry as the underlying crystal symmetry. For example, a tetragonal crystal will normally have a different gap for $\mathbf{k}$ along $\mathbf{c}$ than for $\mathbf{k}$ along $\mathbf{a}$ or $\mathbf{b}$. The variation of $\Delta_{\mathbf{k}}$ at intermediate directions does not need to be simple, but the symmetry cannot be lower than tetragonal. This situation is sometimes referred to as *anisotropic s-wave pairing*: It lacks spherical symmetry but has the full symmetry of the underlying crystal. The term "unconventional pairing" refers to a situation in which the symmetry of the energy gap function $\Delta_{\mathbf{k}}$ is *lower* than that of the underlying crystal.

To address the physics as simply as possible, we return to the original Cooper-pairing argument given in Sec. 3.1, and consider the construction of a two-electron bound state wavefunction from plane waves (or more precisely, Bloch functions with the same $\mathbf{k}$). We retain the pairing of states of equal and opposite momentum. As discussed before (3.1), the required antisymmetry of the two-electron wavefunction can be achieved with either singlet or triplet spin pairing, depending on whether we choose to write the orbital wavefunction as a sum of even or odd functions of $(\mathbf{r}_1 - \mathbf{r}_2)$. In the case treated by Cooper, which led to the BCS model, the singlet was chosen and the orbital part $\Sigma_k \, g(\mathbf{k}) e^{i\mathbf{k}\cdot\mathbf{r}_1} e^{-i\mathbf{k}\cdot\mathbf{r}_2}$ was made symmetric in $(\mathbf{r}_1 - \mathbf{r}_2)$ by choosing $g(-\mathbf{k}) = g(\mathbf{k})$, so that this became a sum of cosine terms

$$\sum_k g(\mathbf{k}) \cos \mathbf{k} \cdot (\mathbf{r}_1 - \mathbf{r}_2) \tag{9.69}$$

The sum was chosen further to form an *s* state by taking $g(\mathbf{k}) = g(k)$, a function only of the *magnitude* of $\mathbf{k}$. This was a natural result of the Cooper approximation that all of the matrix elements $V_{\mathbf{k}\mathbf{k}'}$ had the same value $-V$.

Recalling the definition of $V_{\mathbf{k}\mathbf{k}'}$ from (3.3), namely,

$$V_{\mathbf{k}\mathbf{k}'} = \Omega^{-1} \int V(\mathbf{r}_1 - \mathbf{r}_2) e^{i(\mathbf{k}-\mathbf{k}')\cdot(\mathbf{r}_1 - \mathbf{r}_2)} d(\mathbf{r}_1 - \mathbf{r}_2) \tag{9.70}$$

we see that $V_{\mathbf{k}\mathbf{k}'}$ will in general depend on the *direction* of $\mathbf{k} - \mathbf{k}'$. Therefore, the relation (3.2) cannot be simplified to (3.4), and the expression for g acquires the more general form

$$g(\mathbf{k}) = \frac{\Sigma_{\mathbf{k}'} V_{\mathbf{kk}'} g_{\mathbf{k}'}}{2\epsilon_{\mathbf{k}} - E} \tag{9.71}$$

This implies that $g(\mathbf{k})$ is a function of the *direction* of $\mathbf{k}$ as well as of the energy $\epsilon_{\mathbf{k}}$, and $\Delta(\mathbf{k})$ has the same angular dependence as $g(\mathbf{k})$ since $g(\mathbf{k})$ determines the expectation value $\langle c_{-\mathbf{k}\downarrow} c_{\mathbf{k}\uparrow} \rangle$. Had we chosen the triplet spin pairing, we would have obtained a similar equation, but the $g(\mathbf{k})$ would be an *odd* function of $\mathbf{k}$ to make the orbital function antisymmetric in $(\mathbf{r}_1 - \mathbf{r}_2)$. Without the simplification of isotropy, the solution of (9.71) is no longer trivial. However, the degree of symmetry of $V_{\mathbf{kk}'}$ will be reflected in the symmetry of the solution for $g(\mathbf{k})$ or $\Delta(\mathbf{k})$.

The group theory of this relationship has been analyzed for various crystallographic symmetries by Annett et al.[86] For YBCO, e.g., the situation is complicated by the question of whether the appropriate symmetry is tetragonal or orthorhombic, i.e., whether the superconductivity is dominated by the "planes," which are square insofar as one can treat the **a** and **b** directions as equivalent, or by the "chains" which distinguish the two directions. For the case of tetragonal symmetry with a singlet spin state, Annett et al. identify seven possible symmetries for the gap function, including the nodeless *s* wave having the full tetragonal symmetry of the crystal, and a number of *d*-wave symmetries, including the $(x^2 - y^2)$ symmetry which is suggested in some theoretical models of the relevant interactions. This function is clearly of lower symmetry than tetragonal since it changes sign on the interchange of x and y, which are equivalent in tetragonal symmetry. Hence, it is a clear example of unconventional pairing.

Finally, we remark on the radial, rather than the angular, dependence of the pair state. If the interaction has a strong repulsive core at short distances, and attraction at larger distances, the optimal radial dependence of the pair wavefunction will have a small amplitude for small values of the interparticle distance. This is accomplished automatically by a *d* wave, where the probability amplitude goes to zero as r_{12}^2 at small distances. On the other hand, if the potential is attractive all the way in to $r_{12} = 0$, an *s*-wave solution as found in the BCS case is more favorable.

9.9.2 Pairing Symmetry and Flux Quantization

If, in fact, $\Delta(\mathbf{k}) \propto (k_x^2 - k_y^2) \propto \cos 2\theta$, where θ is measured from the x axis in the *ab* plane, the sign change with angle can introduce an intrinsic phase shift by π into the Josephson effect formalism. As shown by Sigrist and Rice,[87] the Josephson current through a junction between two superconductors (i and j)

---

[86] J. F. Annett, N. Goldenfeld, and S. R. Renn, "The Pairing State of $YBa_2Cu_3O_{7-\delta}$", in D. M. Ginsberg (ed.), *Physical Properties of High Temperature Superconductors II*, World Scientific, Singapore (1990), p. 571.

[87] M. Sigrist and T. M. Rice, *J. Phys. Soc. Jpn.* **61**, 4283 (1992); also *Revs. Mod. Phys.* **67**, 503 (1995).

involving such a d-wave superconductor includes a factor $\cos 2\theta_i$, where θ_i is the angle between the normal to the junction interface and the crystal a axis in superconductor i. Since the Josephson coupling is proportional to $\Delta_i \Delta_j$, this simply reflects the angular dependence of $\Delta(\mathbf{k})$ for the electrons moving normal to the interface, which dominate the tunneling, at least for an ideal planar interface. An s-wave superconductor has no such angular dependence. For simplicity, we consider only cases in which the interface normals are either along the **a** or **b** axes, so that this factor $\cos 2\theta_i$ is either ± 1, corresponding to a phase shift of 0 or π. A particularly simple example, discussed by Sigrist and Rice (and earlier by Geshkenbein et al.[88] in another context), involves a double-junction dc SQUID loop which combines s-wave and d-wave superconductors, as pictured in Fig. 9.12a. Because the two Josephson contacts to the d-wave superconductor are made on interfaces which are rotated by $\pi/2$ in the ab plane, the $\cos 2\theta$ factor for the two junctions must differ by $\cos \pi = -1$. This is equivalent to introducing a phase shift of π in the general relation (6.27) or, alternatively, to introducing a formally *negative* critical current for one of the junctions. The effect on (6.28) of shifting the phase sum by π can be seen to be the same as changing the enclosed flux Φ by $\Phi_0/2$. Accordingly, the lowest-energy state of the ring is now one containing $\Phi_0/2$ of trapped flux, rather than zero flux. Or if the device is subjected to an external field which is swept up and down, the various allowed states of the

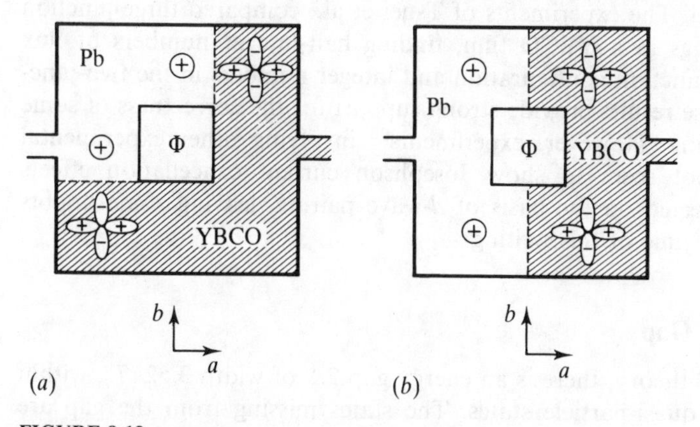

FIGURE 9.12
Superconducting SQUID rings combining films of Pb (with s-wave symmetry) and YBCO (with assumed $d_{x^2-y^2}$ symmetry). The dashed lines represent Josephson junctions between the two metals. In configuration (a), the intrinsic angle-dependent sign of the order parameter effectively introduces a phase difference of π between the two junctions, leading to a half-integral number of flux quanta trapped in the ring. In (b), both junctions are on the same face of the assumed d-wave superconductor, so that there is no such phase difference, and the quantization leads to integral numbers of trapped flux quanta, as expected with conventional superconductors.

[88] V. B. Geshkenbein, A. I. Larkin, and A. Barone, *Phys. Rev.* **B36**, 235 (1987).

ring will be those with a *half-integral fluxoid quantum number*. Assuming that $LI_c \gg \Phi_0$, this implies that the actual *flux* trapped will also be a half-integral multiple of Φ_0, which excludes the possibility of zero flux! As a result, this is sometimes referred to as a type of superconducting paramagnetism rather than diamagnetism.

Although the concept of this experimental test is simple, many spurious effects must be excluded before conclusions can be drawn from raw data. For example, flux pinned in the superconducting material or flux generated by the bias current or the measuring apparatus can give offsets which may act like fractional quanta. The pioneering experiments were carried out by Wollman et al.,[89] who compared the SQUID characteristics of the two configurations shown in Fig. 9.12, finding the expected shift by $\Phi_0/2$ in the response to an externally applied flux. The possibility of errors from undetected stray trapped flux was addressed in subsequent experiments by Tsuei et al.[90] and by Mathai et al.[91] Both experiments used a scanning SQUID microscope not only to measure the flux trapped in the ring under study, but also to scan the entire sample configuration for trapped flux or pinned vortices. The experiments of Mathai et al. compared the two variants of the two-junction SQUID shown in Fig. 9.12—the one proposed by Sigrist and Rice and another in which both junctions were on the same face of the YBCO crystal, so that there is no net phase shift. They found half-quanta of trapped flux in the first case and integral numbers of quanta in the second, as predicted by the *d*-wave pairing model. The experiments of Tsuei et al., compared three-junction and two-junction rings in a YBCO film, finding half-integer numbers of flux quanta in the three-junction configuration and integer numbers in the two-junction rings. All of these results provide strong support for the correctness of some sort of *d*-wave pairing. However, experiments[92] involving other experimental configurations do not seem to show Josephson current cancellation effects which might be expected on the basis of *d*-wave pairing, leaving some doubts to be resolved at the time of this writing.

9.9.3 The Energy Gap

In the isotropic BCS theory, there is an energy gap 2Δ of width $3.52kT_c$, within which there are *no* quasi-particle states. The states missing from the gap are

[89] D. A. Wollman, D. J. Van Harlingen, W. C. Lee, D. M. Ginsberg, and A. J. Leggett. *Phys. Rev.* **71**, 2134 (1993); also see later work by D. A. Wollman, D. J. Van Harlingen, J. Giapintzakis, and D. M. Ginsberg, *Phys. Rev. Lett.* **74**, 797 (1995); D. J. Van Harlingen, *Revs. Mod. Phys.* **67**, 515 (1995).

[90] C. C. Tsuei, J. R. Kirtley, C. C. Chi, L. S. Yu-Jahnes, A. Gupta, T. Shaw, J. Z. Sun, and M. B. Ketchen, *Phys. Rev. Lett.* **73**, 593 (1994).

[91] A. Mathai, Y. Gim, R. C. Black, A. Amar, and F. C. Wellstood, *Phys. Rev. Lett*, **74**, 4523 (1995).

[92] A. G. Sun, D. A. Gajewski, M. B. Maple, and R. C. Dynes, *Phys. Rev. Lett.* **72**, 2267 (1994); P. Chaudhari and S. Y. Lin, *Phys. Rev. Lett.* **72**, 1084 (1994).

"pushed up" above the gap, forming a singular peak in the density of states at the gap edge. As discussed in Chap. 3, this modified density of states, together with the BCS coherence factors, provides the microscopic basis for calculating the electronic specific heat, the *I-V* curves for single-particle tunneling, the magnetic penetration depth λ, the frequency-dependent electromagnetic absorption, the relaxation rate $1/T_1$ for nuclear resonance, etc. The BCS theory explained all of these properties for the classic superconductors with remarkable success. Insofar as the BCS predictions are *not* completely successful for the HTSC, there is an indication of a need for a modified theory. Accordingly, much work has been done to look for deviations in these properties from the BCS predictions. Clear deviations have indeed been found, many of which might be explained in terms of the *d*-wave pairing. A thorough review of all this mass of work is completely beyond the scope of this book; we content ourselves with a brief look at what has been found. For more details and references, the reader is referred to the rather comprehensive and critical review of these issues by Annett et al.[93]

THE GAP WIDTH. Rather than the BCS weak-coupling value of $3.5kT_c$, most data suggest considerably larger values, such as $4kT_c$ to $7kT_c$, for the gap width. At the very least, this implies that the HTSC are not weak-coupling BCS superconductors, but it is possible that very strong coupling could account for such a high $2\Delta/kT_c$ ratio.

THE GAP SYMMETRY. Although the flux quantization experiments give strong support to a *d*-wave pairing scheme, which would account naturally for a change of sign of the Josephson coupling under rotation by $\pi/2$ in the *ab* plane, this behavior is not unique to a specific *d*-wave pairing scheme, such as $d_{x^2-y^2}$. However, support for that specific variety of *d*-wave pairing is provided by angle-resolved photoemission spectroscopy (ARPES) measurements[94] on BSCCO. These experiments have sufficient energy and angular resolution to allow the energy gap to be measured in different directions in *k* space. As would be expected for $d_{x^2-y^2}$ pairing, or something very similar, a gaplike feature is found to develop below T_c along the k_x or k_y direction, but not along directions rotated by $\pi/4$ from them in the *ab* plane.

STATES IN THE GAP. A very large number of experiments have been performed which give information about the density of states but without the angular resolution of the ARPES experiments mentioned above. Almost all these

---

[93] J. F. Annett, N. Goldenfeld, and S. R. Renn, "The pairing state of $YBa_2Cu_3O_{7-\delta}$" in D. M. Ginsberg (ed.), *Physical Properties of High Temperature Superconductors II*, World Scientific, Singapore (1990), p. 571.

[94] B. O. Wells et al., *Phys. Rev.* **B46**, 11830 (1992); Z.-X. Shen et al. *Phys. Rev. Lett.* **70**, 1553 (1993).

experiments suggest that the energy "gap" is not "clean," but has a finite density of states all the way down to zero energy. However, there is limited quantitative consistency between different samples and different experimental techniques on just how great this subgap state density is. Thus, it is not clear whether we are dealing with an intrinsic property, or one which depends on sample perfection. Because of the short coherence lengths in these materials, even single foreign atoms could depress the local gap. Another consequence of the short coherence lengths is that it may be necessary to invoke a "local" density of states, even in ideal crystals. For example, we might assume that the gap would be large in the planes and smaller in other parts of the crystal structure, so that the density of states inferred from a tunneling experiment would represent some sort of average of large and small (or zero) local energy gaps.

On the other hand, if the superconductivity is of the $d_{x^2-y^2}$ pairing type, the energy gap goes continuously to *zero* in k space along the diagonal directions, where $k_x^2 - k_y^2$ is zero. Presumably, the Fermi surface for the electrons in the planes is approximately a cylinder with the axis along k_z. Thus, this nodal line running roughly parallel to the k_z axis would imply a density of states that is finite all the way down to zero energy. Nonetheless, the broad *maximum* in the gap along the x and y directions would be expected to show up as a strong feature in the density of states.

Because of the inevitable doubts as to whether tunneling experiments can avoid being compromised by surface states which are not representative of bulk material, it is attractive to examine electromagnetic properties, which probe a relatively deep surface layer of thickness λ. We examine two cases: energy gap measurements and dc penetration depths.

INFRARED MEASUREMENTS. The far-infrared measurements are aimed at discerning the energy gap, as was done with such impact for the classic superconductors in the early days of BCS. Unfortunately, interpretation of the data has proved to be much less simple with the HTSC because the absorption seems to include several distinct components, rather than the simple Drude absorption which dominates at the much lower frequencies relevant to the small energy gaps of the classic superconductors. Apparently, however, if there is a gap feature in the spectrum, it does not follow the temperature dependence of the BCS gap width, shifting down toward zero frequency as $T \rightarrow T_c$ but, rather, appears to fade out in intensity without a major frequency shift. Recent HREELS (High Resolution Electron Energy Loss Spectroscopy) data show a similar trend. In sum, these spectra cannot be understood in terms of basic BCS, but they have not yet pointed clearly to a specific alternative model.

PENETRATION DEPTH MEASUREMENTS. The detailed study of the temperature dependence of the penetration depth λ has yielded very interesting results. Within the two-fluid picture based on a BCS-like model, the temperature dependence of the superfluid density n_s, which determines the penetration depth by $1/\lambda^2 \propto n_s$, results from the backflow of the normal fluid of quasi-particles, as was discussed in Sec. 3.10.2. At very low temperatures the BCS prediction is that $[\lambda(T) - \lambda(0)]$ should approach zero as $e^{-\Delta/kT}$ as the quasi-particles above the gap freeze out. If instead of a uniform BCS gap Δ there were an $(x^2 - y^2)$ type of gap, it would give rise to a continuum of low-lying excitations, and the exponential freeze-out would be replaced by a linear temperature dependence $n_s \propto (1 - aT)$ at low temperatures. Such a dependence has, in fact, been found in two different types of recent measurements on especially high-quality crystals of YBCO. The first type[95] consists of microwave measurements at $\sim$1 GHz, with a very careful analysis of both the real and imaginary parts of the surface impedance. The second type of measurement involves the muon spin-rotation technique, which probes the penetration depth within the flux lines penetrating the sample in finite applied field, and hence should be independent of surface properties. The data of Sonier et al.[96] at 0.5 Tesla are fitted out to 50 K by the simple expression

$$\frac{1}{\lambda_{ab}^2(T)} = \frac{1}{\lambda_{ab}^2(0)}(1 - aT) \tag{9.72}$$

with $\lambda_{ab}(0) = 1451 \pm 3$ A and $a = 7.2 \pm 0.1 \times 10^{-3}\,\mathrm{K}^{-1}$. The degree of consistency of these two sets of measurements, one of the penetration depth at the surface and the other in the interior, is very impressive. It should be noted, however, that earlier measurements of both types had *not* shown such a linear dependence and were generally thought to be described by a $(1 - t^2)$ dependence, with $t = T/T_c$. According to theoretical calculations,[97] the difference may be due to the effect of impurities in earlier samples, which generally had lower T_c values and less sharp transitions. At higher temperatures the $\lambda(T)$ diverges as $\Delta^{-1} \sim (T_c - T)^{-1/2}$ as $T \to T_c$, so no simple analytical approximation can be expected to hold over the whole range. However, in YBCO, it is found that $n_s \propto (1 - t^2)$ gives a reasonable overall fit, certainly better than the Gorter-Casimir $(1 - t^4)$ dependence. This difference is qualitatively consistent with the more rapid variation observed at very low temperatures.

In summary, there is a general consensus that the variation of $\lambda(T)$ in YBCO at low temperatures is more rapid than would be consistent with a classic clean BCS gap. Although experimental data of different groups are not entirely consistent, there is substantial evidence favoring the linear temperature dependence at low temperatures which would be expected from the *d*-wave pairing.

NMR MEASUREMENTS. One of the triumphs of the BCS theory was the prediction of the rise in the nuclear relaxation rate $1/T_1$ to a "Hebel-Slichter peak"

---

[95] W. N. Hardy et al., *Phys. Rev. Lett.* **70**, 3999 (1993); D. A. Bonn, et al., *Phys. Rev.* **B47**, 11314 (1993).

[96] J. E. Sonier et al., *Phys. Rev. Lett.* **72**, 744 (1994).

[97] P. J. Hirschfeld and N. Goldenfeld, *Phys. Rev.* **B48**, 4219 (1993).

below T_c before an exponential drop-off to zero with the freeze-out of quasi-particles at lower temperatures. (See Fig. 3.9 and the discussion in Sec. 3.9.2.) Such a rise could not be accounted for by any simple two-fluid model since it would imply a density of normal electrons *greater than* that in the normal state. The explanation of the peak within the BCS theory involves two factors: the singular peak in the density of states just above the gap and the BCS coherence factors. The latter distinguish nuclear relaxation from ultrasonic attenuation, which does drop sharply just below T_c, as might have been expected.

When similar NMR (Nuclear Magnetic Resonance) studies were performed on the HTSC, the striking observation was that there is *no* Hebel-Slichter peak in $1/T_1$. Rather, the relaxation rate, which had been decreasing with decreasing T in the normal state above T_c, simply falls more rapidly in the superconducting state. At lower temperatures $1/T_1$ was better described by a power law such as T^3 or $T^{4.5}$ than by the exponential expected for the case of an energy gap. Because there are several nuclear species in the chemical composition of the HTSC, and also Cu atoms at inequivalent sites, comparison of the behavior of different nuclei gives additional information about the spatial variation of the superconductivity within the crystal unit cell. A convenient review of this field was given by Pennington and Slichter.[98]

Reduced to essentials, these experimental results are clearly inconsistent with predictions of the BCS theory in two ways: (1) The absence of the exponential freeze-out of $1/T_1$ at low temperatures indicates the absence of a clean energy gap; the power-law variation is at least qualitatively consistent with the *d*-wave pairing. (2) The absence of the Hebel-Slichter peak suggests a breakdown or modification of the BCS coherence factors, together with the absence of the sharp singularity in the density of states above the energy gap.

CONCLUSION. Despite numerous inconsistencies which need to be resolved, the results of the wide variety of measurements described above indicate that the quasi-particle excitation spectrum of the HTSC is significantly modified from that of the classic BCS superconductors. The gap is wider relative to T_c but is not clean. There is increasing evidence supporting the existence of an intrinsic density of states all the way down to zero energy, as would be expected if the pairing were d wave instead of the BCS *s* wave.

9.9.4 Heavy Fermion Superconductors

As the final topic in this section, we give a brief description of the so-called *heavy fermion superconductors*, an "exotic" type of superconductor discovered before

[98] C. H. Pennington and C. P. Slichter, "Nuclear Resonance Studies of YBCO," in D. M. Ginsberg (ed.), *Physical Properties of High Temperature Superconductors II*, World Scientific, Singapore (1990), p. 269.

the HTSC, which is also believed may have unconventional pairing. Several reviews are available[99] which offer a much more detailed treatment with extensive references to the literature.

Heavy fermion systems always contain ions with f electrons; examples are $CeCu_2Si_2$ and UPt_3. Specific-heat measurements show an electronic specific heat orders of magnitude above the usual value, implying a very narrow energy band with a high density of states, as would occur for carriers with a mass several hundred times the free electron mass. Superconductivity at 0.5 K was discovered in $CeCu_2Si_2$ in 1979 by Steglich et al.[100] Measurements of the specific-heat discontinuity at T_c proved that the superconductivity was indeed coming from the heavy fermions and not from other electrons. This discovery was initially surprising since the addition of magnetic ions containing f electrons usually acts as a pair breaker, which tends to destroy superconductivity. However, in these materials, the f electrons are delocalized and form the carriers which become superconducting.

The superconductivity of some of these materials is complicated by the existence of several distinct superconducting phases in the pressure-temperature plane, which suggests the possibility of unconventional pairing with a complicated order parameter. This idea is also supported by the fact that the low-temperature specific heat appears to have a power-law dependence as T or T^2, depending on the material and the quality of the sample. Such behavior would be expected if the energy gap vanishes on a line or point, respectively, on the Fermi surface, as might be the case with p-wave or d-wave pairing. Because these materials all have critical temperatures of $\sim$1 K or less, they have not been studied with the huge effort devoted to the HTSC materials. Nonetheless, they are obviously of great scientific interest, and may provide one of the few examples of unconventional pairing in superconductors.

[99] See, e.g., L. Taillefer, J. Flouquet, and G. G. Lonzarich, *Physica* **B169**, 257 (1991); G. R. Stewart, *Revs. Mod. Phys.* **56**, 755 (1984); P. Fulde, *J. Phys. F-Met. Phys.* **18**, 601 (1988); P. Fulde, J. Keller, and G. Zwicknagl "Theory of Heavy Fermion Systems," in H. Ehrenreich and D. Turnbull (eds.), *Solid State Physics*, vol. 41, Academic Press, New York (1988), p. 1; N. Grewe and F. Steglich, "Heavy Fermions," in A. Gschneidner and L. Eyring (eds.), *Handbook of Physics and Chemistry of Rare Earths*, vol. 14, North-Holland, Amsterdam (1991), p. 343.

[100] F. Steglich, J. Aarts, C. D. Bredll, W. Lieke, D. Meschede, W. Franz, and J. Schäfer, *Phys. Rev. Lett.* **43**, 1892 (1979).

CHAPTER 10

SPECIAL TOPICS

This chapter is devoted to a brief discussion of several topics which we have skirted earlier to avoid interrupting the more elementary discussions given there. First, we discuss the Bogoliubov equations which govern the spectrum of excitations for spatially inhomogeneous superconductors, treating important examples such as the Anderson dirty superconductor theory and the low-lying excitations in a vortex core which make it "quasi-normal." Then we review the effects of magnetic perturbations in modifying the spectrum of excitations, ultimately producing gapless superconductivity. Finally, we discuss time-dependent superconductivity, as described by the time-dependent GL theory for gapless superconductors. We shall return to time-dependent phenomena in Chap. 11 in the context of relaxation by inelastic electron-phonon processes when an energy gap does exist.

10.1 THE BOGOLIUBOV METHOD: GENERALIZED SELF-CONSISTENT FIELD

In our discussion of the microscopic BCS theory, we considered only pure materials, in which the momentum **k** was a good quantum number, and in which **k** ↑ and −**k** ↓ states were occupied in pairs. In 1959, Anderson[1] showed that a more general prescription, applicable in dirty superconductors as well, was to pair time-

[1] P. W. Anderson, *J. Phys. Chem. Sol.* **11**, 26 (1959).

reversed states. In a dirty metal, the electronic eigenfunctions are some functions $w_n(\mathbf{r})$ which are certainly far from plane waves, but in the absence of magnetic or other time-reversal noninvariant terms in the hamiltonian, each $w_n(\mathbf{r})$ is degenerate with $w_n^*(\mathbf{r})$ if the spin part of the wavefunction is also reversed. Anderson showed that if this pairing of time-reversed states were followed, one could expect the equilibrium properties like T_c, H_c and Δ to be essentially independent of the electronic mean free path.

This result pertains to a superconductor which is still essentially homogeneous on the scale of ξ_0, despite the presence of scattering centers. A more general problem arises if the superconducting order parameter varies in space, e.g., at an interface with another material or in the case of a vortex. To cope with these situations, one may use the Bogoliubov equations, which essentially generalize the ordinary Hartree-Fock equations of the many-electron theory to include the effects of the superconducting *pairing potential* $\Delta(\mathbf{r})$ as well as the ordinary scalar potential $U(r)$. Since de Gennes[2] has given a thorough account of this method, we shall content ourselves with sketching some of the results, largely using his notation and conventions to simplify reference to his discussion.

When there are spatial variations in the hamiltonian, such as scattering centers in the potential $U(\mathbf{r})$ or a variation in $\Delta(\mathbf{r})$ imposed by a vortex core, the plane-wave momentum eigenfunctions characterized by $\mathbf{k}$ used in the original BCS development are no longer appropriate. They must be replaced by suitable position-dependent functions. To find them, one defines a generalization of the Bogoliubov transformation (3.42) by writing

$$\begin{aligned}\Psi(\mathbf{r}\uparrow) &= \sum_n [\gamma_{n\uparrow}\, u_n(\mathbf{r}) - \gamma_{n\downarrow}^*\, v_n^*(\mathbf{r})] \\ \Psi(\mathbf{r}\downarrow) &= \sum_n [\gamma_{n\downarrow}\, u_n(\mathbf{r}) + \gamma_{n\uparrow}^*\, v_n^*(\mathbf{r})]\end{aligned} \tag{10.1}$$

where the Ψ's are annihilation operators for position eigenfunctions rather than for momentum eigenfunctions, as were the $c_{\mathbf{k}\sigma}$ used in Chap. 3. Unlike the $u_\mathbf{k}$ and $v_\mathbf{k}$ of our treatment of the BCS theory in Chap. 3, the u's and v's here are position-dependent eigenfunctions to be determined so as to diagonalize the hamiltonian

$$\begin{aligned}\mathscr{H}_{\text{eff}} = \int \Bigg\{ & \sum_\sigma \Psi^*(\mathbf{r},\sigma)\left[\frac{1}{2m}\left(\frac{\hbar}{i}\nabla - \frac{e\mathbf{A}}{c}\right)^2 + U(r) - \mu\right]\Psi(\mathbf{r},\sigma) \\ & + \Delta(\mathbf{r})\Psi^*(\mathbf{r}\uparrow)\Psi^*(\mathbf{r}\downarrow) + \Delta^*(\mathbf{r})\Psi(\mathbf{r}\uparrow)\Psi(\mathbf{r}\downarrow)\Bigg\}\, d\mathbf{r}\end{aligned} \tag{10.2}$$

[2] P. G. de Gennes, *Superconductivity of Metals and Alloys.* W. A. Benjamin, New York (1966), Chap. 5; reissued by Addison-Wesley, Reading, MA, in 1989.

where $$\Delta(\mathbf{r}) = V\langle\Psi(\mathbf{r}\uparrow)\Psi(\mathbf{r}\downarrow)\rangle = V\sum_n v_n^*(\mathbf{r})u_n(\mathbf{r})(1-2f_n) \tag{10.3}$$

again in close analogy with our discussion in connection with (3.38) and (3.40).[3] The diagonalization requires that u and v satisfy the coupled Bogoliubov equations

$$\mathscr{H}_0 u(\mathbf{r}) + \Delta(\mathbf{r})v(\mathbf{r}) = Eu(\mathbf{r})$$

and
$$-\mathscr{H}_0^* v(\mathbf{r}) + \Delta^*(\mathbf{r})u(\mathbf{r}) = Ev(\mathbf{r}) \tag{10.4}$$

where
$$\mathscr{H}_0 = \frac{1}{2m}\left(\frac{\hbar}{i}\nabla - \frac{e\mathbf{A}}{c}\right)^2 + U(\mathbf{r}) - \mu \tag{10.5}$$

and $U(\mathbf{r})$ includes the ordinary Hartree-Fock averaged Coulomb interaction between electrons as well as the potential of the ion cores and any overall electrostatic potentials.

We note first that if $\Delta = 0$, the equations (10.4) decouple into the forms

$$\mathscr{H}_0 u = Eu$$

and
$$\mathscr{H}_0^* v = -Ev \tag{10.6}$$

so that $u(\mathbf{r})$ and $v(\mathbf{r})$ are the ordinary electron and hole eigenfunctions of the normal state, with energies $\pm E$ relative to the Fermi energy. In general, however, we must seek solutions of the pair of coupled equations (10.4), eventually made self-consistent by computing $\Delta(\mathbf{r})$ from the set of u's and v's by (10.3).

10.1.1 Dirty Superconductors

As a first example, consider the Anderson problem of a dirty, but nonmagnetic, superconductor. Then the normal-state eigenfunctions w_n satisfy

$$\mathscr{H}_0 w_n = \xi_n w_n \tag{10.7}$$

where ξ_n is the eigenvalue measured from the chemical potential μ. In a pure metal, the w_n are Bloch functions with a well-defined $\mathbf{k}$. In general, they are taken as the *exact*, although unknown, solutions in the presence of whatever (elastic) scattering exists. On the assumption that the metal is still homogeneous on the scale of ξ_0, one may take $\Delta(\mathbf{r})$ to be really a constant. In that case, we can satisfy (10.4) by taking both $u_n(\mathbf{r})$ and $v_n(\mathbf{r})$ proportional to $w_n(\mathbf{r})$: That is, we set $u_n(\mathbf{r}) = u_n w_n(\mathbf{r})$ and $v_n(\mathbf{r}) = v_n w_n(\mathbf{r})$, where u_n and v_n are now simply numbers. Then (10.4) becomes

$$(\xi_n - E_n)u_n + \Delta v_n = 0$$

and
$$(-\xi_n - E_n)v_n + \Delta^* u_n = 0 \tag{10.8}$$

---

[3] Note that the one-electron energy term in the first line of (10.2) is explicitly measured with respect to the chemical potential μ. Also, the Fermi function f_n in (10.3) depends only on the energy of the nth eigenfunction, whatever its spatial form.

whose solution requires

$$E_n = (\xi_n^2 + |\Delta|^2)^{1/2} \tag{10.9}$$

as in usual BCS theory. Moreover, when one goes back to find the self-consistent value of Δ, one finds the familiar result that it is determined by

$$\frac{1}{V} = \frac{1}{2}\sum_n \frac{|w_n(\mathbf{r})|^2}{E_n} \tanh \frac{\beta E_n}{2} \tag{10.10}$$

Since we assume the scattering does not change the density of states, this will lead to the same results as (3.50). (Some attention to the normalization of w_n and the definition of V is needed to show this in detail.) Thus, we do not expect much change in T_c or Δ on going from a clean to a dirty specimen; this agrees with experimental fact.

10.1.2 Uniform Current in Pure Superconductors

In our discussion of the critical current in a thin wire, we mentioned in connection with (4.41) that quasi-particle energies were shifted by an amount $\mathbf{v}_s \cdot \mathbf{p}$. This can be seen by using our present methods. If the pairs have center-of-mass momentum[4] $2\mathbf{q}$, we expect

$$\Delta = |\Delta| e^{i2\mathbf{q}\cdot\mathbf{r}} \tag{10.11}$$

If impurity scattering is negligible, the eigenfunctions $u_n(\mathbf{r})$ and $v_n(\mathbf{r})$ become simple plane waves. Then, from (10.3) we see that (10.11) will result if

$$v_{\mathbf{k}}(\mathbf{r}) = V_{\mathbf{k}} e^{i(\mathbf{k}-\mathbf{q})\cdot\mathbf{r}}$$

and

$$u_{\mathbf{k}}(\mathbf{r}) = U_{\mathbf{k}} e^{i(\mathbf{k}+\mathbf{q})\cdot\mathbf{r}} \tag{10.12}$$

Here, $U_{\mathbf{k}}$ and $V_{\mathbf{k}}$ are constants like the $u_{\mathbf{k}}$ and $v_{\mathbf{k}}$ of BCS. Note that with $q \neq 0$ we are no longer pairing time-reversed states. When (10.11) and (10.12) are substituted into the Bogoliubov equations (10.4), they become

$$(\xi_{\mathbf{k}+\mathbf{q}} - E_{\mathbf{k}})U_{\mathbf{k}} + |\Delta| V_{\mathbf{k}} = 0$$

and

$$(-\xi_{\mathbf{k}-\mathbf{q}} - E_{\mathbf{k}})V_{\mathbf{k}} + |\Delta| U_{\mathbf{k}} = 0 \tag{10.13}$$

Solving for the excitation energies $E_{\mathbf{k}}$, we find

$$E_{\mathbf{k}} = \frac{\xi_{\mathbf{k}+\mathbf{q}} - \xi_{\mathbf{k}-\mathbf{q}}}{2} + \left[\left(\frac{\xi_{\mathbf{k}+\mathbf{q}} + \xi_{\mathbf{k}-\mathbf{q}}}{2}\right)^2 + |\Delta|^2\right]^{1/2} \tag{10.14}$$

[4]Note that here we follow de Gennes's convention of assigning the pairs momentum $2\mathbf{q}$, rather than $\mathbf{q}$, as we have done elsewhere. Also, here we take $m^* = 2m$.

Since $\xi_{\mathbf{k}} = (\hbar^2 k^2/2m) - \mu$, we have

$$\frac{1}{2}(\xi_{\mathbf{k}+\mathbf{q}} - \xi_{\mathbf{k}-\mathbf{q}}) = \frac{\hbar^2}{m}\mathbf{k}\cdot\mathbf{q} = \frac{\hbar}{m}\mathbf{p}_{\mathbf{k}}\cdot\mathbf{q} \tag{10.15}$$

Also, so long as $q \ll k_F$, $\xi_{\mathbf{k}+\mathbf{q}} + \xi_{\mathbf{k}-\mathbf{q}} \approx 2\xi_{\mathbf{k}}$. Then, (10.14) can be simplified to

$$E_{\mathbf{k}} = E_{\mathbf{k}}^0 + \mathbf{p}_{\mathbf{k}}\cdot\mathbf{v}_s \tag{10.16}$$

where $\mathbf{v}_s = \hbar\mathbf{q}/m$ is the velocity of the supercurrent and $E_{\mathbf{k}}^0 = (\xi_{\mathbf{k}}^2 + |\Delta|^2)^{1/2}$ is the excitation energy in the absence of a current.

To assure self-consistency, one should use these new shifted energies in the Fermi functions in (10.3). However, at low temperatures the f_n (which are the $f_{\mathbf{k}}$ here) are nearly zero until $E_{\mathbf{k}}$ approaches zero, so that $|\Delta|$ will not decrease much with the current, although the minimum excitation energy

$$E_{\min} = \Delta - p_F v_s \tag{10.17}$$

does. This provides a simple example of the fact that the pair potential Δ is not necessarily the same as the energy gap in the excitation spectrum. In fact, we can obtain a simple concrete realization of *gapless* superconductivity by considering the situation when v_s very slightly exceeds Δ/p_F. Then for a small number of $\mathbf{k}$ values in the direction opposite to $\mathbf{v}_s$, excitations can be made at zero energy just as in the normal state. But, just as in the normal state, the Fermi statistics limit the occupancy of these zero-energy states to $f_k \sim \frac{1}{2}$. In view of (10.3), this means that these few states contribute nothing to sustaining the pair potential, but all the other regions of the Fermi surface contribute more or less normally. When the self-consistent solution is worked out in detail,[5] it is found that there is a small region of v_s above Δ/p_F for which $E_{\min} = 0$ while $\Delta \neq 0$. Thus, there is still a coherent condensed state, with macroscopic quantum properties described by the pair wavefunction $\psi \propto \Delta \propto e^{2i\mathbf{q}\cdot\mathbf{r}}$, and hence we still expect the perfect conductivity property to remain. However, the depairing that sets in for $v_s > \Delta/p_F$ limits the maximum supercurrent to only about 1 percent more than the current when $v_s = \Delta/p_F$. For greater values of v_s, $dJ_s/dv_s < 0$, so the regime is unstable and not easily observed experimentally.

10.1.3 Excitations in Vortex

From the GL theory (Sec. 5.1) we know that in the vortex state of a type II superconductor $\Delta(z, r, \theta)$ has the form $|\Delta(r)|e^{i\theta}$, where $|\Delta(r)|$ rises from zero at the center of the vortex to Δ_∞ at a distance, the major rise occurring in a distance $\sim\xi(T)$. The GL solution for $\Delta(r)$ should be exact at T_c, where $\xi(T) \gg \xi(0)$, so

[5] J. Bardeen, *Rev. Mod. Phys.* **34**, 667 (1962); also, K. T. Rogers, unpublished Ph.D. thesis, University of Illinois, 1960.

that the slow-variation requirement of the GL theory is satisfied. However, it is only a qualitative guide at low temperatures where Δ varies on the scale of $\xi(0)$. Given this rapidly varying Δ, what is the nature of the quasi-particle excitations? This problem was first solved by Caroli, de Gennes, and Matricon;[6] the solution is reviewed in Sec. 5.2 of de Gennes's book. Their solution uses the Bogoliubov equations but makes the approximation of $\kappa \gg 1$, so that the magnetic field is negligible over the core region. Since they assume pure material (so that momentum is conserved), as well as $\kappa \gg 1$, their solution is not strictly applicable to any real material. The calculation was extended to all values of κ in an important paper by Bardeen, Kümmel, Jacobs, and Tewordt,[7] who attempted to obtain self-consistent solutions for $\Delta(r)$ and $h(r)$ using a variational expression for the free energy. Their qualitative conclusions are similar to those of Caroli et al.

At large distances from the center of the vortex ($r \gg \xi$), it is quite a good approximation to use our previous result (10.16) for the shifted energy spectrum due to a uniform velocity field with $v_s = \hbar/2mr = \hbar/m^*r$. As indicated in Sec. 5.5, if this argument is used near the core, it leads to gapless superconductivity inside $r \approx \xi$. Actually, in this region one really must solve the Bogoliubov equations to build in the effects of the rapid spatial variation of $\Delta(r)$. The solutions of Caroli et al. and of Bardeen et al. show that there is indeed a group of low-lying excitations with the wavefunctions $u(\mathbf{r})$ and $v(\mathbf{r})$ localized near the vortex core. The lowest one lies at $\sim\hbar^2/2m\xi^2 \sim \Delta_\infty^2/E_F$. For the classic superconductors, this is $\sim 10^{-4}\Delta_\infty \ll kT_c$, which is effectively gapless. Since the level density is found to correspond roughly to that of a cylinder of normal material of radius $\sim\xi$, this result forms the most microscopic rationale for the concept of the *quasi-normal core* of a vortex. The predicted[8] localized state density in the vortex core of a classic superconductor has been confirmed in beautiful experiments by Hess and coworkers,[9] using an STM (scanning tunneling microscope) to sense the spatially resolved density of states associated with a vortex in the layered superconductor 2H-$NbSe_2$. On the other hand, in the high-temperature superconductors, where $\xi \lesssim 20\,\text{Å}$, the lowest level is comparable with the gap, and there may be only one bound state. Evidence for such a unique bound state in vortices in YBCO has been found by far-infrared absorption spectroscopy experiments by Drew and coworkers.[10]

---

[6]C. Caroli, P. G. de Gennes, and J. Matricon, *Phys. Lett.* **9**, 307 (1964).

[7]J. Bardeen, R. Kümmel, A. E. Jacobs, and L. Tewordt, *Phys. Rev.* **187**, 556 (1969).

[8]For recent calculations, see J. D. Shore, M. Huang, A. T. Dorsey, and J. P. Sethna, *Phys. Rev. Lett.* **62**, 3089 (1989); F. Gygi and M. Schlüter, *Phys. Rev. Lett.* **65**, 1820 (1990).

[9]H. F. Hess, R. B. Robinson, R. C. Dynes, J. M. Valles, and J. V. Waszczak, *Phys. Rev. Lett.* **62**, 214 (1989); H. F. Hess, R. B. Robinson, and J. V. Waszczak, *Phys. Rev. Lett.* **64**, 2711 (1990)

[10]K. Karrai, E. J. Choi, F. Dunmore, S. Liu, H. D. Drew, Qi Li, D. B. Fenner, Y. D. Zhu, and Fu-Chun Zhang, *Phys. Rev. Lett.* **69**, 152 (1992).

10.2 MAGNETIC PERTURBATIONS AND GAPLESS SUPERCONDUCTIVITY

In the previous section, we have seen that the excitation spectrum of a superconductor is modified if it carries a current; if the current is sufficiently strong, the spectrum becomes gapless for a finite current range before superconductivity is destroyed. The origin of this change is that adding the common drift momentum **q** to the paired electrons with initial momenta **k** and −**k** gives them different kinetic energies $\xi_{\mathbf{k}+\mathbf{q}}$ and $\xi_{-\mathbf{k}+\mathbf{q}}$, thus lifting the degeneracy of $\xi_{\mathbf{k}}$ and $\xi_{-\mathbf{k}}$, which had been exact because of time-reversal symmetry. In fact, tracing through the deduction of (10.17) from (10.14), we see that the gap reduction $(\Delta - E_{\min})$ is just equal to half the maximum splitting of the time-reversal degeneracy by the drift momentum. As it stands, this result is applicable only to pure superconductors, in which **k** is a good quantum number. However, as we shall see, the basic idea is very general.

When we considered (Sec. 10.1.1) Anderson's theory of "dirty" superconductors, i.e., nonmagnetic alloys with mean free path $\ell < \xi_0$, we noted that pairing of time-reversed degenerate states led to the same T_c and BCS density of states as that for a pure superconductor. On the other hand, Abrikosov and Gor'kov[11] (AG) showed that *magnetic* impurities (which break the time-reversal symmetry) lead to a strong depression of T_c and a modification of the BCS density of states, so that it becomes gapless for a finite range of concentration below the critical value which destroys superconductivity entirely. Subsequent work by Maki,[12] de Gennes,[13] and others showed that the results of Abrikosov and Gor'kov for the density of states and the depression of T_c could be transcribed to describe the effects of many other pair-breaking perturbations, i.e., those which destroy the time-reversal degeneracy of the paired states. For this transcription to work, there must also be rapid scattering, as in a dirty superconductor, to assure "ergodic" behavior of the electrons. Examples of such perturbations include external magnetic fields, currents, rotations, spin exchange, and hyperfine fields, in addition to magnetic impurities. It was also shown that spatial gradients in the order parameter, such as those induced by proximity to a boundary with a normal metal or in the surface sheath or vortex state of a type II superconductor, have a pair-breaking effect which can induce gapless superconductivity. In fact, gapless superconductivity turns out to be the rule rather than the exception if the transition to the normal state due to the perturbation is of second, rather than first, order.

Since the general theory of these effects is most naturally couched in the Green's function formalism of Gor'kov, which is beyond the scope of this book,

---

[11] A. A. Abrikosov and L. P. Gor'kov, *Zh. Eksperim. i Teor. Fiz.* **39**, 1781 (1960); *Soviet Phys—JETP* **12**, 1243 (1961).

[12] K. Maki, *Prog. Theor. Phys. (Kyoto)* **29**, 333 (1963); **31**, 731 (1964); **32**, 29 (1964); K. Maki and P. Fulde, *Phys. Rev.* **140**, A1586 (1965).

[13] P. G. de Gennes, *Phys. Kondens. Materie* **3**, 79 (1964); P. G. de Gennes and G. Sarma. *J. Appl. Phys.* **34**, 1380 (1963); P. G. de Gennes and M. Tinkham, *Physics* **1**, 107 (1964).

we shall confine our treatment to an outline of some of the major results and their experimental confirmation. For further details, the reader is referred to the reviews by Maki[14] and de Gennes.[15]

10.2.1 Depression of T_c by Magnetic Perturbations

In the AG theory and its extensions to other magnetic perturbations, the pair-breaking strength is characterized by the typical energy difference 2α it causes between time-reversed electrons. Although this pair-breaking energy was simply the constant energy splitting between **k** and $-\mathbf{k}$ electrons in the presence of drift momentum **q** for the pure metal, in the present ergodic case the scattering from one **k** to another is so rapid that the depairing energy $\sim\mathbf{q}\cdot\mathbf{k}$ is constantly changing, tending to average out. The relevant time scale for averaging is the time required for the relative phase of the two time-reversed electrons to be randomized by the perturbation. This is brought out most clearly in de Gennes's version of the theory, in which the effective pair-breaking energy 2α is called $\hbar/\tau_K$, where τ_K is essentially this time. Evidently, an energy difference $\hbar/\tau_K$ operating over a time τ_K produces a phase shift of the order of unity.

This idea may be clarified by an example. Consider the effect of a magnetic field on a particle of dirty superconductor small enough so that no vortex structure is allowed. The leading term in the magnetic perturbation is $(e/mc)\mathbf{p_k}\cdot\mathbf{A}$, so that the difference in its energetic effect on electrons **k** and $-\mathbf{k}$ is $(2e/mc)\mathbf{p_k}\cdot\mathbf{A}$. Since the phase evolves as $e^{-iEt/h}$, the differential rate of change in phase is

$$\frac{d\varphi}{dt} = \left(\frac{2e}{\hbar c}\right)\mathbf{v_k}\cdot\mathbf{A} \tag{10.18}$$

If the scattering time τ is short compared to $(d\varphi/dt)^{-1}$, then the phase change $(d\varphi/dt)\tau$ between collisions is small, and this $d\varphi/dt$ must be integrated over many fragments of trajectory before a phase difference ~ 1 is reached. Since $\mathbf{v_k}$ will change arbitrarily at each scattering event, the phase difference grows as the square root of the number of free paths, by a random-walk process. In time t, the spread in phase will be of the order of $(d\varphi/dt)(\tau t)^{1/2}$. For this to be of order unity, the time τ_K required is given by

$$\frac{1}{\tau_K} = \tau\left\langle\left(\frac{d\varphi}{dt}\right)^2\right\rangle = \frac{1}{3}v_F^2\tau\left(\frac{2e}{\hbar c}\right)^2\langle A^2\rangle = D\left(\frac{2\pi}{\Phi_0}\right)^2\langle A^2\rangle \tag{10.19}$$

---

[14] K. Maki, "Gapless Superconductivity" in R. D. Parks (ed.), *Superconductivity*, vol. II, Dekker, New York (1969), Chap. 18.

[15] P. G. de Gennes, *Superconductivity of Metals and Alloys*, W. A. Benjamin, New York (1966), Chap. 8; reissued by Addison-Wesley, Reading, MA, in 1989.

where $D = \frac{1}{3}v_F l = \frac{1}{3}v_F^2\tau$ is the electronic diffusion constant. We also recall that $2\alpha = \hbar/\tau_K$ gives the connection between the two notational conventions.

The reduced T_c in the presence of such a pair breaker is found to be given by the implicit relation

$$\ln\frac{T_c}{T_{c0}} = \psi\left(\frac{1}{2}\right) - \psi\left(\frac{1}{2} + \frac{\alpha}{2\pi kT_c}\right) \tag{10.20}$$

where our notation is that $T_c = T_c(\alpha)$, $T_{c0} = T_c(0)$, and $\psi(z) = \Gamma'(z)/\Gamma(z)$ is the digamma function. Expanding the digamma function about $\frac{1}{2}$ yields the result

$$k(T_{c0} - T_c) = \frac{\pi\alpha}{4} = \frac{(\pi/8)\hbar}{\tau_K} \tag{10.20a}$$

so that for weak pair breaking, the depression of T_c is linear in α. On the other hand, one finds that superconductivity is completely destroyed (i.e., $T_c = 0$) for

$$2\alpha = \frac{\hbar}{\tau_K} = 1.76kT_c = \Delta_{\text{BCS}}(0) \equiv \Delta_{00} \tag{10.20b}$$

The entire dependence of T_c on the pair-breaker strength α, or conversely, the temperature dependence of the critical pair-breaker strength $\alpha_c(T)$, is depicted in Fig. 10.1. The shaded region indicates the range of parameter values for which gapless behavior is predicted.

As emphasized by Maki and de Gennes, to the extent that all pair-breaking ergodic perturbations are equivalent to magnetic impurities in their effect on T_c, this function $\alpha_c(T)$ should be a *universal function*, which can be applied to any of them. For the cases treated in Chap. 4 using the linearized GL equation (4.56) to

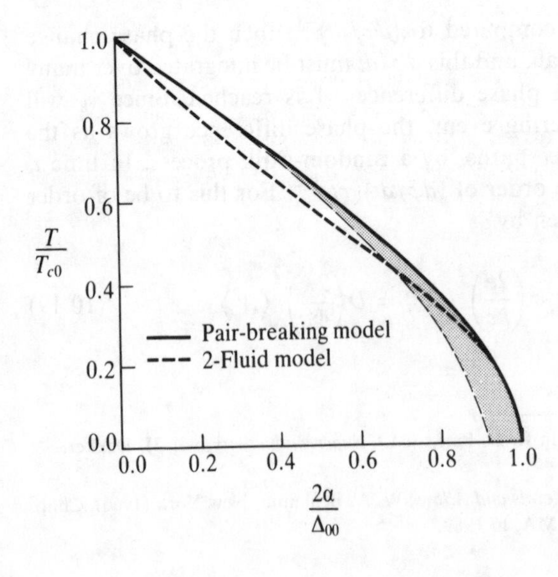

FIGURE 10.1
Universal functional relation between the pair-breaking parameter α and the reduced T_c is shown by solid curve. The shading depicts the gapless region. Values of 2α for various magnetic perturbations are given in (10.22a) to (10.22g). The dashed curve labeled two-fluid is a plot of $2\alpha/\Delta_{00} = (1-t^2)/(1+t^2)$, where $t = T/T_{c0}$. This relation reproduces the results of the GL critical-field calculations building in the two-fluid temperature dependences $\lambda(t) \propto (1-t^4)^{-1/2}$, $H_c(t) \propto (1-t^2)$, and hence $\alpha_c \propto \xi^{-2}(t) \propto \lambda^2 H_c^2 \propto (1-t^2)/(1+t^2)$. [See (10.23).]

find a critical field, it turns out that 2α is the lowest eigenvalue of that equation, written in the form

$$\hbar D\left(\frac{\nabla}{i}-\frac{2e}{\hbar c}\mathbf{A}\right)^2\Delta = 2\alpha\,\Delta \tag{10.21}$$

Note that the result (10.19) obtained above is a special case of this relation since Δ is constant over a small particle. Using our solutions from Chap. 4, we can write down the following cases:

$$\alpha = \frac{DeH}{c} \qquad \text{bulk type II in vortex state} \tag{10.22a}$$

$$\alpha = 0.59\frac{DeH}{c} \qquad \text{surface sheath} \tag{10.22b}$$

$$\alpha = \frac{1}{6}\frac{De^2H^2d^2}{\hbar c^2} \qquad \text{thin film, parallel field} \tag{10.22c}$$

$$\alpha = \frac{DeH}{c} \qquad \text{thin film, perpendicular field} \tag{10.22d}$$

$$\alpha = \frac{2De^2\langle A^2\rangle}{\hbar c^2} \qquad \text{small particle} \tag{10.22e}$$

It is interesting to note that the angular dependence of the critical field of a thin film (4.70) can be obtained by adding (10.22c) and (10.22d), using the appropriate component of the total field in each case, and setting the sum equal to $\alpha_c(T)$. Note also that the critical fields found from (4.56) will agree with those found from (10.20), if the temperature-dependent coherence length is defined by

$$\xi^2(T) = \frac{D\hbar}{2\alpha_c(T)} = D\tau_{Kc}(T) \tag{10.23}$$

In this way, the *linearized* GL equation can be used all the way down to $T = 0$ (for dirty materials for which the present theory is valid) even though the GL theory is normally valid only near T_c. This has the important implication that results of the GL theory often have validity over a wider temperature range than might be expected.

The above pair-breaking effects have all acted on the orbital motion of the electrons. The original AG calculation dealt with the effect of a magnetic impurity coupled to the electron spin by an exchange interaction of the form $J(r)\mathbf{S}\cdot\mathbf{s}$, where $\mathbf{S}$ is the impurity spin and $\mathbf{s}$ is the spin of the conduction electron. Apart from numerical factors, the pair-breaking energy is given by

$$2\alpha \approx \frac{xJ^2}{E_F} \tag{10.22f}$$

where x is the fractional impurity concentration and J is an average over the atomic volume.

There will also be a pair-breaking effect from the effect of an external magnetic field on the electronic spins. The appropriate $d\varphi/dt$ here is $2\mu_B H/\hbar = eH/mc$, and the appropriate scattering time is τ_{so}, the time required for spin flip to occur by spin-orbit coupling associated with the scattering process. If these factors are used in (10.19), one finds

$$2\alpha \approx \frac{\tau_{so} e^2 \hbar H^2}{m^2 c^2} \tag{10.22g}$$

This result applies only when spin-orbit scattering is rapid enough so that $\tau_{so} eH/mc \ll 1$. In the other limit, one may neglect spin-orbit scattering. There is then no random-walk averaging, the pair-breaking energy 2α is simply $2\mu_B H$ and zero-energy excitations can occur when $\mu_B H = \Delta$. But before this field is reached, a first-order transition to the normal state occurs when $\mu_B H = \Delta_{00}/\sqrt{2}$ (at $T = 0$), as pointed out by Clogston and by Chandrasekhar.[16] This limits the critical field of a material with negligible spin-orbit scattering to a value $H_p = \Delta_0/\sqrt{2}\mu_B$. With $\Delta_0 = 1.76kT_c$, this works out to

$$\frac{H_p}{T_c} = 18{,}400\,\mathrm{G/K} \tag{10.24}$$

Many useful type II superconductors have $H_{c2} \gtrsim H_p$, indicating the importance of the randomizing effect of the spin-orbit scattering in reducing the pair-breaking effect of the Zeeman energy of the spins.

When several pair-breaking mechanisms are present, it is usually adequate simply to sum their contributions (10.22) to α in finding the critical condition for the destruction of superconductivity.

10.2.2 Density of States

In interpreting early experimental work on various properties (such as thermal conductivity,[17] microwave absorption,[18] and electron tunneling[19]) of superconducting films in magnetic fields, it was assumed that the density of states could be adequately described by retaining the BCS spectrum with a field-dependent gap $\Delta(H)$. However, the values of $\Delta(H)$ inferred in this way from different sorts of measurements were not really consistent, particularly at low temperatures. The discrepancies were of the sort that could be accounted for if the gap were "fuzzing out," with low-lying excitations coming in before the peak in the spectrum at the gap edge had entirely disappeared.

---

[16] A. M. Clogston, *Phys. Rev. Lett.* **9**, 266 (1962); B. S. Chandrasekhar, *Appl. Phys. Lett.* **1**, 7 (1962)

[17] D. E. Morris and M. Tinkham, *Phys. Rev.* **134**, A1154 (1964)

[18] R. H. White and M. Tinkham, *Phys. Rev.* **136**, A203 (1964).

[19] R. Meservey and D. H. Douglass, Jr., *Phys. Rev.* **135**, A24 (1964)

As noted above, just such a behavior is, in fact, predicted over a finite range of magnetic perturbation strength. The predicted state density for several representative cases is shown in Fig. 10.2, following the computations of Skalski et al.,[20] which appeared at about the same time. In this diagram, energies are normalized to Δ, the so-called gap parameter, which is a measure of the strength of the pairing potential at temperature T in the presence of the pair-breaking strength α. It is *not* the same as the minimum excitation energy, or spectral gap, which we denote Ω_G. This difference is illustrated in Fig. 10.3, which contrasts the dependence of Ω_G and of Δ (at $T = 0$) on α. Also shown is T_c/T_{c0}, the same curve as that presented in Fig. 10.1. Note the sizable departures from the simple BCS theory, in which $\Omega_G(0) = \Delta(0) = 1.76kT_c$, so that all three quantities would scale in the same way. Finally, in Fig. 10.4 we show the temperature dependence of Δ/Δ_{00} for various values of $2\alpha/\Delta_{00}$. (Here, Δ_{00} denotes the value of Δ for $T = \alpha = 0$.) In this figure, the shaded area represents the gapless region, which occurs when $\alpha \geq \Delta$.

The detailed calculations leading to the above results are rather formidable. This enhances the interest in the approach of de Gennes,[21] which deals with the gapless regime where $\Omega_G = 0$ and $\Delta/\alpha \ll 1$. In this regime, he showed that the pair potential Δ can be introduced by a second-order perturbation calculation (although this would have failed in the absence of strong pair

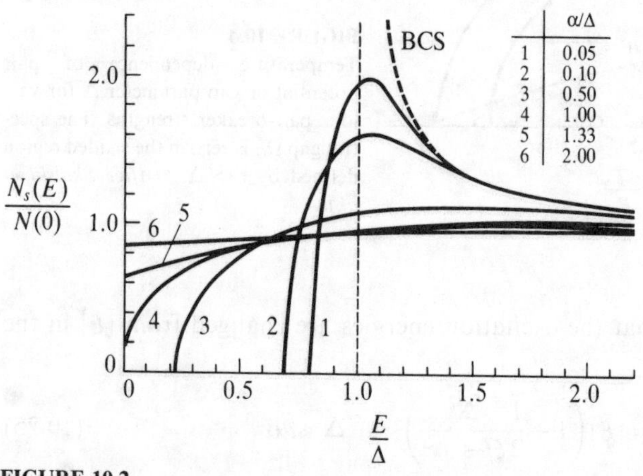

FIGURE 10.2
The density of states as a function of the reduced energy for several values of the reduced pair-breaking strength α/Δ. In this diagram Δ is understood as $\Delta(T, \alpha)$. (*After Skalski et al.*)

---

[20] S. Skalski, O. Betbeder-Matibet, and P. R. Weiss, *Phys. Rev.* **136**, A1500 (1964).

[21] P. G. de Gennes, *Superconductivity in Metals and Alloys*, W. A. Benjamin, New York (1966), p. 265; reissued by Addison-Wesley, Reading, MA, in 1989.

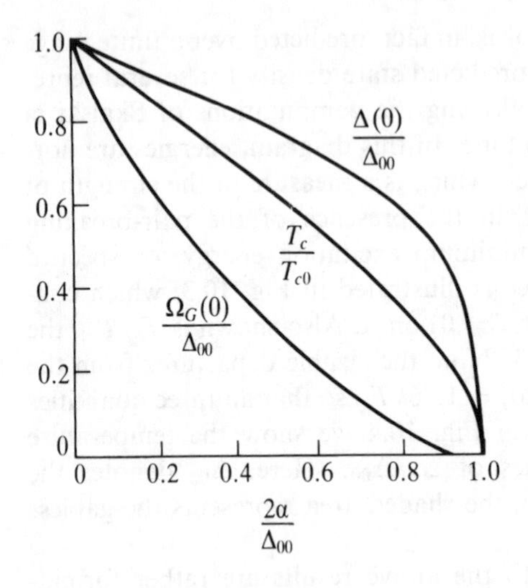

FIGURE 10.3
Decrease of spectral gap Ω_G and gap parameter Δ at $T = 0$ and of transition temperature T_c, with increasing pair-breaking strength α. *(After Skalski et al.)*

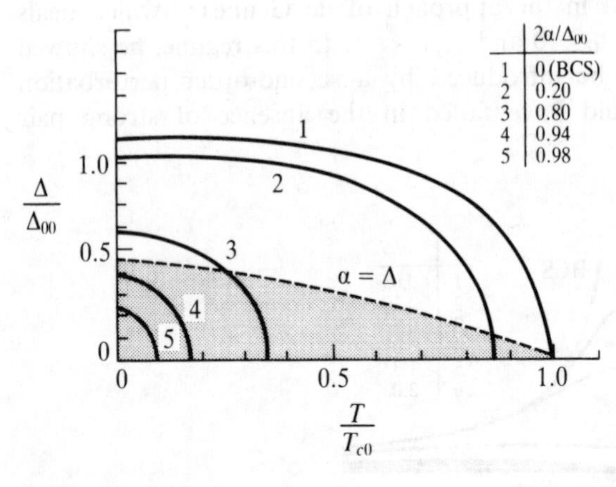

FIGURE 10.4
Temperature dependence of pair potential or gap parameter Δ for various pair-breaker strengths. The spectral gap Ω_G is zero in the shaded region defined by $\alpha > \Delta$. *(After Skalski et al.).*

breaking). The result is that the excitation energies are changed from $|\xi|$ in the absence of Δ to

$$E = |\xi|\left(1 + \frac{1}{2}\frac{\Delta^2}{\xi^2 + \alpha^2}\right) \qquad \Delta \ll \alpha \tag{10.25}$$

Note that for $|\xi| \gg \alpha$, $E \approx |\xi| + \Delta^2/2|\xi| \approx (\Delta^2 + \xi^2)^{1/2}$, which is the BCS result for ordinary superconductors. On the other hand, for $|\xi| \lesssim \alpha$, E approaches $|\xi|(1 + \Delta^2/2\alpha^2)$, which is proportional to $|\xi|$, and hence shows no gap in the spectrum. We can readily compute the density of states from (10.25), finding (to order Δ^2) that

$$\frac{N_s(E)}{N(0)} = \frac{d\xi}{dE} = 1 + \frac{\Delta^2}{2}\frac{E^2 - \alpha^2}{(E^2 + \alpha^2)^2} \qquad \Delta \ll \alpha \tag{10.26}$$

Note that $N_s(E) < N(0)$ for $E < \alpha$, but it remains finite, corresponding to the absence of a gap. On the other hand, for $E > \alpha, N_s(E) > N(0)$. This is analogous to the peaked density of states above the gap in a BCS superconductor, but with an important difference: The energy scale is set by α, not by Δ. Moreover, this theory applies only when Δ is small, i.e., near the transition to the normal state. Thus, the value of α must be nearly $\alpha_c(T)$. This leads to the remarkable conclusion that the energy scale for the deviations of $N_s(E)$ from $N(0)$ is a function only of T, not of the value of Δ, which goes to zero as the transition is approached. In other words, Δ controls only the amplitude of the deviations of $N_s(E)$ from $N(0)$, not their spectral distribution.

The most detailed comparison between theory and experiment is offered by electron-tunneling measurements of the density of states. These were first carried out at sufficiently low temperatures by Woolf and Reif [22] on lead and indium films containing magnetic impurities. They found good agreement with the theoretical predictions in the case of a rare earth impurity (Gd), but with Fe or Mn impurities, the gap seemed to be filled in even more rapidly than predicted. They attributed this difference to the fact that the 4*f* electrons of Gd should closely resemble the localized moments assumed in the AG theory, whereas the 3*d* electrons of Fe and Mn interact more strongly with the conduction electrons, and thus are less localized.

Tunneling experiments using a thin film in a parallel magnetic field offer a cleaner test of the theory since they avoid this question of the degree of localization of the magnetic moment. Moreover, the field can be varied at will, whereas a uniform concentration of impurities is notoriously difficult to achieve, and each concentration requires a different specimen. Careful tunneling experiments in a magnetic field were carried out by Levine[23] and by Millstein and Tinkham.[24] In the latter experiments, e.g., the films were of tin or of tin-indium alloys, of thickness $d \sim 1,000$ Å and with mean free paths from $\ell \sim 300$ to $1,200$ Å. Thus, ℓ was less than, but still comparable with, the coherence length $\xi_0 = 2,300$ Å. The other film in the tunneling sandwich was aluminum with sufficient Mn impurity to keep it in the normal state. Measurements of the density of states were carried out by using ^{3}He cooling at 0.36 K, about $T_c/10$. Even so, the thermal smearing due to the width of the cutoff of the Fermi function leads to quite a sizable difference between the differential conductance (dI/dV) and the true density of states. [For example, comparing (3.79) and (3.80), we see that the normalized differential conductivity $\sigma(V) = G_{ns}(V)/G_{nn}$ measures an average of $N_s(E)/N(0)$ over an energy range of several kT about eV, weighted with the bell-shaped function df/dE.] Thus, when one compares experimental curves with theoretical ones, it is essential first to fold the thermal smearing into the theoretical curve if any real

---

[22] M. A. Woolf and F. Reif, *Phys. Rev.* **137**, A557 (1965).

[23] J. L. Levine, *Phys. Rev.* **155**, 373 (1967).

[24] J. Millstein and M. Tinkham, *Phys. Rev.* **158**, 325 (1967).

accuracy is to be obtained. When this was done, the $\sigma(V)$ curves were found to be in semiquantitative agreement with the transcribed Abrikosov-Gor'kov theory, whereas any attempt to fit the data with a BCS density of states with a modified gap value was completely hopeless.

Still, there was a residual discrepancy, apparently outside experimental error, in that the observed $\sigma(V)$ curves were even more flattened out than the thermally smeared AG curves. Also, it appeared that the films became gapless for fields substantially less than the $(0.91)^{1/2}H_{c\|}$ expected from the theory. A possible explanation for these discrepancies was suggested by the work of Larkin,[25] who had treated the very different idealized case of small superconducting spheres in a magnetic field, under the assumption of specular surface scattering and no volume scattering. He found that the spectrum became gapless at about $0.4H_c$, much *lower* than that found experimentally. Strässler and Wyder[26] then carried out an extension of Larkin's approach, putting in volume scattering. By varying ℓ/ξ_0 from 0 to ∞, they could compute density-of-states curves, such as those shown in Fig. 10.5 for the particular case where the applied field is half the critical field. The limiting case of $\ell = 0$ corresponds to the curve computed by Skalski et al. (Fig. 10.2), whereas the limit $\ell = \infty$ corresponds to Larkin's result. The differences among these curves can be understood qualitatively by observing that without scattering, some electron states are perturbed very strongly by the field and become gapless in low field; with rapid scattering, each electron feels an averaged strength of the perturbation, effectively eliminating the strong effect on any one.

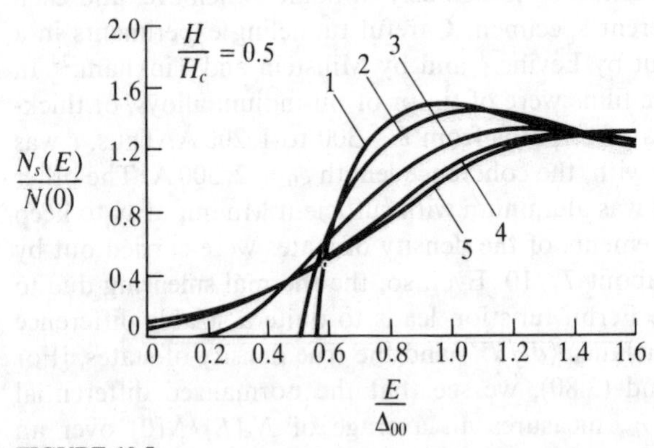

FIGURE 10.5
The density of states as a function of energy E for several values of the mean free path and a magnetic field of half the critical field of the small spheres. [1 : $\ell = 0$; 2 : $\ell = (\pi/10)\xi_0$; 3 : $\ell = \pi\xi_0$; 4: $\ell = 10\pi\xi_0$; 5: $\ell = \infty$.] Curve 1 corresponds to the calculations of Skalski et al. (Fig. 10.2), based on the Abrikosov-Gor'kov theory. Curve 5 corresponds to the limit treated by Larkin. (*After Strässler and Wyder.*)

---

[25] A. Larkin, *Zh. Eksperim. Teor. Fiz.* **48**, 232 (1965) [*Soviet Phys.—JETP* **21**, 153 (1965)].

[26] S. Strässler and P. Wyder, *Phys. Rev.* **158**, 319 (1967).

Returning to the experiments, essentially perfect agreement with the data was obtained if ℓ/ξ_0 for the various films was set equal to values in the range 0.3 to 1, which were within a factor of 2 of independent estimates of this ratio. Considering the uncertainties in these estimates and the idealizations of the theoretical model, such a degree of agreement is quite satisfactory.

In conclusion, it is interesting to note that the temperature dependence of H_c [or, equivalently, $T_c(H)$] found by Millstein and Tinkham was in quantitative agreement with the AG result. One might ask why the finite mean-free-path effects treated by Strässler and Wyder were not evident here. The answer seems to be simply that $T_c(H)$, being an integral property, is less sensitive to details of the model than is the more microscopic $N_s(E)$.

10.3 TIME-DEPENDENT GINZBURG-LANDAU THEORY

In view of the enormous success of the Ginzburg-Landau theory of thermodynamic equilibrium properties of superconductors near T_c, it is natural to seek a time-dependent generalization leading to a differential equation for the space and time dependence of the order parameter Δ. A number of authors[27,28] have attacked this problem, and a critical review of this work has been given by Cyrot.[29] From this work it has emerged that it is very difficult to obtain a nonlinear time-dependent Ginzburg-Landau (TDGL) equation of any generality. As noted by Gor'kov and Eliashberg, this difficulty stems essentially from the singularity in the density of states at the gap edge, which leads to slowly decaying oscillatory responses in the time domain. As we have seen in Sec. 10.2, the presence of magnetic impurities or other pair breakers rounds off the singularity in the BCS density of states, and with sufficient pair-breaking strength, the spectrum becomes gapless. In this gapless regime, one can make an expansion in powers of Δ/α and ω/α, where $\alpha = \hbar/2\tau_K$ is the usual pair-breaking parameter discussed in the previous section. In this way, Gor'kov and Eliashberg obtained a rigorous version of a nonlinear TDGL equation, which is valid for a realistic range of fields and frequencies; but it is, of course, restricted to a gapless superconductor. Schmid had previously obtained equations of a similar form without imposing this restriction but, apparently,[30] they lack rigorous justification except in a gapless regime.

---

[27]See, e.g., A. Schmid, *Phys. Kondens. Mat.* **5**, 302 (1966); C. Caroli and K. Maki, *Phys. Rev.* **159**, 306, 316 (1967); **164**, 591 (1967); E. Abrahams and T. Tsuneto, *Phys. Rev.* **152**, 416 (1966); J. W. F. Woo and E. Abrahams. *Phys. Rev.* **169**, 407 (1968); R. S. Thompson. *Phys. Rev.* **B1**, 327 (1970).

[28]L. P. Gor'kov and G. M. Eliashberg, *Zh. Eksperim. i Teor. Fiz.* **54**, 612 (1968) [*Soviet Phys.—JETP* **27**, 328 (1968)].

[29]M. Cyrot, *Repts. Prog. Phys.* **36**, 103 (1973).

[30]G. M. Eliashberg, *Zh. Eksperim. i. Teor. Fiz.* **55**, 2443 (1968) [*Soviet Phys.—JETP* **29**, 1298 (1969)].

The result of Gor'kov and Eliashberg is conveniently written in the normalized form adopted by Hu and Thompson,[31] namely, the coupled set of equations

$$D^{-1}\left(\frac{\partial}{\partial t}+i\frac{2e\psi}{\hbar}\right)\Delta+\xi^{-2}(|\Delta|^2-1)\Delta+\left(\frac{\nabla}{i}-\frac{2e}{\hbar c}\mathbf{A}\right)^2\Delta=0 \tag{10.27}$$

$$\mathbf{J}=\sigma\left(-\nabla\psi-\frac{1}{c}\frac{\partial\mathbf{A}}{\partial t}\right)+\operatorname{Re}\left[\Delta^*\left(\frac{\nabla}{i}-\frac{2e}{\hbar c}\mathbf{A}\right)\Delta\right]\frac{1}{8\pi e\lambda^2} \tag{10.28}$$

$$\rho=\frac{\psi-\varphi}{4\pi\lambda_{\mathrm{TF}}^2} \tag{10.29}$$

plus the Maxwell equations coupling the scalar and vector potentials φ and $\mathbf{A}$ to the charge and current densities ρ and $\mathbf{J}$. Here D is the normal-state diffusion constant and ψ is the electrochemical potential divided by the electronic charge. Δ is the gap parameter divided by its equilibrium value in the absence of fields $\Delta_0=\pi k[2(T_c^2-T^2)]^{1/2}$, where T is the temperature and T_c is the critical temperature (in the presence of the magnetic impurities); thus, $\Delta=1$ in the absence of fields. The temperature-dependent coherence length is $\xi=\hbar(6D/\tau_s)^{1/2}/\Delta_0$, where τ_s is the spin-flip scattering time, and the temperature-dependent magnetic penetration depth is $\lambda=\hbar c(8\pi\sigma\tau_s)^{-1/2}/\Delta_0$. σ is the normal-state conductivity, and λ_{TF} is the Thomas-Fermi static-charge screening length. It is perhaps appropriate to mention at this point that these equations necessarily assume that heating effects due to the dissipation of energy by the time-varying fields and currents can be neglected. As Gor'kov and Eliashberg remarked, this restricts the applicability of the theory to very near T_c unless there is a very high density of paramagnetic impurities.

Because of the complexity of working with these coupled nonlinear partial differential equations, we shall content ourselves with mentioning a few of the applications which have been made. In their original paper, Gor'kov and Eliashberg first considered the response of a paramagnetic alloy superconductor to a strong variable magnetic field. They worked out an approximate solution for the case of the skin effect with a bulk sample, finding, e.g., the amount of third harmonic radiation generated by the nonlinear surface currents. They then treated the case of a thin film sample of thickness d, subjected to an oscillating field which is the same on both sides of the film and parallel to it. For this simple case, they were able to work out rather exact results, which reduce to intuitively plausible limits for frequencies well above and below a characteristic relaxation rate τ^{-1}. In fact, for $\omega\tau\ll 1$, they found that the order parameter follows the alternating field adiabatically, with $\Delta(t)/\Delta_0=[1-h^2(t)]^{1/2}$, where $h=H/H_{c\|}$, as expected from (4.52) since Δ is proportional to the GL ψ. In the high-frequency limit, one instead finds $h^2(t)$ replaced by the time-average value

---

[31]C. R. Hu and R. S. Thompson, *Phys. Rev.* **B6**, 110 (1972).

$\overline{h^2(t)}$. The characteristic relaxation frequency separating these regimes is found to be $1/\tau = e^2 d^2 D H_{c\|}^2 / 3c^2\hbar$. Comparing this expression with (10.22*c*), and recalling that $2\alpha/\hbar = 1/\tau_K$, we see that this τ is exactly the same as τ_{Kc}. Thus, near T_c, we have

$$\tau = \tau_{GL} = \frac{\pi\hbar}{8k(T_c - T)} \tag{10.30}$$

using (10.20*a*). This is also the relaxation-time expression anticipated in (8.47), in our discussion of fluctuation-enhanced conductivity.

Following the pioneering work of Schmid, and of Caroli and Maki, Thompson and collaborators have applied these equations to an extensive analysis of the dynamic structure of vortices moving in type II superconductors in the resistive mixed state or flux-flow regime, discussed more qualitatively in Sec. 5.5. The Schmid-Caroli-Maki solution was obtained by assuming that there was a uniform electric field associated with a uniformly translating Abrikosov vortex lattice; (10.27) was then solved for Δ, (10.28) was used to compute $\mathbf{J}(\mathbf{r})$, and the ratio of the averaged current density $\langle \mathbf{J} \rangle$ to the assumed uniform $\mathbf{E}$ was used to define an effective flux-flow conductivity. Thompson and Hu showed that for this solution $\nabla \cdot \mathbf{J} \neq 0$, so that charge would build up to generate a nonuniform $\mathbf{E}$ until a new steady-state solution was established. A central feature of this Thompson-Hu solution is a backflow current $\mathbf{J}_b$, which goes through the vortex cores and returns around their sides. This current is lossless, although it flows directly through the core, which illustrates the degree of oversimplification implicit in the "normal" core of Bardeen and Stephen. The order parameter vanishes only along a line at the axis of the vortex, and the presence of low-lying excitations and the absence of an energy gap do not preclude the existence of supercurrents in the core region. One of the quantitative results of the calculation is that the normalized flux-flow resistance $R = \rho_f/\rho_n$ for a high-κ superconductor (containing a large concentration of paramagnetic impurities) should follow a concave-upward curve between $B = 0$ and $B = H_{c2}$, rather than the simple linear relation (5.59). The computed initial slope $dR/d(B/H_{c2})$ is 0.33 and the slope at H_{c2} is computed to be 5.2. Experimental data have this general shape, but there are experimental difficulties in isolating ideal behavior, free from pinning effects.

Probably the most convincing experimental test of the TDGL theory is actually in the measurements of fluctuation conductivity *above* T_c, as outlined in Sec. 8.7. Above T_c, one automatically has a gapless superconductor, but even so there are complications as discussed there unless the so-called Maki terms are suppressed by residual pair-breaking effects of some sort.

10.3.1 Electron-Phonon Relaxation

In many cases, one is interested in understanding the relaxation processes of superconductors which are *not* doped with magnetic impurities, so the Gor'kov-Eliashberg theory is not directly applicable. In the absence of magnetic pair-breakers, one must rely on inelastic phonon-electron interactions to achieve equilibrium between quasi-particles and condensate (by creation and recombination processes) and within the quasi-particle population (by inelastic scattering). Near T_c the characteristic time for both types of processes is the inelastic phonon-scattering time τ_E[32] for electrons in the normal state at T_c, although there will be differences depending on the nature of the disequilibrium being relaxed. The most direct way of estimating this time is from the low-temperature phonon-limited electronic thermal conductivty. (One cannot use the scattering time from low-temperature *electrical* conductivity, since small-angle scattering by low-energy phonons is ineffective in producing electrical resistance, but it *is* effective in relaxing a thermal energy distribution.) A simple, but less reliable, estimate can be made from the room temperature conductivity (where the phonons are equally effective in causing both electrical and thermal resistivity). Assuming a simple Debye phonon spectrum and free-electron Fermi surface, one finds that $\tau_E(T_c)$ varies roughly as $(\Theta_D/T_c)^3$. For tin, it is of the order of 10^{-10} sec; it is much shorter in lead and much longer in aluminum because of their higher and lower T_c values, but all of these times are much longer than the τ_{GL} given by (10.30). The implications of these inelastic relaxation times for superconductors with a well-defined gap are discussed in Chap. 11. There it is shown that they lead to (11.11), which resembles the TDGL equation in form, but with a longer characteristic time scale set by τ_E instead of τ_{GL}.

[32]The subscript E refers to "energy"; it reflects the fact that these inelastic electron-phonon processes relax the energy distribution of the quasi-particles.

CHAPTER 11

NONEQUILIBRIUM SUPERCONDUCTIVITY

11.1 INTRODUCTION

In other chapters, we have discussed examples of superconductors in globally stable equilibrium (e.g., the Meissner state of a superconductor in an external magnetic field) and in metastable equilibrium (e.g., a "persistent" current in a superconducting ring sustaining trapped flux, which occupies a local minimum in free energy but will eventually relax by macroscopic flux motion into the globally stable state with no current). In this chapter, we discuss superconductors in which the electron population is driven *out* of thermal equilibrium, into either a steady-state dynamic equilibrium or a more general time-dependent regime.

A dynamic equilibrium is typically set up when a perturbing source is balanced by relaxation and diffusion to set up a steady state which differs from thermal equilibrium. Typical examples include the conversion of a normal current to a supercurrent at an *NS* interface with associated resistive voltages developed in the superconductor and the stimulation (or weakening) of superconductivity by perturbations such as microwave irradiation that effectively "cool" (or "heat") the electrons.

In the more complex nonequilibrium regimes, both the magnitude and phase of the order parameter $\psi \sim |\Delta(\mathbf{r})|e^{i\varphi(\mathbf{r})}$ vary in time as well as in space. Examples which have received particular attention and which will be discussed below include the dynamic behavior of short, superconducting metallic weak links (Josephson devices) and of phase-slip centers in long, superconducting filaments when $I > I_c$, so that no stable static state is possible.

In this chapter, we focus on a number of illustrative applications of non-equilibrium superconductivity, treated at a relatively elementary level. For a more comprehensive treatment, the reader is referred to the review volume edited by K. E. Gray.[1]

11.2 QUASI-PARTICLE DISEQUILIBRIUM

According to the BCS theory (see Chap. 3), the superconductive ground state is formed of Cooper pairs of electrons with equal and opposite momentum and spin. The energies of single-particle excitations from this state are $E_k = (\Delta^2 + \xi_k^2)^{1/2}$, where ξ_k is the one-electron energy of the state k (relative to the Fermi energy) in the normal state and Δ is the BCS gap parameter. In *thermal equilibrium*, these quasi-particle states are occupied with the probability given by the Fermi function $f_0(E_k/kT) = [1 + e^{E_k/kT}]^{-1}$. This is *not* the case in the non-equilibrium regimes of interest here.

For what follows, it will be important to note that the nature of these excitations changes continuously from electronlike [with effective charge $q_k = (u_k^2 - v_k^2) = \xi_k/E_k \approx 1$] to holelike (with $q_k \approx -1$) as one goes from outside to inside the Fermi surface, i.e., over the range $\sim 2\Delta > \xi_k > -2\Delta$. This fractional charge stems from the fact that the quasi-particle excitations γ_k^* are linear combinations of electron creation and annihilation operators, with probabilities u_k^2 and v_k^2, respectively, as was discussed in Chap. 3. These quasi-particle excitations form the "normal electrons" of a two-fluid model of superconductivity. Their density of states is $N_s(E) = N(0)E/|\xi|$, where $N(0)$ is the usual density of states in the normal metal at the Fermi level. We denote the *actual* occupation numbers of these states by f_k, where in general $f_k \neq f_0(E_k/kT)$, the Fermi function of thermal equilibrium. For the simple spatially uniform case, these f_k determine the magnitude of the gap Δ through the BCS gap equation (3.50), which can be written as

$$\frac{2}{V} = \sum_k \frac{1 - 2f_k}{E_k} = \sum_k \frac{1 - 2f_k}{(\Delta^2 + \xi_k^2)^{1/2}} \tag{11.1}$$

For example, if f_k is set equal to $f_0(E_k/kT)$, one obtains the BCS form of $\Delta(T)$, which goes to zero as $3.07kT_c(1 - t)^{1/2}$ at T_c. [Here $t = T/T_c$.] However, if $f_k \neq f_0(E_k/kT)$, the order parameter Δ will take on a different value, as specified by (11.1).

[1]See review articles in K. E. Gray (ed.), *Nonequilibrium Superconductivity, Phonons, and Kapitza Boundaries*, NATO ASI Series, Plenum, New York (1981).

11.2.1 Energy-Mode vs. Charge-Mode Disequilibrium

To avoid the impractical task of having to deal with all the f_k individually, it is convenient to characterize the departure from thermal equilibrium by introducing two parameters T^* and Q^*, representing the nonequilibrium temperature and quasi-particle charge density, respectively. This scheme is based on the fact that an arbitrary set $\{\delta f_k\}$ of departures from equilibrium, where

$$\delta f_k \equiv f_k - f_0(E_k/kT) \tag{11.2}$$

can be uniquely decomposed as the sum of two orthogonal components, in which (by construction) the two new sets $\{\delta f_k\}$ are, respectively, even and odd with respect to inversion through the local Fermi surface, as illustrated schematically in Fig. 11.1.

The *even* mode has the symmetry corresponding to a change in temperature, which produces more (or fewer) quasi-particles *equally* on both holelike and electronlike branches of the quasi-particle spectrum. According to (11.1), such a change will directly affect the magnitude of the energy gap in the same way that a temperature change does. This even mode was labeled "longitudinal" by Schmid and Schön,[2] and also can be referred to as the "energy" or "temperature" mode. It is excited in pure form by neutral perturbations such as phonons or photons, but also (in combination with the odd mode) by most charged perturbations such as tunneling electrons. It is physically suggestive to parameterize the strength of this longitudinal disequilibrium by defining an effective quasi-particle temperature T^* such that $\Delta_{BCS}(T^*)$ equals the $\Delta(\{f_k\})$ which would result from inserting the actual nonequilibrium distribution in (11.1). To avoid any misunderstanding, we emphasize that this T^* is only a descriptive parameter, *not* a thermodynamic temperature.

Near T_c, where $\Delta \ll kT_c$, one can show by differentiation of (11.1) that the temperature shift δT^* which causes the same (small) change in Δ as a given set $\{\delta f_k\}$ is given

$$\frac{\delta T^*}{T} \equiv \frac{T^* - T}{T} \approx \frac{1}{N(0)} \sum_k \frac{\delta f_k}{E_k} = \int_{-\infty}^{\infty} \frac{\delta f_k}{E_k}\, d\xi_k \tag{11.3}$$

This quantity is similar to the "control function" $-\chi$ introduced by A. Schmid[3] and the function ϕ introduced by Aslamazov and Larkin[4] for the same purpose. To deal with problems involving spatial gradients, one then uses the

---

[2] A. Schmid and G. Schön, *J. Low Temp. Phys.* **20**, 207 (1975).

[3] A. Schmid, *Phys. Rev. Lett.* **38**, 922 (1977).

[4] L. G. Aslamazov and A. I. Larkin, *Zh. Eksp. Teor. Fiz.* **70**, 1340 (1976); Engl. transl. *Sov. Phys. JETP* **44**, 178.

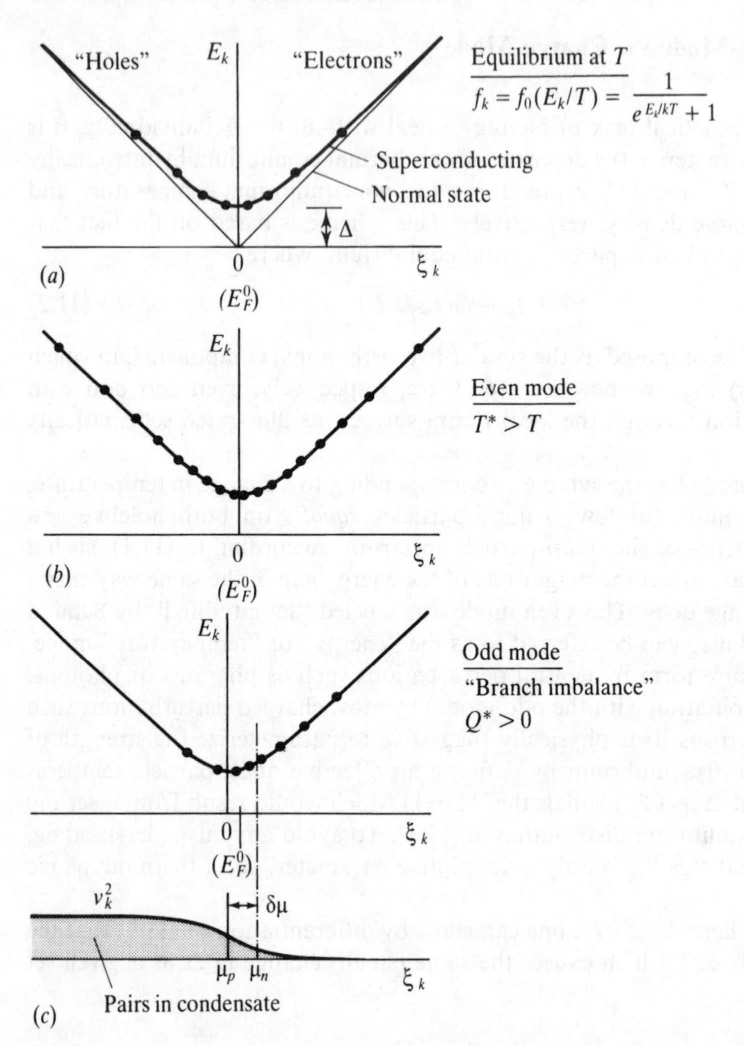

FIGURE 11.1
(*a*) Dispersion curves of excitation energies in normal and superconducting states, with schematic indication of occupation numbers in thermal equilibrium. (*b*) Schematic indication of population with even (or energy) mode excited, with $T^* > T$. (*c*) Schematic indication of population with odd (or charge) mode excited, showing branch imbalance corresponding to $Q^* > 0$ and shift of μ_n and μ_p relative to the equilibrium value E_F^0.

Ginzburg-Landau equation, including this nonequilibrium shift in T^*, which shifts the GL parameter α to $(\alpha + \delta\alpha^*)$.

The *odd* class of disequilibrium is generated only by *charged* perturbations (e.g., electron injection, or conversion of normal current to supercurrent near an interface). It is characterized by a net quasi-particle charge density Q^*, associated with the departures from thermal equilibrium (in which $Q^* = 0$ by symmetry). In units of e, this charge imbalance[5] is

$$Q^* \equiv \sum_k q_k \, \delta f_k = \sum_k \frac{\xi_k}{E_k} \delta f_k \tag{11.4}$$

To maintain overall electrical neutrality, there must be a compensating change in the number of electrons in the condensed BCS ground state. This implies that the electrochemical potentials of the "normal" quasi-particles (μ_n) and of the superconducting pairs (μ_p) must shift in opposite directions from their common equilibrium value. This results in a measurable difference in potential within the selfsame piece of metal, as was demonstrated in a classic experiment of J. Clarke,[6] discussed in Sec. 11.5. Clearly, such an effect is impossible in the normal state, which lacks the extra degree of freedom offered by the presence of the condensate. This odd mode is variously referred to as the "branch imbalance" or "charge" mode of disequilibrium, or as the "transverse" mode in the nomenclature of Schmid and Schön.

11.2.2 Relaxation Times

In order to determine the magnitudes of observable effects stemming from the disequilibrium caused by a given perturbation, we need to know the relaxation time with which equilibrium is approached if the perturbation is halted. According to Schmid and Schön, very near T_c (so that $\Delta \ll kT \approx kT_c$), the relaxation times for the two modes can be written, respectively, as

$$\tau_R^{(L)} = \tau_\Delta = \tau_{T^*} \approx 3.7\tau_E kT_c/\Delta \tag{11.5a}$$

$$\tau_R^{(T)} = \tau_{Q^*} = (4/\pi)\tau_E kT_c/\Delta \tag{11.5b}$$

In these expressions, τ_E is the energy-relaxation or *inelastic* scattering time for an electron at the Fermi surface. This relaxation rate is usually dominated by electron-phonon scattering, the strength of which was computed for many metals by Kaplan et al.[7] This τ_E is the characteristic time for f_k to approach the Fermi function or, equivalently, for δf_k to relax to zero.

The (similar) forms of (11.5*a*, *b*), which predict relaxation times that diverge as $(1-t)^{-1/2}$ near T_c, can be understood qualitatively as follows: τ_E sets the time scale for changing a given δf_k by inelastic scattering, but (near T_c) only processes

---

[5] M. Tinkham and J. Clarke, *Phys. Rev. Lett.* **28**, 1366 (1972); M. Tinkham, *Phys. Rev.* **B6**, 1747 (1972). For an excellent review of related phenomena, see J. Clarke, "Charge Imbalance," in K. E. Gray (ed.), *Nonequilibrium Superconductivity, Phonons, and Kapitza Boundaries*, NATO ASI Series, Plenum, New York (1981), p. 353.

[6] J. Clarke, *Phys. Rev. Lett.* **28**, 1363 (1972).

[7] S. B. Kaplan, C. C. Chi, D. N. Langenberg, J. J. Chang, S. Jaferey, and D. J. Scalapino. *Phys. Rev.* **B14**, 4854 (1976).

involving the fraction $\sim\Delta/kT_c$ of thermally occupied states which lie just above the gap are fully effective in either controlling the gap or relaxing charge imbalance. For example, (11.3) shows that quasi-particles in the $N(0)\Delta$ states with $|\xi_k| < \Delta$, for which $E_k \approx \Delta$, have maximal effect on δT^*. Similarly, (11.4) shows that a quasi-particle injected at a thermal energy $kT \gg \Delta$ contributes a full unit to Q^*, which is relaxed only by an inelastic process which lowers its energy down to near Δ, where $q_k \ll 1$. This approach was used by Tinkham[8] in making the first estimates of the charge relaxation time in interpreting the experiment of J. Clarke before the more rigorous results of Schmid and Schön[9] were available. The interrelationship of τ_{Q^*}, τ_{T^*}, and τ_E was also discussed by Clarke et al.[10]

Although *inelastic* electron-phonon scattering is needed to relax T^* effectively, it is worth noting that Q^* can be efficiently relaxed by *elastic* scattering (as from impurities) *in the presence of gap anisotropy*. The requirement of anisotropy stems from the coherence factor for scattering. This was found in (3.87) to be $(u_k u_{k'} - v_k v_{k'})^2$, where k and k' refer to the initial and final states, respectively, in the scattering process. With the forms of u_k and v_k given in (3.35), this coherence factor is exactly zero for an elastic charge-relaxing scattering process from k on one side of the Fermi surface to k' on the other at the same energy E because then $v_{k'} = u_k$ and $u_{k'} = v_k$. However, since u_k and v_k depend on $\xi_k/E_k = (E_k^2 - \Delta_k^2)^{1/2}/E_k$ and not on E_k alone, if there is gap anisotropy and the scattering connects states of different Δ_k, the cancellation is no longer exact and there is a finite contribution to the relaxation rate $1/\tau_{Q^*}$. As discussed by Clarke,[11] as one approaches T_c, this elastic process is always dominated by the inelastic process discussed earlier.

11.3 ENERGY-MODE DISEQUILIBRIUM: STEADY-STATE ENHANCEMENT OF SUPERCONDUCTIVITY

In this section, we discuss steady-state experimental situations in which, contrary to intuition, the effect of an external perturbation which feeds energy into the system is to *lower* the effective temperature T^* of the superconductor, thus *enhancing* the superconductivity. In the next section, we shall discuss *dynamic* regimes in which the "cooling" (or "heating") is a consequence of the time dependence of the order parameter.

---

[8]M. Tinkham and J. Clarke, *Phys. Rev. Lett.* **28**, 1366 (1972); M. Tinkham, *Phys. Rev.* **B6**, 1747 (1972).

[9]A. Schmid and G. Schön, *J. Low Temp. Phys.* **20**, 207 (1975).

[10]J. Clarke, U. Eckern, A. Schmid, G. Schön, and M. Tinkham, *Phys. Rev.* **B20**, 3933 (1979).

[11]See J. Clarke, "Charge Imbalance," in K. E. Gray (ed.), *Nonequilibrium Superconductivity, Phonons, and Kapitza Boundaries*, NATO ASI Series, Plenum, New York (1981), p. 353.

The basic idea underlying these phenomena is that our expression (11.3) for the nonequilibrium shift in effective temperature can be made to be *negative* (so that superconductivity is strengthened) by a suitable choice of the values of δf_k. This is rather obvious if the δf_k can be made to be predominantly negative by some means of reducing the total number of quasi-particles. But δT^* can be made to be negative in a less obvious way even if $\Sigma(\delta f_k) = 0$, so long as the negative values of δf_k are concentrated in states with lower values of E_k, i.e., in states nearer the gap. Creation of a negative δT^* should not only strengthen superconductivity below T_c, but also should raise T_c itself. The first detailed theoretical discussion of the possibility of such enhancement effects was given by Eliashberg and coworkers.[12]

11.3.1 Enhancement by Microwaves

The first clear experimental demonstration of such effects was the so-called Wyatt-Dayem effect,[13] namely, the enhancement of the dc critical current of a superconducting bridge by microwave radiation. The effect was studied in more detail by Klapwijk and Mooij,[14] who demonstrated that the enhancement occurred in long, narrow strips as well as in bridges, and in addition demonstrated the enhancement of T_c itself. The enhancement of Δ was demonstrated by Kommers and Clarke.[15]

In these experiments, a superconducting film was irradiated with microwave radiation having photon energy $\hbar\omega < 2\Delta$, so that no new quasi-particles are generated. Rather, existing quasi-particles are preferentially elevated from the low-lying states $\Delta \leq E_k \leq (\Delta + \hbar\omega)$, from which *downward* transitions *cannot* be stimulated, to the less populated states higher in energy by $\hbar\omega$. That is, $\delta f_k < 0$ for the low-lying states and $\delta f_k > 0$ for the higher ones, so that (11.3) implies that cooling (in the sense of gap enhancement) occurs despite the net increase in mean energy of the quasi-particle distribution. This contrast illustrates vividly the difference between the parameter T^* and a true thermodynamic temperature.

In addition to this redistribution effect, Chang and Scalapino[16] have pointed out that there is also an actual decrease in the total number of quasi-particles

[12]G. M. Eliashberg, *JETP Lett.* **11**, 114 (1970); *Sov. Phys. JETP* **34**, 668 (1972); B. I. Ivlev and G. M. Eliashberg, *JETP Lett.* **13**, 333 (1971); B. I. Ivlev, S. G. Lisitsyn, and G. M. Eliashberg, *J. Low Temp. Phys.* **10**, 449 (1973).

[13]A. F. G. Wyatt, V. M. Dmitriev, W. S. Moore, and F. W. Sheard, *Phys. Rev. Lett.* **16**, 1166 (1966); A. H. Dayem and J. J. Wiegand, *Phys. Rev.* **155**, 419 (1967).

[14]T. M. Klapwijk and J. E. Mooij, *Physica* **B81**, 132 (1976); T. M. Klapwijk, J. N. van den Bergh, and J. E. Mooij, *J. Low Temp. Phys.* **26**, 385 (1977); J. E. Mooij, N. Lambert, and T. M. Klapwijk, *Solid State Comm.* **36**, 585 (1980).

[15]T. Kommers and J. Clarke, *Phys. Rev. Lett.* **38**, 1091 (1977).

[16]J. J. Chang and D. J. Scalapino, *J. Low Temp. Phys.* **31**, 1 (1978).

because recombination is more rapid at higher energies. This second enhancement effect is typically comparable in importance to the first. Other considerations complicate the quantitative theory. For example, there is a minimum frequency for the enhancement effect to occur. At lower frequencies, the rf current simply serves as a pair breaker, which weakens the superconductivity. The crossover frequency can be shown to be of the order of $1/\tau_E$, i.e., the quasi-particle level width, and this property was used by van Son et al.[17] to determine the dependence of τ_E on the sheet resistance of aluminum films.

Enhancements corresponding to $\delta T^* \sim -0.02T_c$ were obtained experimentally in aluminum films, where τ_E is particularly long, so that departures from equilibrium are easier to sustain than in superconductors having a stronger electron-phonon coupling and a higher T_c. In the particularly interesting case of enhanced superconductivity *above* T_c in the presence of microwave radiation, a question of stability arises since the dependence of δT^* on Δ is usually strong enough to permit two solutions of the gap equation for $\Delta > 0$, in addition to the normal-state solution $\Delta = 0$. This problem was discussed first by Schmid[18] and treated in detail by Eckern et al.[19] In typical examples, they find that the larger of the two nonzero gaps is the more stable, but that there should eventually be a first-order transition to the normal state as one increases the sample temperature. They find a generalized GL free-energy expression [including the stimulation term $\delta T^*(\Delta)$], which plays a role corresponding to that of the usual GL free energy in the analysis of the stability of equilibrium states. This observation allows many of the qualitative insights developed in dealing with equilibrium systems to be brought to bear on non-equilibrium systems as well.

11.3.2 Enhancement by Extraction of Quasi-Particles

Chi and Clarke[20] demonstrated experimentally an enhancement of the energy gap in aluminum films by using a tunnel extraction technique to reduce the quasi-particle density. Although gap enhancements of up to 40 percent were observed very near T_c, this corresponds to a cooling of only $\delta T^* \approx -0.003\ T_c$. Nonetheless, it is a technique well suited for quantitative tests of the enhancement theory because the total extracted current is directly measurable and its energy distribution is known from the well-developed theory of electron tunneling in superconductors.

---

[17]P. C. van Son, J. Romijn, T. M. Klapwijk, and J. E. Mooij, *Phys. Rev.* **B29**, 1503 (1984).

[18]A. Schmid, *Phys. Rev. Lett.* **38**, 922 (1977).

[19]U. Eckern, A. Schmid, M. Schmutz, and G. Schön, *J. Low Temp. Phys.* **36**, 643 (1979).

[20]C. C. Chi and J. Clarke, *Phys. Rev.* **B20**, 4465 (1979).

The qualitative origin of the extraction effect can be seen from the schematic diagram in Fig. 11.2, which depicts the tunneling processes in the conventional semiconductor representation for a bias voltage $eV = |\Delta_2 - \Delta_1|$. With the bias as shown, there is a net flow of quasi-particles from the heavily populated states above the gap in the low-gap superconductor to the lightly populated states of the high-gap superconductor. As a result, $\Sigma_k \delta f_k < 0$ in the low-gap superconductor and its gap is enhanced; the reverse is true for the high-gap material. The effect depends on the difference in the gaps in the two superconductors, which accounts for the smallness of the effect observed by Chi and Clarke, using two Al films whose T_c's differed by only 10 percent.

The full potential of this enhancement technique has been demonstrated more recently by Blamire et al.[21] Using a Nb/AlOx/Al/AlOx/Nb tunneling structure, they were able to obtain T_c enhancement in the Al of *over 100 percent*! For this configuration, the optimum bias voltage is in the range $eV = 2(\Delta_{Nb} \pm \Delta_{Al})$. The key requirements for achieving maximal effects are: (1) a large gap in the outer (Nb) electrodes to allow sweeping out the quasi-particles from a large energy range, (2) thin tunnel barriers to allow a high quasi-particle extraction rate through the surface, (3) a thin Al electrode for a low total thermal regeneration rate in the volume, (4) a slow intrinsic inelastic relaxation in the middle (Al) electrode, and (5) a double-junction configuration to extract both electron and hole excitations. In essence, satisfaction of these conditions allows one to extract

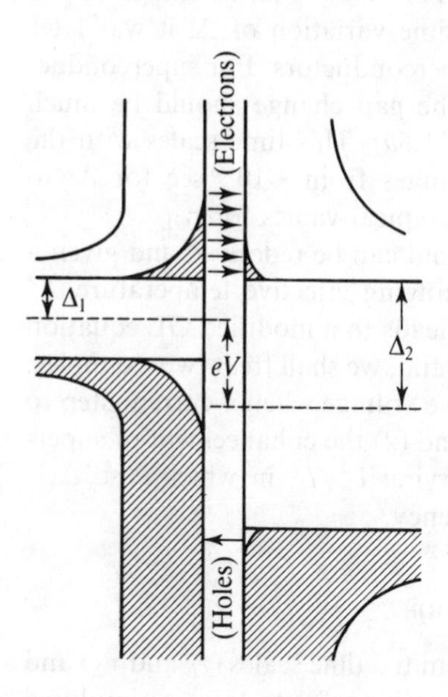

FIGURE 11.2
Schematic diagram of tunnel processes showing net extraction of quasi-particles from the superconductor having the smaller gap and hence a greater density of quasi-particles.

---

[21]M. G. Blamire, E. C. G. Kirk, J. E. Evetts, and T. M. Klapwijk, *Phys. Rev. Lett.* **66**, 220 (1991); see also detailed calculations in D. R. Heslinga and T. M. Klapwijk, *Phys. Rev.* **B47**, 5157 (1993).

most of the low-lying quasi-particles which reduce the gap, and thus maintain the energy gap at a value comparable to its $T = 0$ value even at much higher temperatures. In other words, no matter how much one increases T_c by this technique, $\Delta_{eq}(T = 0)$ remains the upper limit for the energy gap.

11.4 ENERGY-MODE DISEQUILIBRIUM: DYNAMIC NONEQUILIBRIUM EFFECTS

In this section, we discuss *dynamic* regimes in which the cooling (or heating) is a consequence of the time dependence of the order parameter. In this way, we are led to an important form of the time-dependent GL equation, in which the characteristic time is the longitudinal relaxation time τ_Δ $(= \tau_{T^*})$ related to τ_E and defined in (11.5*a*), rather than the much shorter τ_{GL} (see Chap. 10), which is appropriate for gapless superconductors.

When an ideal current step is applied to a superconducting filament, it is initially carried as a normal current, which requires an electric field ρJ. This field accelerates the supercurrent in a characteristic time $\tau_J = \tau_{tr} n/n_s \sim 1/(T_c - T)$, where $\tau_{tr} \sim 10^{-14}$ sec is the transport collision time in the normal state. For a dirty BCS superconductor near T_c, n_s scales with τ_{tr}. It can be shown that τ_J is proportional to, but a factor of 5.79 times shorter than, the characteristic GL time $\tau_{\mathrm{GL}} = \pi\hbar/8k(T_c - T)$, which is $\sim 10^{-12}$ sec for $(T_c - T) = 1$ K. Although it was originally thought that τ_{GL} should govern the time variation of Δ, it was later recognized that this was true only for *gapless* superconductors. For superconductors with a clean gap, Schmid[22] showed that the gap change should be much slower, with a characteristic time τ_Δ given by (11.5*a*). This time scales with the inelastic scattering time τ_E, which (near T_c) ranges from $\sim 10^{-8}$ sec for Al to $\sim 10^{-11}$ sec for Pb, in all cases much larger than typical values for τ_{GL}.

Here we shall show how this result of Schmid can be rederived and given a simple physical interpretation in terms of a changing effective temperature T^* induced by a time-dependent gap, which in turn leads to a modified GL equation including a $\partial\Delta/\partial t$ term. With this formalism in hand, we shall treat two examples: (1) the time delay for the appearance of a resistive voltage when a current step to $I > I_c$ is applied to a superconducting filament and (2) the enhancement of superconductivity in a short, metallic weak link carrying $I > I_c$, in which case Δ is oscillating up and down at the Josephson frequency.

11.4.1 GL Equation for Time-Dependent Gap

The striking contrast between the slow phonon-limited time scales (τ_E and τ_Δ) and the fast electronic time scales (τ_{GL} and τ_J) makes it useful to treat a simplified

---

[22] A. Schmid, *Phys. Kond. Mat.* **8**, 129 (1968).

model for nonequilibrium behavior on the slower τ_E time scales, which treats τ_{GL} and τ_J as instantaneous by comparison. This "slow-motion" regime will then be governed by the ordinary static GL equation, but with an effective temperature T^* which reflects the instantaneous nonequilibrium quasi-particle population. Because of the central role played by the energy gap Δ in this argument, we rewrite the GL equation (4.16) in terms of Δ itself, rather than ψ or $f = \psi/\psi_\infty$, which are proportional to Δ as shown by Gor'kov.[23] The static GL equation then takes the form

$$[\alpha^* + \beta(\Delta^2/T_c^2) - \xi^2(0)\nabla^2]\Delta = 0 \tag{11.6}$$

where $\alpha^* = (T + \delta T^* - T_c)/T_c$ reflects the nonequilibrium population, T is in energy units, $\beta = 7\zeta(3)/8\pi^2 = 0.106$, and $\xi(0)(1 - T/T_c)^{-1/2}$ is the usual coherence length. With these conventions, the equilibrium gap (near T_c) in the absence of gradients and nonequilibrium effects is correctly given by $\Delta_0^2/T_c^2 = -\alpha/\beta = (1 - T/T_c)/\beta$.

We now determine δT^* by inserting into (11.3) the nonequilibrium populations

$$\delta f_k(t) \equiv f_k(t) - f_0[E_k(t)/T] = f_k(t) - f_0\{[\Delta^2(t) + \xi_k^2]^{1/2}/T\} \tag{11.7}$$

which arise because $f_k(t)$, the *actual* occupation number at time t when the instantaneous gap value is $\Delta(t)$, lags behind the value $f_0[E_k(t)]$ that would be correct in thermal equilibrium. If we assume that the time dependence is slower than τ_E, then in the relaxation-time approximation the instantaneous value of f_k at time t will be the equilibrium value at the earlier time $(t - \tau_E)$. This implies that

$$\delta f_k = -\tau_E \frac{\partial f_0(E_k)}{\partial t} = -\tau_E \frac{\partial f_0}{\partial E_k}\frac{\partial E_k}{\partial \Delta}\frac{\partial \Delta}{\partial t} = -\tau_E \frac{\partial f_0}{\partial E_k}\frac{\Delta}{E_k}\frac{\partial \Delta}{\partial t} \tag{11.8}$$

Restricting attention to $E_k \sim \Delta \ll T \approx T_c$ as is appropriate in the GL regime, $(\partial f_0/\partial E_k) = -(4T_c)^{-1}$, and (11.8) becomes

$$\delta f_k = \frac{\tau_E}{4T_c}\frac{\Delta}{E_k}\frac{\partial \Delta}{\partial t} \tag{11.9}$$

Inserting this into (11.3) and integrating over ξ_k, we obtain

$$\delta T^* = (\pi/4)\tau_E(\partial\Delta/\partial t) \tag{11.10}$$

so that (11.6) becomes

$$[(T - T_c)/T_c + \beta(\Delta^2/T_c^2) - \xi^2(0)\nabla^2]\Delta = -(\pi/4T_c)\tau_E\Delta(\partial\Delta/\partial t) \tag{11.11}$$

This takes on a simpler form if written in terms of a normalized order parameter $f \equiv \Delta/\Delta_0(T)$, namely,

$$[1 - f^2 + \xi^2(T)\nabla^2]f = 2\tau_\Delta f(\partial f/\partial t) \tag{11.11a}$$

---

[23] L. P. Gor'kov, *Zh. Eksperim, i. Teor. Fiz.* **36**, 1918 (1959) [*Soviet Phys.—JETP* **9**, 1364 (1959)].

where τ_Δ is just the Schmid-Schön longitudinal relaxation time quoted in (11.5*a*), i.e., the time for the energy mode or Δ to relax. (This normalized order parameter f should not be confused with the occupation numbers f_k used above.)

To check the prefactor of τ_Δ in (11.11*a*), consider the response to a spatially uniform fluctuation of f from its equilibrium value $f = 1$ to $f = 1 + \delta f$. Upon inserting this value into (11.11*a*), we obtain

$$\frac{\partial(\delta f)}{\partial t} = -\frac{(\delta f)}{\tau_\Delta} \tag{11.12}$$

showing explicitly that this τ_Δ does give the exponential relaxation time for a fluctuation in Δ about its equilibrium value.

If the slow-variation approximation is *not* valid, a better[24] approximation is obtained by replacing the instantaneous value of $(\partial f/\partial t)$ in (11.11*a*) by the time-weighted average

$$\frac{1}{\tau_E}\int_{-\infty}^{t}\left(\frac{\partial f}{\partial t'}\right)e^{-(t-t')/\tau_E}\,dt' \tag{11.13}$$

Clearly, this reduces to $\partial f/\partial t$ in the slow-variation limit. In the opposite limit, namely, a discontinuous jump by Δf at some time t_0, this expression reduces (for $t > t_0$) to the plausible form

$$\frac{\Delta f}{\tau_E}e^{-(t-t_0)/\tau_E} \tag{11.13a}$$

rather than to a δ-function at t_0.

11.4.2 Transient Superconductivity above I_c

In 1979, Pals and Wolter[25] measured the time delay τ_d (some nanoseconds) between the application of a current $I > I_c$ to a superconducting aluminum film strip and the appearance of a resistive voltage. They identified this t_d with the time required to destroy superconductivity by driving the energy gap down to zero. A more complete set of experiments, including measurements with picosecond resolution, and a detailed comparison with theory were reported a few years later by Frank et al.[26] Let us use this experiment to illustrate our method of treating such problems and its limitations.

---

[24]See M. Tinkham, "Heating and Dynamic Enhancement in Metallic Weak Links" in K. E. Gray (ed.), *Nonequilibrium Superconductivity, Phonons, and Kapitza Boundaries*, NATO ASI Series, Plenum, New York, (1981), p. 231.

[25]J. A. Pals and J. Wolter, *Phys. Lett.* **70A**, 150 (1979).

[26]D. J. Frank, Ph. D. thesis and Technical Report no. 20 (June 1983), Division of Applied Sciences, Harvard University, Cambridge, MA; D. J. Frank, M. Tinkham, A. Davidson, and S. M. Faris, *Phys. Rev. Lett.* **50**, 1611 (1983); D. J. Frank and M. Tinkham, *Phys. Rev.* **B28**, 5345 (1983).

We consider a superconductor in which a current $I > I_c$ is suddenly applied at $t = 0$. As we did in discussing the equilibrium case in Sec. 4.4, for simplicity we assume that transverse dimensions are small enough $(d < \xi, \lambda)$ to justify a one-dimensional approximation and neglect of magnetic field terms, and also we again assume a spatially homogeneous solution of the form $f = |f|e^{iqx}$. Insofar as the slow-variation approximation holds, we can find the time dependence of f by inserting this assumed form into (11.11*a*), obtaining

$$\frac{\partial f}{\partial t} = \frac{1}{2\tau_\Delta}(1 - f^2 - q^2) \tag{11.14}$$

where q is measured in units of $1/\xi$. [If the time variation is too rapid, the left-hand side should be replaced by (11.13).] Except for initial and final transients of duration $\sim\tau_J \ll \tau_\Delta$, we may reasonably assume that the entire current is carried as a supercurrent, whose density is $j = f^2 q$, using units in which the usual critical current density is $j_c = 2/3\sqrt{3}$. This allows us to replace q^2 in (11.14) by j^2/f^4, obtaining

$$\frac{\partial f}{\partial t} = \frac{1}{2\tau_\Delta}\left(1 - f^2 - \frac{j^2}{f^4}\right) \tag{11.14a}$$

This can be easily integrated numerically to obtain $f(t)$; several representative curves[27] are shown in Fig. 11.3.

If $j < j_c$, this equation implies that f drops initially with a time constant of the order of τ_Δ, approaching the limiting value f_{eq} found in the static equilibrium solution (i.e., for $\partial f/\partial t = 0$), which satisfies the equation $j^2 = f_{eq}^4(1 - f_{eq}^2)$. The limiting equilibrium solution is at $j = j_c = 2/3\sqrt{3}$, where $f = (2/3)^{1/2}$. In the vicinity of this regime, the approach to equilibrium becomes very slow because

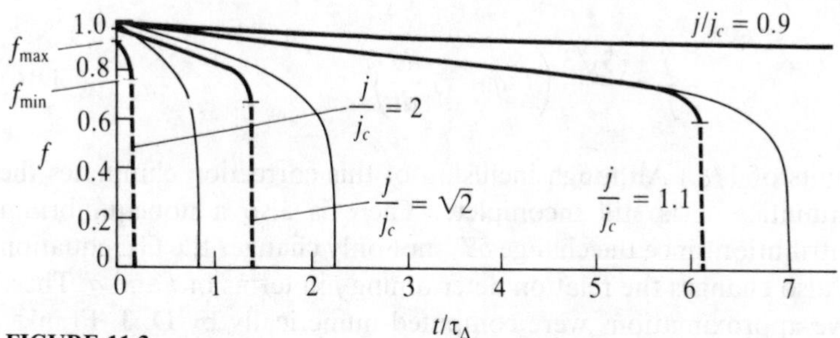

FIGURE 11.3
Time dependence of reduced order parameter $f = \Delta/\Delta_0$ for various current levels. Continuous curves result from the differential equation approximation (11.14*a*); discontinuous curves result from the integro-differential equation obtained by inserting (11.13). (*After Tinkham.*)

[27] M. Tinkham, "Non-Equilibrium Superconductivity" in *Festkörper Probleme XIX*, F. Vieweg, Wiesbaden (1979), p. 363.

the right-hand side of (11.14*a*) approaches zero there. If j is slightly above j_c, the rate of decrease of f slows nearly to zero as f passes near $(2/3)^{1/2}$, then speeds up as f falls farther, with $-\partial f/\partial t$ approaching infinity as $f \to 0$ because the term $-j^2/f^4$ becomes dominant. In fact, if j is substantially above j_c, one can approximate the right-hand side of (11.14*a*) by keeping *only* this term. An elementary integration then yields $f = (1 - 5j^2 t/2\tau_\Delta)^{1/5}$, implying a time delay to the collapse of superconductivity given by $t_d = 2\tau_\Delta/5j^2$ for $j \gg j_c$. In fact, t_d as a function of j can be computed numerically without making this approximation by evaluating the definite integral

$$t_d = 2\tau_\Delta \int_0^1 \frac{f^4\, df}{j^2 + f^6 - f^4} \tag{11.15}$$

Although the above discussion should be qualitatively correct, the slow-variation approximation on which it is based is not self-consistent at either the beginning or the end of the process. At $t = 0_+$ the supercurrent builds up in a time $\sim\tau_J \ll \tau_E$. This implies that df/dt changes in a similarly short time, requiring the use of (11.13) instead of the instantaneous df/dt. With this replacement, one finds that (11.4*a*) can only be satisfied at $t = 0_+$ if f drops discontinuously to some initial value $f_{\max}$, from which it continues to decline. Similarly, as f makes it final plunge toward zero, df/dt is predicted to diverge, again implying rapid change on the τ_E time scale, the need to use the integro-differential form of (11.14*a*), and a discontinuous final drop from some $f_{\min}$ to $f = 0$. These predictions are compared with those of the slow-variation approximation in Fig. 11.3.

These *discontinuous* drops are, in fact, unphysical because any changes of q on the time scale of τ_J will necessarily divert a significant fraction of the current to the normal electron channel in parallel with the supercurrent channel, so that the total current expression is generalized to

$$\frac{j}{j_c} = \frac{3\sqrt{3}}{2}\left(f^2 q + \tau_J \frac{dq}{dt}\right) \tag{11.16}$$

(Again, q is in units of $1/\xi$.) Although inclusion of this correction eliminates the spurious discontinuities, it is still incomplete. There is also a nonequilibrium supercurrent contribution since the change δT^* not only changes the GL equation determining f, it also changes the relation determining j in terms of f and q. These various successive approximations were computed numerically by D. J. Frank[28] and compared with exact numerical solutions of the full Schmid-Schön formalism, as well as with the results of his own experimental measurements. His conclusion was that calculations using the T^* model including both normal current

[28] D. J. Frank, Ph. D. thesis and Technical Report no. 20 (June 1983), Division of Applied Sciences, Harvard University, Cambridge, MA.

and nonequilibrium supercurrent terms gave a semiquantitative agreement with the exact calculations, and with his experimental data.

The experimental results of Frank et al. contain two additional interesting qualitative features. First, by using a picosecond (psec) sampling circuit based on Josephson junctions, Frank et al.[29] were able to resolve an initial voltage pulse lasting ~ 50 psec $\sim \tau_J$ immediately following the current step. This pulse reflects the electric field needed to accelerate the supercurrent to carry the imposed current. Second, they noticed that the breakdown in superconductivity occurred in several steps. They attributed[30] these to the successive nucleation of phase-slip centers (see Sec. 11.6) at different positions along the film strip because of inhomogeneities which caused the local critical current to vary by over 40 percent. When the delay time t_d was plotted vs. the imposed current step normalized to the local critical current, all the data collapsed into a single curve which was well fitted by the theoretical curve, as shown in Fig. 11.4.

11.4.3 Dynamic Enhancement in Metallic Weak Links

Here we discuss the implications of the nonequilibrium populations which result from the variation of the energy gap at the Josephson frequency in a metallic superconducting weak link. For definiteness, consider a short link between

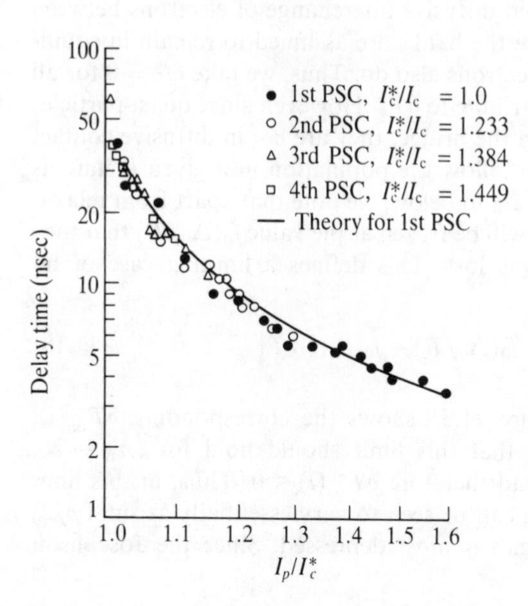

FIGURE 11.4
Delay time measured by Frank and Tinkham for each of four phase-slip centers (PSC) in an In film strip at $T/T_c = 0.994$ as a function of the ratio of the current step to the local critical current I_c^* of the PSC. The solid line is the theoretical prediction for $\tau_E = 148$ psec.

[29] D. J. Frank, M. Tinkham, A. Davidson, and S. M. Faris, *Phys. Rev. Lett.* **50**, 1611 (1983).

[30] D. J. Frank and M. Tinkham, *Phys. Rev.* **B28**, 5345 (1983).

massive superconductors, an idealized form of the variable thickness bridges studied by Octavio et al.[31,32] As shown by Aslamazov and Larkin (see Sec. 6.2.1), in the limit of a short bridge the solution of the static GL equation for the complex gap is the same as that of Laplace's equation for this geometry, namely, a linear interpolation between the boundary values Δ_0 at one bank ($x=0$) and $\Delta_0 e^{i\varphi}$ at the other ($x=L$). This implies that

$$|\Delta(x, \varphi)|^2 = \Delta_0^2\left[1 - \frac{4x}{L}\left(1 - \frac{x}{L}\right)\sin^2\frac{\varphi}{2}\right] \tag{11.17}$$

[See Eq. (6.9).] Now if a voltage V exists between the two banks, the phase φ will advance at the Josephson rate $\omega = d\varphi/dt = 2eV/\hbar$, and (11.17) describes an energy gap which cyclically decreases and returns back to Δ_0 once each cycle, the greatest decrease being in the middle of the bridge. Insofar as quasi-particle energy levels are defined in such a regime, where Δ is a strong function of position and time, the local values of $E_k = (\Delta^2 + \xi_k^2)^{1/2}$ will also pump up and down periodically in time, along with Δ^2. [See Fig. 11.5*a*.] Only if this variation is very slow compared to the inelastic scattering time τ_E can one assume that the occupation numbers f_k are able to relax and stay in instantaneous thermal equilibrium, so that $f_k = f_0[E_k(t)/T]$. For any finite voltage and hence frequency, there will be departures from equilibrium which imply a periodic instantaneous $\delta T^*(t) \neq 0$.

To find $\delta T^*(t)$, we must know $\delta f_k(t)$. This problem is simplified by noting that for $E_k > \Delta_0$, there is a very rapid diffusive interchange of electrons between the short bridge and the banks. Since the banks are assumed to remain in equilibrium, this implies that the bridge electrons also do. Thus, we take $\delta f_k = 0$ for all $E_k > \Delta_0$, and these levels do not contribute to δT^*. However, since quasi-particles in levels with $E_k < \Delta_0$ are trapped in the bridge, they are *not* in diffusive contact with a reservoir, and $\delta f_k \neq 0$. As we follow the population in a given k state as $E_k = (\Delta^2 + \xi_k^2)^{1/2}$ drops from above Δ_0 to below, we note that apart from relaxation effects, the occupation numbers will be frozen at the value $f_0(\Delta_0/T)$ that they had when contact with the bath was lost. This defines a limiting case of no relaxation for which

$$\delta f_k = \delta f_{k,\infty}(t) \equiv f_0(\Delta_0/T) - f_0[E_k(t)/T] \tag{11.18}$$

if $E_k = [\Delta^2(x,t) + \xi_k^2]^{1/2} \leq \Delta_0$. Figure 11.5*b* shows the corresponding $\delta T^*_\infty(t)$; here the subscript reflects the fact that this limit should hold for $\omega\tau_E \to \infty$. Since $E_k \leq \Delta_0$ in (11.18), $\delta f_k \leq 0$, and therefore $\delta T^*_\infty(t) \leq 0$. Thus, in this limit there is a cyclic *cooling* effect, which can be seen to vary essentially as $\sin^2(\varphi/2)$ being greatest at $\varphi = \pi$ when the gap is most depressed. Since the Josephson

[31] M. Octavio, W. J. Skocpol, and M. Tinkham, *IEEE Trans. Magn.* MAG-**13**, 739 (1977).

[32] M. Octavio, W. J. Skocpol, and M. Tinkham, *Phys. Rev.* **B17**, 159 (1978).

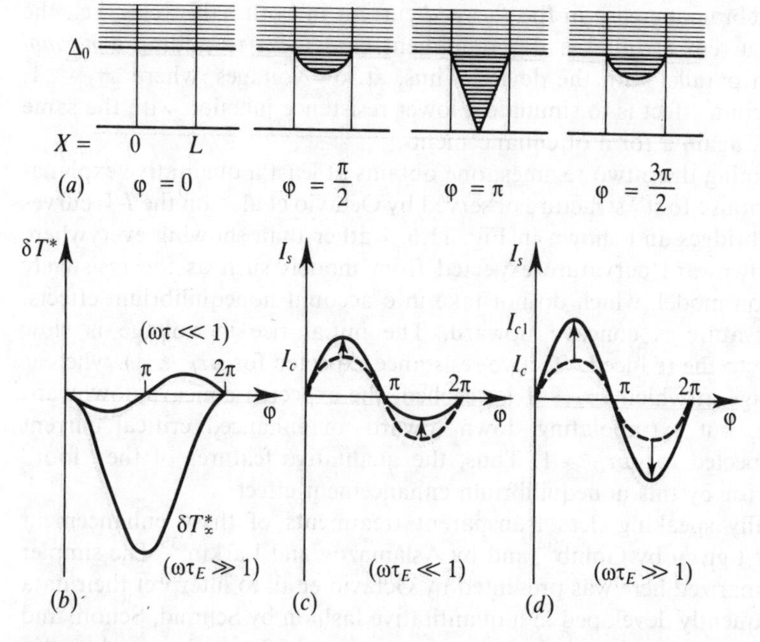

FIGURE 11.5
(*a*) Cyclic depression of the gap as given by (11.17). Quasi-particles in levels above Δ_0 remain in thermal equilibrium by a rapid diffusive exchange with the banks. Those in levels below Δ_0 (heavy shading) can be out of equilibrium. (*b*) Shifts of effective temperature T^* for rapid and slow modulation of Δ. (*c*) $I_s(\varphi)$ for $\omega\tau_E \ll 1$, showing increase in forward current. (*d*) $I_s(\varphi)$ for $\omega\tau_E \gg 1$, showing increase of effective I_c to $I_{c1} > I_c$. In (*c*) and (*d*), equilibrium supercurrents are shown in dashed curves for comparison.

current varies as $I_c(T^*)\sin\varphi$, averaged over a cycle such an effect is qualitatively the equivalent of simply an enhancement of I_{c1} to an "effective" value $I_{c1} > I_c$. [See Fig. 11.5*d*.]

The other simple limiting case occurs when $\omega\tau_E \ll 1$, so that the departures from equilibrium are small. In that case, one can see that

$$\delta f_k = -\tau_E \frac{d}{dt}\,\delta f_{k,\infty} = \tau_E\left(\frac{-\partial f_0}{\partial E}\right)\frac{\partial E}{\partial \Delta}\frac{\partial \Delta}{\partial t} \tag{11.19}$$

From this it follows that

$$\delta T^* \sim \tau_E \frac{\partial \Delta}{\partial t} \sim \tau_E \frac{\partial}{\partial t}(\delta T^*_\infty) \tag{11.20}$$

which varies essentially as $\sin\varphi$, changing sign at $\varphi = \pi$; see Fig. 11.5*b*. Keeping track of the sign, $\delta T^* < 0$, giving *enhancement* of I_c, on the *forward* half-cycle of the Josephson current oscillation $(0 < \varphi < \pi)$; but $\delta T^* > 0$, giving a *reduced reverse* current in the second half-cycle $(\pi < \varphi < 2\pi)$. (See Fig. 11.5*c*.) The net effect is an algebraic increase in the *forward* current in both half-cycles; i.e., the average supercurrent acquires a dc component equivalent to adding a *normal* conductance in parallel with the device. Thus, at low voltages, where $\omega\tau_E \ll 1$, the nonequilibrium effect is to simulate a lower resistance junction with the same critical current, again a form of enhancement.

By combining these two regimes, one obtains at least a qualitative explanation for the curious "foot" structure observed by Octavio et al.[33] on the *I-V* curves of short microbridges and shown in Fig. 11.6. Rather than showing everywhere the concave-downward curvature expected from models such as the resistively shunted junction model, which do not take into account nonequilibrium effects, the initial curvature is concave upward. The initial rise in voltage is slow (corresponding to the reduced effective resistance expected for $\omega\tau_E \ll 1$), whereas when the voltage at which $\omega\tau_E \sim 1$ is reached, the expected concave-downward shape appears, but extrapolating down toward an enhanced critical current $I_{c1} > I_c$, as expected for $\omega\tau_E \gg 1$. Thus, the qualitative features of the "foot" are accounted for by this nonequilibrium enhancement effect.

Historically speaking, less transparent treatments of these enhancement effects were first given by Golub[34] and by Aslamazov and Larkin.[35] The simpler approach summarized here was presented by Octavio et al. to interpret their data and was subsequently developed in a quantitative fashion by Schmid, Schön, and Tinkham.[36] The latter work includes not only the effect of T^* on the gap, but also its direct effect on the supercurrent (as was referred to in the previous section). With the reasonable value $\tau_E = 3 \times 10^{-10}$ sec for tin, this theory accounts quite

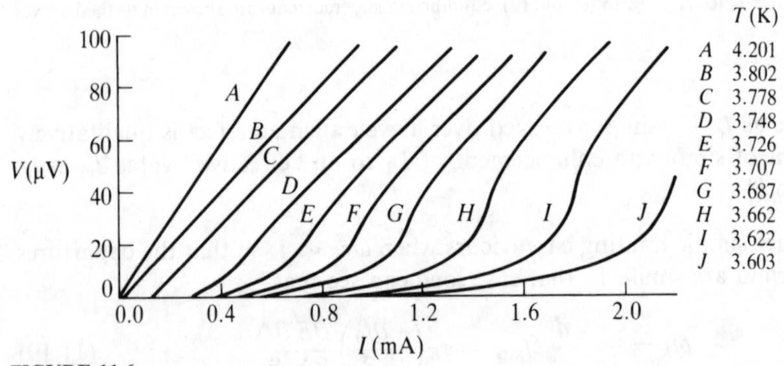

FIGURE 11.6
Low-voltage part of the *I-V* characteristic of tin variable thickness bridge, showing development of "foot" structure as a function of temperature. (*After Octavio et al.*)

---

[33] M. Octavio, W. J. Skocpol, and M. Tinkham, *Phys. Rev.* **B17**, 159 (1978).

[34] A. A. Golub, *Sov. Phys. JETP* **44**, 178 (1976).

[35] L. G. Aslamazov and A. I. Larkin, *Sov. Phys. JETP* **43**, 698 (1976).

[36] A. Schmid, G. Schön, and M. Tinkham, *Phys. Rev.* **B21**, 5076 (1980).

well for the magnitude and temperature dependence $[\sim 1/(T_c - T)]$ of the slope of the foot feature, and also for its high voltage limit $[V_{\max} \approx 3\hbar/e\tau_E \sim 7\,\mu\text{V}]$.

11.5 CHARGE-MODE DISEQUILIBRIUM: STEADY-STATE REGIMES

We begin our discussion of charge-mode or branch-imbalance disequilibrium by reviewing various steady-state examples, including a discussion of Andreev reflection at *NS* interfaces. We then turn in the next section to the dynamic example of the phase-slip centers in long superconducting filaments.

Our quantitative understanding of charge-imbalance or branch-imbalance disequilibrium and its implications stems from an experiment of J. Clarke.[37] He created charge imbalance in a superconducting tin film by tunnel injection of quasi-particles at sufficiently high injection voltage so that they were of a predominantly electronlike (or holelike) character (depending on the sign of the injection voltage). Because of the broad area injection, diffusion plays little role, and the steady-state charge imbalance is determined by balancing the local injection rate $\dot{Q}^*_{\text{inj}}$ against an appropriately averaged relaxation rate Q^*/τ_Q; i.e. $Q^* = \dot{Q}^*_{\text{inj}}\tau_{Q^*}$, and a measureable electrochemical potential difference V between pairs and quasi-particles is established. The quasi-particle potential is sensed by a second normal metal film coupled via a normal tunnel junction to the reverse side of the superconducting film in the injection region. Since the electrochemical potential of the pairs is constant throughout the superconductor, it is convenient to measure V relative to a second probe coupled to the equilibrium superconductor far from the injection region. Tinkham and Clarke showed[38] that the measured potential is proportional to Q^*, and is given by

$$V = \frac{Q^*}{2eN(0)g_{NS}} \tag{11.21}$$

Here $N(0)$ is the usual density of states for electrons of one spin, and g_{NS} is the *measured* normalized conductance of the normal probe junction. Thus the measured V allows the determination of Q^*.

Since the *charge* injection rate $\dot{Q}^*_{\text{inj}}$ differs from the measured injection *current* per unit volume $I_{\text{inj}}/e\Omega$ only by a known[39] function F^*, which is usually near unity, the relaxation time τ_{Q^*} can be determined from the data by using the relation

$$\tau_{Q^*} = \frac{2N(0)e^2\Omega g_{NS}V}{F^* I_{\text{inj}}} \tag{11.22}$$

[37] J. Clarke, *Phys. Rev. Lett.* **28**, 1363 (1972).

[38] M. Tinkham and J. Clarke, *Phys. Rev. Lett.* **28**, 1366 (1972).

[39] J. Clarke and J. L. Paterson, *J. Low Temp. Phys.* **15**, 491 (1974); J. Clarke, U. Eckern, A. Schmid, G. Schön, and M. Tinkham, *Phys. Rev.* **B20**, 3933 (1979).

Clarke's data, analyzed in this way, gave a value of τ_{Q^*} which diverged near T_c as predicted by (11.5*b*), and implied a value of $\tau_E \sim 10^{-10}$ sec, as expected for tin. Subsequently, other materials have been studied, and the effect of impurity scattering has been evaluated. The results generally follow (11.5*b*), but at lower temperatures one must also take into account the additional relaxation mechanism based on gap anisotropy and *elastic* scattering, which was mentioned at the end of Sec. 11.2.2.

In configurations other than the broadside tunnel injection of Clarke's experiment, the nonequilibrium population will diffuse away from the injection point while relaxing. If the geometry is such that the diffusion is essentially one dimensional, this process causes Q^* to decay in space as $e^{-x/\Lambda_{Q^*}}$, where

$$\Lambda_{Q^*} = \sqrt{D\tau_{Q^*}} = \left(\frac{1}{3}v_F \ell \tau_{Q^*}\right)^{1/2} \tag{11.23}$$

is the appropriate diffusion length for a time τ_{Q^*}. The associated potential difference between pairs and quasi-particles decays in the same way since it is proportional to the local value of Q^*. Insofar as the diffusion constant D is known, one can thus infer a value of τ_{Q^*} by measuring the diffusion length Λ_{Q^*}. Since Λ_{Q^*} is typically only a few micrometers, spatially resolved measurements of the potential require microfabrication techniques. The existence of such potential differences between superconducting and normal electrons in the same physical volume near an *SN* interface was demonstrated by Yu and Mercereau.[40]

An approach which avoids the need for microscopic potential probes is that pioneered by Pippard, Shepherd, and Tindall[41] in their study of the resistance of thin *SNS* sandwiches. They consistently observed an extra resistance (above that of the normal layer itself as inferred from its known resistivity and thickness) which appeared to diverge as $T \to T_c$. They argued that this extra resistance arose from the nonequilibrium region in the superconductor in which a quasi-particle current is converted to a supercurrent. Using the concepts outlined here, one sees that the extra resistance corresponds to that of a length Λ_{Q^*} (on either side of the normal layer), after taking into account the fact that again only a fraction F^* of the current is introduced as a quasi-particle current, the remainder being converted to a supercurrent by Andreev reflection (see below) at the interface. This simple interpretation has proved very effective in accounting for measurements of *SNS* resistance values by Hsiang and Clarke.[42]

[40]M. L. Yu and J. E. Mercereau, *Phys. Rev.* **B12**, 4909 (1976).

[41]A. B. Pippard, J. G. Shepherd, and D. A. Tindall, *Proc. Roy. Soc.* **A324**, 17 (1971).

[42]T. Y. Hsiang and J. Clarke, *Phys. Rev.* **B21**, 945, 956 (1980).

11.5.1 Andreev Reflection

When an electrical current passes from a normal to a superconducting metal, a certain fraction is converted to a supercurrent at the interface, and the remainder enters as a nonequilibrium charge Q^*, which relaxes into supercurrent over the charge relaxation distance Λ_{Q^*}, as described above. We now examine how this conversion depends on the nature of the interface between the two metals.

If there is a classic high-barrier tunnel junction at the interface, the fraction F^* of the current delivered as Q^* can be computed by taking into account the charge q_k of each quasi-particle injected into the superconductor, and the rate $\dot{f}_k$ at which they are injected. For example, if $T \approx 0$ and the bias voltage across the junction just exceeds $eV = \Delta$, each quasi-particle is created by injecting an electron right at the gap edge where $E_k = \Delta$. There it contributes its full electronic charge to the electrical current but contributes *zero* to Q^* since $q_k = 0$ in these states because they have an equal mixture of hole and electron character. This implies that the electron directly joins the Cooper pair condensate,[43] and its entire current contribution is converted to supercurrent. For higher bias voltages (or higher temperatures), the injected quasi-particles have nonzero values of q_k, approaching $q_k = \pm 1$ for $E_k \gg \Delta$. Thus, the ratio $F^* = e\dot{Q}^*/I$ ranges from zero for $eV = \Delta$ to unity for eV or $kT \gg \Delta$.

In the opposite limit, in which there is *no* barrier at the interface, the key to understanding the current transfer process is a novel reflection process first described by Andreev[44] (in explaining extra *thermal* resistance at the *NS* interface) and treated in the present context by Artemenko et al.,[45] and by Zaitsev.[46] Normal electrons incident at energies $E \gg \Delta$ pass through the interface depositing essentially all their charge as Q^*, just as in the tunnel case. The major effect is for the electrons incident from the normal metal at energies $E < \Delta$. Upon reaching the interface, they cannot enter as quasi-particles because there are no quasi-particles states in the gap. Instead, they are reflected back into the normal metal *as holes*, thus transferring $2e$ across the interface to the superconducting Cooper pair condensate. (More precisely, the electrons enter evanescent states in the gap which decay into the condensate in a distance $\sim\xi(T)$, which is typically much less than Λ_{Q^*}.) For kT and $eV \ll \Delta$, essentially all incident electrons are Andreev reflected. Thus, each transfers a double charge, giving a differential conductance *twice* that in the normal state, or a normalized conductance value $Y = 2$. (Recall that for the same voltage and temperature conditions, a tunnel

---

[43] More precisely, there is an equal probability that an electronlike quasi-particle is added and that an electronlike quasi-particle is removed and a Cooper pair is added. Thus, on average, half a Cooper pair and zero quasi-particle charge Q^* is added.

[44] A. F. Andreev, *Sov. Phys. JETP* **19**, 1228 (1964).

[45] S. N. Artemenko, A. F. Volkov, and A. V. Zaitsev, *JETP Lett.* **28**, 589 (1978); *Sov. Phys. JETP* **49**, 924 (1979); *Solid State Commun.* **30**, 771 (1979).

[46] A. V. Zaitsev, *Sov. Phys. JETP* **51**, 111 (1980).

junction gives $Y = 0$.) As T is raised, Y falls continuously to unity at T_c. One can define a similar quantity Y^* which gives the normalized rate of Q^* injection. As follows from the discussion above, this $Y^* = 0$ at $T = 0$, where all charge is transferred directly to Cooper pairs and rises to unity at T_c.

More generally, there will be some sort of barrier causing normal reflection at the *NS* interface, e.g., due to some oxide layer there, or because of the different Fermi velocities associated with the different metals. To allow a simple comprehensive treatment for the continuum of possibilities between no barrier and a strong tunnel barrier, Blonder et al.[47] (BTK) introduced a δ-function potential barrier of strength Z at the interface and solved the Bogoliubov equations (see Sec. 10.1) to find the probability for the various outcomes for an electron of energy E incident on the interface, as a function of Z. These various possibilities are shown schematically in Fig. 11.7, and their probabilities are shown in Fig. 11.8 for representative values of Z. Using this approach, they were also able to compute theoretical differential conductance curves for various barrier strengths Z. As shown in Fig. 11.9, these results of the BTK model can account for a wide range of dI/dV curves, ranging from the doubled low-voltage conductance of the pure Andreev case to the classic tunneling characteristic of the *SIS* tunnel junction for large Z. All of these cases reduce to a differential resistance equal to R_n at large voltages. However, integration with respect to V shows that the asymptotic linear I-V curves are displaced, reflecting an "excess current" in the low Z cases where Andreev reflection is significant, but not in the pure tunnel limit. This can be a useful diagnostic indicator of the junction type.

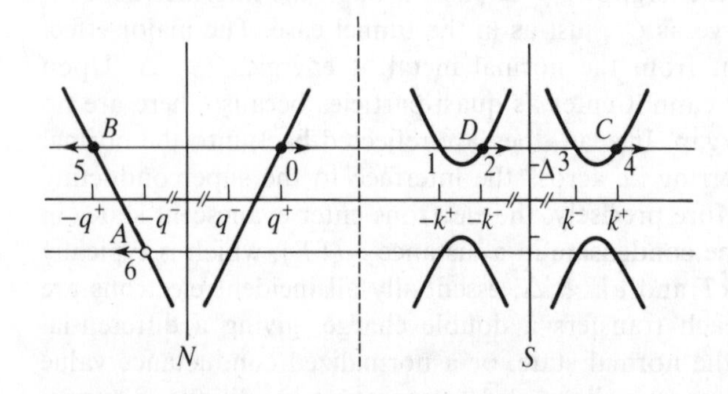

FIGURE 11.7
(*a*) Schematic diagram of energy vs. momentum on the two sides of an *NS* interface. The diagram includes degenerate states both inside and outside the Fermi surface and on both forward and reverse sides of the Fermi sphere. The open circles denote holes; the closed circles, electrons; and the arrows point in the direction of the group velocity, $\partial E_k/\partial k$. This describes an incident electron at (0), along with the resulting transmitted (2, 4) and reflected (5, 6) particles. *A* refers to the Andreev-reflected hole.

[47] G. E. Blonder, M. Tinkham, and T. M. Klapwijk, *Phys. Rev.* **B25**, 4515 (1982).

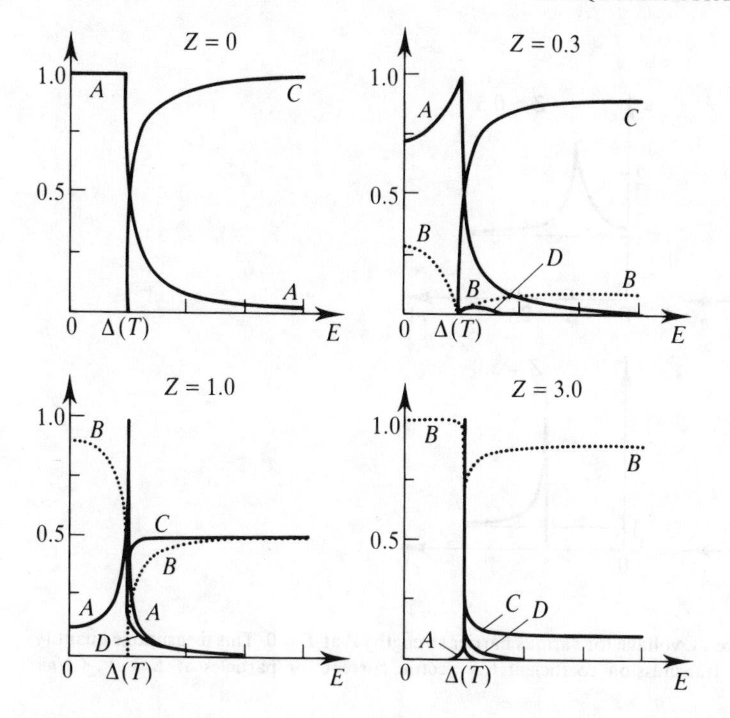

FIGURE 11.8
Plots of transmission and reflection coefficients at an *NS* interface. *A* gives the probability of Andreev reflection, *B* of ordinary reflection, *C* of transmission without branch crossing, and *D* of transmission with branch crossing. (These letters correspond to those in Fig. 11.7.) The parameter *Z* is a measure of the barrier strength at the interface. (*After Blonder et al.*)

11.5.2 Subharmonic Energy Gap Structure

Building on this treatment of an *NS* interface, Klapwijk et al.[48] (KBT) developed a new model, based on multiple Andreev reflections, to account for the subharmonic energy gap structure observed in the *I-V* characteristics of superconducting metallic weak links. They postulate that (for $I > I_c$) the massive superconducting banks, whose pair electrochemical potentials differ by eV, are separated by an effectively normal layer spanning the neck of the constriction weak link.

The concept is illustrated by Fig. 11.10, in which representative particle trajectories are plotted as dotted lines. Consider, e.g., the electron incident from the left in the lower left-hand part of the sketch. After being accelerated by the potential difference eV (represented by the displaced electrochemical potential

[48]T. M. Klapwijk, G. E. Blonder, and M. Tinkham, *Physica* **109**, **110B**, 1657 (1982).

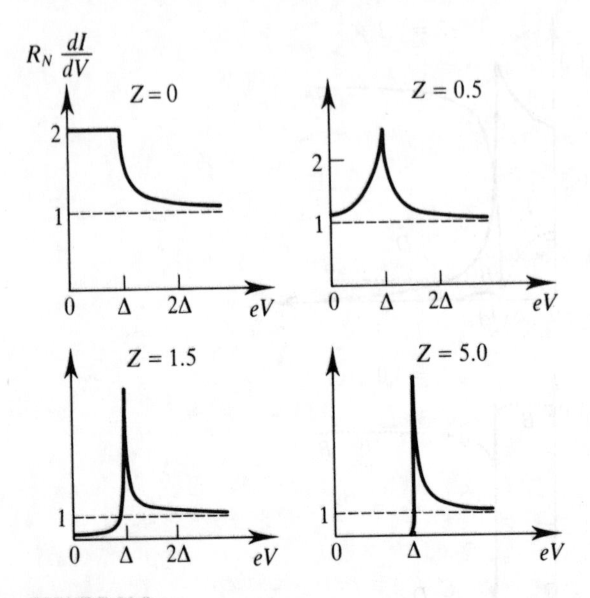

FIGURE 11.9
Differential conductance vs. voltage for various barrier strengths Z at $T = 0$. This measurable quantity is proportional to the transmission coefficient for electric current for particles at $E = eV$. (*After Blonder et al.*)

levels), it arrives at the right electrode with an energy inside the energy gap, causing it to be Andreev reflected as a hole. By energy conservation, the hole state is at an energy found by reflection around the pair chemical potential of the superconductor forcing the reflection, to which charge $2e$ is transferred. Because the particle has switched the sign of its charge, the hole gains another eV by

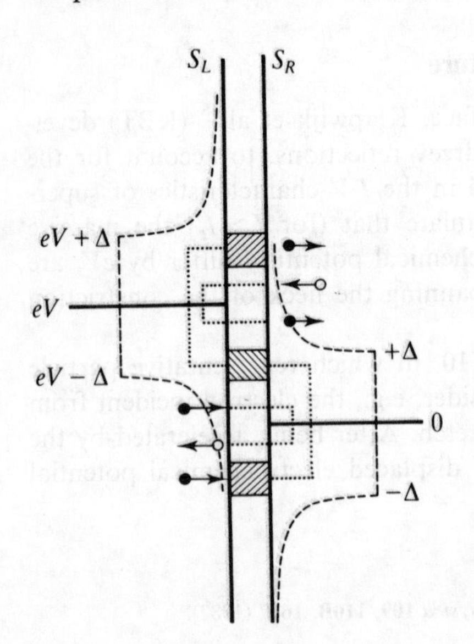

FIGURE 11.10
Semiconductor picture for following trajectories in the symmetric case (where $\Delta_L = \Delta_R = \Delta$) when $eV > \Delta$. The filled circles are electrons, the open circles are holes, and the arrows point in the direction of the group velocity. The Andreev reflection coefficient $A(E)$ is shown in dashed line in place of the density of states to indicate the gap structure. (*After Klapwijk et al.*)

retraversing the normal region. A similar reflection process will then occur at the left interface, again transferring charge $2e$ and causing an electron to emerge outside the gap, where it can now enter the right-hand superconductor. Clearly, by iteration of this process, Andreev reflection provides a way to gain arbitrarily large integral multiples of eV from the electric field, so long as scattering in the normal metal can be neglected. Accordingly, every time the applied voltage V passes through $V_n = 2\Delta/ne$, the $(n+1)$-fold reflection process disappears and the $(n-1)$-fold process opens up, causing a feature on the I-V characteristic at that voltage V_n.

To calculate the actual form of the I-V dependence, one must sum over all allowed trajectories, using the Fermi function to give the probabilities of various starting states in the banks (which are assumed to be in equilibrium), and the energy-dependent probability of the Andreev reflection at each collision with a bank. For simplicity, all phase information about the banks is deliberately suppressed, thus eliminating Josephson effects. This makes the summation quite tractable for the case of no interfacial barrier Z since at each collision with a bank an electron is either Andreev reflected or enters the far bank and remains there. The results obtained are at least semiquantitatively correct. However, with an interfacial barrier, or with scattering in the normal layer, the computation becomes much more complex, requiring a Boltzmann equation approach,[49] but yielding better agreement with the experimental data of Octavio on tin variable thickness bridges.

11.6 TIME-DEPENDENT CHARGE-MODE DISEQUILIBRIUM: PHASE-SLIP CENTERS

It is found experimentally that when the current in a long superconducting filament is increased above I_c, the voltage increases in a series of rather regular steps.[50] The first systematic study appears to have been that of J. D. Meyer[51] on the I-V curves of tin whiskers, and some typical observations from this work are shown in Fig. 11.11. It can be seen that each successive jump adds essentially the same amount to the differential resistance. This suggests that each step represents the appearance somewhere along the filament of an additional similar localized resistive center, across which the time-average pair chemical potential $\bar{\mu}_p$ suffers a discontinuous step increment, $\Delta\bar{\mu}_p$. Since this implies that the phase of the superconducting order parameter is increasing at different rates on the two

---

[49] M. Octavio, M. Tinkham, G. E. Blonder, and T. M. Klapwijk, *Phys. Rev.* **B27**, 6739 (1983); K. Flensberg and J. Bindslev Hansen, *Phys. Rev.* **B40**, 8693 (1989).

[50] See, e.g., W. W. Webb and R. J. Warburton, *Phys. Rev. Lett.* **20**, 461 (1968).

[51] J. D. Meyer and G. v. Minnigerode, *Phys. Lett.* **38A**, 529 (1972); J. D. Meyer, *Appl. Phys.* **2**, 303 (1973). See also the comprehensive review R. Tidecks, *Current-Induced Nonequilibrium Phenomena in Quasi-One-Dimensional Superconductors*, Springer, Berlin (1990).

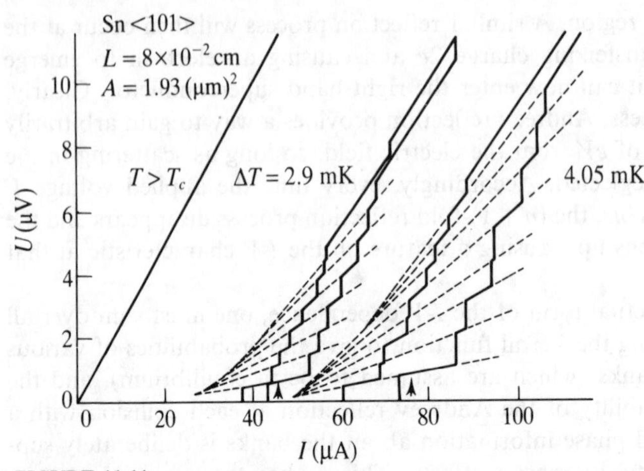

FIGURE 11.11
Current-voltage characteristics of tin "whiskers" showing regular step structures due to successive establishment of phase-slip centers. Here $\Delta T = T_c - T$). (*After Meyer.*)

sides of it, the centers are usually referred to as phase-slip centers (PSC).[52] Similar steps are seen in long superconducting thin film microbridges, and Skocpol et al.[53] used such a bridge with several voltage tabs along its length to verify that each step does indeed arise from the spatially localized region between two adjacent tabs.

The model of the phase-slip center developed by Skocpol, Beasley, and Tinkham (SBT) is shown schematically in Fig. 11.12. To start, it is easiest to imagine that the filament is not perfectly homogeneous, so that its critical current is lower at one point than anywhere else. Then, as the bias current I is raised above this minimum critical current, it is no longer possible to find zero voltage, static superconducting solutions, and a finite electric field appears. This accelerates the supercurrent above the critical velocity corresponding to the critical current, resulting in a collapse of the order parameter so that the entire current must be carried as a normal current; this in turn allows the superconductivity to reappear and the cycle to repeat. The phase difference slips by 2π each time $|\Delta|$ goes to zero in the middle of the PSC. All this action, with its associated strong ac supercurrent, occurs in the "beating heart" of the PSC, a region presumably only of length $\sim\xi(T)$ since that governs the spatial variation of Δ. This regime is reminiscent of the behavior in a short bridge connecting superconducting banks at different voltages, as discussed in Sec. 11.4.3, except that in the present case the *banks* are

[52]The concept of phase slippage in the resistive state of superconducting filaments was introduced by J. S. Langer and V. Ambegaokar, *Phys. Rev.* **164**, 498 (1967), but their calculations (described in Sec. 8.1) focus on the fluctuation-dominated regime near T_c.
[53]W. Skocpol, M. R. Beasley, and M. Tinkham, *J. Low Temp. Phys.* **16**, 145 (1974).

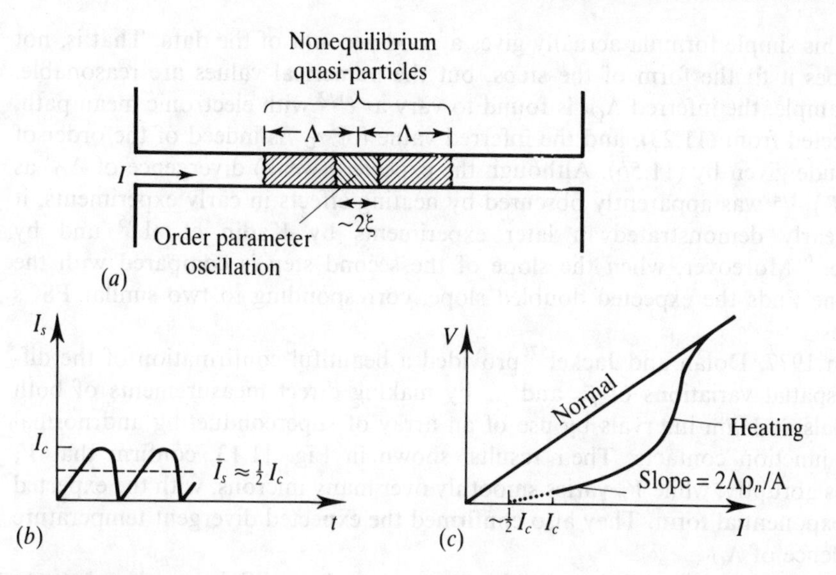

FIGURE 11.12
(*a*) Schematic diagram of a model of the phase-slip center. The oscillation of the gap magnitude occurs in a core length $\sim 2\xi$, whereas the nonequilibrium quasi-particles producing charge imbalance diffuse a distance $\sim\Lambda_{Q^*}$ in either direction before μ_n relaxes to μ_p. (*b*) The oscillatory supercurrent in the core region, with average value $\sim I_c/2$. (*c*) Schematic *I-V* curve of a bridge containing only a single phase-slip center. [*After M. Tinkham,* Revs. Mod. Phys. *46, 587 (1974).*]

also out of equilibrium. The qualitative feature which must be supplied by the theory of charge-imbalance disequilibrium is the connection between the bias current and the voltage across the PSC, and hence the Josephson frequency.

The argument is roughly this: Averaged in time over the Josephson cycle, the supercurrent in the heart of the PSC has some nonzero dc value[54] $\bar{I}_s = \beta I_c$, where $\beta \sim \frac{1}{2}$. The rest of the applied current I, namely, $(I - \bar{I}_s)$, must then be carried as a normal current. To drive this normal current at the center of the PSC requires that $d\bar{\mu}_n/dx \approx e(I - \bar{I}_s)\rho/A$, where A is the cross-sectional area of the filament and ρ is its normal resistivity. Noting that $d\bar{\mu}_s/dx = 0$ on either side of the point at which the phase slippage occurs, this implies also that $d(\bar{\mu}_n - \bar{\mu}_s)/dx \approx e(I - \bar{I}_s)\rho/A$ is the initial slope for an exponential decay of $(\bar{\mu}_n - \bar{\mu}_s)$ over a length Λ_{Q^*} on either side of the heart of the PSC. Integrating over these exponential decays yields the total voltage difference across the PSC, given by

$$V = 2\Lambda_{Q^*}\rho(I - \beta I_c)/A \tag{11.24}$$

with $\beta \sim 1/2$.

[54]This would follow from the cyclic acceleration/quench cycle described above [see Fig. 11.12*b*]. Alternatively, the $T^*(t)$ argument of Sec. 11.4.3 would also account for some net time-average dc supercurrent in the forward direction, which would also scale roughly with I_c.

This simple formula actually gives a good account of the data. That is, not only does it fit the form of the steps, but the numerical values are reasonable. For example, the inferred Λ_{Q^*} is found to vary as $\ell^{1/2}$ with electronic mean path, as expected from (11.23), and the inferred value for τ_{Q^*} is indeed of the order of magnitude given by (11.5*b*). Although the expected (weak) divergence of Λ_{Q^*} as $(T_c - T)^{-1/4}$ was apparently obscured by heating effects in early experiments, it was clearly demonstrated in later experiments by Kadin et al.[55] and by Aponte.[56] Moreover, when the slope of the second step is compared with the first, one finds the expected doubled slope, corresponding to two similar PSCs in series.

In 1977, Dolan and Jackel[57] provided a beautiful confirmation of the different spatial variations of $\bar{\mu}_s$ and $\bar{\mu}_n$ by making direct measurements of both potentials at 1 μm intervals by use of an array of superconducting and normal tunnel junction contacts. Their results, shown in Fig. 11.13, confirm that V_s changes abruptly, while V_n varies smoothly over many microns, with the expected quasi-exponential form. They also confirmed the expected divergent temperature dependence of Λ_{Q^*}.

Since the nonequilibrium currents generated by a PSC extend a distance Λ_{Q^*} on either side, it is clear that PSCs will interact strongly when separated by a distance of less than $\sim 2\Lambda_{Q^*}$. For example, the diffusing normal current from one PSC will reduce the local supercurrent to $I_s(x) = I - I_n(x)$, and hence tend to prevent the occurrence of additional PSCs in the neighborhood of existing ones since the condition for establishment of a PSC is that $I_s(x) > I_c(x)$, *not* $I > I_c(x)$.

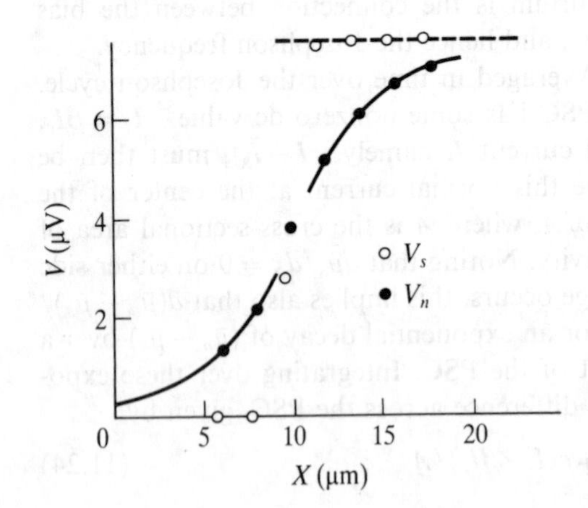

FIGURE 11.13
Spatial variation of superconducting and normal electron potentials, V_s and V_n, measured by tunnel probes near a phase-slip center in a tin film strip. (*After Dolan and Jackel.*)

[55] A. M. Kadin, W. J. Skocpol, and M. Tinkham, *J. Low Temp. Phys.* **33**, 481 (1978).
[56] J. M. Aponte and M. Tinkham, *J. Low Temp. Phys.* **51**, 189 (1983).
[57] G. J. Dolan and L. D. Jackel, *Phys. Rev. Lett.* **39**, 1628 (1977).

Such interaction effects were clearly observed in the clean whisker filaments studied by Meyer and Tidecks,[58] and also by Aponte and Tinkham[59] in thin film microbridges. For a complete treatment, one may also need to take into account the *ac* component in the nonequilibrium currents, which should show an oscillatory spatial behavior[60] as well as exponential decay. Nonetheless, it is of interest to note that the explicit application[61] of the simple dc interaction model to an absolutely uniform filament of length L predicts a succession of steps at increasing currents, such that approximately $(L/2\Lambda_{Q^*})$ steps should occur between I_c and $2I_c$. Thus, the PSC model of SBT does *not* require a spatially inhomogeneous conductor to account for the appearance of steps over a wide range of currents above I_c, although significant inhomogeneity is usually present in real samples.

In another qualitative confirmation of the model, SBT found that Shapiro steps could be induced in the *I-V* characteristic at the voltage corresponding by the Josephson relation to the applied microwave frequency. This result confirms that one is dealing with a dynamic process involving ac supercurrents which can be synchronized with an external signal; all static models, such as the creation of a fully normal region (as in the case of a heating-induced normal "hot spot"[62]) along the filament, are excluded by this experiment.

Kadin[63] extended the PSC measurements in tin film strips to include a parallel magnetic field. He found that the step structure of the *I-V* curve is qualitatively unchanged, but, quantitatively, the steps are much less steep, i.e., the diffusion length Λ_{Q^*} (and hence the implied relaxation time τ_{Q^*}) is shortened. This effect was predicted in the work of Schmid and Schön[64] who found that magnetic perturbations should reduce the relaxation time from the value for τ_{Q^*} given by (11.5*b*), with τ_E being replaced by $(\tau_E\tau_s/2)^{1/2}$ in the limit of $\tau_s \ll \tau_E$. Here τ_s is the magnetic pair-breaking time, given by $\hbar/\tau_s = 1.76kT_c[H/H_{c\parallel}(0)]^2$ in this case. The physical reason for this replacement of τ_E by a geometric mean can be traced to the fact that one needs both a branch-mixing relaxation effect and inelastic scattering to bring quasi-particles down to the low energies where it is fastest. In the cases treated earlier, the electron-phonon process described by τ_E served in both roles, but when $\tau_s < \tau_E$, the pair-breaking perturbation takes over the first. The trend of the data with $H/H_{c\parallel}(0)$ is found to agree well with the Schmid-Schön prediction.

[58] J. D. Meyer and R. Tidecks, *Solid State Comm.* **24**, 643 (1977).

[59] J. M. Aponte and M. Tinkham, *J. Low Temp. Phys.* **51**, 189 (1983).

[60] A. M. Kadin, L. N. Smith, and W. J. Skocpol, *J. Low Temp. Phys.* **38**, 497 (1980).

[61] M. Tinkham, *J. Low Temp. Phys.* **35**, 147 (1979).

[62] W. J. Skocpol, M. R. Beasley, and M. Tinkham, *J. Appl. Phys.* **45**, 4054 (1974).

[63] A. M. Kadin, W. J. Skocpol, and M. Tinkham, *J. Low Temp. Phys.* **33**, 481 (1978).

[64] A. Schmid and G. Schön, *J. Low Temp. Phys.* **20**, 207 (1975).

Finally, reference should be made to a considerable body of theoretical work[65] aimed at providing a more quantitative and rigorous description of the phenomena described above. However, it appears that a rigorous treatment of the problem with realistic parameter values is very difficult and hard to apply to the analysis of experimental data. Accordingly, it is fortunate that the simple picture presented here is able to account quite well for the phenomena observed, despite its evident oversimplification.

[65]For example, L. Kramer and A. Baratoff, *Phys. Rev. Lett.* **38**, 518 (1977); L. Kramer and R. J. Watts-Tobin, *Phys. Rev. Lett.* **40**, 1041 (1978); V. P. Galaiko, *J. Low Temp. Phys.* **26**, 483 (1977); B. I. Ivlev, N. B. Kopnin, and L. A. Maslova, *Zh. Eksp. Teor. Fiz.* **78**, 1963 (1980); B. I. Ivlev and N. B. Kopnin, *J. Low Temp. Phys.* **39**, 137 (1980); **44**, 453 (1981).

APPENDIX 1

UNITS

This book has been written using gaussian cgs units throughout, except for a few formulas that are explicitly quoted in practical units for convenience in applications. This choice conforms to that in the first edition and to that used in much of the physics literature on which the book is based. Moreover, because of the central role of **B** and **H** in superconductivity, it is especially convenient to have $\mathbf{B} = \mathbf{H}$ in vacuum and the natural form $\mathbf{E} = (\mathbf{v}/c) \times \mathbf{B}$ for the electric field associated with a moving flux density.

For readers who are more comfortable with SI or mksa units, we reproduce in Table A.1 a version of the convenient tabular conversion guide given in the appendix of J. D. Jackson, *Classical Electrodynamics*, Wiley, New York, 1975, p. 819. To convert any formula from gaussian to SI units, follow these rules: Symbols for mass, length, time, force, and other quantities that are not specifically of an electromagnetic nature are unchanged. Symbols for electromagnetic quantities listed under "gaussian" in Table A.1 are replaced on both sides of the equation by the corresponding symbols listed under "SI." The reverse transformation can also be made. To illustrate this procedure, we consider a few important examples.

The flux quantum $\Phi_0 = hc/2e$ in gaussian units. Following Table A.1, the left-hand side of this equation becomes $\sqrt{4\pi/\mu_0}\Phi_0$ and the right-hand side becomes $h(1/\sqrt{\mu_0\epsilon_0})(\sqrt{4\pi\epsilon_0}/2e)$. After canceling common factors, one obtains $\Phi_0 = h/2e$, which defines the flux quantum in SI units. Similar manipulations applied to the GL relation (4.20) relating the flux quantum to the product of $H_c\xi\lambda$ leave the form of the equation unchanged, except that the notation H_c must be replaced by B_c because Φ_0 is a quantum of magnetic *flux* or induction, not

TABLE A.1
Conversion table for electromagnetic formulas

Quantity	Gaussian	SI (mksa)
Velocity of light	c	$\frac{1}{\sqrt{\mu_0 \epsilon_0}}$
Magnetic induction or flux density	$\mathbf{B}$	$\sqrt{\frac{4\pi}{\mu_0}}\,\mathbf{B}$
Magnetic field	$\mathbf{H}$	$\sqrt{4\pi\mu_0}\,\mathbf{H}$
Magnetization	$\mathbf{M}$	$\sqrt{\frac{\mu_0}{4\pi}}\,\mathbf{M}$
Charge density (or charge, current, current density, polarization)	ρ (or $Q, I, \mathbf{J}, \mathbf{P}$)	$\frac{1}{\sqrt{4\pi\epsilon_0}}\rho$ (or $Q, I, \mathbf{J}, \mathbf{P}$)
Electric field (or potential, voltage)	$\mathbf{E}$ (or ϕ, V)	$\sqrt{4\pi\epsilon_0}\,\mathbf{E}$ (or ϕ, V)
Displacement	$\mathbf{D}$	$\sqrt{\frac{4\pi}{\epsilon_0}}\,\mathbf{D}$
Conductivity	σ	$\frac{\sigma}{4\pi\epsilon_0}$
Resistance (or impedance)	R (or Z)	$4\pi\epsilon_0 R$ (or Z)
Inductance	L	$4\pi\epsilon_0 L$
Capacitance	C	$\frac{C}{4\pi\epsilon_0}$
Permeability	μ	$\frac{\mu}{\mu_0}$
Dielectric constant	ϵ	$\frac{\epsilon}{\epsilon_0}$

Source: From J. D. Jackson, *Classical Electrodynamics*, Wiley, New York (1975).

field. Similarly, the condensation energy per unit volume $F_n - F_s = H_c^2/8\pi$ becomes $\mu_0 H_c^2/2$ or $B_c^2/2\mu_0$ in SI units, and the London penetration depth $\lambda = \sqrt{mc^2/4\pi n_s e^2}$ becomes $\sqrt{m/\mu_0 n_s e^2}$. The definition of the coherence length ξ is unchanged because it does not involve electromagnetic quantities, but only $\hbar$, m, and the GL coefficient α, all of which are unchanged.

APPENDIX 2

NOTATION AND CONVENTIONS

NOTATION FOR MAGNETIC FIELDS. Because of the ubiquity of magnetic fields in the subject of superconductivity, special notational conventions are common to simplify the discussion. We follow the convention of de Gennes (and others) and use $\mathbf{h}(\mathbf{r})$ to denote the local value of the magnetic induction or flux density, which typically varies on the scale of the penetration depth λ. We reserve the use of $\mathbf{B}$ to denote the value of $\mathbf{h}$ averaged over such microscopic lengths but still capable of varying smoothly over the macroscopic dimensions of the sample.

In *normal* metal or vacuum, of course, there is no microscopic variation of $\mathbf{h}$ (we neglect the Landau diamagnetism and Pauli paramagnetism), so $\mathbf{B} = \mathbf{h}$. In these cases, $\mathbf{B} = \mathbf{H}$, so all three symbols denote equal quantities, and may be used interchangably.

In the *Meissner* state of a massive superconductor, $\mathbf{h}$ is reduced to zero within a penetration depth λ of the surface by supercurrents in the skin layer, as described by the Maxwell equation

$$\text{curl}\,\mathbf{h} = \frac{4\pi \mathbf{J}_{\text{total}}}{c} \tag{A2.1}$$

Hence, $\mathbf{B} = \mathbf{h} = \mathbf{0}$ deep inside. On the other hand, $\mathbf{H}$ is governed by the Maxwell equation

$$\text{curl}\,\mathbf{H} = \frac{4\pi \mathbf{J}_{\text{ext}}}{c} \tag{A2.2}$$

where $\mathbf{J}_{\text{ext}}$ represents a *nonequilibrium* current and excludes currents arising from the equilibrium response of the medium, such as those in the penetration depth described by the London equations. Hence, curl $\mathbf{H} = 0$, and the tangential component H_t is constant through the skin depth, retaining the value of $H_t (= B_t = h_t)$ found outside the sample. If the sample is ellipsoidal, H inside is uniform and everywhere equals the equatorial value of H_t. This H_t will in general exceed the uniform applied field H_a by a factor $(1 - \eta)^{-1}$, where η is the shape-dependent demagnetizing factor of the sample. [See the discussion associated with (2.20).]

In the *intermediate* state of a type I superconductor, which is reached when $H_t = H_c$, the magnitude of $\mathbf{h}$ varies continuously (on the scale of λ) between H_c in the normal lamina and zero in the superconducting ones. $\mathbf{B}$ is the average of this $\mathbf{h}$ over the laminar structure, and it is constant within an ellipsoidal sample. The magnitude of $\mathbf{H}$ must be H_c for coexistence of superconducting and normal regions to be possible. These interrelations are illustrated in Fig. 2.3.

In the *mixed* state of a type II superconductor in a magnetic field above H_{c1}, $\mathbf{h}$ varies on the microscopic scale of the vortex structure, whereas $\mathbf{B}$ is the average of $\mathbf{h}$ over the structure. In the ideal equilibrium case, H is again everywhere equal to H_t at the equatorial surface. In the presence of transport currents and of disequilibrium due to flux pinning, $\mathbf{H}$ will vary because curl $\mathbf{H}$ is no longer zero. This situation is discussed more fully in connection with (5.48) in the text.

NOTATION FOR ELECTRIC FIELDS. For notational symmetry, we also can define a microscopically varying electric field $\mathbf{e}(\mathbf{r})$, whose macroscopic average is $\mathbf{E}$. But because curl $\mathbf{e} = -(1/c)\partial\mathbf{h}/\partial t$, $\mathbf{e}$ is uniform in static situations, and it must be zero in equilibrium. Thus, the distinction between $\mathbf{e}$ and $\mathbf{E}$ arises less frequently than the distinction between $\mathbf{h}$ and $\mathbf{B}$ (which can result from *equilibrium* supercurrents); consequently, we have normally used $\mathbf{E}$ for both to avoid confusion between $\mathbf{e}$ and the electronic charge e. The notation $\mathbf{e}$ is introduced only to describe the electric field distribution about a moving vortex, where the macroscopic average $\mathbf{E}$ is quite different from $\mathbf{e}$ and gives the physically important resistive voltage.

SIGN OF THE ELECTRON CHARGE. Fortunately, the sign convention for the electronic charge is often immaterial since only its square enters into such quantities as the penetration depth or normal conductivity. Also, in the definition of quantities such as the flux quantum $\Phi_0 = hc/2e$ or the Josephson frequency $2eV/h$, it is convenient as well as conventional to take e to be a *magnitude*. However, when the sign of the charge matters, we have chosen to follow de Gennes in adopting the convention that e is the charge of the *electron*, including its sign; i.e., $e = -|e|$. This means that a metal island with n excess electrons on it has a charge ne, not $-ne$, and the supercurrent density is simply $n_s e\mathbf{v}_s$, rather than $n_s(-e)\mathbf{v}_s$. This convention simplifies many expressions, and seems more physical. It has the disadvantage of being contrary to the convention used in such popular textbooks as *Introduction to Solid State Physics* by Kittel and *Solid State Physics* by Ashcroft and Mermin, in which the charge of the electron is written as $(-e)$, so that e *always* refers to a magnitude.

APPENDIX 3

EXACT SOLUTION FOR PENETRATION DEPTH BY FOURIER ANALYSIS

A convenient technique for obtaining an exact expression for the penetrating magnetic field, and hence for the penetration depth, is to apply Fourier analysis to **J** and **A**, and to use (3.101) to obtain a self-consistent solution. Only a one-dimensional Fourier analysis is required since J_x and A_x are functions only of z for the penetration of a magnetic field B_y parallel to a planar surface. Some care is needed in handling the surface, however, since our expressions for the response function $K(q)$ are valid only in an infinite medium. This problem is handled by the mathematical artifice of introducing externally supplied source currents in the interior of the infinite medium to simulate the field applied at a surface.

Consider, e.g., the case in which electrons are assumed to be *specularly reflected* at the surface. If one introduces a current sheet

$$J_{x,\text{ext}} = -\frac{c}{2\pi} B_0\, \delta(z) \tag{A3.1}$$

this introduces a discontinuity $2B_0$ in h_y. This can be taken symmetric about zero, so that h_y switches from $-B_0$ to $+B_0$. Now when the superconductive medium is introduced, its diamagnetic currents will screen out these fields in a length λ (to be determined). Note that electrons passing through this plane at $z = 0$ without scattering have had a past exposure along their trajectory to a vector potential exactly the same as that seen by electrons specularly reflected at the surface in the actual case since $\mathbf{A}(-z) = \mathbf{A}(z)$. (See Fig. A3.1.) Thus, the net supercurrent induced in them should also be the same, and the simulation should be effective.

Having replaced the surface by a current sheet in an infinite medium, we now may proceed to use the response function $K(q)$ worked out for that case. We first note that

$$\nabla^2 \mathbf{A} = -\text{curl curl } \mathbf{A} = -\text{curl } \mathbf{h} = -\frac{4\pi}{c}\mathbf{J}_{\text{total}} = -\frac{4\pi}{c}(\mathbf{J}_{\text{ext}} + \mathbf{J}_{\text{med}})$$

For the qth Fourier component, this becomes

$$q^2 \mathbf{a}(q) = \frac{4\pi}{c}\mathbf{J}_{\text{ext}}(q) - K(q)\mathbf{a}(q)$$

Solving for $\mathbf{a}(q)$, we have the general result

$$\mathbf{a}(q) = \frac{(4\pi/c)\mathbf{J}_{\text{ext}}(q)}{K(q) + q^2} \tag{A3.2}$$

For the current sheet (A3.1), $J_{\text{ext}}(q) = -cB_0/4\pi^2$, and we drop the vector notation since $\mathbf{J}$ and $\mathbf{A}$ have only an x component. Thus,

$$a(q) = \frac{-B_0/\pi}{K(q) + q^2}$$

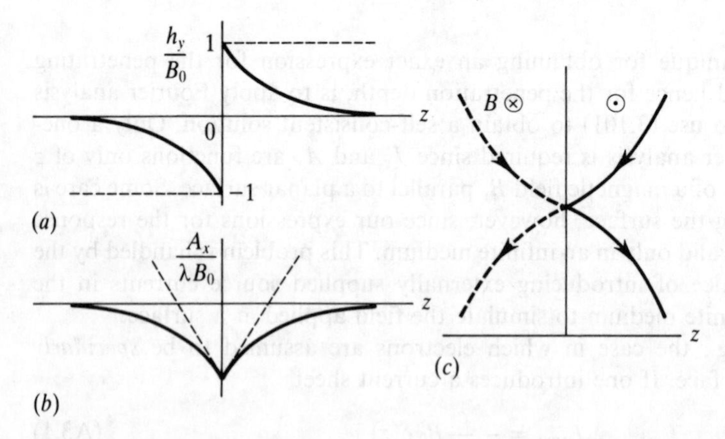

FIGURE A3.1
Simulation of surface with specular reflection by source-current sheet. (*a*) Magnetic field in normal (dashed) and superconducting (solid) states. (*b*) Vector potential in normal (dashed) and superconducting (solid) states. London gauge is used in superconducting state. (*c*) Electron trajectories. The solid curve shows trajectory with specular reflection; the dashed parts show extensions into the other half-space, with current-sheet simulation.

We are more interested in $\mathbf{h} = \text{curl}\,\mathbf{A}$, so that $h_y(q) = iqa(q)$. Integrating over all the Fourier components, we obtain

$$h(z) = \frac{B_0}{i\pi}\int_{-\infty}^{\infty} \frac{qe^{iqz}\,dq}{K(q)+q^2} = \frac{2B_0}{\pi}\int_0^{\infty} \frac{q\sin qz\,dq}{K(q)+q^2} \tag{A3.3}$$

For any $K(q)$, (A3.3) gives the true dependence of h on z, which will not be exactly exponential unless $K(q) =$ constant, as in the London theory. For example, the $h(z)$ computed with the $K(q)$ for either the Pippard or BCS theory actually *changes sign* deep in the interior, where $|h(z)| \ll B_0$.

To get the penetration depth, as usually defined, we integrate (A3.3):

$$\lambda = B_0^{-1}\int_0^{\infty} h(z)\,dz = \frac{2}{\pi}\int_0^{\infty}\int_0^{\infty} \frac{q\sin qz\,dq\,dz}{K(q)+q^2}$$

or

$$\lambda_{\text{spec}} = \frac{2}{\pi}\int_0^{\infty} \frac{dq}{K(q)+q^2} \tag{A3.4}$$

(In carrying out the integration on z, one can replace $\int_0^Z q\sin qz\,dz = 1 - \cos qZ$ by its average value, unity, since as $Z \to \infty$, the oscillatory part effectively averages to zero in the subsequent integration over q.)

Given (A3.4), we can compute λ_{spec} for any model of superconductivity which determines a $K(q)$. For example, in the London theory, $K(q) = 1/\lambda_L^2$. Then

$$\lambda_{\text{London, spec}} = \frac{2}{\pi}\int_0^{\infty} \frac{dq}{\lambda_L^{-2}+q^2} = \lambda_L \tag{A3.5}$$

In the Pippard theory, one has

$$K_p(q) = \frac{1}{\lambda_L^2}\frac{\xi}{\xi_0}\left\{\frac{3}{2(q\xi)^3}[(1+q^2\xi^2)\tan^{-1} q\xi - q\xi]\right\} \tag{A3.6}$$

This is found from (3.117) with $J_p(R,T) = e^{-R/\xi}$ by using the general relation (3.106). If instead one approximates the BCS kernel even more closely by $J(R,T) \approx J(0,T)\exp -[J(0,T)R/\xi_0]$, as discussed in the argument leading to (3.123), the effect is simply to replace ξ_0 by $\xi_0' = \xi_0/J(0,T)$ everywhere in (A3.6), including in the definition (3.121) of ξ. As remarked in Chap. 3, these rather convenient, generalized Pippard forms provide quite a serviceable approximation to the exact numerical results of BCS. However, even with the analytic expression (A3.6) for $K(q)$, numerical integration is required to compute the penetration depth by using (A3.4).

In order to avoid numerical calculations, considerable attention has been given to two limiting cases in which analytic results can be obtained, even though the true situation usually lies in between.

The *local approximation* replaces $K(q)$ for all q by $K(0)$, a constant, thus reducing the problem to the London form, but in general with a modified penetration depth. Using the generalized Pippard approximation

$$K(0,T) = \lambda_L^{-2}\left[1 + \frac{\xi_0}{J(0,T)\ell}\right]^{-1} \tag{A3.7}$$

one finds
$$\lambda(T) = \lambda_L(T)\left[1 + \frac{\xi_0}{J(0,T)\ell}\right]^{1/2} \tag{A3.8}$$

as anticipated in (3.123). This approximation is reasonably well justified in dirty superconductors [if $\ell < \lambda(T)$], in high-temperature superconductors, and even in pure classic superconductors very near T_c where $\xi_0 < \lambda(T)$.

The other approximation is the *extreme anomalous limit*, in which $K(q)$ is replaced for all q values by its asymptotic form for $q \to \infty$, where $K(q) \sim 1/q$. This approximation is reasonably well justified if $\lambda_L \ll \xi_0$ because then the dominant contribution to (A3.4) will come from the q values in which this asymptotic form is valid. Figure A3.2 illustrates the two different approximations to $K(q)$. Since both approximations exceed the true $K(q)$ for some q and never err in the other direction, both will lead to lower bounds to the true value for λ.

Let us now carry out the calculation in the extreme anomalous limit. For complete generality, we write

$$K(q) = \frac{a}{q}$$

where in the Pippard theory $a = 3\pi/4\lambda_L^2\xi_0$, whereas in either the BCS theory or the generalized Pippard theory, ξ_0 is replaced by ξ_0' so that a is increased by a factor of $J(0,T)$. If we introduce the standard notation λ_∞ for the value of λ in this limit, (A3.4) becomes

$$\lambda_{\infty,\text{spec}} = \frac{2}{\pi}\int_0^\infty \frac{dq}{(a/q) + q^2}$$

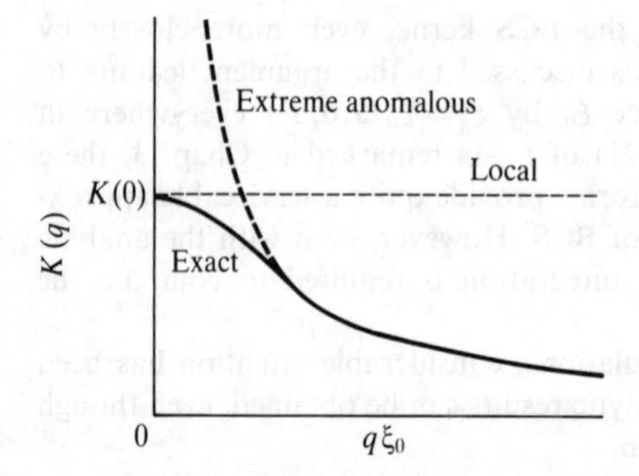

FIGURE A3.2
Schematic comparison of local and extreme anomalous approximations to the exact nonlocal response function $K(q)$.

Making a change of variable to $x = q^3/a$, we obtain

$$\lambda_{\infty,\mathrm{spec}} = \frac{2}{3\pi a^{1/3}} \int_0^\infty \frac{x^{-1/3}\,dx}{1+x} = \frac{4}{3\sqrt{3}a^{1/3}}$$

Inserting the value for a, we have

$$\lambda_{\infty,\mathrm{spec}} = \frac{8}{9}\frac{3^{1/6}}{(2\pi)^{1/3}}(\lambda_L^2\xi_0')^{1/3} = 0.58(\lambda_L^2\xi_0')^{1/3} \tag{A3.9}$$

which has exactly the form anticipated in (3.130) by an elementary argument. Since the BCS correction factor $(\xi_0'/\xi_0)^{1/3} = [J(0,T)]^{-1/3}$ to the simple Pippard form varies smoothly from 1 at $T = 0$ to 0.91 at T_c, it has little effect on the behavior of the result.

If the surface scattering is taken as *diffuse* instead of specular, formulas are obtained that differ only in detail from those given above. In this case, the prescription for handling the surface is simply to cut off the integration over $\mathbf{r}'$ at the surface in the coordinate-space form (3.117) of the response function. The physical reasoning is that electrons coming to $\mathbf{r}$ from the surface do so with no memory of any previous exposure to the field. When this prescription is transcribed into Fourier transform language, it turns out that (A3.4) is replaced by

$$\lambda_{\mathrm{diff}} = \frac{\pi}{\int_0^\infty \ln[1 + K(q)/q^2]\,dq} \tag{A3.10}$$

Although this looks quite different from (A3.4), it actually gives exactly the same result in the local approximation, and for λ_∞ it differs from (A3.9) only in that the factor of 8/9 is missing. Thus, there is little difference in the results for these two different limiting assumptions about the surface scattering.

BIBLIOGRAPHY

While in no way exhaustive, the following list includes many of the standard references for further reading, with a brief indication of their individual features. Within each broad category, these are listed in reverse chronological order. However, it should be noted that many of the older references are classics which are still widely used.

Monographs

Cyrot, M. and D. Pavuna: *Introduction to Superconductivity and High-T_c Materials*, World Scientific, Singapore (1992). A recent book at an introductory level.

Orlando, T. P., and K. A. Delin: *Foundations of Applied Superconductivity*, Addison-Wesley, Reading, MA (1991). A recent book aimed at advanced undergraduates in electrical engineering at MIT. Good source of problems on London electrodynamics. Formulas are in SI units.

Tilley, D. R., and J. Tilley: *Superfluids and Superconductivity* (3d ed.), Institute of Physics, Bristol (1990). An introductory survey which emphasizes the connection between superfluid helium and superconductivity; contains many helpful illustrations.

Tidecks, R.: *Current-Induced Nonequilibrium Phenomena in Quasi-One-Dimensional Superconductors*, Springer, Berlin (1990). An exhaustive discussion of phase-slip-center phenomena.

Phillips, J. C.: *Physics of High T_c Superconductors*, Academic Press, San Diego (1989). An individualistic theoretical perspective.

Likharev, K. K.: *Dynamics of Josephson Junctions and Circuits,* Gordon and Breach, Philadelphia (1986). A classic comprehensive treatment of its subject.

Kittel, C.: *Introduction to Solid State Physics*, 6th ed., Wiley, New York (1986). Chapter 12 gives an introduction to superconductivity.

Wilson, M. N.: *Superconducting Magnets*, Oxford University Press, Oxford (1983). A classic treatment of the design considerations for superconducting magnets by a leading figure.

Barone, A., and G. Paterno: *Physics and Applications of the Josephson Effect*, Wiley, New York (1982). A valuable comprehensive review of the field.

Van Duzer, T., and C. W. Turner: *Principles of Superconducting Devices and Circuits*, North-Holland, Amsterdam (1980). An introduction to practical applications, especially Josephson junction devices. Contains some errors, but has numerous useful graphs and figures.

Huebener, R. P.: *Magnetic Flux Structures in Superconductors*, Springer, New York (1979). A classic review of the field by a leading figure.

Rose-Innes, A. C., and E. H. Rhoderick: *Introduction to Superconductivity*, 2d ed. Pergamon, Oxford (1978). A very readable introductory treatment.

Ashcroft, N. W., and N. D. Mermin: *Solid State Physics*, Holt, Rinehart, and Winston, Philadelphia (1976). Chapter 34 gives an introduction to superconductivity.

Tinkham, M.: *Introduction to Superconductivity* (1st ed.), McGraw-Hill, New York (1975).

Goodstein, D. L.: *States of Matter*, Prentice-Hall, Englewood Cliffs, NJ (1975), pp. 371–411 on superconductivity. Useful discussion emphasizing thermodynamic and statistical aspects.

Campbell, A. M., and J. E. Evetts: *Critical Currents in Superconductors*, Taylor and Francis, London, 1972. A classic treatment of its subject.

Lynton, E. A.: *Superconductivity*, Methuen Monograph, London (1969). A concise survey.

Saint-James, D., G. Sarma, and E. Thomas: *Type II Superconductivity*, Pergamon, New York (1969). An excellent, richly detailed review of this subject, emphasizing theory at the Ginzburg-Landau level.

de Gennes, P. G.: *Superconductivity in Metals and Alloys*, W. A. Benjamin, New York (1966); reprinted by Addison-Wesley, Reading, MA (1989). A classic, excellent, physically motivated treatment of the subject.

Rickayzen, G.: *Theory of Superconductivity*, Wiley, New York (1965). A good general survey.

Schrieffer, J. R.: *Theory of Superconductivity*, W. A. Benjamin, New York (1964). A good account of the theory by one of its founders, including Green's function topics, which are not included in this book.

London, F.: *Superfluids*, vol. I, Wiley, New York (1950). Discusses London theory and its philosophical background. It is still valuable for its thoughtful discussion; a famous footnote on p. 152 predicted fluxoid quantization over a decade before it was discovered experimentally.

Collections

Ginsberg, D. M. (ed.): *Physical Properties of High Temperature Superconductors I, II, III, IV*, World Scientific, Singapore (1989, 1990, 1992, 1994). This ongoing series contains authoritative reviews on topics concerning high-temperature superconductivity.

Bedell, K. S., *et al.* (eds.): *Phenomenology and Applications of High-Temperature Superconductors*, Addison-Wesley, Reading, MA (1992). Proceedings of the Los Alamos Symposium, 1991.

Grabert, H., and M. H. Devoret (eds.): *Single Charge Tunneling*, Plenum, New York (1992). Proceedings of NATO Institute at Les Houches.

Evetts, J. (ed.): *Concise Encyclopedia of Magnetic and Superconducting Materials*, Pergamon, Oxford (1992). Contains authoritative articles by recognized experts on many topics in the science and technology of superconductivity.

Ruggiero, S. T., and D. A. Rudman (eds.): *Superconducting Devices*, Academic Press, New York (1990). Very useful fairly recent survey.

Halley, J. W. (ed.): *Theories of High Temperature Superconductivity*, Addison-Wesley, Reading, MA (1988). A survey of theories proposed in the very early days of high-temperature superconductivity.

Goldman, A. M., and S. A. Wolf (eds.): *Percolation, Localization, and Superconductivity*, Plenum, New York (1984). Proceedings of NATO Institute at Les Arcs, France.

Deaver, B., and J. Ruvalds (eds.): *Advances in Superconductivity*, Plenum, New York (1983). NATO conference proceedings.

Gray, K. E. (ed.): *Nonequilibrium Superconductivity, Phonons, and Kapitza Boundaries*, Plenum, New York (1981). Proceedings of NATO Institute at Maratea, Italy.

Gubser, D. U., T. L. Francavilla, S. A. Wolf, and J. R. Leibowitz (eds.): *Inhomogeneous Superconductors*, AIP Conference Proceedings No. 58, AIP, New York (1980).

Ginzburg, V. L., and D. A. Kirzhnits (eds.): *High Temperature Superconductivity* (1977); translation: Consultants Bureau, New York (1982).

Schwartz, B. B., and S. Foner (eds.): *Small Scale Applications of Superconductivity*, Plenum, New York (1976).

Newhouse, V. L. (ed.): *Applied Superconductivity*, two vols., Academic Press, New York (1975).

Foner, S., and B. B. Schwartz (eds.): *Superconducting Machines and Devices; Large Systems Applications*, Plenum, New York (1974). Proceedings of a summer school.

Parks, R. D. (ed.): *Superconductivity*, two vols., Dekker, New York (1969); reissued by the publisher in 1992. This two-volume treatise, with chapters written by two dozen distinguished authors on their special areas of interest, is the most comprehensive available treatment of the subject as it stood in 1968.

Review articles

Blatter, G., M. V. Feigel'man, V. B. Geshkenbein, A. I. Larkin, and V. M. Vinokur: "Vortices in High-Temperature Superconductors," *Revs. Mod. Phys.* **66**, 1125–1388 (1994). An extremely comprehensive and detailed review of the field.

Averin, D. V., and K. K. Likharev, "Single Electronics: Correlated Transfer of Single Electrons and Cooper Pairs in Systems of Small Tunnel Junctions" in B. L. Altshuler, P. A. Lee, and R. A. Webb (eds.), *Mesoscopic Phenomena in Solids*, Elsevier, New York (1991). A classic reference in its field.

Carbotte, J. P.: "Properties of Boson-Exchange Superconductors," *Revs. Mod. Phys.* **62**, 1027 (1990). A review of the status of detailed calculations.

Tucker, J. R., and M. J. Feldman: "Quantum Detection at Millimeter Wavelengths," *Revs. Mod. Phys.* **57**, 1055 (1985). A review.

Likharev, K. K.: "Superconducting Weak Links," *Revs. Mod. Phys.* **51**, 101 (1979). An often-cited review.

Skocpol, W. J., and M. Tinkham: "Fluctuations Near Superconducting Phase Transitions," *Repts. Prog. Phys.* **38**, 1049 (1975). A review.

Anderson, P. W.: "The Josephson Effect and Quantum Coherence Measurements in Superconductors and Superfluids," in C. J. Gorter (ed.), *Progress in Low Temperature Physics*, vol. 5, Wiley, New York (1967). Notable for its insight into its topic.

Tinkham, M.: "Superconductivity," in C. de Witt, B. Dreyfus, and P. G. de Gennes (eds.), *Low Temperature Physics*, Gordon and Breach, New York (1962), pp. 147–230. A pedagogical presentation at the Les Houches summer school in the early days of BCS.

Bardeen, J.: "Theory of Superconductivity," in S. Flügge (ed.), *Handbuch der Physik*, vol. 15, Springer Verlag, Berlin (1956). A masterful survey of the state of the theory immediately *before* BCS.

Applied Superconductivity Conference proceedings

The proceedings of the biennial Applied Superconductivity Conferences provide an ongoing source of up-to-date surveys of the state of the field. Some of the recent ones are found in:

IEEE Trans. Magn. MAG**25**, no. 2, March 1989.
IEEE Trans. Magn. MAG**27**, no. 2, March 1991.
IEEE Trans. Appl. Supercond. **3**, no. 1, March 1993.

INDEX